The Superworld II

THE SUBNUCLEAR SERIES

Series Editor: **ANTONINO ZICHICHI**, *European Physical Society, Geneva, Switzerland*

Volume 1 was published by W. A. Benjamin, Inc., New York; 2-8 and 11-12 by
Academic Press, New York and London; 9-10, by Editrice Compositori, Bologna; 13-25
by Plenum Press, New York and London.

The Superworld II

Edited by
Antonino Zichichi

European Physical Society
Geneva, Switzerland

PLENUM PRESS • NEW YORK AND LONDON

Library of Congress Cataloging-in-Publication Data

International School of Subnuclear Physics (25th : 1987 : Erice, Italy)
   The superworld II / edited by Antonino Zichichi.
     p. cm. -- (The Subnuclear series ; 25)
   "Proceedings of the Twenty-fifth course of the International School of Subnuclear Physics on the superworld II, held August 6-14, 1987, in Erice, Sicily, Italy"--T.p. verso.
   Includes bibliographical references.
   ISBN-13: 978-1-4684-7469-5   e-ISBN-13: 978-1-4684-7467-1
   DOI: 10.1007/978-1-4684-7467-1
   1. String models--Congresses. 2. Particles (Nuclear physics)--Congresses. I. Zichichi, Antonino. II. Title. III. Title: Superworld two. IV. Title: Superworld 2. V. Series: Subnuclear series ; v. 25.
QC794.6.S85I57 1987
539.7'2--dc20   90-6750
                                                       CIP

Proceedings of the Twenty-Fifth Course of the International School
of Subnuclear Physics on The Superworld II,
held August 6–14, 1987, in Erice, Sicily, Italy

© 1990 Plenum Press, New York
Softcover reprint of the hardcover 1st edition 1990

A Division of Plenum Publishing Corporation
233 Spring Street, New York, N.Y. 10013

PREFACE

During August 1987, a group of 76 physicists from 51 laboratories in 22 countries met in Erice for the 25th Course of the International School of Subnuclear Physics. The countries represented were: Austria, Bulgaria, Canada, Chile, China, Colombia, Czechoslovakia, France, Federal Republic of Germany, Greece, Hungary, India, Italy, Lebanon, The Netherlands, Poland, Portugal, Spain, Sweden, Switzerland, United Kingdom, and the United States of America.

The School was sponsored by the European Physical Society (EPS), the Italian Ministry of Public Education (MPI), the Italian Ministry of Scientific and Technological Research (MRSI), the Sicilian Regional Government (ERS), and the Weizmann Institute of Science.

This is the 25th anniversary of the School and, for the second time, the programme has been mainly devoted to the Superworld. Needless to say that the Superworld appears to be, at present, very far from the experimental axis. Nevertheless, the Superworld is a fascinating field of modern physics: we ought to know what boils in the heads of our theoretical colleagues, keeping in mind that the source of basic truth is, and will remain, experimental physics.

Relevant news in experimental physics was scarce in the past year and the most interesting results have been reported. The future has also been presented with LEP, Gran Sasso, HERA: projects to become operative by 1990; and ELOISATRON as the driving force for Europe to keep a central role in Subnuclear Physics.

I hope the reader will enjoy this book as much as the students enjoyed attending the lectures and the discussion sessions, which are the most attractive features of the School. Thanks to the work of the Scientific Secretaries, the discussions have been reproduced as faithfully as possible. At various stages of my work I have enjoyed the collaboration of many friends whose contributions have been extremely important for the School and are highly appreciated. I thank them most warmly. A final acknowledgement to all those who, in Erice, Bologna, Rome and Geneva, helped me on so many occasions and to whom I feel very much indebted.

Antonino Zichichi

CONTENTS

THE GLORIOUS DAYS OF PHYSICS

CLOSING LECTURE

CLOSING CEREMONY

THEORETICAL LECTURES

SIGMA-MODELS AND STRINGS

M.T. Grisaru

Physics Department
Brandeis University
Waltham, MA 02254, USA

INTRODUCTION

The modern covariant approach to string theory is based on the Polyakov ansatz: The dual model scattering amplitude is given by an expression

$$\sum_{topologies} \int [d\gamma_{\mu\nu}][dX^m] \quad exp[\tfrac{1}{2} \int d\zeta \, d\tau \sqrt{\gamma}\gamma^{\mu\nu}\partial_\mu X_m \partial_\nu X^m]$$

$$\cdot V(k_1,s_1)V(k_2,s_2)\cdots V(k_n,s_n) \tag{1}$$

where $\int [d\gamma_{\mu\nu}][dX^m]$ represents functional integration over all possible embeddings of the two-dimensional world-sheet with coordinates (ζ,τ) into D-dimensional space-time with coordinates X^m, and over all possible (gauge-inequivalent) metrics on the world-sheet, and the sum is over all topologies of the world-sheet. The V's are vertex functions $V(k_i,s_i) = \int d\zeta_i d\tau_i V(k_i,s_i,X_i)$ for emission of particle species s_i with momentum k_i [1].

The above expression can be interpreted as the computation of correlation functions (Green's functions) of composite operators $V(\cdots,X)$ in a two-dimensional field theory described by the action

$$S = \frac{1}{2}\int d\zeta \, d\tau \sqrt{\gamma}\gamma^{\mu\nu}\partial_\mu X_m \partial_\nu X^m \tag{2}$$

with scalar fields $X^m(\zeta,\tau)$ and a nondynamical (gauge-)field $\gamma_{\mu\nu}$. Many of the remarkable features of the string follow from the fact that after gauging away the metric, $\gamma_{\mu\nu} \to \eta_{\mu\nu}$ this is a free field theory and, more importantly, that it is a conformally invariant field theory.

The action is invariant under *general coordinate transformations* ($\sigma^\mu = \zeta,\tau$)

$$
\begin{aligned}
\delta\sigma^\mu &= \xi^\mu(\sigma) \\
\delta\gamma_{\mu\nu} &= -\xi^\rho \partial_\rho \gamma_{\mu\nu} - \partial_\mu \xi^\rho \gamma_{\rho\nu} - \partial_\nu \xi^\rho \gamma_{\mu\rho} \\
&= -(\nabla_\mu \xi_\nu + \nabla_\nu \xi_\mu)
\end{aligned} \tag{3}
$$

The Superworld II
Edited by A. Zichichi
Plenum Press, New York, 1990

(∇ = covariant derivative), while the $X^m(\sigma)$ transform as world scalars. The action is also invariant under *Weyl rescaling*

$$\delta\gamma_{\mu\nu} = 2\Lambda(\sigma)\gamma_{\mu\nu}$$
$$\delta X^m = 0 \tag{4}$$

There exists an old theorem which asserts that a theory invariant under general coordinate transformations and Weyl transformations, when restricted to flat space ($\gamma_{\mu\nu} = \eta_{\mu\nu}$) will be invariant under the transformations of the conformal group (Poincaré transformations, dilatations, and conformal boosts). The matter energy-momentum tensor, defined by

$$T_{\mu\nu} = \frac{1}{\sqrt{\gamma}}\frac{\delta S}{\delta\gamma^{\mu\nu}}\Big|_{\gamma_{\mu\nu}=\eta_{\mu\nu}} = \frac{1}{2}\partial_\mu X_m \partial_\nu X^m - \frac{1}{4}\eta_{\mu\nu}\partial_\rho X_m \partial^\rho X^m \tag{5}$$

is conserved and traceless:

$$\partial_\mu T^{\mu\nu} = T_\mu{}^\mu = 0 \tag{6}$$

The tracelessness is a consequence of the conservation of the dilatation current $\Delta_\mu = \sigma^\nu T_{\mu\nu}$.

In light-cone coordinates $\sigma_\pm = \tau \pm \zeta$ we have, under general coordinate transformations,

$$\delta\gamma_{++} = -2\nabla_+\xi_+ \quad , \quad \delta\gamma_{--} = -2\nabla_-\xi_- \tag{7}$$

In general, any two-dimensional metric can be written as

$$\gamma_{\mu\nu} = \nabla_\mu v_\nu + \nabla_\nu v_\mu + \mathcal{T}_{\mu\nu} \tag{8}$$

where $\mathcal{T}_{\mu\nu} = 0$ for a surface with trivial topology and otherwise corresponds to a Teichmuller deformation. On a zero genus surface, which is all we consider here, we can therefore gauge to zero γ_{++}, γ_{--}. We define this as *conformal gauge*. We take $F_{++} = \gamma_{++}$, $F_{--} = \gamma_{--}$ as gauge-fixing functions. (The remaining component $\gamma_{+-} = e^\phi\eta_{+-}$ could also be gauged to η_{+-} by Weyl rescalings but we don't do that because a) the transformation in general has anomalies and b) in any event it does not lead to any new Faddeev-Popov ghosts.) The gauge variation of the gauge-fixing functions is $\delta F_{\pm\pm} = \nabla_\pm\xi_\pm$ and this leads to a Faddeev-Popov action which, together with the Polyakov action in conformal gauge give the total quantum action

$$S = \int d^2\sigma[\partial_+ X_m \partial_- X^m + b^{++}\nabla_+ c_+ + b^{--}\nabla_- c_-] \tag{9}$$

with ghosts $c_\pm$ and antighosts $b^{\pm\pm}$.

It is convenient to rewrite the action so that it makes sense for any metric. One way to interpret this is to imagine that one has done a background-quantum splitting of the world-sheet metric $\gamma_{\mu\nu} \to \gamma_{\mu\nu} + h_{\mu\nu}$ and gauge-fixed the quantum part $h_{\mu\nu}$ only, maintaining the background reparametrization invariance. The action is then

$$S = \frac{1}{2}\int d^2\sigma\sqrt{\gamma}\gamma^{\mu\nu}[\partial_\mu X_m \partial_\nu X^m + c^\rho\nabla_\mu b_{\nu\rho}] \tag{10}$$

where c^ρ, $b_{\nu\rho}$ are a contravariant vector and a covariant symmetric traceless tensor, respectively. The corresponding expression for the energy-momentum tensor is

$$T_{\mu\nu} = \frac{1}{2}\partial_\mu X_m \partial_\nu X^m + \frac{1}{2}c^\rho \partial_{(\mu} b_{\nu)\rho} + \partial_{(\mu} c^\rho b_{\nu)\rho} - traces \tag{11}$$

The above action is still Weyl invariant and the energy-momentum tensor is traceless. However, quantum corrections destroy the Weyl invariance and give rise to a nonzero trace. To see this one computes the one-loop *unrenormalized* effective action, by integrating out X^m and the ghosts *in the presence of the background metric*. Using dimensional regularization this can be done so as to maintain Weyl invariance. But the result must be renormalized and subtracting out the divergences unavoidably leads to a breaking of the Weyl invariance for the *renormalized* effective action. To be explicit, one finds

$$\Omega_{un} = \int d^n\sigma \sqrt{\gamma}\frac{D-26}{24\pi}\left[\frac{1}{\epsilon}R^{(2)} + \frac{1}{4}R^{(2)}\frac{1}{\Box}R^{(2)}\right] \tag{12}$$

where $R^{(2)}$ is the world-sheet curvature corresponding to the metric $\gamma_{\mu\nu}$. Under the Weyl rescaling in (4) we have, in $n = 2 - \epsilon$ dimensions

$$\sqrt{\gamma} \rightarrow e^{n\Lambda}\sqrt{\gamma}$$
$$\sqrt{\gamma}R \rightarrow e^{-\epsilon\Lambda}\sqrt{\gamma}[R - 2(1-\epsilon)\Box\Lambda + (1-\epsilon)\epsilon(\partial_\mu\Lambda)^2] \tag{13}$$

and one can check that the variation of the first term in (12) cancels that of the second term. On the other hand, the renormalized effective action Ω_{ren} is obtained by removing the first, divergent term in (12) and Weyl invariance is lost. Its variation can be written as

$$\delta\Omega_{ren} = \int d^2\sigma \sqrt{\gamma}\Lambda T_{ren} \tag{14}$$

where T_{ren} is the trace of the *renormalized* energy-momentum tensor, $\sqrt{\gamma}T_{ab,ren} = \delta\Omega_{ren}/\delta\gamma^{\mu\nu}$ r

$$T_{ren} = \frac{D-26}{24\pi}R^{(2)} \tag{15}$$

We observe that this *trace anomaly* is given by minus the coefficient of the pole term $1/\epsilon$ in (12). This is a general feature. If we start with a theory which is Weyl invariant in two dimensions it is possible to regularize the quantum corrections so that the unrenormalized effective action is still Weyl invariant. A simple way to achieve this is to insert in the quantum Lagrangian suitable dependence on the

determinant of the two-dimensional metric. The trace anomaly can be obtained either from the renormalized effective action or, more simply, from the coefficient of the divergence.

When the trace anomaly is not zero the string, which classically has only D degrees of freedom X^m, acquires an extra degree of freedom, the so-called Liouville mode, described by the trace of the world-sheet metric. Although attempts have been made to incorporate this into string dynamics, the generally accepted philosophy is that it should not be present and that for consistency one should require that the renormalized energy-momentum tensor be traceless.

For the string propagating in a flat background the trace anomaly vanishes for $D = 26$. We discuss now propagation in more general backgrounds. The natural assumption is that these backgrounds can be described by fields associated with the particles present in the dual model amplitudes. In particular, in the presence of metric, antisymmetric tensor, dilaton and tachyon fields $g_{mn}(X)$, $B_{mn}(X)$, $\Phi(X)$, $V(X)$ the action in (2) gets generalized to

$$S = \frac{1}{2} \int d^2\sigma \sqrt{\gamma} \{ [\gamma^{\mu\nu} g_{mn} + \epsilon^{\mu\nu} B_{mn}] \partial_\mu X^m \partial_\nu X^n - \frac{\alpha'}{2} \Phi R^{(2)} + 4\alpha' V \} \qquad (16)$$

The string action is now that of a generalized σ-model [2] . It is still classically Weyl invariant for $\alpha' = 0$, and furthermore invariant under transformations in the target manifold: Reparametrizations $X^m \to \xi^m(X)$, $g_{mn} \to g_{mn} - \nabla_{(m}\xi_{n)}$, scale transformations of the fields, and gauge transformations of the antisymmetric tensor field.

As an interacting quantum theory the above action is renormalizable. Quantum corrections lead to divergences that require introducing counterterms

$$g_{mn} \quad \to \quad g_{mn} + \frac{1}{\epsilon} T_{mn}^{(1)} + \frac{1}{\epsilon^2} T_{mn}^{(2)} + \cdots$$

$$B_{mn} \quad \to \quad B_{mn} + \frac{1}{\epsilon} C_{mn}^{(1)} + \frac{1}{\epsilon^2} C_{mn}^{(2)} + \cdots$$

$$\Phi \quad \to \quad \Phi + \frac{1}{\epsilon} \Psi^{(1)} + \frac{1}{\epsilon^2} \Psi^{(2)} + \cdots \qquad (17)$$

(from now on we drop any reference to the tachyon). The β-functions of the theory are related to the coefficients of the simple pole $1/\epsilon$ terms. They are functions of the background fields. The anomaly in the trace of the energy-momentum tensor is now given by a sum of terms with coefficients related to these β-functions. Requiring the vanishing of the trace anomaly leads to constraints on the background fields which can be interpreted as their equations of motion. Much of the activity in this field has centered on the computation of the β-functions and comparison with the equations of motion directly obtainable from the string scattering amplitudes. So far this interpretation has been completely justified but, although very compelling reasons have been given for its correctness no convincing proof exists as yet. Before discussing details we describe the other σ-models of interest.

On the two-dimensional world-sheet it is possible to introduce fermions. For the present discussion it is sufficient to work in conformal gauge and use the light-cone coordinates introduced above. We deal with right- and left-handed spinors denoted by λ, χ. The σ-model of interest has the form

$$S = \frac{1}{2} \int d^2\sigma \{ (g_{mn} + B_{mn}) \partial_+ X^m \partial_- X^n + i\lambda^J [\partial_- \lambda^J + A^J_{-K} \lambda^K]$$

$$+ i\chi^M [\partial_+ \chi^M + B^M_{+N} \chi^N] + U_{JK,MN}(X) \chi^M \chi^N \lambda^J \lambda^K \} \qquad (18)$$

Here the right- and left-handed fermions may be in representations of different internal symmetry groups, the two-dimensional vectors A, B gauge these symmetries, and U transforms covariantly.

Under suitable conditions the above action possesses (p, q) supersymmetry, i.e. is invariant under the action of p right-handed and q left-handed supersymmetry charges Q_i, $\tilde{Q}_j$. To find these conditions one postulates supersymmetry transformations

$$\delta X^m = f^m_{iM}(X)\epsilon^i \chi^M + \tilde{f}^m_{jJ}(X)\tilde{\epsilon}^j \lambda^J \tag{19}$$

and corresponding transformations for the spinors, and requires that the action is invariant, and also that the supersymmetry algebra is satisfied (the commutator of two supersymmetry transformations is a translation). The analysis is straightforward though algebraically complicated. The following cases of interest emerge:

a) Introduce a D-dimensional Yang-Mills field A_m, frame fields $e_m{}^a(X)$ with $e_m{}^a e_{na} = g_{mn}$ and spin-connections

$$
\begin{aligned}
\omega_m^{(\pm)ab} &= \omega_m{}^{ab} \mp H_{mnr} e^{na} e^{rb} \\
H_{mnr} &= \frac{1}{2}(\partial_m B_{nr} + \partial_n B_{rm} + \partial_r B_{mn})
\end{aligned} \tag{20}
$$

and identify

$$
\begin{aligned}
A_-^{JK} &= A_m^{JK} \partial_- X^m \\
B_+^{ab} &= \omega_m^{(+)ab} \partial_+ X^m \\
U_{ab}^{JK} &= \frac{1}{2} F_{ab}^{JK}
\end{aligned} \tag{21}
$$

where F_{mn} is the field strength of the gauge field A_m. Here, the internal symmetry group acting on the χ^M has been identified with the Lorentz group and the χ's become D-dimensional vectors. The action is

$$
\begin{aligned}
S = \frac{1}{2}\int d^2\sigma \{ &(g_{mn} + B_{mn})\partial_+ X^m \partial_- X^n + i\lambda^J[\partial_- \lambda^J + A_m^{JK}\partial_- X^m \lambda^K] \\
&+ i\chi^a[\partial_+ \chi^a + \omega_m^{(+)ab}\partial_+ X^m \chi^b] + \frac{1}{4}F_{ab}^{JK}\lambda^J \lambda^K \chi^a \chi_-^b \}
\end{aligned} \tag{22}
$$

This action has at least $(0,1)$ supersymmetry with $\delta X^m = e_a{}^m \epsilon \chi^a$, and will have $(0, q)$ supersymmetry if the target manifold is a complex manifold supporting $q - 1$ complex structures $J_a^{(r)m}$, $r = 1, q-1$ which, together with the frame field $e_a{}^m$ can be identified with the coefficients f^m_{iM} in the transformation laws (19).

For each r the complex structures are covariantly constant tensors

$$\nabla_m^{(+)} J_a{}^n \equiv \partial_m J_a{}^n + \Gamma_{mr}^{(+)n} J_a{}^r + \omega_{ma}^{(+)b} J_b{}^n = 0 \tag{23}$$

where

$$\Gamma_{mn}^{(\pm)r} = \Gamma_{mn}^{r\,(Christoffel)} \pm H_{mn}^r \tag{24}$$

They satisfy the additional conditions (with $J_n{}^m = J_a{}^m e_n{}^a$)

$$
\begin{aligned}
J_{mn} &= -J_{nm} \\
F_{mn}^{JK} J_r{}^m J_s{}^n &= F_{rs}^{JK} \\
J_n^{(r)m} J_r^{(s)n} + J_n^{(s)m} J_r^{(r)n} &= -2\delta^{rs}\delta_r^m
\end{aligned} \tag{25}
$$

The dimension of the space must be even. Furthermore, unless the manifold is parallelizable (in which case the antisymmetric tensor field strength H_{mnp} is the parallelizing torsion) $q = 0, 1, 2, 4$ only. To see this one determines that the holonomy group $h(\omega^{(+)})$ must commute with the covariantly constant complex structures defined above and one proves that it must be trivial if $q > 4$.

This case describes the heterotic string in a background, with the dimension of the target manifold $D = 10$. So does the next one.

b) In the example above the χ's transform as vectors under tangent space transformations of the target manifold. One can choose them instead to transform as spinors. The case of interest is when the manifold is the manifold for the heterotic string in light-cone gauge, of dimension $D = 8$, so that X^m transforms as a vector (8_v) under $SO(8)$ while χ^α transforms as a right-handed Weyl spinor (8_s). The action can be written as

$$S = \frac{1}{2} \int d^2\sigma \{ (g_{mn} + B_{mn})\partial_+ X^m \partial_- X^n + i\lambda^J[\partial_- \lambda^J + A_m^{JK}\partial_- X^m \lambda^K]$$
$$+ i\chi^\alpha[\partial_+ \chi^\alpha + \frac{1}{4}\omega_m^{(+)ab}(\gamma_{ab})_{\alpha\beta}\partial_+ X^m \chi^\beta] + \frac{1}{8}F_{ab}^{JK}(\gamma^{ab})_{\alpha\beta}\lambda^J \lambda^K \chi^\alpha \chi^\beta \} \quad (26)$$

The action has $(0, q)$ supersymmetry if there exist a set of q covariantly constant spinors $\eta_\alpha^{(r)}$ and therefore $q(q-1)/2$ covariantly constant tensors

$$J_{ab}^{(rs)} = \eta_\alpha^{(r)}(\gamma^{ab})^{\alpha\beta}\eta_\beta^{(s)} \quad (27)$$

that are complex structures, and if the additional condition is satisfied

$$J^{ab}R_{abcd}^{(+)} = 0 \quad (28)$$

where $R^{(+)}$ is the curvature associated with the connection in (24). Again if the holonomy group is nontrivial q cannot be greater than four. By triality this model is equivalent to the one above, when the latter is written in light-cone gauge.

c) Models with (p, q) supersymmetry can be obtained by putting additional restrictions on the examples above. Starting with the case a), one requires that

$$R_{abcd}^{(+)} = R_{cdab}^{(-)} \quad (29)$$

and

$$\omega_m^{(-)ab} = A_m^{\ ab} \quad (30)$$

thus relating the spin connection to the gauge connection. The action becomes

$$S = \frac{1}{2} \int d^2\sigma \{ (g_{mn} + B_{mn})\partial_+ X^m \partial_- X^n) + i\chi^a[\partial_+ \chi^a + \omega_m^{(+)ab}\partial_+ X^m \chi^b]$$
$$+ i\lambda^a[\partial_- \lambda^a + \omega_m^{(-)ab}\partial_- X^m \lambda^b] + \frac{1}{2}R_{abcd}^{(-)}\lambda^a \lambda^b \chi^c \chi^d \} \quad (31)$$

(plus that of additional free fermions). This gives a model with at least $(1, 1)$ supersymmetry. To have (p, q) supersymmetry one needs complex structures satisfying conditions similar to the ones given above.

8

d) The supersymmetric σ-models can be written more conveniently in superspace. For systems with $(0,1)$ supersymmetry we introduce a left-handed spinorial coordinate on the world-sheet, $\bar{\theta}$, and scalar and spinor superfields

$$
\begin{aligned}
\Psi^m(\sigma,\bar{\theta}) &= X^m(\sigma) + \bar{\theta}\chi^m(\sigma) \\
\Lambda^J(\sigma,\bar{\theta}) &= \lambda(\sigma) + \bar{\theta}F(\sigma)
\end{aligned}
\tag{32}
$$

We also introduce a spinorial derivative

$$
\bar{D} = \frac{\partial}{\partial\bar{\theta}} + \bar{\theta}\partial_-
\tag{33}
$$

The action for the σ-model can be written as

$$
S = \int d^2\sigma d\bar{\theta}\{(g_{mn} + B_{mn})\bar{D}\Psi^m\partial_+\Psi^n + \Lambda^J[\bar{D}\Lambda^J + A_m^{JK}\bar{D}\Psi^m\Lambda^K]\}
\tag{34}
$$

If the theory has $(1,1)$ supersymmetry, with $\omega^{(+)} = \omega^{(-)}$, it can be rewritten in terms of $(1,1)$ superfields. We introduce a second spinorial coordinate, θ, and a superfield

$$
\psi^m(\sigma,\theta,\bar{\theta}) = \Psi^m + \theta\Lambda^m
\tag{35}
$$

The action becomes

$$
S = \int d^2\sigma d^2\theta (g_{mn} + B_{mn})D\psi^m\bar{D}\psi^n
\tag{36}
$$

If the model has $(2,2)$ supersymmetry the target manifold is a Kahler manifold and is best described by introducing yet another pair of θ's and chiral superfields $\phi^\mu(\theta,\bar{\theta})$ which are complex coordinates on the manifold. The action is simply

$$
S = \int d^2\sigma d^2\theta d^2\bar{\theta} K(\phi^\mu,\bar{\phi}^\nu)
\tag{37}
$$

where K is the Kahler potential. The metric is given by

$$
g_{\mu\bar{\nu}} = \frac{\partial^2 K}{\partial\phi^\mu\partial\bar{\phi}^\nu}
\tag{38}
$$

If the manifold is hyperKahler, the σ-model has $(4,4)$ supersymmetry.

e) A different kind of σ-model is obtained by considering the superstring Green-Schwarz action in the presence of background supergravity fields. We consider here the heterotic case, with a ten-dimensional target manifold. The system is described by ten space-time coordinates X^m as before, and by a set of right-handed world-sheet spinors λ^J, but instead of the left-handed world-sheet spinors χ^a we have a set of ten-dimensional Majorana-Weyl spinors θ^α, $(\alpha = 1,\ldots,16)$ that, together with the X^m can be interpreted as the coordinates of ten-dimensional superspace. In the absence of any background the Green-Schwarz action is

$$
S = \int d^2\sigma[V_-^a V_+^a + V_-^a(\partial_+\theta^\alpha\Gamma_{\alpha\beta}^a\theta^\beta) - V_+^a(\partial_-\theta^\alpha\Gamma_{\alpha\beta}^a\theta^\beta)
$$

where
$$
+ \lambda^J\partial_-\lambda^J]
\tag{39}
$$

$$V_{\pm}^a = \partial_{\pm} X^a + \partial_{\pm}\theta^\alpha \Gamma_{\alpha\beta}^a \theta^\beta \tag{40}$$

The action is invariant under ten-dimensional supersymmetry transformations

$$\delta X^a = \epsilon^\alpha \Gamma_{\alpha\beta}^a \theta^\beta \quad , \quad \delta\theta^\alpha = \epsilon^\alpha \tag{41}$$

and under a two-dimensional fermionic symmetry

$$\delta_\kappa \theta^\alpha = V_-^a \Gamma_{\alpha\beta}^a \kappa_+^\beta \quad , \quad \delta_\kappa X^a = -\delta\theta^\alpha \Gamma_{\alpha\beta}^a \theta^\beta \tag{42}$$

provided the Virasoro constraint $V_-^2 = 0$ is satisfied. Here κ is a two-dimensional self-dual vector and a ten-dimensional spinor. The spinors λ^J are invariant under these transformations.

The propagation of the Green-Schwarz superstring in a supergravity and super-Yang-Mills background turns it into a σ-model as follows [3] : We introduce first the supervielbein $E_M^A(X,\theta)$, a scalar (dilaton) superfield $\phi(X,\theta)$ a super-two-form field $B_{MN}(X,\theta)$, and a super-Yang-Mills potential $A_M(X,\theta)$, functions of local coordinates

$$Z^M = (X^m, \theta^\mu) \tag{43}$$

In *flat* superspace the supervielbein has the form

$$E_M^A = \begin{pmatrix} e_m{}^a = \delta_m^a & e_m{}^\alpha = 0 \\ e_\mu{}^a = \delta_\mu^\alpha \Gamma_{\alpha\beta}^a \theta^\beta & e_\mu{}^\alpha = \delta_\mu^\alpha \end{pmatrix} \tag{44}$$

while the only nonvanishing component of B is $B_{\alpha a} = 1/2\Gamma_{a\alpha\beta}\theta^\beta$. We define now

$$V_{\pm}^a = \partial_{\pm} Z^M E_M^A \tag{45}$$

It is easy to check that in flat superspace it reduces to the expression in (40).

The superfields we have introduced are required to satisfy the torsion and curvature constraints of ten-dimensional supergravity coupled to super-Yang-Mills (and, as a consequence of the Bianchi identities, the field equations). Under these conditions the action

$$S = \int d^2\sigma \{ [\phi V_-^a V_+^a + \partial_- Z^N \partial_+ Z^M B_{MN}] + \lambda^J [\partial_- \lambda^J - A_-^{JK} \lambda^K] \} \tag{46}$$

with

$$A_- = \partial_- Z^M A_M \tag{47}$$

has local ten-dimensional supersymmetry, ten-dimensional Yang-Mills invariance, and still has local two-dimensional κ-symmetry, given this time by

$$\delta_\kappa Z^M E_M^a = 0 \quad , \quad \delta_\kappa Z^M E_M^\alpha = V_-^a \Gamma_a^{\alpha\beta} \kappa_{+\beta} \tag{48}$$

provided the supergravity superfields transform as scalars, while the Yang-Mills potential and the λ's undergo a gauge transformation with parameter

$$\Lambda = \delta_\kappa Z^M A_M \tag{49}$$

The consistency of the procedure requires that space-time is ten-dimensional and that the background satisfies the classical field equations already at the classical level, whereas for the other σ-models the dimensionality of the target manifold and the background equations of motion emerge only at the quantum level. The quantization of the Green-Schwarz action (in flat space or a background) requires gauge-fixing the κ invariance, a task that has turned out to be difficult except in light-cone gauge where, under suitable simplifying assumptions the action reduces to that of (26).

QUANTIZATION

The σ-models we have described (except the Green-Schwarz action) can be quantized as ordinary two-dimensional field theories of fields X^m, λ, χ and the corresponding effective action can be computed perturbatively by separating the Lagrangian into a quadratic part and interactions and developing ordinary Feynman rules. This procedure has the serious drawback that it does not manifestly preserve the symmetries of the target manifold (reparametrization invariance, etc.). One uses instead the background-field method and a particular quantum-background splitting, using normal coordinates, which leads to Feynman rules and an effective action where the symmetries are manifest [4]. For simplicity we describe the procedure for a bosonic σ-model given by the action

$$S = \frac{1}{2} \int d^2\sigma \, g_{mn} \partial_\mu \phi^m \partial^\mu \phi^n \tag{50}$$

where the ϕ's are coordinates of a manifold with metric g_{mn}. The same procedure however works in the presence of additional terms, and also in superspace. The action is invariant under reparametrizations $\phi^m \rightarrow \phi^m + f^m(\phi)$, $g_{mn} \rightarrow g_{mn} - \nabla_{(m} f_{n)}$.

The usual background-field method splits

$$\phi^m(\sigma) = \phi_{cl}^m(\sigma) + \pi^m(\sigma) \tag{51}$$

and treats π^m as a quantum field, expanding the action in powers of π^m and computing an effective action that depends on ϕ_{cl}^m. However π^m does not transform as a vector under reparametrizations and consequently the manifest invariance is lost. One introduces instead, as a quantum field, $\xi^m(\sigma)$, the tangent vector to the geodesic joining the points with coordinates ϕ_{cl}^m and $\phi_{cl}^m + \pi^m$ whose length is equal to the length of the geodesic. The geodesic is parametrized by $\phi^m(s)$ with $\phi^m(0) = \phi_{cl}^m$, $\phi^m(1) = \phi_{cl}^m + \pi^m$ and the tangent vector is $\xi^m(s) = d\phi^m/ds$, satisfying the geodesic equation

$$\frac{d^2\phi^m}{d^2 s} + \Gamma^m_{np} \frac{d\phi^n}{ds} \frac{d\phi^p}{ds} = 0 \tag{52}$$

Since

$$\phi^m(1) = \sum \frac{1}{N!} \frac{d^N\phi^m}{ds^N}\big|_{s=1} \tag{53}$$

repeated differentiation of the geodesic equation gives the relation between π^m and ξ^m.

$$\pi^m = \xi^m - \frac{1}{2}\Gamma^m_{np}\xi^n\xi^p - \frac{1}{3}\Gamma^m_{npq}\xi^n\xi^p\xi^q - \cdots \tag{54}$$

a nonlinear redefinition of the quantum field. Here Γ^m_{np} is the Christoffel connection while the higher coefficients are derivatives thereof defined by

$$\Gamma^m_{p_1\cdots p_{N+1}} = \partial_{(p_{N+1}}\Gamma^m_{p_1\cdots p_N)} - N\Gamma^q_{(p_1p_2}\Gamma_{p_3\cdots p_{N+1})q} \tag{55}$$

In principle, given any function of ϕ^m one can insert the expression in terms of ϕ^m_{cl} and π^m and rearrange the Taylor series as a power series in ξ^m, with coefficients that are guaranteed to be covariant. Instead, the following algorithm is much more efficient [5]:

For a given point with coordinates ϕ^m_{cl} one can choose *normal coordinates* so that the Christoffel connection and all its derivatives vanish at that point. In normal coordinates $\pi^m = \xi^m$ and therefore *in this coordinate system* the σ-model Lagrangian has an ordinary Taylor expansion

$$\mathcal{L}(\phi + \xi) = \sum \frac{1}{N!}(\partial_{k_1\cdots k_N}L(\phi))\xi^{k_1}\cdots\xi^{k_N} \tag{56}$$

The problem is to rewrite this in an arbitrary coordinate system.

The N-th order term in the expansion of the Lagrangian can be rewritten as

$$\mathcal{L}^{(N)} = \frac{1}{N!}\int[\partial^{\sigma_1}_{m_1}\cdots\partial^{\sigma_N}_{m_N}\mathcal{L}(\phi(\sigma))]\xi^{m_1}(\sigma_1)\cdots\xi^{m_N}(\sigma_N)d\sigma_1\cdots d\sigma_N \tag{57}$$

where we have defined the functional derivative

$$\partial^\sigma_m = \frac{\delta}{\delta\phi^m(\sigma)} \tag{58}$$

We note that

$$\int d\sigma\xi^m(\sigma)\partial^\sigma_m\xi^n(\sigma') = 0 \tag{59}$$

since this takes the directional derivative of $\xi^n(\sigma')$ along the straight line (in normal coordinates) joining ϕ^m_{cl} and $\phi^m_{cl} + \xi^m$. Therefore,(57) can be rewritten as

$$\mathcal{L}^{(N)} = \frac{1}{N!}\int d\sigma_1\xi^m_1(\sigma_1)\partial^{\sigma_1}_{m_1}\int d\sigma_2\xi^m_2(\sigma_2)\partial^{\sigma_2}_{m_2}\cdots$$
$$\cdots\int d\sigma_N\xi^m_N(\sigma_N)\partial^{\sigma_N}_{m_N}\mathcal{L}(\phi(\sigma)) \tag{60}$$

However this expression is valid in arbitrary coordinates since the operator

$$\int d\sigma\xi^m(\sigma)\partial^\sigma_m \tag{61}$$

maps scalars into scalars. To make this into a useful statement we note that in arbitrary coordinates we can replace (61) by

$$\Delta = \int d\sigma\xi^m(\sigma)D^\sigma_m \tag{62}$$

and (59) by

$$\Delta\xi^l(\sigma) = 0 \tag{63}$$

where the functional covariant derivative is defined by its action on a vector

$$D^\sigma_m V^p(\phi(\sigma')) = [\partial_m V^p(\phi(\sigma)) + \Gamma^p_{mq}(\phi(\sigma))V^q(\phi(\sigma))]\delta(\sigma - \sigma') \tag{64}$$

The background field expansion of the action can be obtained now by repeated application of the operator Δ, using (63) and the following relations which are easily derived

$$
\begin{aligned}
\Delta T_{k_1 \cdots k_n}(\phi(\sigma)) &= T_{k_1 \cdots k_n; m} \xi^m(\sigma) \\
\Delta(\partial_\mu \phi^m(\sigma)) &= D_\mu \xi^m(\sigma) \\
\Delta(D_\mu \xi^m(\sigma)) &= R^m_{pqr} \partial_\mu \phi^p \xi^q \xi^r(\sigma)
\end{aligned}
\tag{65}
$$

Using the algorithm, it is easy to obtain the following terms in the expansion of the action:

$$
\begin{aligned}
S^{(0)} &= \frac{1}{2} \int g_{mn} \partial_\mu \phi^m \partial^\mu \phi^n \\
S^{(1)} &= \int g_{mn} \partial_\mu \phi^m D^\mu \xi^n \\
S^{(2)} &= \frac{1}{2} \int R_{mnpq} \partial_\mu \phi^m \partial^\mu \phi^q \xi^n \xi^p + \frac{1}{2} \int g_{mn} D_\mu \xi^m D^\mu \xi^n \\
S^{(3)} &= \frac{1}{6} \int R_{mnpq;r} \partial_\mu \phi^m \partial^\mu \phi^q \xi^n \xi^p \xi^r + \frac{2}{3} \int R_{mnpq} \partial_\mu \phi^m D^\mu \xi^q \xi^n \xi^p \\
S^{(4)} &= \frac{1}{24} \int (R_{mnpq;rs} + 4 R^t_{npm} R_{trsq}) \partial_\mu \phi^m \partial^\mu \phi^q \xi^n \xi^p \xi^r \xi^s \\
&\quad + \frac{1}{4} \int R_{mnpq;r} \partial_\mu \phi^m D^\mu \xi^q \xi^n \xi^p \xi^r + \frac{1}{6} \int R_{mnpq} D_\mu \xi^m D^\mu \xi^q \xi^n \xi^p
\end{aligned}
\tag{66}
$$

etc. This form is useful for deriving Feynman rules for the quantum field ξ^m and calculating the effective action up to two-loop order. Higher order terms are straightforward to obtain.

We note for further reference that exactly the same expansion is valid for the superspace model with $(1,1)$ supersymmetry described by (36). One simply has to replace the fields ϕ, ξ by corresponding superfields and the world-sheet partial derivatives ∂_μ by corresponding spinorial derivatives.

We turn now to a discussion of the perturbative evaluation of the background effective action and its renormalization, using again the bosonic case above as a simple example.

As a first step we simplify the part of the action quadratic in the quantum fields ξ^m by referring them to tangent frames on the manifold:

$$
\xi^a = e^a{}_m(\phi) \xi^m
\tag{67}
$$

so that

$$
D_\mu \xi^a = \partial_\mu \xi^a + \omega_m^{ab} \partial_\mu \phi^m \xi^b
\tag{68}
$$

The quadratic part of the action becomes then

$$
\int d^2\sigma D_\mu \xi^a D^\mu \xi^a
\tag{69}
$$

from which one obtains the free propagator (by separating out the ordinary derivative term) or a covariant propagator (the inverse of the operator $D_\mu D^\mu$). The other terms in the expansion of the action can also be written in terms of ξ^a with tangent space indices. The resulting Feynman diagrams one obtains contain vertices where the quantum fields interact with themselves, with coefficients that depend on the background fields and can be regarded as generalized coupling constants.

Feynman diagrams can be computed in standard fashion, in momentum space, with propagators $p^{-2}\delta^{ab}$. They must be regularized, both in the ultraviolet and the infrared. For the UV it is convenient to use dimensional regularization in $n = 2 - \epsilon$ dimensions, and to handle the infrared divergences one can introduce a mass or use IR subtraction techniques.

In the background-field method two kinds of renormalization are required. One must renormalize the quantum fields and Green's functions containing external quantum field lines, and one must renormalize the background fields and the background effective action. For ordinary field theories the procedure is fairly well understood, and the relevant issues have also been dealt with for the present situation [6]. For the purpose of computing the divergences of the background effective action the following procedure has proven useful: Compute Feynman diagrams with only external background lines using dimensional regularization. At any loop order subtract *sub*divergences using BPHZ techniques (i.e. for each divergent one-loop subintegral subtract out the pole part; then, for each divergent two-loop subintegral including the one-loop subtraction subtract the pole part, etc). This is equivalent to introducing counterterms for the renormalization of the quantum fields. One is left in the end with an overall divergence which is a local function of the background fields and their momenta and can be removed by a counterterm. In the renormalizable theories we are considering such a counterterm has the same form as one of the couplings in the original action. The β-function corresponding to such couplings is related to the coefficient of the $1/\epsilon$ part of the counterterm.

We summarize now some of the results. We consider first the bosonic model. Starting with the action (66) one computes the divergences of the background effective action at any loop order and finds that they can be cancelled by replacing g_{mn} by

$$g_{mn}^B = g_{mn} + \sum_{n=1}^{\infty} \frac{1}{\epsilon^n} T_{mn}^{(n)}(g) \tag{70}$$

The corresponding β-function is given by

$$\beta_{mn}(g) = \epsilon g_{mn} + (1 + \lambda \frac{\partial}{\partial \lambda}) T_{mn}^{(1)}(\lambda^{-1} g)|_{\lambda=1} \tag{71}$$

At the one- and two-loop level one finds

$$\beta_{mn}^{(1)} = \frac{1}{2\pi} R_{mn}$$
$$\beta_{mn}^{(2)} = -\frac{2}{3(4\pi)^2} R_{p(qr)(m} R_{n)}{}^{pqr} \tag{72}$$

Recently, calculations up to two-loop order for the σ-model with torsion, i.e. including coupling to the antisymmetric tensor field, have been presented [7]. Starting with the classical action

$$S = \frac{1}{2} \int d^2\sigma [g_{mn} \partial_\mu \phi^m \partial^\mu \phi^n + \epsilon^{\mu\nu} B_{mn} \partial_\mu \phi^m \partial_\nu \phi^n] \tag{73}$$

one performs a normal coordinate expansion as before. One finds

$$S^{(2)} = \frac{1}{2}\int d^2\sigma[g_{mn}\bar{\nabla}^\mu\xi^m\bar{\nabla}_\mu\xi^n + (\eta^{mn}+\epsilon^{mn})\bar{R}_{mnpq}\partial_\mu\phi^m\partial_\nu\phi^q\xi^n\xi^p]$$

$$S^{(3)} = \frac{1}{3}\int d^2\sigma[\epsilon^{\mu\nu}H_{mnp}\xi^m\bar{\nabla}_\mu\xi^n\bar{\nabla}_\nu\xi^p$$
$$+(\eta^{\mu\nu}\bar{R}_{m(np)q}+\epsilon^{\mu\nu}\bar{R}_{m[np]q})\partial_\mu\phi^n\bar{\nabla}_\nu\xi^p\xi^m\xi^q$$
$$+\frac{1}{2}(\eta^{\mu\nu}+\epsilon^{\mu\nu})(\bar{R}_{mnpq;r}-\bar{R}_{mnp}{}^s H_{qrs})\partial_\mu\phi^m\partial_\nu\phi^q\xi^n\xi^p\xi^r] \tag{74}$$

where

$$\bar{\nabla}^\mu\xi^m = \nabla^\mu\xi^m + \epsilon^{\mu\nu}H^m{}_{np}\partial_\nu\phi^n\xi^p$$

$$H_{mnp} = \frac{1}{2}(\nabla_m B_{np} + cyclic)$$

$$\bar{R}_{mnpq} = R_{mnpq} + H_{mn[q;p]} + H^s{}_{p[n}H_{sq]m} \tag{75}$$

The calculation of the divergences is straightforward except for the handling of the $\epsilon^{\mu\nu}$ in dimensional regularization, which has caused a certain amount of confusion in the literature. The correct prescription appears to be that, for example in minimal subtraction, one should keep it in two dimensions throughout the calculation and use all the two-dimensional identities it satisfies. (However, other prescriptions may correspond to a change in renormalization scheme.) One finds the gravitational and antisymmetric tensor β-functions

$$\beta^{(1)}_{mn} = \frac{1}{2\pi}(R_{mn}+H_{mpq}H^{pq}_n)$$

$$\bar{\beta}^{(1)}_{mn} = \frac{1}{2\pi}H_{mnp}{}^{;p} \tag{76}$$

and

$$\beta^{(2)}_{mn} = -\frac{1}{(4\pi)^2}[\frac{2}{3}\bar{R}_{pqr(m}\bar{R}_{n)}{}^{pqr}+\frac{4}{9}\bar{\nabla}_m H_{pqr}\bar{\nabla}_n H^{pqr}$$
$$+2\bar{R}_{p(mn)q}H^{prs}H^q{}_{rs}+\frac{4}{3}H_{pq(m}\bar{\nabla}_{n)}H^{prs}H^q{}_{rs}]$$

$$\bar{\beta}^{(2)}_{mn} = \frac{1}{(4\pi)^2}[\frac{2}{3}\bar{R}_{p(qr)[m}\bar{R}_{n]}{}^{pqr}$$
$$-2\bar{R}_{p[mn]q}H^{prs}H^q{}_{rs}-\frac{4}{3}H_{pq[m}\bar{\nabla}_{n]}H^{prs}H^q{}_{rs}] \tag{77}$$

We turn now to supersymmetric σ-models and consider in particular cases with $N = 1$ $(1,1)$ and with $N = 2$ $(2,2)$ supersymmetry. The calculations are most efficiently carried out in superspace, using supergraphs. We start by considering the gravitational β-function for the σ-model described by the action in (36) with $N = 1$ superfields $\psi^m(\sigma,\theta)$. We set the antisymmetric tensor to zero. As we have already mentioned the background-field expansion takes exactly the same form as for the bosonic expansion in (66). Using supergraphs and supersymmetric dimensional regularization one finds a one-loop result

$$\beta^{(1)}_{mn} = \frac{1}{2\pi}R_{mn} \tag{78}$$

15

There are no two- or three-loop contributions to the β-function. At the four-loop level we find [8]

$$\beta_{mn}^{(4)} = \frac{\zeta(3)}{3(4\pi)^4} T_{(mn)} \tag{79}$$

where

$$
\begin{aligned}
T_{mn} &= 2R_{pmnq;rs}(R^{quvs}R_{(uv)}{}^{p}{}^{r} + R^{quvp}R_{uv}{}^{s}{}^{r}) + 4R_{npqr;[ms]}R^{q(up)v}R_{vu}{}^{r}{}^{s} \\
&+ 3(R_{mpqr;s}R^{ruvs}R_n{}^{pq}{}_{v;u} + R_{mpqr;s}R^r{}_u{}^{pv}R_n{}^{us}{}_{v}{}^{;q} + 2R_{mpqr;s}R^r{}_u{}^{sv}R_n{}^{up}{}_{v}{}^{;q}) \\
&+ (2R_{pqrs;m} - R_{prqs;m})R^s{}_v{}^{pu}R^{vqr}{}_{u;n} \\
&- 12R_{pqrm}R_{nuv}{}^{p}(R^r{}_{st}{}^{u}R^{vstq} + R^r{}_{st}{}^{v}R^{quts})
\end{aligned} \tag{80}
$$

We discuss next the $N = 2$ σ-model. The target manifold must then be a Kahler manifold. As already indicated, it is described by chiral superfields $\phi^{\mu}(\sigma, \theta, \bar{\theta})$ which are complex coordinates on the manifold, and an action

$$S = \int d^2\sigma d^2\theta d^2\bar{\theta} K(\phi^{\mu}, \bar{\phi}^{\bar{\nu}}) \quad \theta = \theta_{\alpha} , \quad \bar{\theta} = \bar{\theta}_{\alpha} \quad \alpha = 1, 2 \tag{81}$$

Because of the chirality constraint $\bar{D}_{\alpha}\phi^{\mu} = 0$ a normal coordinate expansion does not seem convenient and one resorts to a linear splitting $\phi = \phi_{cl} + \xi$ for doing the background field expansion. Using $N = 2$ supergraphs one finds divergences at the one- and four-loop level which can be removed by counterterms that can be viewed as additions to the Kahler potential

$$
\begin{aligned}
S + \Delta S &= \int d^2\sigma d^2\theta d^2\bar{\theta} [K + \frac{1}{\epsilon}\Delta K^{1-loop} + \frac{\zeta(3)}{6(4\pi)^4\epsilon}\Delta K^{4-loop} \\
&+ higher - order \ poles]
\end{aligned} \tag{82}
$$

$$
\begin{aligned}
\Delta K^{1-loop} &= -tr ln K_{\mu\bar{\nu}} \\
\Delta K^{4-loop} &= R_{\kappa\bar{\lambda}\mu\bar{\nu}}R^{\sigma\bar{\lambda}\tau\bar{\nu}}R_{\sigma}{}^{\kappa}{}_{\tau}{}^{\mu} - R_{\kappa\bar{\lambda}\mu\bar{\nu}}R^{\mu\bar{\nu}\sigma\bar{\tau}}R_{\sigma\bar{\tau}}{}^{\kappa\bar{\lambda}}
\end{aligned} \tag{83}
$$

where

$$K_{\mu\bar{\nu}} = \frac{\partial^2 K}{\partial\Phi^{\mu}\partial\bar{\Phi}^{\bar{\nu}}} \equiv \partial_{\mu}\partial_{\bar{\nu}}K \tag{84}$$

To this order the β-function is given by [9]

$$\beta_{\mu\bar{\nu}} = -\frac{1}{2\pi}R_{\mu\bar{\nu}} - \frac{4\zeta(3)}{3(4\pi)^4}\partial_{\mu}\partial_{\bar{\nu}}\Delta K^{4-loop} \tag{85}$$

The result agrees with that obtained in the $N = 1$ case when the latter is restricted to a Kahler manifold.

In some recent work we have looked at the $N = 2$ five-loop situation where, a priori, one can expect new contributions. However, it turns out that although divergences are present they are such that by a change in the subtraction scheme one can eliminate any contribution to the β-function [10].

So far we have not discussed the dilaton β-function. In principle it can be determined by direct calculation of quantum corrections on a curved world-sheet.

However this turns out not to be necessary. It can be shown that renormalization group methods allow one to determine β^Φ from a knowledge of the other β-functions and the renormalization of composite operators such as $\partial\phi\partial\phi$ [11]. Alternatively, the same goal can be achieved by consistency requirements and certain plausible identifications, as we describe now [12]:

In general, the requirement of vanishing β-functions is slightly stronger than that of conformal invariance. The correct condition is

$$\beta_{mn} + \nabla_{(m}V_{n)} = 0 \tag{86}$$

for some vector V_n. At the one loop level let us write the condition for the vanishing of the purely gravitational β-function

$$R_{mn} + \nabla_{(m}V_{n)} = 0 \tag{87}$$

By taking the curl and the divergence of this equation and using the Bianchi identities it is possible to show that

$$V_m = \frac{1}{2}\nabla_m\Phi \tag{88}$$

for some scalar field Φ. The condition (87) becomes

$$R_{mn} - \nabla_m\nabla_n\Phi = 0 \tag{89}$$

which is exactly the same equation that would express the vanishing of the one-loop gravitational β-function including dilaton contributions. Clearly, one should identify the scalar above with the dilaton field.

The analysis can be carried out to the two-loop level, and again one obtains an integrability condition for V_m which allows one to identify it with the gradient of the dilaton field giving

$$R_{mn} + \frac{1}{2}R_{mpqr}R_n{}^{pqr} - \nabla_m\nabla_n\Phi = 0 \tag{90}$$

Furthermore, taking the divergence of this equation, using the Bianchi identities and using the equation itself one find that to this order

$$\nabla_m\left(-R + 2\nabla^2\Phi + (\nabla\Phi)^2 - \frac{1}{4}R_{pqrs}R^{pqrs}\right) = 0 \tag{91}$$

When integrated this leads to

$$-R + 2\nabla^2\Phi + (\nabla\Phi)^2 - \frac{1}{4}R_{pqrs}R^{pqrs} = const. \tag{92}$$

and one can argue that the constant is zero.

It is reasonable to identify this equation with the condition for vanishing of the dilaton β-function and to identify the left hand side with β^Φ.

This procedure can be generalized at any loop order, in an iterative approach in powers of α'. Starting with

$$R_{mn} + T_{mn} + \nabla_{(m}V_{n)} = 0 \tag{93}$$

one can conclude that $V_m = 1/2\nabla_m\Phi$ provided

$$\nabla_n T_m{}^n + \nabla_n\Phi T_m{}^n = \frac{1}{2}\nabla_m T \tag{94}$$

for some scalar function T. If this integrability condition is satisfied one solves for T and obtains

$$\beta^\Phi = -R + 2\nabla^2\Phi + (\nabla\Phi)^2 + T \tag{95}$$

Therefore, provided one can satisfy (94) and carry out the algebra, the dilaton β-function can be determined without any new loop calculations.

While the above identification of β^Φ is plausible it does not replace the more rigorous methods mentioned above. However, for the $N = 2$ case we have found another way to determine it, and we describe now the procedure [13]:

The action for the $N = 2$ σ-model can be rewritten in terms of N=1 superfields by defining

$$\theta_{1\alpha} = \theta_\alpha + \bar\theta_\alpha \qquad \theta_{2\alpha} = \theta_\alpha - \bar\theta_\alpha \tag{96}$$

and integrating out $\theta_{2\alpha}$. We obtain, using $\int d\theta_\alpha = \partial_\alpha \sim D_\alpha$

$$\int d^2x\, d^2\theta_1 g_{\mu\bar\nu} D_{1\alpha}\psi^\mu D_1^\alpha \bar\psi^{\bar\nu} \quad , \qquad g_{\mu\bar\nu} = K_{\mu\bar\nu} \tag{97}$$

where we have used the definition

$$\psi^\mu(\theta_1) = \phi^\mu(\theta,\bar\theta)|_{\theta_2=0} \tag{98}$$

and the chirality condition $\bar D\phi = 0$. To this, we could add the dilaton term

$$\int d^2x\, d^2\theta_1 R^{(2)}\Phi(\psi) \tag{99}$$

where $R^{(2)}$ is the $N = 1$ superspace world-sheet curvature. Radiative corrections to this term should come from corresponding radiative corrections to the N=2 theory. However, in the N=2 theory all radiative corrections are corrections to the Kahler potential, $K \to K + \Delta K$ and these only seem to give rise to corrections to the Kahler metric, $g_{\mu\bar\nu} \to g_{\mu\bar\nu} + \Delta g_{\mu\bar\nu}$. (In principle in the N=2 theory one can write a superpotential term $\sim \int d^2\theta R\Phi$, where R is the N=2 world-sheet curvature, but it does not get renormalized, according to standard arguments.)

One's first reaction is to conclude that for the N=2 theory the dilaton β-function vanishes. This however is not possible, because if one starts with the N=1 theory, computes the two-loop $\beta^\Phi(g_{ij}) \sim R(g_{ij})$, and restricts to a Kahler manifold one certainly does not get zero. Since one has no reason to believe that the nonrenormalization theorem (of F-terms) is violated, the answer must lie elsewhere.

We proceed as follows: To determine the dilaton β-function we must stay on a curved world-super-sheet, i.e. keep some background supergravity present, and do the calculations in a manner that respects local N=2 supersymmetry. In particular, in dimensional regularization, we should include coupling to supergravity in $2 - \epsilon$ dimensions. In that case, the N=2 action can be written (in superconformal gauge) as

$$\int d^2x\, d^4\theta e^{\epsilon(\sigma+\bar\sigma)} K(\phi,\bar\phi) \tag{100}$$

18

where σ is the (chiral) compensator of N=2 supergravity. Now, if we reduce to N=1 as above, we obtain

$$\int d^2x d^2\theta_1 e^{\epsilon\sigma_1}\{g_{\mu\bar{\nu}}D_{1\alpha}\psi^\mu D_1^\alpha\bar{\psi}^{\bar{\nu}} + \frac{\epsilon}{2}[D^2\sigma_1 - \frac{\epsilon}{4}D_{1\alpha}\sigma_1 D_1^\alpha\sigma_1]K(\psi,\bar{\psi})\} \qquad (101)$$

Here

$$\sigma_1 = \sigma|_{\theta_2=0} \qquad (102)$$

and the second term in (14) is a dilaton-like term since

$$D^2\sigma_1 - \frac{\epsilon}{4}D_{1\alpha}\sigma_1 D_1^\alpha\sigma_1 \equiv R^{(2)} \qquad (103)$$

Note that *at $\epsilon = 0$ the dilaton term disappears.*

However, if we perform the same reduction on $K + \Delta K$ we find

$$\int d^2x d^2\theta_1 e^{\epsilon\sigma_1}[(g + \Delta g)_{\mu\bar{\nu}}D_{1\alpha}\psi^\mu D_1^\alpha\bar{\psi}^{\bar{\nu}} + \frac{\epsilon}{2}R^{(2)}\cdot(K + \Delta K)] \qquad (104)$$

and the $R^{(2)}$ term survives if ΔK has higher order poles. In particular since β^Φ is related to the first order pole of the coefficient of $R^{(2)}$ *the dilaton β-function is determined by the second order $1/\epsilon^2$ poles of ΔK.* Furthermore, by the renormalization group pole equations, the $1/\epsilon^2$ $(L + 1)$-loop ΔK is determined by the $1/\epsilon$ L-loop ΔK. Thus, at any loop order we have $\beta^{\Phi(L+1)}$ determined essentially by $\beta_{\mu\bar{\nu}}^{(L)}$:

$$\beta^{\Phi(L+1)} = -\partial_\eta\sum_{L'}\Delta K_1^{(L-L')}(g_{\rho\bar{\sigma}} - \eta L'\beta_{\rho\bar{\sigma}}^{(L')})|_{\eta=0} \qquad (105)$$

where $\Delta K_1^{(L-M)}$ is the $1/\epsilon$ pole part of the $(L - M)$-loop correction to the Kahler potential.

SIGMA-MODELS AND STRINGS

In this section we discuss the relationship between σ-models and strings. With the vanishing of the β-functions interpreted as equations of motion for the massless string fields one derives a corresponding Lagrangian from which these equations can be obtained. On the other hand, starting with dual model S-matrix elements it is possible to construct an effective low-energy Lagrangian that gives tree amplitudes corresponding to these S-matrix elements and compare the two. No disagreement has been found so far.

In his original paper Lovelace [14] approached the subject as follows: Consider (in the present context) a two-dimensional field theory defined on a cylinder (the world-sheet of the closed string) and conformally map the cylinder into the z-plane so that $\tau = -\infty$ becomes the origin, while $\tau = +\infty$ becomes the point at infinity. Quantization becomes radial quantization: The Hamiltonian is the dilation operator and time ordering is radial ordering. One can develop perturbation theory in the standard way and using operator product expansions one obtains a set of Feynman rules where the graphs (including loops) of the conventionally quantized field theory are tree graphs that turn out to be identical to those that represent scattering in the dual model.

Lovelace considers the two-dimensional bosonic model of (16). The ultraviolet divergences of the usual approach can be related to divergences that occur in the dual-model amplitudes, i.e. to situations when one of the internal lines of a dual-model tree graph is on the mass-shell. In particular, the β-function corresponds to tree graphs where on one side of the on-shell propagator one has a tadpole amplitude, i.e. an amplitude corresponding to a particle going into the vacuum. Furthermore, for renormalizable theories, the mass of the on-shell state must be zero. Thus, the β-function can be related to a dual model tadpole amplitude for a massless state.

At this point one has a method for computing β-functions in two-dimensional field theories: Start with the Virasoro-Shapiro formula, or some other dual-model S-matrix formula, and compute the amplitude for a massless particle to go into N other particles, each at zero momentum. The result represents the σ-model β-function associated with that particular coupling (field) describing the massless particle. The zero-momentum particles can be associated with various other couplings of the σ-model and N represents the order of perturbation theory. In particular, the requirement that the β-function vanish can be seen as equivalent to the requirement that the amplitude for the massless state to go into the vacuum is zero. This of course means that one sits at a minimum of the effective potential for the massless state i.e. that its classical equations of motion are satisfied. This is the link between σ-model β-functions and equations of motion for the states of the string. Although the method may be too complicated to carry out in other cases it provides, as far as I know, the only real derivation of this link (see also the paper of Fridling and Jevicki [15]).

Other arguments exist, which are very persuasive but do not provide a real derivation. In one way or another, one claims that the σ-model indeed corresponds to the propagation of strings in nontrivial backgrounds which can be thought of as condensates of the massless states of the string and one tries to derive an effective Lagrangian describing these condensates. Alternatively the requirement of conformal invariance is seen as imposing conditions on the background which are interpreted as equations of motion.

Consider again the Polyakov expression for the bosonic string in the presence of background gravity and, possibly, other fields:

$$\Gamma = \sum_{topologies} \int \sqrt{g(X)}[d\gamma_{\mu\nu}][dX^m] exp[\frac{1}{2} \int d^2\sigma \sqrt{\gamma}\gamma^{\mu\nu}[g_{mn}\partial_\mu X^m \partial_\nu X^n + \cdots] \quad (106)$$

(including a factor of $\sqrt{g}$ in order to have reparametrization invariance of the functional measure). By writing $g_{mn}(X) = \eta_{mn} + h_{mn}(X)$ and expanding in powers of $h_{mn}(X)$ one certainly makes contact with the scattering amplitude for gravitons by choosing

$$h^{mn} = \partial X^m \partial X^n e^{ik.X} \quad (107)$$

so that one can view the σ-model as some kind of generating function for the string S-matrix. Instead [16], observe that in the functional integral one can separate out a "zero-mode"

$$X^m = X_0^m + \bar{X}^m(\sigma, \tau) \quad (108)$$

where X_0^m is independent of the world-sheet coordinates. One also expands the metric as

$$g_{mn}(X) = g_{mn}(X_0 + \tilde{X}) = g_{mn}(X_0) + g_{mn,p}(X_0)\tilde{X}^p + \cdots \tag{109}$$

and the functional integration

$$[dX] = [dX_0][d\tilde{X}] \tag{110}$$

(In practice one would use a normal coordinate expansion in the separation.) Therefore one can rewrite Γ as

$$\Gamma = \int dX_0 \sqrt{g(X_0)} \mathcal{L}(g_{mn}(X_0),\ \partial_p g_{mn}(X_0), \cdots) \tag{111}$$

and $\mathcal{L}$ can be interpreted as an effective Lagrangian for the metric and other background fields as functions of the coordinates X_0^m of the embedding space.

To attempt to determine $\mathcal{L}$ one first goes to conformal gauge and integrates over $\tilde{X}$. This amounts to computing the σ-model effective action. In general of course this cannot be done explicitly. However, if one is sitting at a zero of β_{mn} the result of this integration can be written in the form

$$\int [d\tilde{X}] exp[\frac{1}{2} \int \cdots] = exp[-W] \tag{112}$$

where (cf. (12))

$$W = \chi\Phi + \beta^{\Phi} \int d^2\sigma \sqrt{\gamma}[\frac{1}{\epsilon}R^{(2)} + \frac{1}{4}R^{(2)}\frac{1}{\Box}R^{(2)}] \tag{113}$$

Here χ is the Euler number of the surface, and the β^{Φ}, calculated to low orders of perturbation theory, can be interpreted as the dilaton β-function

$$\beta^{\Phi} = \frac{D-26}{24\pi} + \frac{1}{8\pi}[(\nabla\Phi)^2 - 2\nabla^2\Phi - R - \frac{1}{3}H^2] + \cdots \tag{114}$$

(We have included contributions, up to two-loops, from the ghosts as well as the dilaton and antisymmetric tensor fields.)

Subtracting out the divergent part, and integrating over (conformally inequivalent metrics) leads to the action

$$\begin{aligned}
\Gamma &= \int dX^0 \sqrt{g} e^{\chi\Phi} \beta^{\Phi} \\
&= \int dX^0 \sqrt{g} e^{\chi\Phi}[-R + (\nabla\Phi)^2 - 2\nabla^2\Phi - \frac{1}{3}H_{mnp}H^{mnp} + \cdots]
\end{aligned} \tag{115}$$

where we have taken D equal to the critical dimension. Clearly this is the correct action to describe the massless states of the string. Furthermore, to this order, it is easy to check that the gravitational and antisymmetric tensor equations of motion that follow from this action are identical to the requirement that the corresponding β-functions vanish. We discuss now attempts to verify this to higher order.

There are two aspects that one must consider. First, given the explicit gravitational and antisymmetric tensor β-functions that have been calculated and interpreting them as equations of motion, one can use an integrability condition to determine the corresponding Lagrangian. Second, starting from dual model scattering amplitudes, one can determine effective Lagrangians for the massless fields. In comparing the two results one must keep in mind that the requirement of vanishing β-functions is slightly stronger than that of conformal invariance and that one should only require (86). Correspondingly, the determination of an effective Lagrangian from the S-matrix is ambiguous because of the invariance of the S-matrix under field redefinitions. In particular, it only makes sense to talk about an 'on-shell' Lagrangian since any terms proportional to the equations of motion (i.e. β-functions themselves) do not contribute to the S-matrix. The comparisons should be made modulo these ambiguities.

For simplicity consider again the case when only dilaton and gravitational fields are present. On the basis of general arguments one knows the form of the low-energy effective action:

$$S_{eff} = \int dX \sqrt{g} e^{\Phi} \mathcal{L}(g_{mn}, \nabla \Phi) \tag{116}$$

and the corresponding equations of motion are

$$\begin{aligned}
\frac{\delta S_{eff}}{\delta g^{mn}} &= e^{\Phi} W_{mn} + \frac{1}{2} g_{mn}(e^{\Phi}\mathcal{L} + 2B) = 0 \\
\frac{\delta S_{eff}}{\delta \Phi} &= e^{\Phi}\mathcal{L} + A = 0
\end{aligned} \tag{117}$$

where B and A are total derivatives. (Contributions to B come from terms in the Lagrangian which explicitly contain the Ricci tensor.)

Let us assume that these equations are equivalent to the vanishing of the corresponding β-functions. Noting that the gravitational β-function β_{mn} does not contain terms proportional to g_{mn} (this follows, for example, from the normal coordinate expansion; the dependence on the background fields is only through the curvature) we conclude that

$$A = 2B \tag{118}$$

and the equations of motion become

$$\begin{aligned}
W_{mn} &= 0 \\
A &= -e^{\Phi}\mathcal{L}
\end{aligned} \tag{119}$$

On the other hand, invariance of the action under diffeomorphisms gives

$$\nabla^n W_{mn} + \frac{1}{2}\nabla_m \mathcal{L} = 0 \tag{120}$$

where we have also used the equations of motion. Obviously from a knowledge of W_{mn} one can reconstruct the Lagrangian [17].

Under the assumption that the vanishing of β_{mn} gives the same information as the gravitational equations of motion it is reasonable to expect a relation of the form

$$W_{mn} = (1 + N)_{mn}{}^{pq} \beta_{pq} \tag{121}$$

where $N = O(\alpha')$. Substituting into (120) we have therefore

$$\nabla^n \beta_{mn} = -\frac{1}{2}\nabla_m \mathcal{L} - \nabla^n (N\beta)_{mn} \tag{122}$$

Therefore, in an iterative calculation in powers of α', if we drop terms proportional

22

to β_{mn} on the right hand side of (122) this equation allows us to determine the Lagrangian up to terms proportional to β_{mn}, i.e. proportional to the equations of motion, which are irrelevant for the on-shell $\mathcal{L}$.

Clearly, this method for determining the Lagrangian is very similar to that discussed in the previous section for determining the dilaton β-function and certainly suggests identifying the two.

These methods have been used for determining the Lagrangian that corresponds to the four-loop gravitational β-functions computed in the supersymmetric models discussed in the previous sections. Although the algebra is cumbersome, it is perfectly straightforward. One finds, corresponding to the expression in (80) the following result for the on-shell gravitational effective action, representing higher-order modifications of the Einstein-Hilbert action [18,19]

$$S = \int dX \sqrt{g} [-R + \frac{\alpha'^3 \zeta(3)}{8} R^{hrsp} R^q_{rs}{}^k (R_{hmnk} R_p{}^{mn}{}_q + \frac{1}{2} R_{hkmn} R_{pq}{}^{mn})] \qquad (123)$$

We turn now to strings and describe the manner in which an effective Lagrangian is derived from a dual model amplitude by considering as an example the Type II closed superstring scattering amplitude [20]. The string contains of course an infinite number of particles but an expansion in powers of momentum divided by the Planck mass allows us to obtain the Lagrangian for the massless particles with the Regge slope $\alpha' \sim M_P^{-1}$ as the only dimensionful parameter.

Gauge invariance and Lorentz invariance fixes the starting point: The Lagrangian consists first of the Einstein Lagrangian for the graviton, a term quadratic in the antisymmetric tensor field strength, and a term quadratic in the gradient of the dilaton field. Furthermore, invariance under a constant shift of the dilaton field together with a scale transformation of the other fields fixes

$$\mathcal{L}_{2pt} = \frac{1}{2\kappa^2} R - \frac{1}{6} e^{-2c\Phi} H_{mnp} H^{mnp} - \frac{1}{2} (\nabla \Phi)^2 \qquad (124)$$

where c is a constant to be determined later. This part of the Lagrangian fixes the two-point functions. (To compare with (115) one should rescale the fields.)

One must examine now the three-point amplitudes, and see if a corresponding $\mathcal{L}_{3pt}$ must be added to the Lagrangian. The three-point amplitude for particles of momenta k_1, k_2, k_3 can be written as

$$A_{3pt} = g[(k_2 \Theta_1 k_2) tr(\Theta_2 \Theta_3^T) + (k_3 \Theta_2 \Theta_3^T \Theta_1 k_2) + (k_1 \Theta_3 \Theta_2^T \Theta_1 k_2)]$$
$$+ cyclic\ permutations \qquad (125)$$

where Θ_{mn} is a polarization tensor, and the notation describes matrix multiplication, e.g. $(k\Theta k) \equiv k_m \Theta_{mn} k_n$. By choosing Θ to be symmetric traceless, antisymmetric, or transverse diagonal one obtains the amplitude for graviton, antisymmetron, or dilaton.

The corresponding three-point Lagrangian would have to contain two derivatives, and be constructed out of the Riemann curvature, the field strength and the gradient of the dilaton. However there are no such terms and one concludes that (124) already accounts for the three-point amplitude. The latter can be used however to determine the relation between the various coupling constants, and one finds $c = \kappa/\sqrt{2}$ and $2\kappa = g(2\alpha')^2$.

One examines next the four-point function which can be written in the following form

$$A_{4pt} = -2g^2 \frac{\Gamma(-\frac{s}{8})\Gamma(-\frac{t}{8})\Gamma(-\frac{u}{8})}{\Gamma(1+\frac{s}{8})\Gamma(1+\frac{t}{8})\Gamma(1+\frac{u}{8})}$$
$$\cdot \frac{1}{16}[t_{abcdefgh}\tilde{f}_1^{ab}\tilde{f}_2^{cd}\tilde{f}_3^{ef}\tilde{f}_4^{gh}] \otimes [t_{mnpqrsuv}f_1^{mn}f_2^{pq}f_3^{rs}f_4^{uv}] \quad (126)$$

where

$$f^{mn} = k^m\rho^n - k^n\rho^m \quad (127)$$

and ρ^m, ρ^n are the left- and right-moving parts of Θ^{mn} (corresponding to left- and right-moving operators in the oscillator formalism). Finally $t_{mn\ldots}$ is defined by

$$t_{mnpqrsuv}f_1^{mn}f_2^{pq}f_3^{rs}f_4^{uv} = \frac{1}{2}[tr(f_1f_2f_3f_4) - \frac{1}{4}tr(f_1f_3)tr(f_2f_4)] + cyclic \quad (128)$$

One expands now the ratio of Γ functions

$$\frac{\Gamma(-\frac{s}{8})\Gamma(-\frac{t}{8})\Gamma(-\frac{u}{8})}{\Gamma(1+\frac{s}{8})\Gamma(1+\frac{t}{8})\Gamma(1+\frac{u}{8})} = -\frac{2^9}{stu} - 2\zeta(3) + \cdots \quad (129)$$

The $(stu)^{-1}$ term corresponds to particle exchange involving terms in the Lagrangian we have already considered so that any new contribution, corresponding to a contact interaction, comes from the second term and involves four derivatives (the ellipses represent terms with a higher number of derivatives). Since the Riemann tensor at the linearized level is

$$R^{ab}{}_{cd} = -\kappa\partial^{[a}\partial_{[c}h^{b]}{}_{d]} = \kappa\tilde{f}^{ab} \otimes f_{cd} \quad (130)$$

it is easy to identify the corresponding correction to the Einstein Lagrangian

$$\mathcal{L}_{4pt,grav} = \frac{\zeta(3)}{3\cdot2^7\kappa^2}e^{-\frac{3\kappa}{\sqrt{2}}\Phi}t^{abcdefgh}t^{mnpqrsuv}R_{abmn}R_{cdpq}R_{efrs}R_{ghuv} \quad (131)$$

This expression agrees with the result obtained from the four-loop β-function calculation and is perhaps the strongest direct evidence for the consistency of the picture we have described.

The remaining terms depending on the antisymmetric tensor and dilaton can also be found [20]. It is appropriate to remark at this point that they, and the corresponding field equations, are much easier to obtain this way, than from a direct σ-model loop calculation, thus supporting Lovelace's original assertion that one should use dual models to determine β-function rather than the other way around.

REFERENCES

[1] All relevant information about strings can be found in M.B. Green, J.H. Schwarz and E. Witten, *Superstring theory*, Cambridge University Press (1987).

[2] For a recent review of σ-models see C.M. Hull 'Lectures on non-linear sigma models and strings' in *Super Field Theories*, Proceedings of the 1986 Workshop in Vancouver, Canada, Plenum Press (1987).

[3] J.J. Atick, A. Dhar and B. Ratra, Phys. Lett. 169B (1986) 54.

[4] L. Alvarez-Gaume, D.Z. Freedman and S. Mukhi, Ann. Phys. 134 (1981) 85.

[5] S. Mukhi, Nucl. Phys. B264 (1986) 640.

[6] P.S. Howe, G. Papadopoulos and K.S. Stelle, Princeton IAS preprint (1986).

[7] R.R. Metsaev and A.A. Tseytlin, Phys. Lett. 191B (1987) 354; C.M. Hull and P.K. Townsend, Phys. Lett. 191B (1987) 115; D. Zanon, Phys. Lett. 191B (1987) 363; D.R.T. Jones, Phys. Lett. 192B (1987) 391.

[8] M.T. Grisaru, A.E. van de Ven and D. Zanon, Nucl. Phys. B277 (1986) 409.

[9] M.T. Grisaru, A.E. van de Ven and D. Zanon, Nucl. Phys. B277 (1986) 388.

[10] M.T. Grisaru, D. Kazakov and D. Zanon, Nucl. Phys. B287 (1987) 189.

[11] A.A. Tseytlin, Phys. Lett. 178B (1986) 34; G. Curci and G. Paffuti, Nucl. Phys. B286 (1987) 399.

[12] C.G. Callan, I.R. Klebanov and M.J. Perry, Nucl. Phys. B278 (1986) 78.

[13] M.T. Grisaru and D. Zanon, Phys. Lett. 184B (1987) 209.

[14] C. Lovelace, Nucl. Phys. B273 (1986) 413.

[15] B.E. Fridling and A. Jevicki, Phys. Lett. 174B (1986) 75.

[16] E.S. Fradkin and A.A. Tseytlin, Nucl. Phys. B261 (1985) 1.

[17] D. Zanon, Phys. Lett. 186B (1987) 309.

[18] M.T. Grisaru and D. Zanon, Phys. Lett. 177B (1986) 347.

[19] Q-Han Park and D. Zanon, Phys. Rev. D35 (1987) 4038.

[20] D.J. Gross and J.H. Sloan, Nucl. Phys. B291 (1987) 41.

DISCUSSION I

- Miele:

In order for a theory to be conformally invariant is it sufficient and necessary to have $T^\mu_\mu = 0$?

- Grisaru:

At a classical level one way to answer the question, if one has a symmetry which in this case is conformal invariance, is to check the conservation of the corresponding currents. For scale transformations one constructs a charge Q that generates this symmetry such that $\delta(\text{field}) = [Q, \text{field}] = (\text{conformal weight}) \times$ field. The corresponding current is $J_\mu = x^\nu T_{\mu\nu}$. Then one uses $\partial^\mu T_{\mu\nu} = 0$ and then it follows $\partial^\mu J_\mu = 0$ if $T^\mu_\mu = 0$.

- Diaz:

In the functional integral à la Polyakov one sums over the topologies of the different surfaces. What is the meaning of conformal invariance when one deals with the different topologies being summed up?

- Grisaru:

Conformal invariance is a statement that the two-dimensional theory does not depend on the size of the world-sheet one is working with. In this sense it really does not matter whether the world-sheet has the topology of a sphere or a torus or any higher genus surface. Another way of saying is that locally the energy-momentum tensor whose trace characterizes the conformal invariance of the theory, has the same form regardless of the topology of the surface.

- Diaz:

To be concrete, does it mean that when one is summing over all topologies one can rescale the size of the torus and that of the sphere by different amounts? Or does one has to rescale all the different genus surfaces by the same amount?

- Grisaru:

Conformal invariance is a statement that the functional integral does not depend on the size of the Riemann surface. That is one takes the contribution from the sphere to get a number which is independent of the size of the sphere; then one adds the contribution from the torus and so on.

Diaz:

Should corrections be computed from string theory or sigma model approach?

– *Grisaru:*

One should compute the corrections both from string theory and the sigma model approach to see if the answer coincides. My personal experience based on beta-function computations for the sigma-model is that in the long run it will be easier to compute these corrections from string theory.

– *Kiritsis:*

What does it mean to compute the beta-function from string theory?

– *Grisaru:*

From the dual model one can compute the different scattering amplitudes and from them obtain the effective action and hence the classical equations of motion for the massless string fields. We believe that these equations should be the same as those obtained from the vanishing of the beta-function of a two dimensional sigma-model.

– *Liu:*

Is the vanishing of the beta-function the consequence of conformal invariance or the global scale invariance?

– *Grisaru:*

The vanishing of the beta-functions are the consequence of local scale invariance. For global scale invariance, you only can conclude that certain integrals involving beta-functions are zero. The statement that beta-functions should be zero is not quite the right statement. The correct statement is that if the theory is locally scale invariant then, for example

$$\beta_{mn} + \nabla_{(m} V_{n)} = 0$$

This extra term expresses the difference between the local scale invariance and global scale invariance requirements.

– *Liu:*

β–functional equal to zero corresponds to classical field equation for string theory. Can you also get quantum effect from non–linear σ–model approach?

– *Grisaru*

This is the issue still being debated. Let me just for the general audience explain what the question is. If one computes the beta-functional, let us say, associated with $g_{mn}(x)$, which is supposed to describe the gravity in 10-dimensions, one finds that the beta-functional is equal to the Ricci tensor in 10-dimensions plus some other terms. The beta-functional being zero says $R_{nm} = 0$. Everybody recognizes that this is the Einstein's equation in 10-dim. That is very nice, because you expect the graviton to satisfy Einstein's equation. This is what one would call the classical equation of motion for the string. This is correlated with tree dual diagrams, if you like, there are no loops, no $\hbar$, so to speak. The question is where in this business one should see quantum effects? Now, if you are in normal QFT, you say look, I have my classical Lagrangian which gives me the classical equation of motion, and the quantum corrections in a certain sense don't change the equation of motion, they do other things that are important when I compute the scattering amplitude and so on. This is one point of view, the classical equation of motion of the string are given by setting beta-functional to zero. Indeed there are loop corrections to string scattering amplitudes. They should be computed in a separate way. The thing that causes some confusion has to do with the fact that when one computes these beta-functionals, they take account of properties that are local in world sheet. Now people have asked questions whether topology will affect the beta-functional, and people have computed effects that are coming from these things. Effects of this kind could be incorporated in these fields equations as some kind of quantum effects. That is one point of view. There is also a point of view that some of these effects should be used to cancel some of the sigma-model higher loop correction to the field equations. But my point of view is that there is a distinction between the classical equation of motion which you get by studying the local properties, and whatever you get which is due to global properties which are associated indeed with quantum effects of the string, but these should be calculated in a second step, separate from obtaining these equations of motion.

– *Kugo:*

How can you deal with massive modes in the framework of sigma-model?

– *Grisaru:*

How to treat massive modes appearing as external lines is not clear in the sigma-model approach, unfortunately. This is because the incorporation of such massive modes in the sigma-model seems to lead to an unrenormalizable model and one simply does not know how to treat it.

– *Toppan:*

I would like to know what happens if the string theories are formulated away from the critical dimensions and we loose conformal invariance. Is it possible to give meaning to these theories?

– Grisaru:

If you are not in the critical dimension and hence loose the conformal invariance, you will get an extra mode, the Liouville mode, as in the original paper on strings by Polyakov. He has tried to look at the theories in which this mode is present. I really don't know much about this work. I sort of fell into the trap of believing the standard lore in string theory which is to keep the conformal invariance, but I think it is premature to claim, certainly the membrane people would agree, that conformal invariance has to be a main principle of string theory.

– Liu:

I just want to make one comment about the question asked by Dr. Toppan. Global scale invariance implies only certain moments of $T^\mu_\mu(x)$ are zero, only conformal invariance implies $T^\mu_\mu(x) = 0$. See J. Polchinski.

DISCUSSION II

– *Quackenbush:*

What is the motivation for the addition of the antisymmetric B_{mn} field in the sigma-model action?

– *Grisaru:*

In string theory the analysis of the field content reveals the presence of an antisymmetric tensor field. In fact, in D=10 supergravity, it is crucial for the presence of supersymmetry. Also, since it is a renormalizable term in 2 dimensions, there is no reason to leave it out.

– *Martinelli (comment):*

Such a term will arise in higher loop calculations.

– *Grisaru:*

That is not true, it will not be generated if we don't have it from the beginning, unless we have fermions or some other sources of $\varepsilon^{\mu\nu}$

– *Villasante:*

Could you please explain the difference between the different beta-functions you mention in your talk?

– *Grisaru:*

Basically all the different β_{mn} functions are related to each other by adding a term:

$$\tilde{\beta}_{mn} \; = \; \beta_{mn} \; + \; \nabla_{(m} V_{n)}$$

The question is really which one is computed when one looks at the divergences of the effective action, or looks at the renormalization of various composite operators. To do the latter one takes $T_{\mu\nu}$ and defines:

$$T_{\mu\nu}^{ren} \; = \; Z^i O_{i\mu\nu}$$

where O_i are all the composite operators of the same dimension. From these one gets beta functions which according to Tseytlin (Ref. 11) should be compared with the string equations of motion.

– *Villasante:*

If auxiliary fields are used in the 10-D supergravity multiplet, shouldn't one include in the two dimensional sigma-model an extra contribution which reproduces the quantum contribution of these auxiliary fields?

– *Grisaru:*

Interesting question, yet the answer is not clear. The only connection one can make for sure is between a 2-D sigma model and the 10-D on-shell effective theory. In that sense one would imagine the auxiliary fields have been eliminated by means of the field equations, and so the answer one gets should be the same. However, at this moment I cannot think of any extra terms to add to the 2-D sigma model.

– *Villasante:*

If one would know what are the auxiliary fields in 10-D supergravity, how could one incorporate them in the 2-D sigma-model?

– *Grisaru:*

There is a common belief that in 10-D supergravity one needs an infinite number of auxiliary fields. Assuming this is true, the question is how could one include an infinite number of auxiliary fields in the sigma-model. Presumably one way to do it is to include some non local term in the 2-D sigma-model. At this point I lack the imagination to know how to do it.

–*Villasante:*

In superstrings the auxiliary fields are now here in sight. If one would carry such a program, would there still be any relation between sigma-model and strings?

– *Grisaru:*

Perhaps they would fit much better in string field theory formulations than in sigma models. Certainly in string field theory there is room for those fields.

– *Liu:*

Can you explain the connection between condensed matter physics and 2D-sigma-model. Do you know of any condensed matter physics system which has a global scale invariance but not local scale invariance?

– *Grisaru:*

For those condensed matter systems which can be described by 2-D field theories one knows that some of the parameters that appear in critical phenomena are connected to the beta-function of the corresponding field theory. Some condensed matter systems when expressed as 2-D field theories do exhibit the same type of conformal invariance properties as the string. I do not have an answer for the second part of the question.

– Kiritsis (comment):

Referring to the second part of the previous question I would like to mention that Mack and Luscher in 1976 proved that in a quantum field theory in 2-D, scale invariance implies conformal invariance at the quantum level, under very mild assumption.

– Grisaru:

In general, theories that have local scale invariance and Poincare invariance have conformal invariance, at least at the classical level.

– Kiritsis:

True also at the quantum level.

– Quackenbush:

Yesterday you discussed the sigma-model which contained the antisymmetric tensor field and the $\varepsilon^{\mu\nu}$ tensor. How does one handle this tensor when using dimensional regularization?

– Grisaru:

The correct procedure appears to be to keep $\varepsilon^{\mu\nu}$ in 2-D and to do all the momentum integrals in n-dimensions. That is, when one encounters $\varepsilon^{\mu\nu}$, one should use the two-dimensional identities:

$$\varepsilon^{\mu\nu}\varepsilon^{\rho\sigma} = \eta^{\mu\sigma}\eta^{\nu\rho} - \eta^{\mu\rho}\eta^{\nu\sigma}$$

Tseytlin in a recent paper claims however one can use any rules one wants, i.e. one can dimensionally continue the right hand side, and this amount to a change $\nabla_{(m}V_{n)}$.

– Quackenbush:

What is the effect of the $\varepsilon^{\mu\nu}$ symbol on 2-D anomalies?

– Grisaru:

For heterotic string type theories, that is theories which deal differently with left and right moving fermions, one can produce Lorentz or gravitational anomalies.

– *Martinelli:*

Has the computation of the beta-function for the 2-D sigma been done using dimensional reduction?

– *Grisaru:*

The rule of keeping $\varepsilon^{\mu\nu}$ in 2-D is the rule of dimensional reduction. The paper of T. Jones (Ref. 7) makes essentially this point. He points out that in a superfield calculation the $\varepsilon^{\mu\nu}$ tensor never shows up explicitly and doing the computation in components, using dimensional reduction, should give the same answer as that obtained using superfields.

– *Martinelli:*

To higher loops, normally the beta-function picks up renormalization scheme dependent terms, therefore it is not possible to attach any physical meaning to it. What happens when one computes it for the sigma-model containing the $\varepsilon^{\mu\nu}$ tensor term.

– *Grisaru:*

The feeling is that the scheme dependence only affects the off shell contributions. The scheme dependence is equivalent to making a field redefinition in the 10-D fields. And so it will not give incorrect information of the S-matrix. If one compares the beta-function to what one gets from the dual model S-matrix, the scheme dependence is equivalent to the ambiguities in going from the S-matrix to the Lagrangian.

– *Martinelli:*

What happens if you use dimensional reduction and you are dealing with two fields, one a gauge field, the other the ε-scalar? Do you get two beta functions?

– *Grisaru:*

Our fields are two-dimensional scalars, not vectors, so the question of ε-scalars doesn't come up. It might if one quantizes world-sheet vectors.

– *Kugo:*

You said that the beta-function of the dilaton gives an effective action S of graviton, anti–symmetric tensor and dilaton. However, according to Lovelace's interpretation the beta function β^Φ must give the dilaton tadpole, which is $\delta S/\delta\Phi$ but not the action S itself. How can both statements be consistent?

– Grisaru:

In Lovelace's context, he gets the renormalization group beta functions, from various tadpole amplitudes and in his explicit example he looks at tachyon equations. In general $\beta^\Phi = 0$ and $\delta S/\delta \Phi = 0$ can be reconciliated because of the distinction between β and $\tilde{\beta}$ (Tseytlin, Ref. 11) and because in fact $S = e^\Phi \beta^\Phi$.

– Morelli:

Could you illustrate the equivalence between Feynman loop diagrams and dual diagrams?

– Grisaru:

Lovelace takes a 2-D theory, does radial quantization, and computes diagrams in perturbation theory using the standard Gell-Mann-Low formula, but instead of using Wick's theorem he uses Operator Product Expansions. For example, in ϕ^3 theory, instead of the Wick contractions in $< \phi^3(x)\phi^3(y)\phi^3(z)\ldots >$ he uses Operator Product Expansions

$$\phi^3(y)\phi^3(z) = \sum G_n(y - z)\phi^{(n)}(z)$$

where $\phi^{(n)}(z)$ are composite operators. He then takes a Mellin transform, and what he gets is a series of terms which can be interpreted as diagrams that look exactly like dual model diagrams.

NOT THE STANDARD SUPERSTRING REVIEW

M.J. Duff

Theoretical Physics Division
CERN
1211 Geneva 23, Switzerland

The word "orthodoxy" not only no longer means being right; it practically means being wrong.

> G.K. Chesterton
> 1874-1936

1. – SUPERSTRING ORTHODOXY

Each week the CERN preprint library receives about 60 papers on theoretical particle physics of which about 25-30 are devoted to superstrings or related topics. So as I put pen to paper I feel the burden of having to review one half of theoretical physics. To make matters worse, just as the number of superstring publications is mushrooming, so are the number of superstring review talks. There now exist many excellent reviews, some of them by experts who have been studying superstrings for 15 years[1]. Well, there are two respects in which this review may differ from previous ones. First, to paraphrase Shelley[*], I never was attached to that great sect whose doctrine was the Dual Resonance Model and who pioneered string theory in the late 60's and early 70's in its original incarnation as a theory of hadrons. I entered the subject comparatively recently via the route of supersymmetry and for me ten-dimensional superstrings were the natural extension of the rather different theoretical developments that had been taking place in supergravity theories and Kaluza-Klein unification via extra space-time dimensions[2].

The second, and more important, difference is the degree of scepticism with which I view much of the current superstring orthodoxy. Accordingly, I have chosen the above quote of Chesterton's as my theme. Of course, I doubt that he was referring to superstrings; more probably had in mind religious belief. Since there is as yet no shred of experimental confirmation of superstrings, however, the comparison is perhaps not altogether inappropriate.

In order not to be misunderstood, let me say straight away that I share the conviction that superstrings are the most exciting development

----

[*] The poet, not the physicist.

The Superworld II
Edited by A. Zichichi
Plenum Press, New York, 1990

in theoretical physics for many years, and that they offer the best promise to date of achieving the twin goals of a consistent quantum theory of gravity and a unification of all the forces and particles of Nature. Where I differ is the degree of emphasis that I would place on the unresolved problems of superstrings, and the likely time scales involved before superstrings (or something like superstrings) make contact with experimental reality. To illustrate what I mean, let us consider the three boasts of superstrings as a Theory of Everything:

1. Superstrings provide a consistent (finite) quantum theory of gravity.
2. They are likely to explain the standard model with no free parameters.
3. No other theory (neither points, nor membranes) would work.

Whereas, the harsh realities are:
1. <u>Finiteness</u>?: Still no proof after 15 years.
2. <u>Uniqueness</u>?: Disappears in going from ten dimensions to four. Now have <u>billions</u> of consistent four-dimensional strings with desired theoretical properties ($N=1$ supersymmetry, chiral, anomaly-free, modular invariant, conformally invariant...) with more or less any gauge group of rank $\leqslant 22$.
3. <u>No other theory</u>?: Now exists a <u>supermembrane</u> in <u>eleven</u> dimensions which yields a superstring in ten dimensions upon dimensional reduction.

I draw attention to the finiteness problem (1) above and in Section 5, not because I think superstrings are necessarily infinite, but because of an attitude pervading the orthodox string community that finiteness is so obvious that we do not need to prove it! At the recent Nobel Symposium[3], for example, almost everybody spoke about strings but almost nobody spoke about finiteness. It was taboo.

Much more serious, in my opinion, is the "vacuum degeneracy" problem (2) which often receives short shrift in superstring review talks. It is particularly ironic that superstrings should founder on the uniqueness problem, since, as discussed in Section 3, it was the apparent uniqueness of ten-dimensional superstrings (following from the Green-Schwarz anomaly-cancellation paper) which triggered the 1984 superstring revolution.

Finally, the maximum space-time dimension permitted by supersymmetry is $D = 11$, whereas superstrings exist in $D = 10$. The recent discovery of a supersymmetric theory of a three-dimensional extended object in $D = 11$ may resolve this old puzzle. We do not yet know whether this "supermembrane" is consistent at the quantum level, but the orthodox claim that only strings can be quantum consistent now looks much less certain.

Before giving a verdict on superstrings' claim to be the Theory of Everything, let us examine what such a theory might be and why we need it.

2. - A THEORY OF EVERYTHING?

Our experimental colleagues tell us that the standard $SU(3) \times SU(2) \times U(1)$ model with three families of quarks and leptons is still in good shape, although there is some uncertainty about the top quark. Moreover, the Higgs bosons necessary for symmetry-breaking have yet to be observed. Nevertheless, the standard model is remarkably robust. There are as yet no signs of proton decay, no signs of monopoles and no signs of supersymmetry. In short, there is no empirical evidence at all that we should go beyond the standard model. So why bother? Although the standard $[SU(3) \times SU(2) \times U(1)]$ model is probably the truth, and although grand unified theories [SU(5), SO(10), E_6,...] are possibly the truth, none can claim to be the whole truth. There are still too many unanswered question. For example:

* Why chiral fermions?
* How many families?
* What dictates the Higgs sector?
* Why SU(3) × SU(2) × U(1)?

By "chiral" we mean that we put the left-handed quarks and leptons into doublets of the weak-isospin and the right-handed into singlets. It is true that with these assignments we avoid the otherwise disastrous chiral anomalies in triangle graphs provided we group the quarks and leptons into families, but this logic does not tell us how many families there are; and whereas the gauge principle beautifully pins down the gauge boson sector and its coupling to fermions, the scalar Higgs sector remains quite arbitrary. Nor does the gauge principle alone tell us which gauge group we should start from.

A Theory of Everything must also include the force of gravity whose typical energy scale is given by the Planck mass

$$M_P = \sqrt{\frac{hc}{G}} \sim 10^{-5} \text{ gms} \sim 10^{19} \text{ GeV} \qquad (2.1)$$

where G is Newton's constant of gravitation. But if we now worry about gravity, the mystery deepens:

* How to incorporate gravity?
* The ultra-violet problem?
* Why is the cosmological constant $\Lambda \simeq 0$?
* Why is M_W/M_P so small?

It seems unlikely that the fundamental theory will be obtained merely by grafting gravity, as an afterthought, on to one's favourite GUT theory. Indeed, all the current theoretical indications are that gravity is actually at the root of everything. Moreover, the success in formulating the theory of strong, weak and electromagnetic forces relied crucially on the criterion of renormalizability, i.e., the ability to absorb into the masses and coupling constants those infinities which arise from ultra-violet divergent integrals in radiative corrections. Yet Einstein's theory of general relativity, the cornerstone of classical gravity, is at the quantum level <u>non-renormalizable</u>! (just because Newton's constant G has dimensions). Thirdly, there is the problem of the cosmological constant Λ. Originally Einstein suggested that the field equations of gravity should be obtained by equating geometry (the Einstein tensor $G_{\mu\nu}$) to matter (the energy- momentum tensor $T_{\mu\nu}$), i.e.,

$$G_{\mu\nu} = 8\pi G T_{\mu\nu} \qquad (2.2)$$

where (μ,ν = 1,2,3,4). But he later realized that the principle of general covariance did not rule out a term proportional to the metric tensor $g_{\mu\nu}$, i.e.,

$$G_{\mu\nu} + \Lambda g_{\mu\nu} = 8\pi G T_{\mu\nu}. \qquad (2.3)$$

Upper bounds on the constant of proportionality Λ, known as the cosmological constant, can be put by studying the rate of expansion of the Universe, and the result is one of the most accurate of experimental statements:

$$\Lambda < 10^{-120} M_P^2. \qquad (2.4)$$

37

Yet theoretically the most natural value seems to be $\Lambda \sim M_P^2$ or perhaps else $\Lambda \sim M_W^2$ where M_W = W-boson mass $\sim 10^2$ GeV; no one knows why $\Lambda \sim 0$. Fourthly, we have the hierarchy problem: why is $M_W/M_P \sim 10^{-17}$ so small?

There are even more fundamental questions one might ask: questions so fundamental that their answer is usually taken for granted but which have recently been brought into focus by superstrings, namely:
* Are ultra-violet divergences inevitable?
* Is field theory right?
* Are the fundamental objects points? Why not extended objects: strings, membranes...?
* Why four apparent space-time dimensions?

Ultra-violet divergences have been with us for so long now that we forget the horror with which they were first greeted. Indeed, Dirac maintained to the end that they were only artefacts of our feeble approximations to Nature and that the fundamental theory should be finite. Field theory has been very successful at low energies but should we expect the field concept to persist beyond the Planck scale? The non-renormalizability of quantum gravity suggests that it does not. Indeed, as we shall see, this is one of the main reasons for introducing extended objects. And finally why does space-time appear to have four dimensions, as opposed to some other number?

Murray Gell-Mann reminds us that a Theory of Everything requires three ingredients: (a) the initial conditions of the Universe; (b) the fundamental laws; (c) the parameters peculiar to our specific Universe. Ingredient (a) is the subject of cosmology which is beyond the scope of this review, but we shall be asking whether superstrings provide not only ingredient (b) but also ingredient (c). For example, no one would expect a Theory of Everything to predict the radius of the earth from its fundamental laws alone, but do we expect it to predict things like the fine-structure constant or the electron-muon mass ratio? We shall consider this problem in Section 8, but let us first understand the excitement of superstrings.

3. - THE SUPERSTRING REVOLUTION

Despite the success of the standard model's electroweak unification, and the possibility of grand electronuclear unification schemes, a Theory of Everything remained only a dream so long as gravity was neglected. Consequently, the arrival of local supersymmetry, a Bose-Fermi symmetry which predicted gravity, completely changed the psychology of theorists. Moreover, the fact that supersymmetry also places an upper limit on the dimension of space-time

$$D \leqslant 11 \tag{3.1}$$

raised hopes of a superunification involving the higher space-time dimension ideas, originally put forward by Kaluza and Klein in the 1920's. Up until the summer of 1984 various proposals were put forward combining supersymmetry and the Kaluza-Klein idea[2] but none with complete success. Those based on conventional field theory (e.g., eleven-dimensional supergravity) suffered from various problems, not least of which was the traditional objection to a non-renormalizable theory of gravity. Those based on string theory seemed better from this point of view but had problems of their own. The realistic-looking strings appeared to suffer from inconsistencies (anomalies akin to the triangle anomalies of the standard model) while the anomaly-free strings did not appear realistic.

In particular, they seemed to prefer to live in ten space-time dimensions rather than undergoing a spontaneous compactification to four space-time dimensions as demanded by the Kaluza-Klein idea. This was the sorry state of affairs until the September 84 superstring revolution:

1. Green and Schwarz[4] discovered that the gravitational and Yang-Mills anomalies of ten-dimensional superstrings all cancel provided the gauge group is either $SO(32)$ or $E_8 \times E_8$;

2. Gross, Harvey, Martinec and Rohm[5] discovered the heterotic (hybrid) string with the above gauge groups;

3. Candelas, Horowitz, Strominger and Witten[6] discovered that the $E_8 \times E_8$ heterotic string admits a "spontaneous compactification" from ten to four dimensions on a six-dimensional Calabi-Yau manifold. The resulting four-dimensional theory resembles a GUT theory based on the group E_6. In particular, there are chiral families of quarks and leptons.

To understand the hysteria which greeted the Green-Schwarz discovery, one has to realize that theorists love uniqueness. They like to think that the Theory of Everything will one day be singled out on the grounds that it is the only possible theory. In other words, all rival theories will be eliminated not merely because they disagree with experiment but because they are internally inconsistent. Anomalies are just the sort of inconsistency one looks for. Already in the four-dimensional standard model the cancellation of triangle anomalies narrows down the choice of quark and lepton representations. Anomaly cancellation in the ten-dimensional superstring is much more dramatic. Here, and for the first time ever, the Yang-Mills gauge groups were pinned down to either $SO(32)$ or $E_8 \times E_8$. $SO(32)$ is just the rotation group in 32 dimensions and E_8 is the largest of the "exceptional" group in Cartan's classification of Lie groups. The subscript 8 refers to the rank of the group. Both $SO(32)$ and $E_8 \times E_8$ have rank 16, which means that we need 16 quantum numbers to specify a particle multiplet. Both groups have dimension 496, which means that there are 496 massless gauge bosons. [Compare with $SU(2)$ which has rank one (isospin) and dimension 3 or or $SU(3)$ which has rank two (isospin and strangeness) and dimension 8.] Of course, $E_8 \times E_8$ and $SO(32)$ are enormous groups but since both contain $SU(3) \times SU(2) \times U(1)$ as a subgroup, one might optimistically hope to find the standard model after some spontaneous symmetry breaking. For example: E_8 contains E_6, E_6 contains $SO(10)$; $SO(10)$ contains $SU(5)$; and $SU(5)$ contains $SU(3) \times SU(2) \times U(1)$.

Before discussing the heterotic string, its compactification to four dimensions and the subsequent phenomenology let us first recall what strings are.

4. - WHY STRINGS ?

Normally one thinks of fundamental objects as being points in space which sweep out one-dimensional world-lines in space-time as in Fig. 1a. But suppose instead we consider the fundamental objects to be one-dimensional strings which sweep out two-dimensional world-sheets in space-time. Strings can either be "open" so that their ends move freely in space-time as in Fig. 1b or "closed" so that their ends are joined to form a closed-loop as in Fig. 1c. In either case, the string may carry an intrinsic orientation. Two open strings may interact to form a third open string but they may also form a closed string. So all theories with open strings necessarily contain closed strings as well, though the converse is not true.

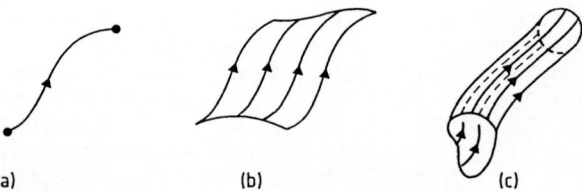

Fig. 1   a) Point-particle sweeps out world-line in space-time with trajectory $X^\mu(\tau)$.

b) Open string sweeps out world-sheet in space-time with trajectory $X^\mu(\tau,\sigma)$. The boundary conditions are $X'^\mu = 0$ for $\sigma = 0$ and $\sigma = \pi$.

c) Closed string sweeps out world-sheet in space-time with trajectory $X^\mu(\tau,\sigma)$. The boundary conditions are $X^\mu(\tau,\sigma) = X^\mu(\tau,\sigma+\pi)$.

These strings were originally invented in the 1960's by Nambu[7] to explain the Veneziano[8] spectrum of hadrons. The string has vibrational modes and each of these modes represents a particle in the spectrum. So a string describes an infinite number of elementary particles: a finite number of massless particles corresponding to the zero-eigenvalue modes and an infinite tower of massive particles corresponding to the non-zero eigenvalue modes. The mass M of each particle in the spectrum is related to its spin J by

$$J = \alpha_0 + \alpha' M^2 \qquad (4.1)$$

where the intercept α_0 is given by 1 for open strings and 2 for closed. So open strings include massless spin-1 particles and closed strings massless spin-2. The slope parameter α' with dimensions of (mass)$^{-2}$ sets the fundamental scale for string theories. In these early attempts to relate strings to hadrons one assumed that $(\alpha')^{-\frac{1}{2}} \sim 1$ GeV.

It was not until the 1970's, however, that Yoneya[9] and Scherk and Schwarz[10] abandoned the hadron interpretation and instead interpreted the massless spin-2 particle as the graviton! In this picture we take $(\alpha')^{-\frac{1}{2}} \sim 10^{19}$ GeV i.e., the Planck scale. Similarly, the massless spin-1 particle of the open string was identified with the Yang-Mills gauge boson. At such enormous energy scales, the effects of "stringiness" are important but at much lower energy scales these may be approximated by an ordinary quantum field theory. In particular the low-energy effective Lagrangians will possess general covariance and Yang-Mills gauge invariance and are given by the familiar Einstein and Yang-Mills Lagrangians plus an infinite sequence of corrections suppressed by powers of E/M_p:

$$\mathcal{L}_{\text{GRAVITY}} = \sqrt{-g}\ R + 0(\alpha') \qquad (4.2)$$

$$\mathcal{L}_{\text{YANG-MILLS}} = -\frac{1}{4} F^{\mu\nu}\ F_{\mu\nu} + 0(\alpha') \qquad (4.3)$$

Physically, a string describes a two-dimensional surface embedded in a D-dimensional space-time. Bosonic strings may thus be described by a two-dimensional field theory in which the commuting bosonic fields are the space-time co-ordinates $X^\mu(\tau,\sigma)$ with τ and σ the temporal and spatial co-ordinates on the two-dimensional world-sheet. The index μ ranges over 1 to D. We may also introduce anticommuting world-sheet spinors $\lambda(\tau,\sigma)$. We speak of a "superstring" if there is a supersymmetry which rotates X^μ into λ and vice-versa. These can be classified by n = 0,1,2 or 4, where n is the number of world-sheet supercharges. We also demand world-sheet reparametrization invariance under

$$\tau \to \tau'(\tau,\sigma) \qquad \sigma \to \sigma'(\tau,\sigma) \tag{4.4}$$

so that the laws of physics are the same for all world-sheet "observers". In addition to such local transformations, we also require invariance under "global diffeomorphisms" of the world sheet, i.e., transformations that are not connected to the identity. This invariance, known as "modular invariance" is more difficult to describe but proves to be a crucial constraint on the allowed string theories. Finally we demand Weyl invariance, i.e., invariance under the conformal rescaling

$$\gamma_{ij} \to \Omega \, \gamma_{ij} \tag{4.5}$$

where $\gamma_{ij}(\tau,\sigma)$ is the metric that describes the two-dimensional world-sheet geometry and Ω is an arbitrary function of σ and τ. Classically, it is relatively easy to satisfy the above requirements but quantum-mechanically it is a different story.

In 1974 it was shown that there are "Weyl anomalies"[11], i.e., the classical conformal invariance (4.5) displayed by various classical matter systems in interaction with gravity, breaks down at the quantum level. This implies, in particular, that the regularized energy-momentum tensor $\theta^i_{\ j}$ acquires a non-vanishing trace which in the case of two dimensions is proportional to the two-dimensional scalar curvature R[12]. For a single scalar field[13]

$$\theta^i_{\ i} = \frac{1}{24\pi} R \tag{4.6}$$

and, for a Majorana fermion[13]

$$\theta^i_{\ i} = \frac{1}{48\pi} R \tag{4.7}$$

In 1981, Polyakov[14] demonstrated that when applied to the two-dimensional string world-sheet, these Weyl anomalies accounted for the previously known "critical dimension" of string theories: the critical dimension D_c being the dimension in which the Weyl anomalies cancel. When a theory with the above symmetries is quantized we must also include a contribution to $\theta^i_{\ i}$ from the Faddeev-Popov ghosts which Polyakov calculated to be -26 times that of a single scalar in the case of the diffeomorphism ghosts and 11 times in the case of the supersymmetry ghosts. Anomaly-free theories are then obtained by having different numbers of bosons and fermions on the world sheet. The known open strings and their critical dimensions are shown in Table 1.

The n = 0 case is just what used to be called the Veneziano model back in the days of dual resonance models. From (4.1) we see that it exhibits an undesirable tachyon ($M^2 < 0$) at J = 0. The beauty of the Type I superstring is that it exhibits space-time supersymmetry (N=1) which eliminates the tachyon. Moreover, as shown by Green and Schwarz[4] in 1984, its chiral anomalies vanish only in the case of SO(32). Hence the revolution. The n = 2 and n = 4 cases are the superstrings which, for obvious reasons, dare not speak their name. (To obtain their critical dimension we also require two scalar ghosts for each world sheet vector.)

In a suitable gauge the equations of motion of the string co-ordinate

TABLE 1 - Open strings

NAME	n	D_c	COMMENTS
Veneziano	0	26	SO(n), Sp(n) and SU(n) as possible gauge groups. No fermions. Tachyon.
Type I Superstring	1	10	Fermions. Space-time supersymmetry. No tachyons. Chiral. Only SO(32).
?	2	2	Cannot explain D = 4.
?	4	−2	Nonsense.

is just

$$(\frac{\partial^2}{\partial\tau^2} - \frac{\partial^2}{\partial\sigma^2}) \; X^\mu(\tau,\sigma) = 0 \tag{4.8}$$

In the case of the closed string the solution may therefore be written as a sum of left and right moving waves

$$X^\mu(\tau,\sigma) = X_L^\mu(\tau-\sigma) + X_R^\mu(\tau+\sigma) \tag{4.9}$$

and X_L^μ and X_R^μ are then described by the same mathematics as an open string. Symbolically we have

$$\text{CLOSED} = (\text{OPEN})_L \; \times \; (\text{OPEN})_R \tag{4.10}$$

The resulting closed strings and their critical dimensions are shown in Table 2 (where we have excluded the unrealistic cases D_c=+2 and −2).

The case 26 × 26 yields what used to be called the Shapiro-Virasoro model in the old days, whereas 10 × 10 yields the so-called Type IIA or Type IIB strings depending of whether the handedness of the space-time fermions is opposite or equal. In these cases we can still say with confidence that the space-time dimension is either 26 or 10. Much more mysterious is the heterotic (hybrid) string discovered by Gross, Harvey, Martinec and Rohm[5] in 1984. This idea came as a rude shock to traditional Kaluza-Klein enthusiasts who were used to treating the extra space-time

TABLE 2 - Closed Strings

NAME	DESCRIPTION	D_c	COMMENTS
Shapiro-Virasoro	26 × 26	26	Gravity but no Yang-Mills. Tachyon. No fermions.
Heterotic	26 × 10_L	10	N = 1 supergravity plus $E_8 \times E_8$ or SO(32) Yang-Mills. Chiral.
Type IIA	10_R × 10_L	10	N = 2a supergravity. Non-chiral.
Type IIB	10_L × 10_L	10	N = 2b supergravity. Chiral.

dimensions as real, albeit compactified. Here we have a theory with 26 left-moving dimensions but only 10 right-moving dimensions. What on earth can this mean?

As we shall see in Section 7, the best way to regard this theory is as a ten-dimensional theory in which the extra 16 left-moving dimensions are interpreted as the internal symmetry of $E_8 \times E_8$ or $SO(32)$. We then still have the luxury of treating the surviving ten dimensions in the usual way. (However, even this luxury will be denied to us when we come to the "asymmetric orbifolds".)

There are thus five consistent ten-dimensional superstrings[*], one open and four closed. The question naturally arises why there should be five consistent superstrings. Should not the Theory of Everything be unique? One intriguing suggestion, which I do not have time to discuss, is that the fundamental theory is actually the closed <u>bosonic</u> string[15], and that the other <u>closed</u> superstrings: heterotic $SO(32)$, <u>heterotic</u> $E_8 \times E_8$, Type IIA and Type <u>IIB</u> are somehow different vacuum states of this one fundamental string. I shall not pursue this idea further here since it has been discussed at length elsewhere[16]. Even if it were true, however, we would still be left with the problem of which vacuum to choose, a problem which becomes even more severe in four dimensions. This will be the subject of Section 7.

After compactification to four dimensions, the superstring spectrum looks like Fig. 2. There are infinite towers of bosons and fermions of increasingly high mass and spin and a finite number of massless particles with maximum spin-two. Since the mass gap is 10^{19} GeV, however, these massive particles are way beyond the scope of any current or foreseeable accelerators. If string theory is to explain the real world, we must look for the particles we actually observe amongst the massless sector of the theory. Particles like the electron or W-boson must then acquire their small masses (small compared to the Planck scale $M_W \ll M_P$) through some symmetry breaking effect, probably at the quantum level. In fact, explaining where this other small scale in nature comes from remains one of the biggest unsolved problems. Nevertheless all the right ingredients

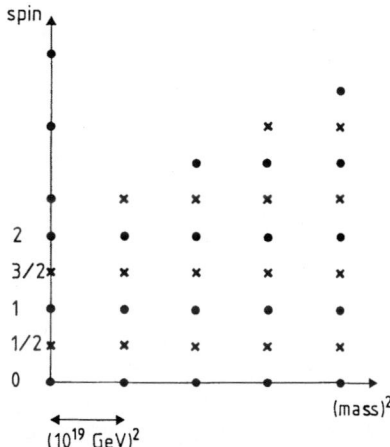

Fig. 2   The superstring mass spectrum. Bosons are denoted by dots and fermions by crosses.

----

[*] There is also the $SO(16) \times SO(16)$ string which has no space-time supersymmetry but is nevertheless tachyon-free, at least classically.

for a Theory of Everything are present in the massless sector. We have a single massless spin-2 particle, identified with the graviton and its spin-3/2 superpartner, the gravitino (assuming N=1 supersymmetry remains unbroken). In addition there are massless spins 1, ½ and 0 which (if we are lucky) will be the gauge bosons, quarks, leptons and Higgs bosons of the standard model.

At low energies, therefore, these massless string states constitute a gravity-matter system which is in principle capable of describing low-energy particle-physics phenomenology and standard general relativity but which, taken in isolation, would suffer from the usual breakdown of quantum gravity at the Planck scale. But it is just at the Planck scale where the <u>massive</u> string states can no longer be neglected. This phenomenon is at the root of the claims that strings solve the ultra-violet problem. It is to the issue of finiteness that we now turn.

5. - ARE STRINGS FINITE ?

Consider a scattering process in which two incoming strings collide, coalesce, and then split into two outgoing strings as in Fig. 3. Now imagine two world-sheet observers A and B and their corresponding equal time lines. According to observer A, the strings meet at the point marked a whereas according to observer B they meet at the point b. Since the laws of physics should be the same for both observers by virtue of the reparametrization invariance of (4.4), the point at which the strings meet cannot be objectively specified and is without physical significance. In contrast to conventional field theory based on point particles there is no concept of an "interaction-point", i.e., string diagrams have no sharp corners! This has far-reaching consequences when we consider radiative corrections to such scattering processes. Whereas the one-loop field theory diagrams of Fig. 4 would yield ultra-violet divergences owing to the sharp corners (i.e., from products of field operators at the same space-time point), the corresponding string diagram of Fig. 5 (of which

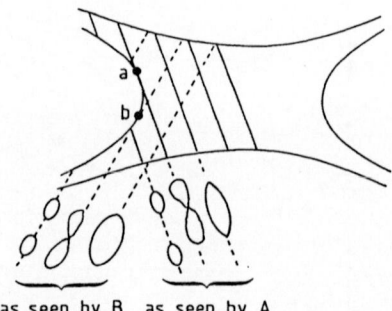

as seen by B   as seen by A

Fig. 3 The scattering of strings as seen by two different world-sheet observers A and B.

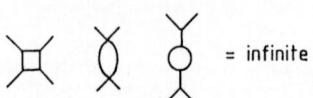

Fig. 4 Feynman diagrams for one-loop corrections to four-particle scattering in field theory.

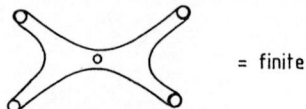

Fig. 5 The corresponding process in string theory.

44

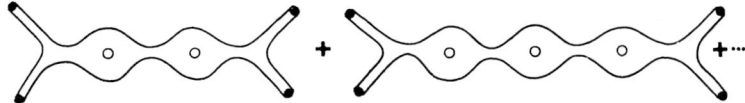

Fig. 6   Higher loop corrections.

there is only one) should be finite; a fact borne out by explicit one-loop calculations.

If we go beyond one loop as in Fig. 6, the same hand-waving argument about no sharp corners still applies. Unfortunately the highly sophisticated mathematical apparatus (involving the theory of Riemann surfaces and its super-extension) necessary to check finiteness beyond one loop has still not been worked out. Why then are orthodox string theorists so confident that superstrings are finite? I put this question to a superstring expert at the recent Nobel Symposium and was told "Experts don't ask that question"!

To be charitable, what he meant (I think) is that those who have worked with superstrings for many years have developed a feeling for the subject which enables them to make educated guesses about the outcome of calculations still too complicated to carry out. No one could argue with that. Conjecture and refutation is the way that science progresses, after all. But since the claim of a consistent (finite) quantum gravity is the single most important boast of superstrings, this should not be a matter for complacency. Yet there is an attitude pervading the orthodox string community that finiteness is so obvious that we do not need to prove it[*]. I suggest that the finiteness question is one that all of us, experts and non-experts alike, should continue to ask since the future of theoretical physics hinges on the outcome.

6. – ORTHODOX PHENOMENOLOGY

Up until recently, superstring phenomenology was based on the Calabi-Yau approach of Ref. 6 and looked like Fig. 7. Aside from the question of why we pick the $E_8 \times E_8$ heterotic string and not one of the others (another question the experts do not like to ask), there are several other questionable features. Must we first take the field-theory limit in ten dimensions before compactifying to four dimensions? Is a Calabi-Yau manifold the only way to go from 10 dimensions to 4? Is the only way to achieve $SU(3) \times SU(2) \times U(1)$ via E_6? Can one prove that the extra $U(1)$'s or $SU(2)$'s are inevitable? Do we understand the mechanism for supersymmetry breaking? The answer to all these questions is "No"!

The main objection to this orthodox phenomenology, however, is that we now understand, much better than we did three years ago, how to arrive at a four-dimensional string theory. In order to appreciate this we shall briefly review the Kaluza-Klein idea, the Frenkel-Kac idea, the Narain idea and the ideas of orbifolds and four-dimensional strings.

---

*) Two physicists exempt from this criticism are Stanley Mandelstam[17] and John G. Taylor[18]. Indeed, J.G. Taylor even claims to have proved finiteness in a series of papers largely ignored by the orthodox community. He surely deserves a response. By contrast, A. Casher[19] claims that there is no consistent finite string theory.

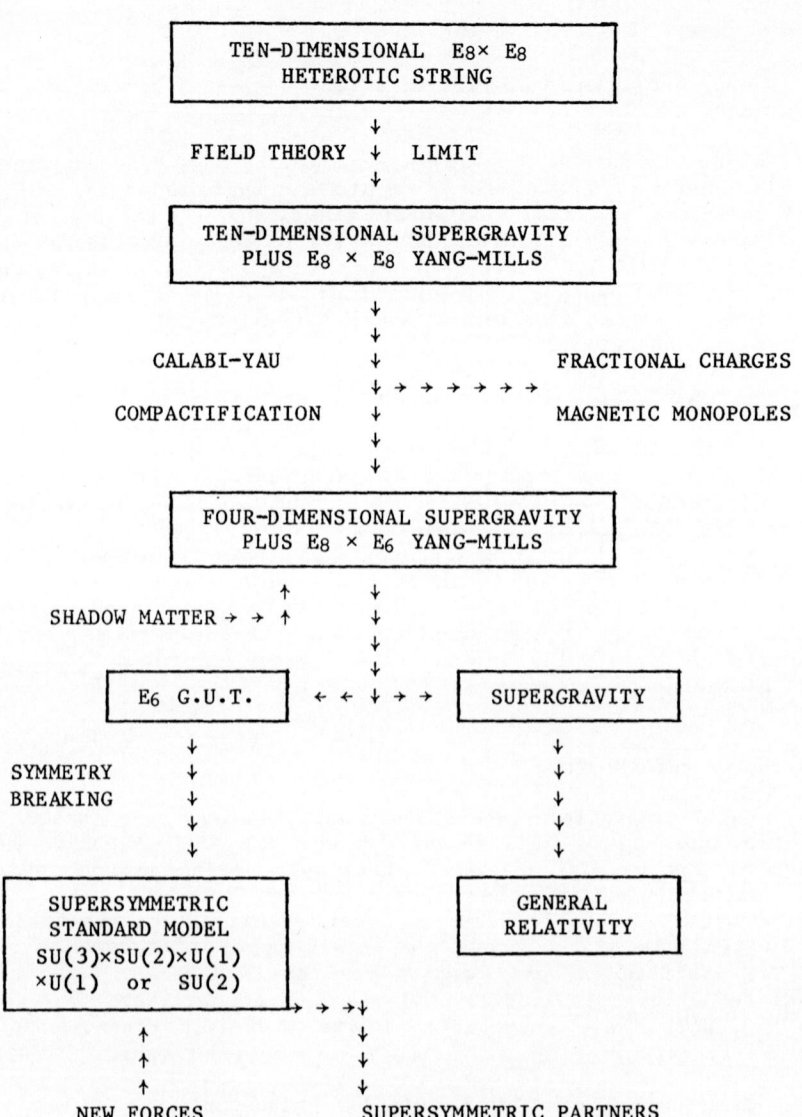

Fig. 7 Orthodox phenomenology

7. - FOUR-DIMENSIONAL SUPERSTRINGS

The Kaluza idea

By 1919, Maxwell's theory of electromagnetism had been around for some time and Einstein had recently formulated his general theory of relativity. According to Einstein the laws of physics should be the same to all observers and not merely the same to those in relative uniform motion, as he had earlier proposed in his special relativity. This is the principle of general co-ordinate invariance. In this picture, the gravitational field is described by a symmetric second-rank tensor, $g_{\mu\nu}(x)$ where the indices μ and ν run over the four dimensions of space-time ($\mu, \nu = 1, 2, 3, 4$). Moreover, this tensor admits the geometrical interpretation of a metric, i.e., it describes the infinitesimal proper distance ds between points in a __curved__ space-time[*]:

$$ds^2 = g_{\mu\nu}(x) \ dx^\mu dx^\nu. \tag{7.1}$$

Note that in curved space-time, the metric is a field: it itself depends on the space-time co-ordinates $x^\mu = (ct, x, y, z)$, just like Maxwell's electromagnetic vector-potential $A_\mu(x)$.

Of course, very little was known about the strong and weak nuclear forces at this time so it was natural, in searching for a unified field theory, to try and combine gravity with electromagnetism. This Kaluza[20] was able to do by the ingenious device of postulating an extra, fifth dimension for space-time. Kaluza considered Einstein's theory of gravity alone but with a five-dimensional line element

$$d\hat{s}^2 = \hat{g}_{MN}(z) dz^M dz^N \tag{7.2}$$

where the indices M, N now run over 1 to 5 and z^M are the co-ordinates of a five-dimensional space-time. Thus

$$z^M = (x^\mu, y) \tag{7.3}$$

where x^μ are the usual co-ordinates of space-time and $y = x^5$ is the extra fifth co-ordinate. We put hats on ds and g_{MN} to distinguish them from the four-dimensional quantities of (7.1). Kaluza then suggested the following four-dimensional interpretation

$$
\begin{array}{lllll}
\hat{g}_{\mu\nu} = g_{\mu\nu}(x) & \text{TENSOR} & \text{SPIN-2} & \text{GRAVITON} & \\
\hat{g}_{\mu 5} = A_\mu(x) & \text{VECTOR} & \text{SPIN-1} & \text{PHOTON} & (7.4) \\
\hat{g}_{55} = \phi(x) & \text{SCALAR} & \text{SPIN-0} & \text{DILATON} &
\end{array}
$$

i.e., the $\mu\nu$ components have the four-dimensional interpretation of a tensor which we identify with the usual gravitational field; the $\mu 5$ components have the four-dimensional interpretation of a vector which (and this is the crucial step) we identify with Maxwell's electromagnetic vector potential; and the 55 components have the four-dimensional interpretation of a scalar which is a new (and in 1919, rather embarrassing) field. All this was before the invention of quantum field theory but nowadays we would make the identification of these fields with particles, namely a spin 2 graviton, a spin 1 photon, and a spin 0 dilaton (so called because, in Klein's interpretation to be discussed below, the field ϕ measures the size of the fifth dimension).

Of course, it is not enough to call $g_{\mu 5}$ by the name photon, one must also check that it obeys the correct equations and here we see the Kaluza-Klein miracle at work: Einstein's pure gravity equations in d = 5 yielded not only the correct gravity equations for $g_{\mu\nu}(x)$ in d = 4 but also the correct Maxwell's equations for $A_\mu(x)$ and, as an extra bonus, the Klein-Gordon equation for $\phi(x)$. In other words, Maxwell's theory of electromagnetism was seen to be a consequence of Einstein's theory of gravity, given that one was willing to buy this extra fifth dimension. In particular, conservation of electric charge (gauge invariance) was a consequence of conservation of energy momentum (general co-ordinate invariance).

The Klein idea

Attractive though Kaluza's idea was, it contained a rather unsatisfactory ad hoc assumption. Although the indices were allowed to run over 1 to 5, the metric $\hat{g}_{MN}$ depended on only the four space-time co-ordinates x^μ. For no very good reason, the dependence on the fifth co-ordinate y had been suppressed. An attempt to remedy this defect was made in 1926 by Oskar Klein[21] who suggested that the fifth dimension should have a different "topology" from the other four. In other words, whereas

$$-\infty < x^\mu < \infty \qquad\qquad (7.5)$$

the fifth co-ordinate is periodic·

$$0 \leqslant my \leqslant 2\pi \qquad\qquad (7.6)$$

and the fifth dimension has been "compactified" to a circle of radius m^{-1}. Mathematically we say that space-time has topology $R^4 \times S^1$ (the circle S^1 is a special case of a k-dimensional sphere S^k), which means physically that if we set out in the fifth direction we will always return to our starting point.

It is difficult to conceptualize a Universe with $R^4 \times S^1$ topology, but a simpler model with $R^1 \times S^1$ is provided by the hose-pipe analogy. Consider a long, thin hose-pipe as in Figure 8. At large distances it looks like a one- dimensional line (topology R^1), but closer inspection reveals that at each point on the line there is a little circle (topology S^1). So it was that Klein suggested a little circle (S^1) at each point of four-dimensional space-time (R^4).

The point of all this is that we can now retain the dependence on the fifth co-ordinate without giving up the four-dimensional interpretation. All we have to do is perform a Fourier expansion

$$\hat{g}_{MN}(x,y) = \Sigma_{n=-\infty}^{\infty} \hat{g}_{MN}^{(n)}(x)e^{inmy}. \qquad\qquad (7.7)$$

The n = 0 modes in this expansion are just the massless particles that Kaluza had already discussed, namely the graviton, the photon and the dilaton of (7.4). But now we also have an infinite tower of n ≠ 0 modes. Substituting this expression (7.7) into the five-dimensional Einstein equations reveals that these extra modes, which satisfy

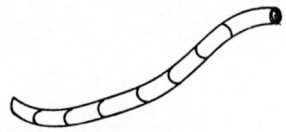

Fig. 8 The hose-pipe analogy. At large distances the hose-pipe looks like R; closer inspection reveals it to be $R \times S^1$.

$$\hat{g}_{MN}(n) = \hat{g}_{MN}(-n)^*$$ (7.8)

admit the four-dimensional interpretation of charged, massive particles with masses

$$m_n = nm$$ (7.9)

and electric charges

$$q_n = n\kappa m$$ (7.10)

where

$$\kappa = \sqrt{G}.$$ (7.11)

Note that the masses are quantized in units of m, the inverse radius of the circle, and that the charges are quantized in units of a fundamental charge

$$e = \kappa m.$$ (7.12)

Although this five-dimensional Kaluza-Klein theory is only a simple model, it illustrates three important points that survive in more realistic models:

(1) The electric charges of the elementary particles are quantized in units of a fundamental charge (a well-known, but hitherto unexplained, empirical fact).

(2) The gravitational constant $\sqrt{G}$ and the electric charge e are proportional to one another, with the constant of proportionality given by the size of the extra compact dimensions, as in (7.2).

(3) This means that with e of order unity, we are forced to take the radius of the extra dimensions to be of the very small; the Planck size

$$m^{-1} \sim 10^{-33} \text{ cm}$$ (7.13)

which is satisfactorily consistent with the fact that these extra space-time dimensions are not apparent in our everyday lives. But this in turn means that we are forced to take the fundamental mass m to be Planck size, i.e., 10^{19} GeV.

We may summarize the Kaluza-Klein philosophy as follows: what we perceive to be internal symmetries in d = 4 (electric charge, colour, charge conjugation, etc.) are really space-time symmetries in d > 4 (general covariance, parity, etc.). For a comprehensive review, see Ref. 2).

The Frenkel-Kac idea

Let us now extend the idea to r extra dimensions. If we choose these extra dimensions to be a torus T^r, i.e., a product of r circles

$$T^r = \underbrace{S^1 \times S^1 \times \dots S^1}_{r \text{ times}},$$ (7.14)

then the four-dimensional gauge group is just Abelian

$$G = \underbrace{U(1) \times U(1) \times \dots U(1)}_{r \text{ times}}$$ (7.15)

It might appear at first sight, therefore, that the torus is not a very realistic choice for describing groups containing the SU(3)×SU(2)×U(1) of the standard model. However, our Kaluza-Klein discussion has so far dealt only with ordinary field theory. As we shall now describe, the story with string theory has an interesting "twist" to it.

The reason is that the string can wind around the circles of the extra dimensions. For a single circle, one identifies

$$X \equiv X + 2\pi \, Rn \qquad\qquad (7.16)$$

where R is the radius of the circle and the integer n is the number of times the string wraps around the circle. When this winding number is non-zero, there are extra topological "soliton" states in the theory in addition to the usual Kaluza-Klein states. For a specific value of the radius R, moreover, some of the spin-1 states are massless and the Abelian Kaluza-Klein U(1) is enlarged to a non-Abelian SU(2).

One may generalize the argument to r circles and one finds that for specific tori the Kaluza-Klein group $[U(1)]^r$ is enlarged to any simply-laced non-Abelian group of rank r[22]. The simply-laced groups of rank r are the orthogonal groups SO(2r), the unitary groups SU(r+1), and the exceptional groups E_r (r=6,7,8). One may also obtain any product of smaller simply-laced groups provided the total rank is r. Thus in string theory, the humble torus (as Abdus Salam calls it) can nevertheless give rich symmetries big enough to contain SU(3)×SU(2)×U(1).

In the case of the compactification of the left-moving heterotic string from 26 to 10 dimensions, the requirements of modular invariance force the T^{16} tori to be such as to yield the very special simply-laced rank 16 groups $E_8 \times E_8$ or SO(32). This is how the heterotic string was first constructed[5].

The Narain idea

We have seen how the heterotic string has 26 left-moving dimensions and 10 right-moving dimensions and how, by compactification of 16 left-movers on T^{16}, we arrive at a string theory with the interpretation of 10 space-time dimensions and internal gauge groups $E_8 \times E_8$ or SO(32).

The next important development in string compactification came from K.S. Narain[23], a young Indian physicist working at the Rutherford Laboratory. Although in detail it is technically complicated, the broad idea is so simple and yet so far-reaching that one wonders why no one had thought of it before. Instead of going from 26 to 10 with the left-movers leaving the right movers alone, why not go straight away from 26 to 4 on the left with T^{22} and go from 10 to 4 on the right with T^6 so as to obtain a four-dimensional string theory directly? This is precisely what Narain did. Of course, one must check that the tori are so chosen as to comply with the requirements of modular invariance. In this way we arrive at a string theory with the interpretation of four space-time dimensions and an internal symmetry of any simply-laced group with rank

$$r = \text{rank } G = 22 \qquad\qquad (7.17)$$

Some comments are now in order. First this is a compactification of the string theory itself. There is no need to go first to 10 dimensions, take the field-theory limit, and then compactify the field theory from 10 to 4 as in Section 6. Secondly, by including the right movers in the compactification we have lost the uniqueness of $E_8 \times E_8$ or SO(32). Now we can have a four-dimensional theory with any simply-laced group of rank 22, of which there are thousands! For example we could have SO(44), a group which neither contains $E_8 \times E_8$ as a subgroup nor is contained in it! Thirdly, the four-dimensional theory so obtained allows $22 \times 6 = 132$ free parameters.

In Narain's original construction, the resulting four-dimensional theory was still far from phenomenology. In particular there was $N = 4$ supersymmetry and no chiral fermions. However, a more realistic class of theories with $N = 1$ supersymmetry and chiral fermions may be obtained by appealing to the idea of "orbifolds" (an idea first applied in the context of going from 10 to 4 in the ten-dimensional heterotic string[24]). Instead of taking the simple tori T^{22} or T^6 we consider $T^{22}/\mathcal{G}$ and $T^6/\mathcal{G}$ where $\mathcal{G}$ is a discrete group. One may even consider "asymmetric orbifolds"[25] where we consider $T^{22}/\mathcal{G}_L$ and $T^6/\mathcal{G}_R$ and where the left-moving and right-moving discrete groups are different: $\mathcal{G}_L \neq \mathcal{G}_R$. Now we must give up the luxury of interpreting the theory as a ten-dimensional string compactified to four dimensions, since there is no such space-time interpretation! The resulting models are then best described by the title "four-dimensional superstrings".

Fermionic Formulations

In Section 4, we saw that the string is described by boson fields $X^\mu(\tau,\sigma)$ and fermion fields $\lambda(\tau,\sigma)$ on the two-dimensional world sheet. It is a peculiar property of two dimensions (where spin is not defined) that we may swap one boson for two Majorana fermions and vice versa. This does not spoil the conformal invariance since from (4.6) one boson contributes to the conformal anomaly by the same amount as two Majorana fermions. This process of "fermionization" is, however, possible only when the bosonic co-ordinate is periodic. We may thus express the compactified X^μ_L as a left-handed Majorana fermion and the compactified X^μ_R as a right-handed Majorana fermion. If so desired, we may dispense completely with the idea of extra space-time dimensions! All we have is a conformally invariant set of bosons and fermions on the world sheet. Such fermionic formulations have proved very popular recently and have led to many new four-dimensional string theories[26].

8. - CONCLUSIONS

The result of all this activity is that we may now construct four-dimensional superstrings satisfying all the theoretically desired properties: conformal invariance, unbroken $N = 1$ supersymmetry, chiral fermions and modular invariance. But the uniqueness of the ten-dimensional strings is nowhere to be seen in four dimensions. We may have any simply-laced (and many non-simply laced) gauge groups with rank $\leqslant 22$. (Thus, to paraphrase Shelley[*], we can in particular incorporate a grand unified theory with the rank 18 group $[SU(4)]^6$.) We may have many difficult numbers of families and many different quark-lepton representations. The authors of Ref. 27, who used the device of even self-dual Lorentzian lattices of dimension 88, pointed out that there was a lower bound of about 10^{1500} such lattices each giving a different four-dimensional superstring! Not all such lattices give rise to consistent theories, however, so this number does not accurately reflect the number of four-dimensional superstrings[27]. In fact, in the absence of an exhaustive classification, we do not yet know just how many there are but it surely runs into billions. This number becomes even larger when we include the four-dimensional superstrings that come from the Type II
----

[*] The physicist, not the poet. See the talk by Prof. Glashow at this school.

construction of Section 4. Contrary to what was originally suspected one may, in fact, construct four-dimensional Type II superstrings with realistic gauge groups and chiral fermions[28]. Unfortunately, although the possibilities are much more tightly constrained than in the heterotic formulation, none of them seems to give realistic fermion representations[29].

For the time being, therefore, the phrase "superstring-inspired phenomenology" can only mean sifting through these billions of heterotic models in the hope of finding one that is realistic. The trouble with this needle-in-a-haystack approach is that even if we found a model with good phenomenology, we would be left wondering in what sense this could be described as a prediction of superstrings.

We have ended on a pessimistic note. Since this review deliberately set out to highlight the unresolved problems of string theory, let me say on the positive side how truly amazing it is that superstrings, motivated as they were by the desire for a consistent quantum theory of gravity, should have come anywhere close to the standard model. The moral I would draw at this stage is that we have seen some remarkable breakthroughs but that this is no time for complacency. We must understand the outstanding problems of finiteness, of supersymmetry breaking, of the cosmological constant and of gauge hierarchy.

Embracing perhaps all of these is the crucial question of vacuum degeneracy. There are good reasons for believing that the billions of four-dimensional superstring models are not billions of different theories but rather billions of different vacua of the same theory. But this will not help the phenomenology of superstrings one iota unless we can find some principle for lifting this degeneracy. Some string theorists put their faith in non-perturbative effects, others in string field theory, others in both. Some cosmologists, on the other hand, accept vacuum degeneracy as a fact of life. They argue that the Universe has really billions of different vacua and that we just happen to be living in one of them with $SU(3) \times SU(2) \times U(1)$, three families, etc. This view is very difficult, if not impossible, to refute but is certainly rather a pessimistic one for the future of scientific inquiry. As Murray Gell-Mann points out, if it is really true that we have to give not only the fundamental laws, but also hundreds of parameters peculiar to our specific Universe then physics will have been reduced to an environmental science like botany.

Perhaps the vacuum degeneracy problem par excellence is the question of why we appear to live in four space-time dimensions. Superstrings have as yet offered no clue. What about supermembranes?[29] This is the subject of my second lecture[30].

REFERENCES

1. J.H. Schwarz - "Superstrings - A Progress Report", Preprint CALT-68-1417 (1987).
2. M.J. Duff, B.E. W. Nilsson and C.N. Pope - Physics Reports 130 (1986) 1.
3. "Unification of the Fundamental Interactions", Proceedings of the Second Nobel Symposium on Elementary Particle Physics, Marstrand, Sweden, June 1986. (Eds. L. Brink et al., Physica Scripta T15, 1987).
4. M.B. Green and J.H. Schwarz - Phys.Lett. 149B (1984) 117.
5. D. Gross, J. Harvey, E. Martinec and R. Rohm - Phys.Rev.Lett. 54 (1985) 502, Nucl.Phys. B256 (1985) 253.
6. P. Candelas, G. Horowitz, A. Strominger and E. Witten - Nucl.Phys.

B258 (1985) 46.

7. Y. Nambu — in "Proceedings of the International Conference on Symmetries and Quark Models", Wayne State University (Gordon and Breach, 1970).

8. G. Veneziano — Nuovo Cimento 57A (1968) 190.

9. T. Yoneya — Progr.Theor.Phys. 51 (1974) 1907.

10. J. Scherk and J. Schwarz — Nucl.Phys. B81 (1974) 118.

11. D.M. Capper and M.J. Duff — Nuovo Cimento 23A (1974) 173.

12. S. Deser, M.J. Duff and C.J. Isham — Nucl.Phys. B111 (1976) 45.

13. M.J. Duff — Nucl.Phys. B125 (1977) 334.

14. A.M. Polyakov — Phys.Lett. 103B (1981) 207, 211.

15. P.G.O. Freund — Phys.Lett. 151B (1985) 387;
    M.J. Duff, B.E.W. Nilsson and C.N. Pope — Phys.Lett. 163B (1985) 343;
    A. Casher, F. Englert, H. Nicolai and A. Taormina — Phys.Lett. 162B (1985) 121;
    M.J. Duff, B.E.W. Nilsson, C.N. Pope and N.P. Warner — Phys.Lett. 171B (1986) 170;
    F. Englert, H. Nicolai and A. Schellekens — Nucl.Phys. B274 (1986) 315;
    W. Lerche, D. Lust and A. Schellekens — Phys.Lett. 181B (1986) 71;
    A. Chamseddine, M.J. Duff, B.E.W. Nilsson, C.N. Pope and D.A. Ross — Phys.Lett. 193B (1987) 444.

16. M.J. Duff — in Proceedings of the Fourth Capri Symposium, May 1986 (Ed. F. Buccella, World Scientific).

17. S. Mandelstam — Proceedings of the Niels Bohr Centennial Conference, Copenhagen (May 1985).

18. J.G. Taylor — CERN Preprint TH. 4596 (1986).

19. A. Casher — CERN Preprint TH. 4738 (1987).

20. Th. Kaluza — Sitzungsber.Preuss.Akad.Wiss. Berlin Math.Phys. K1 (1921) 966.

21. O. Klein — Nature 118 (1926) 516.

22. I. Frenkel and V.G. Kak — Inv.Math. 62 (1980) 23:
    P. Goddard and D. Olive — in "Workshop on Vertex Operators in Mathematics and Physics", Berkeley (1983).

23. K.S. Narain — Phys.Lett. 169B (1986) 41.

24. L. Dixon, J. Harvey, C. Vafa and E. Witten — Nucl.Phys. B261 (1985) 678 and B274 (1986) 285.

25. K.S. Narain, M.H. Sarmadi and C. Vafa — Harvard Preprint HUTP-86/A089 (1986).

26. H. Kawai, D.C. Lewellen and S.H.H. Tye — Phys.Rev.Lett. 57 (1986) 1832; Cornell University Preprint CLNS 86/75 (1986);
    I. Antoniadis, C. Bachas and C. Kounnas — LBL Preprint LBL-22709 (1986).

27. W. Lerche, D. Lüst and A.N. Schellekens — CERN Preprint TH. 4590 (1986).

28. L. Castellani, R. D'Auria, F. Gliozzi and S. Sciuto — Phys.Lett. 168B (1986) 47;
    R. Bluhm, L. Dolan and P. Goddard, Rockefeller Preprint RU-B86 (1987);
    H. Kawai, D.C. Lewellen and S.H.H. Tye — Cornell Preprint CLNS 87/760 (1987).

29. L. Dixon, V. Kaplunovsky and C. Vafa — SLAC Preprint SLAC-PUB-4282 (1987).

30. E. Bergshoeff, E. Sezgin and P.K. Townsend — Phys.Lett. 189B (1987) 75.

31. M.J. Duff — CERN Preprint TH. 4797 (1987).

DISCUSSION

– *Morelli:*

For a low energy effective theory from superstrings you have shown the zero mass spectrum of the theory and the supersymmetries which it contains. Is there a way of getting a low energy Lagrangian directly from the superstring Lagrangian?

– *Duff:*

To get the exact low energy Lagragian, you have to integrate out all the massive modes; no one knows how to do that yet. What people usually do is to determine from the spectrum what symmetries we have, then to construct the most general Lagrangian consistent with these symmetries. That is all we can do. In fact, one of the criticisms of the Calabi-Yau phenomenology is that the Lagrangian is, to a certain extent, guessed instead of derived.

– *Morelli:*

In this way you have shown that the spectrum (i.e. the free part of the low energy theory) exhibits supersymmetry (or other symmetries). How can you argue that the same holds for the interaction terms of the low energy theory?

– *Duff:*

If I have a massless gravitino, which means I have an unbroken supersymmetry, then the only consistent Lagrangians I can construct are those whose both free and interaction parts are invariant under the corresponding supersymmetry transformations, so there is no question that these Lagrangians have to have that symmetry. What is not clear is if the symmetry will uniquely determine the Lagrangians. To come back to your original question, can you derive the low energy Lagrangian from the string theory directly without guessing it? The answer is that we probably can but we do not know how to do it, yet.

– *Liu:*

Why do you not have $N > 5$ world sheet supersymmetry?

– *Duff:*

It turns out that $N > 5$ world sheet supersymmetry is not consistent with conformal symmetry, but I cannot reproduce the proof here.

– Miele:

I would like to know if it is possible to consistently go beyond the supermembrane?

– Duff:

Yes, you can go beyond membranes. In fact you can have objects with p-spatial dimensions called "p-branes". Tomorrow I shall tell you which values of p are consistent with which space-time dimensions, assuming you want them to be supersymmetric.

– Miele:

Is there a proof of unitary for superstrings?

– Duff:

That it is a good question. I believe there is, but I cannot give you a reference to it.

– Miele:

If you believe in unitary, how can you explain that when you supersymmetrize the Chern-Simons term you get ghosts?

– Duff:

In supersymmetrizing the Chern-Simons term, you need to find terms in your Lagrangian which are quadratic in the Riemann tensor. Now it has been known for many years that if you have more than two derivatives in your Lagrangian, you will get ghosts. However if you have infinitely many higher derivative terms as string theory does (we have terms cubic in Riemann-tensor, quartic in Riemann-tensor etc.), you will find in any finite order perturbation theory, that these ghost poles never actually appear.

– Quackenbush:

Would you discuss the general features of orbifold compactification?

– Duff:

I don't want to pre–empt Professor Kounnas, and I believe that this will be the topic of his lecture.

– Lewellen:

Why insist on a geometrical picture of particles as external objects, when it is the infinite dimensional symmetries behind these theories which is the truly necessary ingredient? The current trend in string theories has led us to models which look less and less like geometrical objects.

– Duff:

I am sympathetic to this point of view, but for the moment we still need the geometric picture as a crutch to discover the necessary underlying symmetries. It is perfectly possible that we will be able to abandon this picture and replace it with a more abstract principle. Remember also that the work you are describing relies heavily on conformal invariance, something we do not have for supermembranes.

FROM SUPERSPAGHETTI TO SUPERRAVIOLI

M.J. Duff

Theoretical Physics Division
CERN
1211 Geneva 23, Switzerland

ABSTRACT

There are four fundamental extended objects with d > 2 world-volume dimensions and D-dimensional space-time supersymmetry: (d=3,D=11), (d=6, D=10), (d=4,D=6) and (d=3,D=4). A simultaneous dimensional reduction of the world-volume and the space-time then yields four sequences of super extended objects which include the four classical Green-Schwarz super-strings (d=2,D=10,6,4 and 3) as special cases. We discuss which of these models is likely to be quantum consistent and finish with some speculations.

1. - INTRODUCTION

Many of the supergravity theories that we used to study a few years ago are now known to be merely the field theory limit of an underlying string theory[1]. For example, N = 2a supergravity in ten dimensions is just the field theory limit of the Type IIA superstring. What are we to make, therefore, of supergravity theories which cannot be obtained from strings such as N = 1 supergravity in eleven dimensions[2],[3]? This is a particularly puzzling example since it is well known that upon dimensional reduction to ten dimensions, it yields the above-mentioned N = 2a theory. Indeed, if supersymmetry allows D ≤ 11, why do strings stop at D = 10? Another mystery concerns the dual formulation of N = 1 supergravity in ten dimensions in which the three-form field strength $H_{\mu\nu\rho}$ is swapped for a seven-form field strength $\tilde{H}_{\mu\nu\rho\sigma\lambda\kappa\tau}$[4]. As a field theory it seems every bit as good. In particular, provided we couple it to $E_8\times E_8$ or $SO(32)$ Yang-Mills, it is just as anomaly free as its three-form counterpart[5]. Why then does only the $H_{\mu\nu\rho}$ version correspond to the field theory limit of a string?

In the last few weeks we have begun to see a glimpse of an answer to some of these questions, thanks to the work of Bergshoeff, Sezgin and Townsend[6] on "supermembranes". In particular they have constructed a theory which describes an extended object with two space and one time dimensions moving in an eleven-dimensional supergravity background. We describe this theory in Section 2 and then show in Section 3 how to derive from it the Type IIA superstring in ten dimensions by a simultaneous dimensional reduction of the world-volume and the space-time [Duff, Howe, Inami and Stelle[7]].

Inspired by earlier work of Hughes, Liu and Polchinski[8], Bershoeff, Sezgin and Townsend[6] pointed out that a covariant Green-Schwarz action for other extended objects with one temporal and p spatial dimensions is possible whenever there is a closed (p+2)-form in superspace. As described in Section 4, the four fundamental "p-branes" are then given by p = 2 in D = 11, p = 5 in D = 10, p = 3 in D = 6 and p = 2 in D = 4 [Achúcarro, Evans, Townsend and Wiltshire[9]]. Applying the above-mentioned simultaneous dimensional reduction k times, we find four sequences of (p-k)-branes in (D-k) dimensions which include the well-known Green-Schwarz superstrings in D = 10,6,4 and 3.

Of course, of these four superstrings, only the D = 10 case is consistent at the quantum level and this raises the crucial question of which of the super p-branes, if any, can be quantum-mechanically consistent. We do not yet know an answer to this question, but in Section 5 we discuss some of the objections which have been raised in the literature against membranes and, whenever possible, provide a refutation. In particular, we discuss the cancellation of the vacuum energy between the bosons and fermions in the eleven-dimensional super-membrane [Duff, Inami, Pope, Sezgin and Stelle[10)]*).

Section 6 is reserved for some wilder speculations concerning the heterotic five-brane, supermembranes on the seven-sphere and super-membranes as "singletons".

2. – THE D = 11 SUPERMEMBRANE

To describe the coupling of a closed d = 3 membrane to a D = 11 supergravity background, let us introduce world-volume co-ordinates $\hat{\xi}^{\hat{i}}$ ($\hat{i}$ = 1,2,3) and a world-volume metric $\hat{\gamma}_{\hat{i}\hat{j}}(\hat{\xi})$ with signature (-,+,+). The target space is a supermanifold with superspace co-ordinates $\hat{z}^{\hat{M}} = (\hat{x}^{\hat{m}}, \hat{\theta}^{\hat{\mu}})$ where $\hat{m}$ = 1,...,11 and $\hat{\mu}$ = 1,...,32 with space-time signature (-,+,...,+). We also define $\hat{E}_{\hat{i}}{}^{\hat{A}} = (\partial_{\hat{i}}\hat{z}^{\hat{M}})\hat{E}_{\hat{M}}{}^{\hat{A}}(\hat{z})$ where $\hat{E}_{\hat{M}}{}^{\hat{A}}$ is the supervielbein and $\hat{A} = (\hat{a}, \hat{\alpha})$ is the tangent space index ($\hat{a}$ = 1,...,11 and $\hat{\alpha}$ = 1,...,32). The action is then given by[6)]

$$S = \int d^3\hat{\xi} \left\{ \frac{1}{2} \sqrt{-\hat{\gamma}} \, \hat{\gamma}^{\hat{i}\hat{j}} \, \hat{E}_{\hat{i}}{}^{\hat{a}} \, \hat{E}_{\hat{j}}{}^{\hat{b}} \, \eta_{\hat{a}\hat{b}} - \frac{1}{2} \sqrt{-\hat{\gamma}} \right.$$
$$\left. - \frac{1}{6} \varepsilon^{\hat{i}\hat{j}\hat{k}} \, \hat{E}_{\hat{i}}{}^{\hat{A}} \, \hat{E}_{\hat{j}}{}^{\hat{B}} \, \hat{E}_{\hat{k}}{}^{\hat{C}} \, \hat{A}_{\hat{C}\hat{B}\hat{A}} \right\} \tag{2.1}$$

Note that there is a Wess-Zumino term involving the super three-form $\hat{A}_{\hat{A}\hat{B}\hat{C}}(\hat{z})$ and also a world-volume cosmological term. In addition to world-volume diffeomorphisms, target space superdiffeomorphisms, Lorentz invariance and three-form gauge invariance, the action (2.1) is invariant under a fermionic gauge transformation[6)] whose parameter

----
*) For further detailed discussion, see the conference talk by K.S. Stelle[11)].

$\hat{\kappa}^{\hat{\alpha}}(\hat{\xi})$ is a 32-component space-time Majorana spinor and a world-volume scalar. This is a generalization to the case of membranes of the symmetry discovered by Siegel[12] for the superparticle and Green and Schwarz[13] for the superstring in the form given by Hughes, Liu and Polchinski[8].

To facilitate a discussion of this fermionic symmetry, it is convenient to eliminate the world-volume metric as an independent variable. In this way we avoid having to discuss the rather complicated transformation rule for the metric. The action (2.1) then takes on its Nambu-Goto form

$$S = \int d^3\hat{\xi} \left[\sqrt{-\det \hat{E}_{\hat{i}}{}^{\hat{a}} \hat{E}_{\hat{j}}{}^{\hat{b}} \eta_{\hat{a}\hat{b}}} - \frac{1}{6} \varepsilon^{\hat{i}\hat{j}\hat{k}} \hat{E}_{\hat{i}}{}^{\hat{A}} \hat{E}_{\hat{j}}{}^{\hat{B}} \hat{E}_{\hat{k}}{}^{\hat{C}} \hat{A}_{\hat{C}\hat{B}\hat{A}} \right]$$

$$(2.2)$$

It is invariant under the transformation[6]

$$\delta \hat{z}^{\hat{a}} \equiv \delta \hat{z}^{\hat{M}} \hat{E}_{\hat{M}}{}^{\hat{a}} = 0$$

$$\delta \hat{z}^{\hat{\alpha}} \equiv \delta \hat{z}^{\hat{M}} \hat{E}_{\hat{M}}{}^{\hat{\alpha}} = \hat{\kappa}^{\hat{\beta}} (1 + \hat{\Gamma})_{\hat{\beta}}{}^{\hat{\alpha}}$$

$$(2.3)$$

where

$$\hat{\Gamma}_{\hat{\beta}}{}^{\hat{\alpha}} = \frac{1}{6\sqrt{-\hat{g}}} \varepsilon^{\hat{i}\hat{j}\hat{k}} \hat{E}_{\hat{i}}{}^{\hat{a}} \hat{E}_{\hat{j}}{}^{\hat{b}} \hat{E}_{\hat{k}}{}^{\hat{c}} (\hat{\Gamma}_{\hat{a}\hat{b}\hat{c}})_{\hat{\beta}}{}^{\hat{\alpha}} \quad (2.4)$$

In (2.4) $\hat{g}_{\hat{i}\hat{j}}$ is the metric on the world-volume induced from the bosonic metric on superspace,

$$\hat{g}_{\hat{i}\hat{j}} = \hat{E}_{\hat{i}}{}^{\hat{a}} \hat{E}_{\hat{j}}{}^{\hat{b}} \eta_{\hat{a}\hat{b}} \qquad (2.5)$$

In order for (2.2) to be invariant under this transformation, it is necessary that the background supergeometry be constrained. The constraints found in Ref. 6) are

60

$$\hat{T}_{\hat{a}\hat{\beta}}{}^{\hat{c}} = -i\,(\hat{\Gamma}^{\hat{c}})_{\hat{a}\hat{\beta}} \quad ; \quad \hat{F}_{\hat{a}\hat{\beta}\hat{c}\hat{d}} = i\,(\hat{\Gamma}_{\hat{c}\hat{d}})_{\hat{a}\hat{\beta}} \quad (2.6)$$

$$\hat{F}_{\hat{a}\hat{\beta}\hat{\gamma}\hat{\delta}} = \hat{F}_{\hat{a}\hat{\beta}\hat{\gamma}\hat{d}} = 0$$

$$\hat{T}_{\hat{a}\hat{b}\hat{c}} = \eta_{\hat{b}\hat{c}}\,\hat{\Lambda}_{\hat{a}} \quad ; \quad \hat{F}_{\hat{a}\hat{b}\hat{c}\hat{d}} = (\hat{\Gamma}_{\hat{b}\hat{c}\hat{d}})_{\hat{a}}{}^{\hat{\beta}}\,\hat{\Lambda}_{\hat{\beta}} \quad (2.7)$$

Although these equations are not the standard equations of on-shell D = 11 supergravity in superspace[14),15)], they are equivalent to them. That is to say, by suitable redefinitions of the superconnections and parts of the supervielbein, we may set $\hat{\Lambda}_{\hat{\alpha}}$, $\hat{T}_{\hat{\alpha}\hat{\beta}}{}^{\hat{\gamma}}$ and $\hat{T}_{\hat{a}\hat{b}}{}^{\hat{c}}$ to zero[7)],

$$\hat{T}_{\hat{a}\hat{b}}{}^{\hat{c}} = \hat{T}_{\hat{a}\hat{\beta}}{}^{\hat{\gamma}} = \hat{F}_{\hat{a}\hat{b}\hat{c}\hat{d}} = 0 \qquad (2.8)$$

Equations (2.8) are just the field equations of D = 11 supergravity as may be checked using the Bianchi identities. At this classical level, at least, all solutions of D = 11 supergravity are possible backgrounds for the supermembrane. We shall return to this point in Section 6.

The other important point to make about the Siegel invariance, is that

$$\hat{\Gamma}^2 = \mathbb{1} \qquad (2.9)$$

on using (2.5), and hence $\frac{1}{2}(1\pm\Gamma)$ act as projection operators. This will allow us to gauge away one half of the fermionic degrees of freedom, i.e., there is a total of $16 \div 2 = 8$ on-shell. The number of bosonic degrees of freedom is also $8 = 11-3$ owing to the diffeomorphisms on the three-dimensional world-sheet.

3. – SUPERSTRINGS FROM SUPERMEMBRANES BY SIMULTANEOUS DIMENSIONAL REDUCTION

It is well known that $N = 2a$ supergravity in ten dimensions $(g_{mn}, A_m, \Phi; \psi_m, \chi; A_{mnp}, A_{mn})$ may be obtained by dimensional reduction from $N = 1$ supergravity in eleven dimensions $(\hat{g}_{\hat{m}\hat{n}}, \hat{\psi}_{\hat{m}}, \hat{A}_{\hat{m}\hat{n}\hat{p}})$. On the other hand, $N = 2a$ supergravity is also the field-theory limit of the Type IIA superstring. The purpose of this section is to derive the Type IIA superstring from the supermembrane of Section 2 by a dimensional reduction of the world-volume from three to two dimensions and, simultaneously, a dimensional reduction of the space-time from eleven to ten[7].

To see how the dimensional reduction works, let us first focus our attention on the purely bosonic sector for which the action (2.1) reduces to

$$
S = \int d^3\hat{\xi} \left[\frac{1}{2} \sqrt{-\hat{\gamma}} \; \hat{\gamma}^{\hat{i}\hat{j}} \; \partial_{\hat{i}} \hat{x}^{\hat{m}} \; \partial_{\hat{j}} \hat{x}^{\hat{n}} \; \hat{g}_{\hat{m}\hat{n}}(\hat{x}) \; - \frac{1}{2} \sqrt{-\hat{\gamma}} \right.
$$
$$
\left. + \frac{1}{6} \, \varepsilon^{\hat{i}\hat{j}\hat{k}} \; \partial_{\hat{i}} \hat{x}^{\hat{m}} \, \partial_{\hat{j}} \hat{x}^{\hat{n}} \, \partial_{\hat{k}} \hat{x}^{\hat{p}} \, \hat{A}_{\hat{m}\hat{n}\hat{p}}(\hat{x}) \right]
$$

$$(3.1)$$

Varying with the respect to the metric $\hat{\gamma}_{\hat{i}\hat{j}}$ yields the embedding equation

$$
\hat{\gamma}_{\hat{i}\hat{j}} = \hat{g}_{\hat{i}\hat{j}} \equiv \partial_{\hat{i}} \hat{x}^{\hat{m}} \, \partial_{\hat{j}} \hat{x}^{\hat{n}} \, \hat{g}_{\hat{m}\hat{n}}(\hat{x})
\tag{3.2}
$$

while varying with respect to $\hat{x}^{\hat{m}}$ yields the equation of motion

$$
\frac{1}{\sqrt{-\hat{g}}} \, \partial_{\hat{i}} \left(\sqrt{-\hat{g}} \, \hat{g}^{\hat{i}\hat{j}} \, \partial_{\hat{j}} \hat{x}^{\hat{m}} \right) + \hat{\Gamma}_{\hat{n}\hat{p}}{}^{\hat{m}} \, \partial_{\hat{i}} \hat{x}^{\hat{n}} \, \partial_{\hat{j}} \hat{x}^{\hat{p}} \, \hat{g}^{\hat{i}\hat{j}}
$$
$$
= \frac{1}{6} \hat{F}^{\hat{m}}{}_{\hat{n}\hat{p}\hat{q}} \, \partial_{\hat{i}} \hat{x}^{\hat{n}} \, \partial_{\hat{j}} \hat{x}^{\hat{p}} \, \partial_{\hat{k}} \hat{x}^{\hat{q}} \frac{\varepsilon^{\hat{i}\hat{j}\hat{k}}}{\sqrt{-\hat{g}}}
$$

$$(3.3)$$

where $\hat{F}_{\hat{m}\hat{n}\hat{p}\hat{q}}$ is the field-strength of $\hat{A}_{\hat{m}\hat{n}\hat{p}}$

$$\hat{F}_{\hat{m}\hat{n}\hat{p}\hat{q}} \equiv 4 \, \partial_{[\hat{m}} \hat{A}_{\hat{n}\hat{p}\hat{q}]} \; . \tag{3.4}$$

We now make a two-one split of the world-volume co-ordinates

$$\hat{\xi}^{\hat{\iota}} = (\xi^{\iota} , \, \mathcal{S}) \qquad \iota = 1,2 \tag{3.5}$$

and a ten-one split of the space-time co-ordinates

$$\hat{x}^{\hat{m}} = (x^{m} , \, y) \qquad m = 1, \ldots, 10 \tag{3.6}$$

in order to make the partial gauge choice

$$\mathcal{S} = y \; . \tag{3.7}$$

which identifies the eleventh space-time dimension with the third dimension of the world-volume. The dimensional reduction is then effected by demanding that

$$\partial_{\mathcal{S}} x^{m} = 0 \tag{3.8}$$

and

$$\partial_{y} \hat{g}_{\hat{m}\hat{n}} = 0 = \partial_{y} \hat{A}_{\hat{m}\hat{n}\hat{p}} \; . \tag{3.9}$$

A suitable choice of ten-dimensional variables is now given by

$$\hat{g}_{\hat{m}\hat{n}} = \Phi^{-2/3} \begin{bmatrix} g_{mn} + \Phi^{2} A_{m} A_{n} & \Phi^{2} A_{m} \\ \Phi^{2} A_{n} & \Phi^{2} \end{bmatrix} \tag{3.10}$$

$$\hat{A}_{\hat{m}\hat{n}\hat{p}} = (\hat{A}_{mnp} , \, \hat{A}_{mny})$$

$$= (A_{mnp} , \, A_{mn})$$

From (3.2), the induced metric on the world-sheet is now given by

$$\hat{g}_{\hat{i}\hat{j}} = \Phi^{-2/3} \begin{bmatrix} g_{ij} + \Phi^2 A_i A_j & \Phi^2 A_i \\ \Phi^2 A_j & \Phi^2 \end{bmatrix} \qquad (3.11)$$

where

$$g_{ij} \equiv \partial_i x^m \partial_j x^n g_{mn} \quad , \quad A_i \equiv \partial_i x^m A_m \ . \qquad (3.12)$$

Note that

$$\sqrt{-\hat{g}} = \sqrt{-g} \qquad (3.13)$$

Substituting these expressions into the field equations (3.3) yields in the case $\hat{x}^{\hat{m}} = x^m$

$$\frac{1}{\sqrt{-g}} \partial_i \left(\sqrt{-g}\, g^{ij} \partial_j x^m \right) + \Gamma_{np}{}^m \partial_i x^n \partial_j x^p g^{ij}$$

$$= \frac{1}{2} F^m{}_{np} \partial_i x^n \partial_j x^p \frac{\varepsilon^{ij}}{\sqrt{-\hat{g}}} \qquad (3.14)$$

where F_{mnp} is the field-strength of A_{mn}

$$F_{mnp} \equiv 3\, \partial_{[m} A_{np]} = \hat{F}_{mnpy} \qquad (3.15)$$

In the case $\hat{x}^{\hat{m}} = y$, (3.3) is an identity, as it must be for consistency. But (3.14) is just the ten-dimensional string equation of motion derivable from the action

$$S = \int d^2y \left[\frac{1}{2} \sqrt{-\gamma}\, \gamma^{ij} \partial_i x^m \partial_j x^n g_{mn} + \frac{1}{2} \varepsilon^{ij} \partial_i x^m \partial_j x^n A_{mn} \right]$$

$$(3.16)$$

Comparing with (3.1), we see that the overall effect is to reduce the eleven-dimensional membrane to a ten-dimensional string, to replace the three-form by a two-form in the Wess-Zumino term and to eliminate the world-volume cosmological constant. Note that the other ten-dimen-

64

sional bosonic fields A_{mnp}, A_m and Φ have all decoupled. They have not disappeared from the theory, however, since their coupling still survives in the fermionic θ sector, to which we shall turn shortly. First, we make some remarks.

As is well known, the dimensional reduction (3.9) corresponds to a Kaluza-Klein compactification of space-time on a circle in which one discards all the massive modes. The difference from conventional Kaluza-Klein is that by identifying the eleventh space-time dimension with the third dimension on the world-volume as in (3.7), the world-volume is also compactified on the same circle. The condition (3.8) means that we are discarding the massive world-sheet modes at the same time. By retaining all the U(1) singlets but only the U(1) singlets, these truncations are guaranteed to be consistent[3] with the membrane equations of motion and, as we shall see, with the equations of motion of the background fields. As an extra check on consistency, we have been careful to substitute the Kaluza-Klein ansatz into the equations of motion rather than directly into the action. The signal for consistency is that the $\hat{x}^{\hat{m}} = y$ component of the field equations (3.3) is an identity. Having established consistency one may then, if so desired, substitute directly into the action (3.1) and integrate over ρ.

To include the fermions, it is now straightforward to repeat the above procedure in superspace. The Kaluza-Klein ansatz for the $N = 1$, $D = 11$ supervielbein is

$$\hat{E}_{\hat{M}}{}^{\hat{A}} = \begin{bmatrix} \hat{E}_M{}^\alpha & \hat{E}_M{}^\alpha & \hat{E}_M{}^{11} \\ \hat{E}_y{}^a & \hat{E}_y{}^\alpha & \hat{E}_y{}^{11} \end{bmatrix} \tag{3.17}$$

$$= \begin{bmatrix} E_M{}^\alpha & E_M{}^\alpha + A_M \chi^\alpha & \Phi A_M \\ 0 & \chi^\alpha & \Phi \end{bmatrix} \tag{3.18}$$

where $E_M{}^A = (E_M{}^a, E_M{}^\alpha)$ is the $N = 2a$, $D = 10$ supervielbein, A_M the superspace $U(1)$ gauge field and Φ and χ^a are superfields whose leading components are the dilaton and the dilatino respectively. In writing (3.18), we have made a partial $D = 11$ local Lorentz gauge choice to set $\hat{E}_y{}^a = 0$. For the superspace three-index potential $\hat{A}_{\hat{M}\hat{N}\hat{P}}$ we have

$$\hat{A}_{MNP} = A_{MNP} \qquad (3.19)$$

$$\hat{A}_{MNy} = A_{MN} \qquad (3.20)$$

All of the $D = 10$ superfields $E_M{}^A$, χ^α, A_M, Φ, A_{MN}, A_{MNP} are taken to be independent of y. Note also that ten-dimensional spinor indices run from 1 to 32 so that $\hat{\alpha}$ and α can be identified. With $\hat{z}^{\hat{M}} = (z^M, y)$, we also impose $\partial_\rho z^M = 0$ and $y = \rho$ as before. Substituting the ansätze (3.18), (3.19) and (3.20) into (2.2) yields the action for a Type IIA superstring coupled to a supergravity background

$$S = \int d^2\varsigma \left[\Phi \sqrt{-\det E_i{}^a E_j{}^b \eta_{ab}} - \frac{1}{2} \varepsilon^{ij} \partial_i z^M \partial_j z^N A_{NM} \right]$$

$$(3.21)$$

Purely for convenience in superspace calculations, we have omitted an overall factor of $\Phi^{-2/3}$ in the ansatz (3.17); the factor of Φ in (3.21) can be removed by a suitable rescaling of the supervielbein. To find the fermionic symmetry of the dimensionally-reduced action (3.21), one substitutes the Kaluza-Klein ansätze into (2.3). It is straightforward to show that

$$\delta\hat{z}^\alpha = \delta z^\alpha \equiv \delta z^M E_M{}^\alpha = \kappa^\beta (1 + \Gamma)_\beta{}^\alpha \qquad (3.22)$$

and that

$$\hat{\Gamma}_\beta{}^\alpha = \Gamma_\beta{}^\alpha = \frac{1}{2\sqrt{-g}} \varepsilon^{ij} E_i{}^a E_j{}^b (\Gamma_{ab}\Gamma_{11})_\beta{}^\alpha \qquad (3.23)$$

Since (2.2) is invariant under (2.3) when the $D = 11$ field equations are satisfied, it follows that the reduced action (3.21) will be invariant under (3.23) if the $N = 2a$, $D = 10$ supergravity field equations are satisfied. This is because the compactification of the $N = 1$, $D = 11$ field theory on a circle is known to yield the $N = 2a$,

D = 10 field theory. Note that all of the N = 2a supergravity fields are now coupled, including A_{mnp}, A_m and Φ which decoupled from the purely bosonic sector. The transformation (3.22) can be recast into the Green-Schwarz[13] form by introducing

$$\lambda^{i\,\alpha} = \frac{1}{2} \frac{\varepsilon^{ij}}{\sqrt{-g}} \; E_j{}^\alpha \; \kappa^\beta \left(\Gamma_\alpha \Gamma_{11} \right)_\beta{}^\alpha \qquad (3.24)$$

In conclusion, we have succeeded in deriving [for the first time*)] the action of the Type IIA superstring coupled to an N = 2a, D = 10 supergravity background starting from the action of the super-membrane coupled to the background of N = 1 supergravity in D = 11. The dimensional reduction corresponds to a compactification of both the space-time and the world-volume on the same circle and then discarding the massive modes. Classically, this is equivalent to letting the membrane tension κ_2 tend to infinity and the radius of the circle R tend to zero in such a way that the string tension $\kappa_1 = [\alpha']^{-1}$

$$\kappa_1 = 2\pi R \kappa_2 \qquad (3.25)$$

remains finite.

4. - OTHER DIMENSIONS, OTHER BRANES

So far we have discussed a p = 2 extended object in D = 11 dimensions, but super p-extended objects can exist in D dimensions whenever there is a closed (p+2)-form in superspace*). This follows from the Siegel invariance of the Green-Schwarz action[6]:

$$S = \int d^{p+1} \mathcal{S} \left\{ \frac{1}{2} \sqrt{\gamma} \gamma^{ij} E_i{}^a E_j{}^b \eta_{ab} - \frac{(p-1)\sqrt{\gamma}}{2} \right.$$

---
*) The Type IIB action is given in Ref. 16).

$$+ \frac{1}{(p+1)!} \varepsilon^{i_1 i_2 \cdots i_{p+1}} E_{i_1}{}^{A_1} E_{i_2}{}^{A_2} \cdots E_{i_{p+1}}{}^{A_{p+1}} A_{A_{p+1} \cdots A_2 A_1} \Big\} \quad (4.1)$$

This closure is equivalent to the identity

$$\left(d\bar{\theta} \Gamma_a d\theta \right) \left(d\bar{\theta} \Gamma^{a b_1 \cdots b_{p-1}} d\theta \right) = 0 \qquad (4.2)$$

for a commuting spinor $d\theta$.

Achúcarro, Evans, Townsend and Wiltshire[9] have recently classified those values of p and D, and the number of space-time supersymmetries N for which this identity is valid. They find that this is
possible only when the number of on-shell p-dimensional Bose and Fermi
degrees of freedom are equal. There are thus four "fundamental" super
p-branes as shown in the table. All four have the minimal N = 1
supersymmetry and the D = 10 and D = 6 cases are chiral. The D = 10
case couples to the dual form of D = 10 supergravity discussed in
Section 1.

Table 1

The fundamental super p-branes.

p	D	Bose+Fermi	Algebra
2	11	8 + 8	$\mathbb{O}$
5	10	4 + 4	$\mathbb{H}$
3	6	2 + 2	$\mathbb{C}$
2	4	1 + 1	$\mathbb{R}$

However, starting from these fundamental p-branes in D dimensions
other (p-k)-branes in (D-k) dimensions may be obtained by repeating k
times the simultaneous dimensional reduction of Section 3. Thus we
make a (D-k)-k split of the target space co-ordinates

$$\hat{x}^{\hat{m}} = \left(x^m, y^\alpha \right) \qquad \alpha = 1 \ldots k \qquad (4.2)$$

----
[*] We confine our attention to closed p-branes. This is required in
D = 11. The existence of open supermembranes in D < 11 is less
clear.

68

and a (p+1-k)-k split of the world-volume co-ordinates

$$\hat{\xi}^{\hat{\imath}} = (\xi^{i}, \rho^{\alpha}) \qquad \alpha = 1 \dots k \qquad (4.3)$$

in order to make the partial gauge choice

$$\rho^{\alpha} = y^{\alpha} \qquad (4.4)$$

which identifies the extra k space-time dimensions with the extra k dimensions of the world-volume. The dimensional reduction is then effected by demanding that the remaining co-ordinates x^{m} be independent of ρ^{α} and the target-space background fields be independent of y^{α}. This procedure now results in four sequences of super p-branes as shown in the figure.

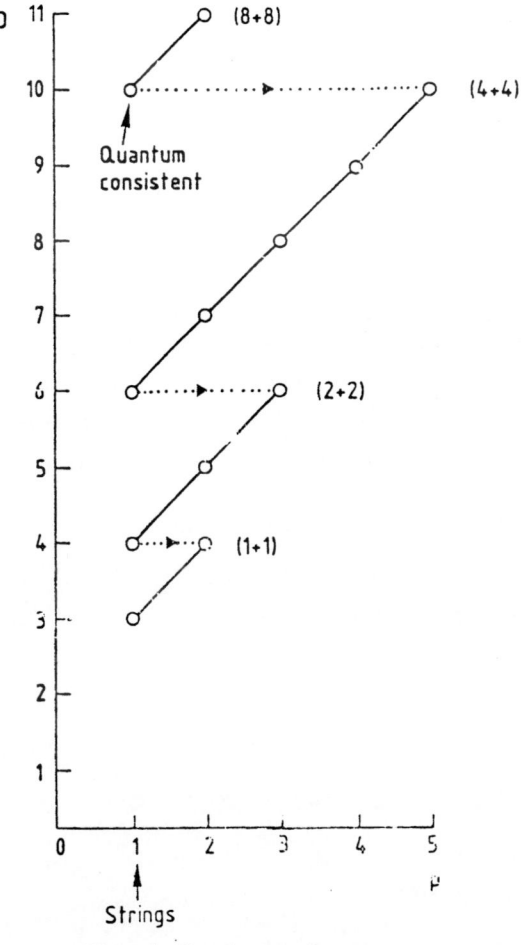

Fig. 1 The Brane Scan

Remarkably, for p = 1 we recover the well-known four classical Green-Schwarz superstrings in D = 10,6,4 and 3 and these are now seen to be but special cases of a more general framework of four sequences of super p-brane with (8+8), (4+4), (2+2) and (1+1) degrees of freedom respectively. This immediately suggests a connection with the four-division algebras $\mathbb{R}$, $\mathbb{C}$, $\mathbb{H}$ and $\mathbb{O}$, a connection which can in fact be made more precise[9),17)]. In the p = 1 case, of course, we know that only the D = 10 superstring is quantum consistent. Note that the only other member of this $\mathbb{O}$ sequence is the eleven-dimensional supermembrane.

All the points on the scan represent N = 1 theories except for the p = 1 superstring case where we find N = 2. Two world-sheet dimensions are also special in that one can separate modes into left- and right-movers. This permits a truncation to N = 1. The horizontal arrows in the Table represent a "duality transformation" in which each of these N = 1 string theories is swapped for its dual theory in which the three-form field strength is replaced by a (p+2)-form field strength. Note that there is no dual version of N = 1, D = 11 supergravity in which the four-form field strength is replaced by a seven-form[18]. (This means that one cannot <u>simultaneously</u> dualize the four-form and three-form field strengths of N = 2a, D = 10 supergravity, but one can do either separately[*)]. There appear to be no corresponding N = 2 membranes, however.)

Finally, we note that one may travel vertically downwards in the figure by making an ordinary dimensional reduction of the space-time only. The theories so obtained need not have a closed (p+2)-form since Chern-Simons terms will arise from the dimensional reduction[10].

5. - OBJECTIONS TO QUANTUM MEMBRANES

In the literature one may find statements to the effect that strings provide the only consistent quantum theories of extended

----
*) H. Nishino, private communication.

objects.   Below we list some of these objections and, whenever possible, provide the refutation.

Fermionic Symmetry

The Siegel symmetry of the Green-Schwarz superstring is essential for ensuring the correct count of boson and fermion degrees of freedom and hence its consistency. It has been claimed that this fermionic symmetry cannot be generalized to membranes. But, as we have already seen in Section 2, it can! This first objection is thus rather easy to refute.

Triality and SO(8)

Another argument in favour of superstrings and, in particular, ten space-time dimensions, invokes the appearance of the group SO(8), with its unique triality properties, as the invariance group of the space of physical variables. Thus the eight bosonic variables corresponding to the transverse 10-2 = 8 dimensions are matched by the 8 = 16÷2 fermionic variables remaining after the Siegel symmetry has been taken into account. But as we have seen in Section 2, the same balance occurs for the supermembrane in eleven dimensions: here we have 11-3 = 8 physical bosonic variables, and again 16÷2 = 8 fermionic variables after taking account of the Siegel symmetry.

Ghosts and Lorentz Invariance

Clearly, any consistent theory must be free from ghosts. Both strings and membranes apparently suffer from the existence of an infinity of ghosts, owing to the negative sign of the kinetic term of the time co-ordinate x^0. In the case of strings, this problem is known to be solved in the critical dimension. The proof[19),20),21)] invokes the conformal symmetry, which in two dimensions is infinite dimensional. In more than two dimensions the conformal group is only finite dimensional, and is in any case unlikely to be relevant for membranes[22)]. To address the ghost problem for membranes, one can proceed in one of two ways. In the first, covariant, approach one looks for an infinite dimensional symmetry that will substitute for the Virasoro algebra. Little has been achieved in this direction so far, although it is interesting to note that Hoppe[23] has demonstrated the existence of infinite parameter symmetries of the membrane Hamiltonian. An alternative approach is to go to a physical gauge, where ghost-freedom is manifest but Lorentz invariance is not. The problem now is to demonstrate that Lorentz invariance is nevertheless preserved in the quantization. For strings, this is a problem owing to the existence of conformal anomalies which spoil the closure of the Lorentz algebra. Hence the need for a critical dimension, in which the conformal anomalies cancel. For membranes, it is difficult to see how a critical dimension could arise in this sense, when there is no obvious candidate for an anomalous symmetry at least in the case of the bosonic membrane. In particular, there is no classical conformal invariance. Moreover, there are no diffeomorphism anomalies in three dimensions. In fact it seems that many of the special features of string theory are needed in order to cure the problems of living in two dimensions. For a Hamiltonian formulation of the bosonic membrane, see Refs. 24)-27), and of the supermembrane Refs. 10) and 28).

Massless Particles

An important question for any membrane theory is whether massless particles are present in the spectrum. In the case of the super-membrane, consistency presumably requires the occurrence of the graviton, gravitino and antisymmetric tensor of eleven-dimensional supergravity in the spectrum[*]. For the purely bosonic membrane, Kikkawa and Yamasaki have concluded that the open membrane does not admit massless spin-one or spin-two particles in any integer dimension[29]. Their argument involved computing the intercept of the classical mass/spin relation induced by one-loop vacuum energy effects by analogy with the string calculations of Brink and Nielsen[30].

Consider a bosonic p-brane with Nambu-Goto action

$$ S = K_p \int d\tau \, d^p\sigma \, \sqrt{\det \partial_i x^m \partial_j x^n \eta_{mn}} \qquad (5.1) $$

----
[*]Or, perhaps, the massless particles to which they give rise after spontaneous compactification of the space-time. See Section 6.

Then the membrane tension κ_p has dimensions

$$[\kappa_p] = M L^{1-p} T^{-1} \tag{5.2}$$

To obtain a classical ($\hbar$ independent) relation between the angular momentum J

$$[J] = M L^2 T^{-1} \tag{5.3}$$

of a rotating p-brane and its rest mass, using only κ_p and the velocity of light c

$$[c] = L T^{-1} \tag{5.4}$$

Kikkawa and Yamasaki[29] argue that we must have

$$J = A \kappa_p^{-1/p} (cm)^{(p+1)/p} \tag{5.5}$$

and that this gives rise in the quantum theory to a spin-mass relation

$$J = A \kappa_p^{-1/p} (cm)^{(p+1)/p} + B \hbar \tag{5.6}$$

where A and B are dimensionless numbers.

This certainly reduces in the case of the bosonic string (p=1), to the familiar Regge formula (A=1, $\kappa_1^{-1}=\alpha'$)

$$J = \alpha' m^2 + \alpha_0 \tag{5.7}$$

To compute α_0, Brink and Nielsen[30] noted that the intercept α_0 arises from the Casimir energy and that the mass squared is shifted by an amount

$$\Delta m^2 = \frac{1}{\alpha'} \frac{D-2}{2} \sum_{n=1}^{\infty} n \tag{5.8}$$

$$\tag{5.9}$$

$$= -\frac{1}{\alpha'} \frac{D-2}{24}$$

73

on using ξ function regularization for which $\zeta(-1) = -1/12$ where $\zeta(s) = \sum_{n=1}^{\infty} n^{-s}$. Thus we obtain from (5.7) a massless (m=0) spin one (J=1) particle when D = 26. As noted by Kikkawa and Yamasaki[29], it is a miracle that the Casimir term happens to be a rational number so that J is made an integer by adjusting the dimension of space-time. In the case of the closed bosonic string, the Casimir energy is doubled (one considers two open strings back-to-back) and we find a massless spin two particle

$$\Delta m^2 = \kappa_1 (D-2) \sum_n n \qquad (5.10)$$

Similar miracles occur for the Neveu-Schwarz formulation of the open and closed superstrings where we find massless particles when D = 10.

To examine the case for membranes, Kikkawa and Yamasaki consider rigid rotator solutions since these maximize the angular momentum for a given energy. Assuming that the Casimir energy is independent of the membrane tension κ, they argue that the one-loop result will be exact since this corresponds to the $\kappa \to \infty$ limit. Perhaps not surprisingly they find that there are no miracles and that the open membrane (p=2) does not admit massless spin one or spin two particles in any integer dimension.

The question naturally arises as to whether the inclusion of fermions would cure this problem. However, in order to follow the method of Ref. 30), it would be necessary to have a Neveu-Schwarz-Ramond (NSR) formalism for the supermembrane; i.e., a formalism which exhibits world-volume supersymmetry. The supermembrane theory of Ref. 6) is written in a Green-Schwarz formalism, in which space-time super-symmetry is manifest. It is not clear at present whether an NSR formalism exists for the supermembrane, but we draw attention to the following problem. As discussed in Section 3, the supermembrane, by contrast to the superstring, requires a cosmological term in the action. In three-dimensional supergravity, the cosmological term arises from eliminating an auxiliary field S which appears both linearly and quadratically in the action. However, there is no super-invariant quadratic in S that does not also involve the Einstein-Hilbert gravitational action. This means that the Nambu-Goto action cannot be supersymmetrized. This problem was implicit in the paper of Howe and Tucker[31] (we are grateful to P. Howe for drawing this to our attention), and is further discussed in Ref. 32).

Although an NSR calculation of the shift in the intercept is thus not possible, it is gratifying to find that the vacuum energy can be zero in the Green-Schwarz formalism [Duff, Inami, Pope, Sezgin and Stelle[10)]. Owing to the intrinsic non-linearity of the membrane equations, there is no clever gauge which linearizes the equations of motion for $p > 1$. Consequently, in order to analyze the self-energy question, the best we can do at the moment is to pick a stable classical solution and to quantize the fluctuations about this background. Stable solutions are not easy to find, and instead of considering a rotating membrane in Minkowski space as in Ref. 29), we considered in Ref. 10) a space-time with topology $R^9 \times S^1 \times S^1$. With such a topology there is a stable classical solution with a closed toroidal membrane stretched around the torus. In this case, the mode expansions are particularly simple. For the bosonic membrane, one finds a (mass)^2 shift arising from the normal ordering of the oscillators which is proportional to (D-3), and is given by an Epstein ζ-function

$$\Delta m^2 = K_2 (D-3) \sum_{n,m} \left[(m \ell_2 R_2)^2 + (n \ell_1 R_1)^2 \right]^{1/2} \quad (5.11)$$

where R_1 and R_2 are the radii of the two circles and ℓ_1 and ℓ_2 are integers characterizing the winding numbers of the membrane around the two circles. The beauty of the supermembrane is that this vacuum energy is cancelled mode by mode by an equal and opposite contribution from the fermions, just as in the Green-Schwarz superstring. This is encouraging for the prospect of the theory's spectrum containing massless states, but one would not expect to find them in the toroidal background considered here but rather in the classical background that corresponds to the expected leading Regge trajectory.

Linear Trajectories?

On the subject of the Regge trajectory, an apparent paradox arises. According to the dimensional argument (5.6) of Kikkawa and Yamasaki[29)], the Casimir energy of the membrane (p=2) induces a quantum shift, not in m^2 as for the string (p=1) but rather in $m^{3/2}$. Whereas our analysis[10)] yielded a shift in m^2 for the same reason as for the string, namely the operator ordering of terms quadratic in oscillators. The resolution of the paradox is that the toroidal compactification introduces a length scale R just as in Section 3, where the membrane tension κ_2 is related to the string tension κ_1 by

$$K_1 = 2\pi R K_2 \qquad (5.12)$$

The introduction of a length scale means that the dimensional arguments of Kikkawa and Yamasaki no longer apply, and linear trajectories are possible even for membranes. Thus, we may replace (5.6) by

$$J = A K_p^{-1} c^2 m^2 R^{1-p} + B\hbar \qquad (5.13)$$

which also reduces to the familiar formula when p = 1. In fact it is tempting to speculate that consistency may require space-time compactification when p > 1. See Section 6.

Non-Renormalizability

The Green-Schwarz action for super p-branes is power-counting non-renormalizable except in the case of strings (p=1)[*]. Of all the objections to membranes, this is in our opinion the most difficult to refute. One suggestion is to follow Weinberg[34] and appeal to asymptotic safety. Another suggestion is to include couplings to all the modes of the membrane in the non-linear σ-model and not just the massless ones. In the case of strings this would also render the σ-model non-renormalizable (since higher rank tensors would require more derivatives to form an invariant) and yet presumably leads to a perfectly sensible theory. The hope of this suggestion is that the problem resides not with the membrane per se but only in the truncation to the massless modes which is consistent for strings but perhaps not for membranes (in analogy with the consistency problems that arise in Kaluza-Klein theories). A third suggestion is based on the observation that certain three-dimensional O(N) σ-models do become renormalizable in the large N limit[35].

A fourth (wild?) idea which I favour personally is that although the theory is non-renormalizable and non-finite for generic choices of the background supergravity fields, there may exist a very special choice of vacuum which renders the σ-model finite. This idea is

----

[*]One thus expects to generate curvature corrections to the Nambu-Goto action. These are typically of the form $\kappa^{1-\ell}\sqrt{-\gamma}\, R^{\ell(p+1)/2}$ at ℓ loop order with only ℓ even for even p. Note that we never induce an Einstein-Hilbert term, however. For this reason, we do not follow recent atempts to interpret space-time itself as the world-volume of a membrane[33].

further discussed in Section 6. Even if this problem is solved, we still have to worry about constructing membrane scattering amplitudes and whether such amplitudes are finite.

6. - WILD SPECULATIONS

Looking at the brane-scan of the figure, the most interesting p-branes seem to be the supermembrane in D = 11 and the super five-brane in D = 10. In this section we make some speculations concerning their properties, beginning with the five-brane.

A Heterotic Five-Brane?

In Section 1 we remarked that there are two anomaly-free ten-dimensional N = 1 supergravity theories coupled to either $E_8 \times E_8$ or SO(32): the version with the three-form field strength and the version with the seven-form field strength. Since the former arises as the field-theory limit of the superstring, it is natural to conjecture that the latter is related to the super five-brane. However, this would require a coupling of the super five-brane to background Yang-Mills fields which, to my knowledge, has not yet been achieved. Of course, the heterotic string construction[36] is peculiar to a two-dimensional world-sheet which permits a separation into left- and right-moving modes, one corresponding to the 26-dimensional bosonic co-ordinates and the other to the ten-dimensional super co-ordinates. For the five-brane we have no such split and no conformal invariance to guide us. However, a world-volume with one time and (4k+1) space co-ordinates does have something in common with one time and one space in that both permit a real self-duality condition. In the case of the five-brane it is left- and right-handed three-forms rather than the left- and right-handed one-form of the world-sheet. Perhaps this construction or else one involving chiral fermions on the world-volume may permit the required Yang-Mills coupling. If such a five-brane were quantum consistent, it would be every bit as appealing from the point of view of chirality and gauge groups as the heterotic string.

Of course, this five-brane lies on the same (quaternionic) sequence as the superstring in six dimensions which is known to be quantum inconsistent. From this point of view a more promising candidate for a consistent quantum theory is the supermembrane in D = 11 which is the only other member of the (octonionic) sequence containing the superstring in ten dimensions, which is known to be quantum consistent.

Supermembrane in D = 11: Hidden Symmetries?

It is interesting to note that the three-eight split

$$SO(1,10) \supset SO(1,2) \times SO(8) \qquad (6.1)$$

implied by the embedding of the three-dimensional world-volume of the supermembrane in eleven-dimensional space-time had previously been invoked in Refs. 37) and 38) to exhibit the hidden SO(16) symmetry of D = 11 supergravity, where the 128 bosonic degrees of freedom may be assigned to the coset $E_{8(+8)}/SO(16)$. One cannot help wondering what rôle these symmetries will play for the supermembrane, not to mention the Kac-Moody extension E_9 and the Lorentzian algebra E_{10}[37].

Another mathematical curiosity, already noted by string theorists, is that the number of disconnected components of the D = 10 diffeomorphism group is 992, which is just the dimension of G×G for G = SO(32) or $E_8 \times E_8$. But this, in turn, is given by the number of exotic eleven-spheres S^{11}[39].

Spontaneous compactification: 11 = 4+7

One of the most appealing features of D = 11 supergravity, which was swept aside in the wake of the superstring revolution, was the rather natural way in which four space-time dimensions were singled out. This is because the field equations admit a ground state of (D=4 non-compact Einstein space-time) × (D=7 compact Einstein space) in which the four-index field strength F_{mnpq} of Section 2 acquires a non-zero VEV proportional to ε_{mnpq} when the indices range over 1, 2, 3, 4 and zero otherwise[40]. This introduces a length scale into the theory. Since, at the classical level, all solutions of D = 11 supergravity are possible backgrounds for the supermembrane, this four-dimensional feature persists for the supermembrane. The classical field equations of D = 11 supergravity also permit D = 11 Minkowski space, of course, but we have already seen in Section 5 that the quantum membrane may somehow prefer space-time compactification and the accompanying length scale. Fujikawa and Kubo[41] have also suggested, for different reasons, that space-time compactification is a natural phenomenon in the membrane theory. It is worth recalling that the reason why we live in four dimensions remains of the greatest unsolved mysteries of string theory.

Of course, $11 = 7+4$ too[42] and an equally acceptable solution of the D = 11 field equations is to set the dual of F_{mnpq} proportional to $\varepsilon_{mnpqrst}$ when the indices ranger over 1,2,3,4,5,6,7 and zero other-wise[40]. This yields a ground state of (D=7 non-compact Einstein space-time) × (D=4 compact Einstein space). Although locally both solutions seem to be on an equal footing, globally they are quite different[3] since the volume form on a compact manifold is closed but not exact, i.e., the four-index field strength cannot be written as the curl of any non-singular three-form potential. This has an interesting consequence for the supermembrane when we recall that we may rewrite the Wess-Zumino term as an integral over a four-manifold whose boundary is the three-dimensional world-sheet. Thus (dropping hats), the bosonic part of the action is

$$S = \int_{\partial M^4} d^3\xi \sqrt{-g} + \int_{M^4} d^4\xi \, \varepsilon^{ijk\ell} \, \partial_i X^m \partial_j X^n \partial_k X^p \partial_\ell X^q \cdot F_{mnpq} \quad (6.2)$$

In the case (D=7 space-time) × (D=4 compact space) the VEV of F_{mnpq} and hence the curvature of the compact four-manifold will be <u>quantized</u> by virtue of the non-trivial topology (non-vanishing fourth Betti number). It remains to be seen what the implications are, but it is curious to note that it is the four-dimensional space-time choice which does <u>not</u> give rise to this quantization. A possible resolution of this old dilemma is given below.

<u>Supermembranes on the Seven-Sphere</u>

An important special case of the above compactifications corresponds to (d=4 anti-de Sitter space-time) × S^7 since this gives rise to a four-dimensional supergravity with maximum (N=8) supersymmetry and local SO(8) invariance[3]. The symmetry of the vacuum is given by the supergroup OSp(4/8). Thus the supermembrane provides a niche for the gauged extended supergravities[43] which had no place in superstring theory[1].

Apart from group manifolds, S^7 is the only compact manifold which admits an absolute parallelism[*] and consequently one is tempted to draw an analogy between supermembranes on S^7 and strings on group manifolds[**].

Now the non-linear σ-models describing strings on group manifolds are known to be conformally invariant and finite in D > 26 provided we satisfy the correct critical dimension formula. The observation that the theories are not finite in D > 26 "even" for flat backgrounds is true but irrelevant. In fact, one can show that the σ-models on space-time × G correspond to a free field theory. Thus, as discussed in Section 5, might the observation that the supermembrane σ-model is not renormalizable or finite "even" for flat backgrounds be equally irrelevant? Might the theory be finite on a very special background? Might that background be (d=4 AdS) × S^7?

Is the Supermembrane a Singleton?

Owing to its anti de Sitter subgroup, the supergroup OSp(4/8) admits the strange "singleton" representations which have no analogue in the Poincaré group and no immediate field theory interpretation. Owing to the N = 8 supersymmetry they form an ultrashort N = 8 supermultiplet with maximum spin-$\frac{1}{2}$[46]. The multiplet consists of eight spin-$\frac{1}{2}$ fermions and eight spin-0 bosons which transform according to the 8_s and 8_v representations of SO(8). (Thus the fermions have only half the naive number of degrees of freedom.) Although we are dealing

----

[1] For better or worse, this also means a fond "hello again" to anti-de Sitter space (AdS)[44].

[*] A totally antisymmetric three-index torsion for which the corresponding Riemann tensor vanishes.
[**] We note in this connection the recent discovery of loop algebras and superalgebras based on S^7[45].

with the four-dimensional anti de Sitter group SO(2,3), one cannot write down a four-dimensional action for these singletons. However, as discussed by Fronsdal[47], we can write down a <u>three-dimensional</u> action where the three-dimensional integration is over the boundary of anti-de Sitter space, which has signature (-,+,+) and topology $S^1 \times S^2$. The N = 8 case is discussed by Sezgin and Nicolai[48].

But 8_s fermions and 8_v scalars on a three-dimensional world-volume with signature (-,+,+) is just what we obtain after gauge-fixing the supermembrane action! Just as for the singleton, the fermions have only half the naive number of degrees of freedom owing to the Siegel symmetry. Moreover, we have already seen from (6.2) how the world-volume of the membrane is naturally regarded as the boundary of a four-dimensional manifold. Can the supermembrane be regarded as the singleton of OSp(4/8) whose world-volume is the boundary of anti-de Sitter space? If so, this would definitely favour 11 = 4+7 and not 11 = 7+4 too!

In fact, it will not have escaped the reader's attention that OSp(4/8) is also the vacuum symmetry of the (d=4 AdS) $\times$ S^7 background of D = 11 supergravity. Indeed, the above singleton representations appear in the harmonic expansion of the D = 11 fields on S^7, but can be gauged away[49],[3]. This is suggestive of a form of bootstrap in which the field theory gives rise to the membrane which, in turn, yields the field theory. [Singleton representations also occur for the super-groups OSp(6,2/4) and SU(2,2/4) whose anti-de Sitter groups are SO(2,6) and SO(2,4)[50]. This is discussed in Ref. 50) where an attempt is made to relate singletons to superstrings. The dimensions of the boundaries of the corresponding space-times are d = 6 and d = 4 which, from the table, are also the world-volume dimensions of the quaternionic and complex supermembranes. In these cases, however, the bootstrap idea seems no longer to apply.]

Finally, there is one other feature which membranes and singletons have in common: they were both invented by Dirac[51],[52] at about the same time!

Supermembrane-Inspired Phenomenology?

No concrete experimental predictions so far (much like super-strings).

ACKNOWLEDGEMENTS

Much of the work described in this lecture was carried out in collaboration with Paul Howe, Takeo Inami, Chris Pope, Kelly Stelle and
Ergin Sezgin. I would also like to thank E.S. Fradkin and Paul Townsend
for stimulating discussions on supermembranes.

REFERENCES

1) Schwarz, J.H., ed. "Superstrings: the first fifteen years" (World Scientific, 1985).

2) Cremmer, E., Julia, B. and Scherk, J. Phys. Lett. $\underline{76B}$, 409 (1978).

3) Duff, M.J., Pope, C.N. and Nilsson, B.E.W., Physics Reports $\underline{130}$, 1 (1986).

4) Chamseddine, A., Phys. Rev. $\underline{1724}$, 3065 (1981).

5) Gates, S.J. and Nishino, H., Phys. Lett. $\underline{157B}$, 157 (1985).

6) Bergshoeff, E., Sezgin, E. and Townsend, P.K., Phys. Lett. $\underline{189B}$, 75 (1987).

7) Duff, M.J., Howe, P.S., Inami, T. and Stelle, K.S., Phys. Lett. $\underline{191}$, 70 (1987).

8) Hughes, J., Liu, J. and Polchinski, J., Phys. Lett. $\underline{B180}$, 370 (1986)

9) Achucarro, A., Evans, A., Townsend, P.K. and Wiltshire, D.L., DAMTP preprint (1987).

10) Duff, M.J., Inami, T., Pope, C.N., Sezgin, E. and Stelle, K.S., preprint IC/87/74; CERN preprint TH.4731/87 (1987).

11) Stelle, K.S., to appear in Proceedings of the 4th Quantum Gravity Seminar, Moscow (May 1987).

12) Siegel, W., Phys. Lett. $\underline{128B}$, 397 (1983).

13) Green, M.B. and Schwarz, J.H., Phys. Lett. $\underline{136B}$, 367 (1984).

14) Cremmer, E. and Ferrara, S., Phys. Lett. $\underline{91B}$, 61 (1980).

15) Brink, L. and Howe, P., Phys. Lett. $\underline{91B}$, 384 (1980).

16) Grisaru, M.T., Howe, P., Mezincescu, L., Nilsson, B.E.W. and Townsend, P.K., Phys. Lett. $\underline{162B}$, 116 (1985).

17) Sierra, G., Class. Quantum Grav. $\underline{4}$, 227 (1987).

18) Nicolai, H., Townsend, P.K. and van Nieuwenhuizen, P., Lett. Nuovo Cim. $\underline{30}$, 315 (1981).

19) Brower, R.C., Phys. Rev. $\underline{D6}$, 1655 (1972).

20) Goddard, P. and Thorn, C.B., Phys. Lett. $\underline{40B}$, 235 (1972).

21) Goddard, P., Goldstone, J., Rebbi, C. and Thorn, C.B., Nucl. Phys. $\underline{B56}$, 109 (1973).

22) Rivers, R.J., Phys. Rev. $\underline{D9}$, 2920 (1974).

23) Hoppe, J., Aachen preprint, PITHA 86/24 (1986) and Ph.D. Thesis, MIT (1982).

24) Collins, P.A. and Tucker, R.W., Nucl. Phys. B112, 150 (1976).

25) Inamoto, T., University of Tokyo preprint (1987).

26) Savvidi, G.K., Yerevan preprint EFN-972(22)-87 (1987).

27) Laziev, M.E. and Savvidi, G.K., Yerevan preprint EFN-957(7)-87 (1987).

28) Bergshoeff, E., Sezgin and Tanii, Y., ICTP Trieste preprint IC/87/107 (1987).

29) Kikkawa, K. and Yamasaki, M., Progr. Theor. Phys. 76, 1379 (1986).

30) Brink, L. and Nielsen, H.B., Phys. Lett. 45B, 332 (1973).

31) Howe, P. and Tucker, R., J. Phys. A10, L155 (1977).

32) Higashijima, K., Uematsu, T. and Yu, Y.-Z., Tokyo preprint UT-423 (1984).

33) Gibbons, G.W. and Wiltshire, D.L., DAMTP preprint (1987).

34) Weinberg, S., in General Relativity; An Einstein Centenary Survey, Eds. Hawking, S.W. and Israel, W. (CUP, 1979).

35) Aref'eva, Ya., Ann. Phys. 117, 393 (1979).

36) Gross, D.J., Harvey, J.A., Martinec, E. and Rohm, R., Nucl. Phys. B256, 253 (1985).

37) Duff, M.J., CERN preprint TH.4124/85 (1985), to appear in Quantum Field Theory and Quantum Statistics. Essays in honour of E.S. Fradkin's 60th birthday, Eds. Batalin, H.I., Isham, C. and Vilkovisky G. (Adam Hilger, London, 1987).

38) Nicolai, H., Phys. Lett. B187, 316 (1987).

39) Killingback, T.P., Class. Quantum Grav. L71, (1987).

40) Freund, P.G.O. and Rubin, M.A., Phys. Lett. 97B, 233 (1980).

41) Fujikawa, K. and Kubo, J., Stony Brook preprint ITP-SB-87-36 (1987).

42) Pilch, K., van Nieuwenhuizen, P. and Townsend, P.K., Nucl. Phys. B242, 377 (1984).

43) de Wit, B. and Nicolai, H., Phys. Lett. 108B, (1982) 285.

44) Freedman, D.Z, in "Supersymmetry and its Applications", eds. Gibbons, G.W., Hawking, S.W. and Townsend, P.K., (C.U.P., 1986).

45) Englert, F., Sevrin, A., Troost W., van Proeyen A. and Spindel, Ph., University of Leuven preprint KUL-TF-87/12 (1987).

46) Freedman, D.Z. and Nicolai, H., Nucl. Phys. B237, 342 (1984).

47) Fronsdal, C., Phys. Rev. D12, 3819 (1975), Phys. Rev. D26, 1988 (1982).

48) Nicolai, H. and Sezgin, E., Phys. Lett. <u>143B</u>, 389 (1984).

49) Sezgin, E., Phys. Lett. <u>138B</u>, 57 (1984).

50) Gunaydin, M., Nilsson, B.E.W., Sierra, G. and Townsend, P.K., Phys. Lett. <u>176B</u>, 45 (1986).

51) Dirac, P.A.M., Proc. Roy. Soc. <u>A268</u>, 57 (1962).

52) Dirac, P.A.M., J. Math. Phys. <u>4</u>, 901 (1963).

DISCUSSION

– *Martinelli:*

We start from quantum gravitation which is not renormalizable, then go to string theory to cure this problem, then do membranes as a generalization of string theories. At this point we discover that the non-linear sigma–model associated with membranes is not renormalizable. What do we gain from this? We don't know exactly what happens for the non–linear sigma–model beyond 2 dimensions, but there are hints that it exists as a theory even if it is not perturbatively renormalizable. For example, in old studies by Symanzik and Parisi, it is suggested that the non-linear sigma–model in more than 2 dimensions works more or less like a linear sigma–model with the same properties. The difference is that the fixed point is no longer at zero coupling, but at some other point. You can in any case define a theory in the limit that the cutoff goes to infinity. This may well happen for gravitation itself, in the sense that beyond perturbation theory, the Einstein Lagrangian may contain effects that cannot be studied loop by loop. Would you comment on this?

– *Duff:*

In answer to your first question, it is perfectly true that, if in attempting to cure the problems of a nonrenormalizable theory of gravity, we swap it for a theory which is itself nonrenormalizable, then we haven't gained anything. So, unless there is some way around the nonrenormalizability of the 3–dimensional membrane sigma–model, we are in trouble. My attitude at the moment is that since everything else is working out so well, we should continue with the supermembrane until something goes wrong. As far as finiteness is concerned, there are two issues which must be addressed (and this is also true for strings): one is the finiteness of the sigma–model and the other is the finiteness of string scattering amplitudes. The issue of a renormalizable sigma–model occurs already at tree level in strings, as Mark Grisaru mentioned in his lectures. The finiteness or conformal invariance of the sigma–model is an issue even before we start constructing string loop amplitudes. When it comes to the membrane, there are two separate questions. One is the renormalizability of the 3–dimensional sigma–model. In addition, there is the question of the finiteness of membrane scattering amplitudes. As far as I can tell, these are separate issues. What is the solution to the problem of the nonrenormalizability of the 3–dimensional sigma-model? One possibility is the one you have mentioned: that although it is power–counting nonrenormalizable, we will be able to make sense of it in the way suggested by Symanzik, Parisi, and others. In fact, there is a paper by Aref'eva which claims that in 3–dimensions, but in no more than 3–dimensions, the $O(N)$ sigma–model is actually renormalizable in the large N limit. As an observation, even the string sigma-model is nonrenormalizable if one includes coupling to the massive states. This is because the massive states are higher spin states, and so require more than 2 derivatives to couple to the sigma–model. Yet, we know that the string still makes sense. So it is not obvious that the fact that we have the luxury of a renormalizable theory for the purely massless modes is crucial. Another possible way out is the "asymptotic safety" suggested by Weinberg as a way of curing the renormalizability problem of gravity. Now, a nonrenormalizable theory is not bad simply because it has divergences, which can be removed by the addition of a counterterm. They are bad because you generally require infinitely many counterterms, and therefore infinitely many undetermined parameters and so the theory has no predictive power. If in very special circumstances you could show that there are only a finite number of parameters in spite of the nonrenormalizability, Weinberg showed that this would be equally acceptable. The fourth possibility is the one I suggested in my talk. In this respect, it is not fair to compare the situation to that of general relativity. In general relativity, you are given a Lagrangian and either it's finite or it's not; you have no control over the situation. In the case of the sigma–model, you have many sigma-models depending on your choice of background, so it is perfectly conceivable that some are good and some are bad. The other remark that I might make is that no one has yet probed the consequences of the world volume fermionic Siegel symmetry. No one knows what effects this will have on the divergence structure.

– Liu:

In your membrane Lagrangian, do you have a $^{(3)}R$ term?

– Duff:

We do not have an Einstein–Hilbert action; we do not have the curvature term. This is because we want to have, as closely as possible, the Nambu-Goto action, so the metric appears only algebraically in the action; there is no kinetic term. In fact, it would be bad if we had such a term because without it, we have 11 bosonic coordinates, of which I can gauge away 3 using the world–volume diffeomorphisms, leaving 8 physical boson coordinates. That matches the 8 physical fermion coordinates. If we were to include an Einstein–Hilbert term, we would no longer have the freedom to gauge away 3 of the coordinates because the diffeomorphism group would be used up in compensating for the world volume graviton. As a result, we would have 11 degrees of freedom. Moreover, one of those 11 would have the wrong sign for its kinetic term, and we would be back with the problem of ghosts. For this reason we do not have an Einstein–Hilbert term. One consequence of not having such a term, but of having a cosmological constant term, is that we can't have any supersymmetry on the world–volume. This is because in 3–dimensional supergravity, we have an auxilliary field s which enters only algebraically into the Lagrangian in the form

$$L \sim As^2 + Bs$$

Now when you vary the Lagrangian with respect to s you find s is equal to a constant and that then gives you the cosmological constant. However you cannot have a term quadratic in s without also having the Einstein–Hilbert term because they belong to the same supermultiplet. So without the curvature term, we cannot make the cosmological term supersymmetric. That is why we don't have a Neveu–Schwarz formalism for the membrane, but only a Green–Schwarz formalism.

– *Liu:*

Are there only 4 p–branes, and can there be others in more than 11 dimensions?

– *Duff:*

You have the 4 fundamental p–branes and you have the others which follow from the simultaneous dimensional reduction. All the p–branes which you obtain must satisfy the criterion of the existence of a Green–Schwarz action which has space–time supersymmetry and world volume Siegel symmetry, so there are no others and they cannot be constructed in greater than 11 dimensions.

– *Gourdin:*

I have been troubled by the absence of a link between theory and experiment. For supersymmetry, I realize that the discovery of super particles would be a positive indication. However, while superstrings are very beautiful mathematical construction, I don't see a clear relation with experiment. What theoretical progress is necessary for theory to produce simple tests such as a particle spectrum or a unitary S–matrix?

– *Duff:*

All the string theories will give you a unitary S–matrix. As far as connecting with experiment, there is one outstanding problem which must first be resolved, and that is the problem of vacuum degeneracy. At the moment we have billions of 4–dimensional superstring models, and there are good reasons for believing that these are not billions of different theories, but rather billions of different vacuum states of the same theory. The problem is that, as far as we can tell at the moment, they are degenerate in energy, so we have no criteria for saying which is the right one. In order to make any contact with experiment, it is necessary to have some criteria for singling out one (or possibly a small number) of these vacua as the vacuum state in which we live, and finding some principle for rejecting the others. At the moment, we don't have such a principle, but it doesn't mean that one does not exist. It may be that you have to go to nonperturbative effects, which we don't understand very well. It may be that you have to go to string field theory. There are all sorts of possibilities. Perhaps I should add that some people actually don't mind this situation. They say that the Universe may have billions of vacuum states, and we may just be living in one which just happens to have the symmetry group $SU(3) \times SU(2) \times U(1)$, three families, etc; Now, of course, there is no argument against this point of view, but it is a bit pessimistic because it means that we'll never be able to make any predictions.

– *Balug:*

What are the boundary conditions in membrane theory and how do you induce interactions?

– *Duff:*

All membranes I have shown you correspond to closed membranes. The supersymmetry of the action requires that you have a closed membrane in 11 dimensions; I don't know whether you can construct open membranes in less than 11 dimensions. Regarding interactions, we don't yet know how to construct them. For strings, you can write down a functional integral which instructs you to integrate over all geometries and sum over all topologies, so you might imagine that there is something similar for membranes. However, unlike strings, where conformal invariance means that you only have to worry about different topologies, each membrane geometry is different. The other problem is that mathematicians have not succeeded in classifying all possible 3 dimensional topologies.

– *Zichichi:*

In the case of strings, there are open and closed strings, why are membranes only closed?

– *Duff:*

The requirement of $N = 1$ supersymmetry in $D = 11$ forces the membrane to be closed. I have no intuitive explanation, I am afraid. For $D < 11$ the situation is not so clear, but all known cases are also closed.

– *Toppan:*

In string field theory, conformal invariance ensures that when we perform covariant quantization, the negative–norm states are cancelled. What is the corresponding mechanism for membrane theory?

– Duff:

There is a belief that strings are special because the $D = 2$ conformal group is infinite-dimensional, allowing us to kill the infinity of ghosts arising from the time-like coordinate. While this issue has not been settled rigourously for the membrane, what seems to be happening is that when you fix the gauge of your 3–dimensional membrane action by picking something like the light–cone gauge, there remains an infinite dimensional symmetry, which is a subgroup of the 3–dimensional diffeomorphism group. That infinite parameter symmetry seems to play the same role as the infinite parameter symmetry of the string plays in killing ghosts. It is the fact that you have reparameterization invariance which is essentially doing the job. Another way of looking at the problem is to go to a physical gauge, where there are obviously no ghosts. The price you pay, however, is that you no longer have manifest Lorentz invariance, and you have to show that closure of the Lorentz algebra is not lost. This is an important calculation which has not yet been carried out; you can nevertheless try to anticipate what could go wrong because we know what goes wrong for the string. For the string, there is a symmetry which is violated at the quantum level, creating an anomaly and spoiling the closure of the algebra. In the case of the membrane, as far as we can tell, there is no anomalous symmetry, with the possible exception of the Siegel symmetry. It is a very interesting problem which no one has yet looked into, but it is the only possible source of an anomaly that we can see at the moment. Based on that, I would guess that there is no problem with either Lorentz invariance or ghosts.

– Kiritsis:

What kind of 3–manifolds you are considering when you write the action for your 3–dimensional nonlinear sigma–model? I am asking this question because you have a Wess–Zumino term in the action, and it is known that in this case topology plays an important role even at the classical level?

– Duff:

Apart from the fact that it is a closed manifold, there is no further restriction.

– Kiritsis:

Then because of the presence of the Wess–Zumino term, are there any restrictions that topology can give you on the possible theories?

– Duff:

We haven't investigated this much yet, but we do have some indication of something interesting. We recently found solutions of the membrane equations of motion in which the 11–dimensional space–time corresponds to the product of 4–dimensional anti-de Sitter space–time and a 7–dimensional internal space, and in which the 3-dimensional membrane is sitting out at the boundary of the anti–de Sitter space-time. That is why we have singleton representations. It has been known for many years that these singleton representations do not admit an ordinary 4–dimensional Lagrangian even though they are representations of the 4–dimensional anti–de Sitter group. Rather, they only admit a formulation in which you write a Lagrangian at the boundary of anti de Sitter space. We've proven that for our particular configuration, the Lagrangian is the corresponding Lagrangian of the singletons. This is an example of the importance of the Wess–Zumino term and the sorts of membrane configurations that you find when you can solve the field equations.

– Kiritsis:

You mentioned that you don't have any anomolous symmetries with the possible exception of the Siegel symmetry. Would you discuss this further?

– Duff:

This is a crucial question, and if anyone is looking for a problem, this would be a good one. I think the problem is that the Siegel symmetry is an on-shell symmetry and I don't know how you go about looking at the conservation of the corresponding quantum current. There is probably some way of doing it but I don't see what it is at the moment. But clearly, that is what we have to do: to check whether the Siegel symmetry is consistent at the quantum level. Who knows, there might be an anomaly that will give us a critical dimension.

– Quackenbush:

In $D = 10$ string theory, we know that the theory is only consistent if the gauge group is limited to $E_8 \times E_8$ or $SO(32)$ (and to groups of rank < 22 for other 4-dimensional string theories). Which gauge groups are allowed in the $D = 11$ membranes?

– Duff:

The short answer is that we don't know. However, the $D = 11$ theory is the maximal supergravity theory in 11 dimensions, and as such, you cannot couple any matter to it. This was one of the phenomenological difficulties of 11–dimensional supergravity. You can generate gauge groups from the compactification, and seven extra dimensions are enough to accomodate $SU(3) \times SU(2) \times U(1)$ (that was proven by Witten). But these are not gauge groups you put in; they are ones you get out from spontaneous compactification. Now, an interesting question is whether there is any analog of the Frenkel–Kac mechanism for membranes which would give you an even bigger group than you would expect from naive Kaluza–Klein reasoning. Suppose we try the more brute force approach of having Yang–Mills fields already in the theory. Then we know in the case of 10–dimensional strings, as you pointed out, that we can couple a Yang–Mills multiplet to the superstring, and provided that it is $E_8 \times E_8$ or $SO(32)$, we get a quantum consistent theory. Now a very interesting question is whether the 10–dimensional 5–brane, which couples to the dual formulation of 10–dimensional supergravity, also allows you to couple in Yang–Mills. At the level of a field theory, the dual formulation of 10–dimensional supergravity is every bit as good as the original one, and in fact, as Gates and Nishino have shown, it is just as anomaly free as the three-form version, provided you couple it to $SO(32)$ or $E_8 \times E_8$ Yang-Mills. So if you could couple Yang–Mills to this theory, and assuming that it is quantum consistent, you would have a theory which from the group theory point of view is every bit as attractive as the heterotic string. Now, the heterotic string relies for its construction on the division between left-moving and right-moving modes, which is something very particular to two space–time dimensions (p=1). However, there is an observation that in (2 mod 4)–dimensional Minkowski space, you have a real self-duality condition. In the case of strings, it is self-dual one forms. In the case of 5–branes, it is self-dual three forms. Whether there is some heterotic 5–brane which exploits this self-duality and allows you to couple in Yang–Mills is another unsolved and interesting question. Alternatively, you could try putting extra fermions on the world-volume and couple in Yang–Mills that way. If you want to do things the conventional way, that seems an interesting theory to try. I should also have mentioned that the 10–dimensional 5–brane and the 6–dimensional 3–brane are chiral which is also good from the phenomenological point of view.

– Kugo:

If you are in trouble in the presence of an Einstein–Hilbert term, then you may be in trouble already. In this connection, I would like to remind you of the so-called "induced gravity" in four dimensions, where the graviton pole is generated dynamically, irrespective of whether the Einstein–Hilbert term is included in the original Lagrangian. This is in fact a Nambu–Goldstone phenomenon: in any theory which is invariant under general coordinate transformations in d–dimensions, the global general linear transformations $M_j^i \in GL(d)$ must be a symmetry even after the gauge is fixed. If the metric $g_{ij}(x)$ develops into the flat Minkoski metric η_{ij}, then the vacuum expectation value of

$$[M_j^i, g_{k\ell}(x)] \; = \; \delta_k^i g_{j\ell}(x) - \delta_\ell^i g_{kj}(x)$$

reads

$$< 0|[M_j^i, g_{k\ell}(x)]|0 > = \delta_k^i \eta_{j\ell} - \delta_\ell^i \eta_{kj}$$

which implies that the symmetric part of $GL(d)$ is spontaneously broken. This means that the graviton appears as a "Goldstone tensor".

– Duff:

The action involves an integral over the three dimensional world volume, and so if we were to put in a membrane tension parameter (which I have been setting to 1), we see that it has to have the dimensions of $(mass)^3$ or $(length)^{-3}$. What that means is that on dimensional grounds, you will never induce an Einstein–Hilbert Lagrangian. The first term you might possibly induce would be at two loops and which would involve the inverse of the tension and the cube of the Riemann tensor. On dimensional grounds, I can never generate anything linear in R. Of course I could go on generating terms higher in curvature, which is, of course, just the nonrenormalizability problem.

– Singh:

Regarding yesterday's lecture, you showed the spectrum of superstring states in 4–dimensions on a graph of spin J versus $(mass)^2$. Is this mass the mass before or after symmetry breaking?

– Duff:

In the diagram, I assumed that we had already gone to four dimensions, and where, if we had a massless gravitino, we also had an unbroken supersymmetry. In that case, the $(mass)^2$ corresponds to the 4–dimensional string prior to supersymmetry breaking. In that regime, all the particles that we would identify as those of the standard model (if we were fortunate enough to get the standard model) would be massless. Clearly since the particles of the standard model are not massless, we have to look for some symmetry breaking mechanism which will give us small masses for the W–boson, the electron, and so on, and which will break supersymmetry. So the mass in the diagram you refered to is the mass before such symmetry breaking.

– Singh:

This also refers to yesterday. The renormalization group equation for alphas places a limit on the number of quark families. Since the strings have an infinite number of spin 1/2 particles, is this an inconsistency?

– Duff:

We don't want to get superstrings models with more than four families because then we're in trouble for the reason you've mentioned. The problem is that the superstrings does not predict the number of families. Of these billions of vacuum states, some of them have three families, some have many more than three families. The number is something which no one has yet predicted.

– Miele:

In the spectrum of possible *P*-branes, are the theories on any given sequence simply a single theory with different formulations?

– Duff:

This is something one might think about, as we derived the 10–dimensional superstring from the 11–dimensional membrane and indeed, all the theories can be derived from the 4 fundamental p–branes at the top of each sequence. So there is a sense in which all the "secondary" theories are coming from the 4 "primary" ones. In the reduction process which we use to generate these secondary theories, we are really performing a truncation; there are fewer states in the spectrum of the theories as you go down each sequence. In that sense, it is difficult to see how they could be describing the same dynamics. However, this subject is so new, and we know so little about it that it would be foolhardy to make any concrete statements about which theories are equivalent and which are not. There may be some grander scheme in which all these theories will appear to be different facets of just one theory.

– Schuler:

You mentioned a connection between the membrane and the boundary of space–time. Could you explain this a bit further?

– Duff:

There is a solution of the membrane equations of motion in which the eleven dimensions are the product of a 4–dimensional space–time and a 7–dimensional internal space, and for which the membrane occupies the boundary of space-time, so it's out there at infinity. That's the solution we've found and we're currently trying to interpret it.

– Martinelli:

A comment: phenomenology excludes the existence of more than 3 or 4 families where the neutrino is always light or is heavy but stable. I believe there are no limits if you admit the existence of heavy neutrinos which are unstable.

THE NEW LOOP SPACE INDEX THEOREMS AND STRING THEORY*

Paul WINDEY

*Department of Physics and Lawrence Berkeley Laboratory,
University of California, Berkeley, CA 94720, USA
and
Laboratoire de Physique Théorique et Hautes Energies
Université de Paris VI,
4 pl. Jussieu, F-75252 Paris CEDEX 05, FRANCE* [†]

1   Introduction

These lectures are concerned with index theorems as seen from the point of view of field theory: not with the various uses of index theorems in field and string theory —like the study of anomalies—, but merely with the use of supersymmetric quantum field theory to prove (or maybe more accurately to derive) index theorems. Of course we are eventually more interested by the physical consequences of the theorems than by their proofs, but I would like to convince you that in the process of understanding the structure of physical theories —in this case field and string theories— one often recovers deep mathematical results and sometimes discovers exciting new ones. As we will see it is quite remarkable that one of the most profound result of the last twenty years in mathematics, the Atiyah-Singer index theorem [1][2], has an extremely simple rewriting in terms of quantum mechanics language. But perhaps the most surprising fact is that the field theoretical approach leads naturally to extensions of this theorem to infinite dimensional spaces, namely the space of loops of any given compact closed manifold. In fact the story I will try to present has three very different versions and I am going to concentrate mainly on one of them which stems from the study of the Dirac-Ramond operator introduced long time ago in the study of string theory (see for example [3]). This operator is the generalization to loop space of the Dirac operator and we are going to compute its index just as easily as one computes the index of the

---

*This work was supported in part by the Director, Office of Energy Research, Office of High Energy and Nuclear Physics, Division of High Energy Physics of the U.S. Department of Energy under Contract DE-AC03-76SF00098 and in part by the National Science Foundation under grant PHY85-15857.

Dirac operator [4,5,6,7,8,9]. The second version of the story, due to Schellekens and Warner [10][11], came independently from the study of a generating function for all the field theory anomalies coming from string theory. Finally there is the mathematical version of our tale which came from the desire to extend an old result of Atiyah and Hirzebruch. This was done by Ochanine [12] and Landweber and Stong [13] who named the theory elliptic cohomology —for more references on the mathematical literature the interested reader may consult the entire volume of the Princeton conference [14]). It is interesting to point out that their very sophisticated analysis was originally motivated by a conjecture made by Witten [15] in the context of Kaluza-Klein theories. The hope is that these new loop space index theorems will shed some light on the structure of string theory and two dimensional field theory.

Few years ago it was noticed that all anomalies are governed by one version or another of the Atiyah-Singer index theorem. It was also noticed that a very simple way to derive this theorem was provided by supersymmetric quantum mechanics [16][17][18]. It was then natural to ask whether first quantized string theory would lead to a simple generalization of the Atiyah-Singer theorem to loop space which would be relevant for string theory and provide a tool to analyze the properties of the spectrum of the Dirac-Ramond operator. In field theories index theorems govern the spectrum of massless point particles and link it with the topology of the configuration space M (space-time) where these evolve. Similar theorems in string theory should then probe the structure of the loop space of M (the configuration space of the string) and relate its topological properties with the string spectrum. More precisely such index theorems should give at least two broad type of constraints on string theories. Firstly the constraint must be related to the traditional field theory anomalies. Indeed one might expect to recover at once all the known anomalies of the field theories which are implicitly contained in a given string theory. This is the problem solved, independently of any loop space index considerations, by Schellekens and Warner [10][11] in a very insightful paper where they essentially got (for reasons which will become clear later) the index formulae we are going to derive. The second type of constraint concerns the definition of the index itself and should be the equivalent for loop space to the conditions that a manifold must satisfy to admit a spin structure. Finally one can hope that an index theorem for string theory would reveal some new and unexpected structures and certainly a lot of work remains to be done in that direction.

The first part of these lectures (Sections 2 and 3) will review in detail the use of supersymmetric quantum mechanics to derive the Atiyah-Singer index theorem and some of its consequences, mainly the fixed point theorems. The second part of these lectures presents the extension of these results to loop space. We simply look for the analogue of the Dirac operator in string theories, *i.e.* the Dirac-Ramond operator, and compute its index. At first sight the result doesn't look like any known index formula, but a careful analysis reveals that it is simply a character valued index, properly extended to the infinite dimensional case. The expression for the index is valid only if the Green-Schwarz anomaly cancellation occurs [19]. Otherwise, the index is not well defined. This anomaly comes as an obstruction to the index being a well behaved modular function. This is the analogue of the spin structure alluded to above. When that condition is satisfied the general index formula takes the form of a generating function for the index of the Dirac operator associated with each supersymmetric state of the spectrum of the string. In Section 4 we give the precise definition of the index of the Dirac-Ramond operator in loop space. It is done in the framework of a $N = 1/2$ supersymmetric sigma model, *i.e.* in the Ramond sector of a gauged fixed heterotic string, with a supersymmetric right moving sector, embedded in an arbitrary external background gravitational field. This case corresponds to the simplest extension of the Dirac operator to an infinite dimensional loop space. The derivation of the index theorem for this operator is presented in Section 5 and some of its relations to the theory of modular forms are pointed out. In Section 6 we generalize our treatment to include the coupling of the original $N = 1/2$ system to a left moving sector with the four boundary conditions corresponding to the spin structures of the torus. Each of the four resulting topological invariants is simply expressed as the ratio of two theta functions or equivalently in terms of the Jacobi elliptic functions. We briefly explain how to interpret these modular invariants as the natural loop space generalizations of the Euler number and Hirzebruch signature. A simple modification of the left-moving sector would lead to equivalent index theorems for the Dirac-Ramond operator twisted by a coupling to external gauge fields. The above mentioned consistency condition $\mathrm{Tr}\, R^2 = 0$ would simply become $\mathrm{Tr}\, R^2 - \mathrm{Tr}\, F^2 = 0$. Finally we give in Section 7 a simple proof due to Witten of the extension of the Atiyah-Hirzebruch theorem to twisted Dirac operator on manifolds with an S^1 action. This result follows from the loop space index theorem described above. For the convenience of the reader we have included in an appendix the basic definitions of the elliptic and theta functions as well as a summary of the main steps our computations.

2 The Atiyah-Singer Index Theorem

The analytical index of an elliptic operator Q is defined to be

$$I \equiv \mathrm{Ind}\, Q = \dim \ker Q - \dim \mathrm{coker}\, Q \ . \tag{2.1}$$

The computation of such an index is in principle a problem in analysis. What the Atiyah-Singer theorem proves is that this integer is a topological invariant. The best known example in particle physics is the one of the Dirac operator; we will also see that in a sense it is the most general. In that case the analytical index defined above is the difference between the number of massless fermions of positive and negative chiralities and the topological invariant is an integral over space-time of a polynomial function of the gauge and gravitational fields curvature. For example it is well known that for an electron on a sphere this index is zero in the absence of external gauge field but equal to one if the electron moves in the field of a magnetic monopole of minimal charge. The topological invariant is simply the integral (suitably normalized) of the magnetic flux over the sphere. Conversely one can conclude from the theorem that this particular integral must be an integer since it is simply equal to the difference of two integers. The topological invariant is called a topological index and the A-S theorem simply states that it is equal to the analytical index defined above. This gives a relation between local data which depend on the solution of a differential equation with the global properties of the manifold.

Let us start by analyzing the features shared by all analytical indices. Consider an operator Q acting on an Hilbert space $\mathcal{H}$ and such that $H = Q^\dagger Q$ is semi-positive definite. Moreover there is an operator $(-1)^F$ (we will call it the fermion parity operator) which anticommutes with Q and provides a Z_2 grading of $\mathcal{H}$. They obey the algebra:

$$H = Q^\dagger Q \qquad \text{with} Q^\dagger = -Q \tag{2.2}$$

$$\text{and} \qquad Q(-1)^F + (-1)^F Q = 0. \tag{2.3}$$

The operator $(-1)^F$ splits the Hilbert space $\mathcal{H}$ into $\mathcal{H} = S_+ \oplus S_-$, such that $(-1)^F$ has $+1$ (-1) eigenvalue on S_+ (respectively S_-.) The operator Q maps these subspaces into each other: $Q : S_+ \to S_-$ It is easy to see that any eigenstate $|\psi_+\rangle \in S_+$ of H with non zero eigenvalue is paired with an eigenstate $Q|\psi_+\rangle = |\psi_-\rangle \in S_-$ with the same eigenvalue. Obviously the pairing does not hold for the states annihilated by Q (respectively $Q^\dagger$) $i.e.$ for states belonging to the finite dimensional kernel of Q ($Q^\dagger$). These states, which are called the zero energy states, satisfy $H = 0$. The pairing implies the homotopy invariance of the index of Q for any smooth deformation will continously rearrange the spectrum of Q. In particular, some states will leave or reach the zero energy level. But the pairing forces them to do that in pairs with opposite eigenvalues of $(-1)^F$. It is clear that the index can be formally rewritten as $I = \text{Tr}(-1)^F|_{H=0}$ or taking into account the pairing, one can equally well compute the index through:

$$I = \text{Tr}(-1)^F e^{-\tau H} \tag{2.4}$$

since only the zero energy states contribute to the trace. Of course in computing (2.4) some care should be exercised, since the sum is now over an infinite dimensional space . The basis of the field theory proof of the index theorem is the remarkable observation by Witten [20] that the algebra just described is realized in any supersymmetric theory where Q is the symmetry generator and H the Hamiltonian. So to any such theory one can associate an index. The index of any operator Q satisfying the relations (2.2) and (2.3) can be computed through path integral techniques [21] provided that the second quantized expression of the supersymmetric charge coincide with Q and the Hamiltonian with H.

We are now ready to give the by now standard supersymmetric derivation [17][16]of the Atiyah-Singer index theorem[2]. This derivation is based on the identification of the Hilbert space of simple supersymmetric quantum mechanical systems with the classical elliptic complexes in differential geometry, like the spin or the de Rham complexes. The supersymmetric subset of states corresponds to the cohomology groups associated with these complexes. The elliptic differential operators acting on these complexes are identified with the supersymmetry generator of the theory, while the respective involutions are seen to be equivalent to the fermion parity operator of the corresponding quantum mechanical system. In the case of the Dirac operator mentioned at the beginning of this section we have to find a quantum mechanical system whose supersymmetric charge is the Dirac operator and such that the fermion parity operator is γ_5. The states of the Hilbert space are chiral space time fermions. They are called "bosonic" or "fermionic" depending on their chirality. In an other example we will see that one can construct a system whose supersymmetric charge $Q = d - d^*$, where d is the usual exterior differentiation and d^* its adjoint. In that case the Hilbert space of the theory will consist of the space of exterior forms on a compact manifold M, supersymmetric states will be in one to one correspondence with the space of harmonic forms and the involution is $(-1)^p$ with p an integer corresponding to the degree of a form. The Atiyah-Singer index gives then the Euler-Poincaré characteristic of the manifold. Let's briefly review this case. Let Λ^p, $p = 0, 1, ..., n$, be the space of differential forms of degree p on a compact closed manifold M. One defines as usual the exterior derivative $d_p : \Lambda^p \to \Lambda^{p+1}$ and its adjoint $d_p^* : \Lambda^{p+1} \to \Lambda^p$. This defines the de Rham complex:

$$0 \to \Lambda^0 \overset{d_0}{\to} \Lambda^1 \overset{d_1}{\to} \cdots \overset{d_{n-1}}{\to} \Lambda^n \to 0, \tag{2.5}$$

in which the composition of two successive operators is zero. The space of harmonic $p-$forms ($d\omega = d^*\omega = 0$) is the kernel of the laplacian operators defined by

$$\Delta_p = d_{p-1} d_{p-1}^* + d_p^* d_p : \Lambda_p \to \Lambda_p. \tag{2.6}$$

This kernel is easily identified with the cohomology group H_p (i.e the finite dimensional vector space of closed $p-$forms modulo the exact ones) through the well known Hodge decomposition theorem. Its dimension is the pth Betti number. The space Λ^* of all exterior forms plays the role of $\mathcal{H}$ and $(-1)^F = (-1)^p$ splits it into the space of even forms Λ^+ and Λ^- the space of odd forms. The operator $Q = d - d^* : \Lambda^+ \to \Lambda^-$. If the Euler number is defined to be

$$\chi = \sum_{p=0}^{n} (-1)^p \dim H_p = \sum_{p=0}^{n} \dim \ker \Delta_p, \tag{2.7}$$

then it follows from the above definitions that $\chi = \text{Ind } Q$. Instead of taking the index with respect to $(-1)^p$ there is for orientable manifold another involution which anticommutes with $d - d^*$. We first define the Hodge duality operator: $* : \Lambda^p \to \Lambda^{n-p}$ which corresponds to the contraction with $\epsilon_{\mu_1 \cdots \mu_n}$. We then choose $(-1)^F = (-1)^{\frac{p(p-1)}{2}} * \equiv \tau$. The Hodge duality gives an isomorphism $* : H^p(M, \mathbf{R}) \simeq H^{n-p}(M, \mathbf{R})$ and $b_p = b_{n-p}$. So if the splitting of the complex is defined with respect to τ we can have non zero contribution only from the middle dimension. The index is now called the signature of M.

These are the indices we are going to study. Notice once again that the definitions we have given correspond to analytical indices. The essence of the Atiyah-Singer index theorem is the equality between these analytical indices and topological invariants. Roughly speaking we will see that the integers just defined are given by the integral over the manifold of certain characteristic classes. We will comment upon these only after we have derived the theorem.

We now turn to our principal task which is to determine which supersymmetric quantum mechanical system gives after quantization the algebra (2.2) and (2.3) which corresponds to one of the above complexes. The case of the Dirac operator is of primary interest to us. Let consider a supersymmetric quantum mechanical system with the following (anti-)commutation relations among the dynamical variables:

$$[ip_\mu, x^\nu] = \delta_\mu^\nu \tag{2.8}$$

$$\{\psi^\mu, \psi^\nu\}_+ = 2\delta^{\mu\nu}, \tag{2.9}$$

where x^μ and p^μ are respectively the position and momentum operator in a flat space M ($\mu = 1 \cdots n$), while ψ^μ correspond to Dirac matrices. If we choose $Q = ip^\mu \psi_\mu$, $H = p^2$ and $(-1)^F = \gamma_5 = i^{-n/2} \frac{1}{n!} \epsilon_{\mu_1 \cdots \mu_n} \psi^{\mu_1} \ldots \psi^{\mu_n}$, then the relations (2.2) and (2.3) are satisfied. The above operators appear after second quantization of the Lagrangian:

$$L = \frac{1}{4} \partial_t x^\mu \partial_t x^\mu + \frac{1}{4} \psi^\mu \partial_t \psi^\mu \tag{2.10}$$

This Lagrangian is invariant under the supersymmetry transformations

$$\delta x^\mu = \epsilon \psi^\mu \qquad \text{and} \qquad \delta \psi^\mu = -\epsilon \partial_t x^\mu \,, \tag{2.11}$$

where ϵ is an infinitesimal anticommuting parameter. We now generalize the above to include coupling to an arbitrary background gravitational field $g_{\mu\nu}$ describing an arbitrary spin manifold M. Following [17] we will use superfield notations and sometimes suppress the indices when no confusion is possible. Let $X^\mu = x^\mu + \theta \psi^\mu$ where θ is the Grassmann variable associated with the time t. Defining $D = \theta \partial_t - \partial_\theta$ we have:

$$S = \int^\tau dt\, d\theta\, \frac{1}{4} g_{\mu\nu}(X) \partial_t X^\mu D X^\nu \quad \text{with} \tag{2.12}$$

$$g_{\mu\nu}(X) = g_{\mu\nu}(x) + \theta \psi^\lambda \partial_\lambda g_{\mu\nu}(x) \quad . \tag{2.13}$$

In this form the action is explicitly supersymmetric. The transformation of a field is given by its commutator with $\epsilon Q = \epsilon(\theta \partial_t + \partial_\theta)$. In components we have

$$L \sim g_{\mu\nu}(x) \partial_t x^\mu \partial_t x^\nu + g_{\mu\nu}(x) \psi^\mu (\partial_t \psi^\nu + \partial_t x^\alpha \Gamma^\nu_{\alpha\beta} \psi^\beta), \tag{2.14}$$

which leads through canonical quantization to $Q = \not{D}$ where $\not{D}$ is the usual Dirac operator on spinors. To compute the index of the Dirac operator we just have to follow the steps explained at the beginning of this section and compute $I = \text{Tr}(-1)^F e^{-\tau H}$. With the usual correspondence between the path integral and the canonical formalism this is very easily done. We refer the reader to [17] for the details and just mention the main logical steps of the computation as well as some formulae which will turn out to be useful in the sequel. Firstly notice that the $(-1)^F$ factor implies computing the partition function with periodic boundary conditions in time for both the x and the ψ fields. Secondly, since the index is a topological invariant independent of τ — it is just an integer,— we can perform a small τ expansion of the action (around the constant solutions of the field equations), keep only the terms of lowest order in the background metric, and nevertheless get exact result. Let $x^\mu = x_0^\mu + \xi^\mu$ and $\psi^\mu = \psi_0^\mu + \zeta^\mu$. Using Riemann normal coordinate expansion about x_0 we have

$$g_{\mu\nu}(x_0) = \delta_{\mu\nu} \tag{2.15}$$

$$\partial_\lambda g_{\mu\nu}(x_0) = 0 \tag{2.16}$$

$$\partial_\rho \partial_\lambda g_{\mu\nu}(x_0) = -\frac{1}{3}(R_{\mu\rho\nu\lambda} + R_{\mu\lambda\nu\rho}) \,. \tag{2.17}$$

The component lagrangian becomes:

$$L \sim -\partial_t \xi^\mu \partial_t \xi^\mu + \mathcal{R}_{\mu\nu} \partial_t \xi^\mu \xi^\nu + \partial_t \zeta^\mu \zeta^\nu \tag{2.18}$$

where $\mathcal{R}_{\alpha\beta} = \frac{1}{2} R_{\mu\nu\alpha\beta} \psi_0^\mu \psi_0^\nu$. This reduces the problem of proving the index theorem to the computation of $\det(\partial_t - \mathcal{R})$ —properly normalized. A special role is played by the zero modes which essentially provide the volume integration. The result is the well known Atiyah-Singer index for the Dirac operator:

$$I = \int_M \left(\frac{i}{2\pi}\right)^{n/2} \det^{-1/2}\left[\left(\frac{1}{2}\mathcal{R}\right)^{-1} \sinh \frac{1}{2}\mathcal{R}\right] \,, \tag{2.19}$$

where now $\mathcal{R}^\mu{}_\nu = \frac{1}{2}R^\mu{}_{\nu\alpha\beta}dx^\alpha dx^\beta$. Notice the substitution of dx^μ for ψ_0^μ coming from the integration over the fermionic zero modes. To obtain the appropriate polynomial in dimensions $2n$, one simply expands the integrand in powers of the curvature two form and keep the terms degree n. A simple modification of the above leads to the index of a Dirac operator coupled to external gauge fields $A_\mu(x)$. One simply introduces new fermionic operators η^a and $\bar\eta^a$ with the commutation relations $\left[\bar\eta^a, \eta^b\right] = \delta^{ab}$. The Hilbert space is now the tensor product of spinors with antisymmetric tensors of arbitrary rank in internal space. The new covariant Dirac operator corresponds to the generator of supersymmetry of original Lagrangian (2.12) modified by the addition of the following term:

$$L_\eta = \int d\theta \bar{N} D_A N + i\alpha \bar{N} N \,, \tag{2.20}$$

with the fermionic superfield $N = \eta + \theta\phi$ and $D_A = D + DX^\mu A_\mu(X)$. Here α is a Lagrange multiplier. Elimination of the auxiliary field ϕ leads to:

$$L_\eta = \bar\eta(\partial_t + \partial_t x^\mu A_\mu)\eta - \frac{1}{2}\bar\eta F_{\mu\nu}\psi^\mu\psi^\nu\eta + i\alpha\bar\eta\eta \,. \tag{2.21}$$

Notice that since the generator of supersymmetry is quadratic in the η fields, they may be chosen to be either periodic or antiperiodic in time. One can then evaluate the character valued index $I(\alpha) = \mathrm{Tr}(-1)^F e^{-\tau H + i\alpha N_\eta} = \sum_k I_k e^{-i\alpha k}$, where N_η is the number operator for the η field. The last equality results from $[N_\eta, H] = 0$. It is obvious that I_k is the index of the Dirac operator coupled to an antisymmetric tensor of rank k in internal space. The computation is performed (with antiperiodic boundary conditions for η) by expanding around $A_\mu(x_0)$, and computing a simple fermion determinant $\det(\partial_t - \frac{1}{2}F_{\mu\nu}\psi_0^\mu\psi_0^\nu)$. The result is:

$$I(\alpha) = \int_M \left(\frac{i}{2\pi}\right)^{n/2} \det(1 + e^{F - i\alpha}) \det{}^{-1/2}\left[\left(\frac{1}{2}\mathcal{R}\right)^{-1} \sinh \frac{1}{2}\mathcal{R}\right] \,, \tag{2.22}$$

where $F = 1/2 F_{\mu\nu}dx^\mu dx^\nu$. Notice that the above implies: $\mathrm{Tr}\, e^{-\bar\eta\omega\eta} = \det(1 + e^\omega)$ for any antisymmetric matrix ω. Similarly $\mathrm{Tr}(-1)^F e^{-\bar\eta\omega\eta} = \det(1 - e^\omega)$. Along the same lines we have $\mathrm{Tr}\, e^{\frac{1}{4}\gamma\mathcal{R}\gamma} = (2)^{n/2} \det(\cosh \mathcal{R}/2)^{\frac{1}{2}}$ while $\mathrm{Tr}(-1)^F e^{\frac{1}{4}\gamma\mathcal{R}\gamma} = (2i)^{n/2} \det(\sinh \mathcal{R}/2)^{\frac{1}{2}}$ These equations can of course be checked directly. Finally we give the result corresponding to the following particular choice of the η system: $F_{\mu\nu} = -\frac{1}{2}R^\alpha{}_{\beta\mu\nu}\gamma_\alpha\gamma^\beta$. Computing the path integral with respectively periodic or anti-periodic boundary conditions gives:

$$\chi = \int_M \left(\frac{i}{2\pi}\right)^{n/2} \det{}^{-1/2}\left[\left(\frac{1}{2}\mathcal{R}\right)^{-1} \sinh \frac{1}{2}\mathcal{R}\right] \mathrm{Tr}\, \gamma_5 e^{\frac{1}{4}\gamma\mathcal{R}\gamma} \tag{2.23}$$

$$= \int \frac{1}{(4\pi)^{n/2}(n/2)!} \epsilon^{\mu_1 \cdots \mu_n} \mathcal{R}_{\mu_1\mu_2} \cdots \mathcal{R}_{\mu_{n-1}\mu_n} \,, \tag{2.24}$$

$$\mathrm{sign}(M) = \int_M \left(\frac{i}{2\pi}\right)^{n/2} \det{}^{-1/2}\left[\left(\frac{1}{2}\mathcal{R}\right)^{-1} \sinh \frac{1}{2}\mathcal{R}\right] \mathrm{Tr}\, e^{\frac{1}{4}\gamma\mathcal{R}\gamma}$$

$$= \int \left(\frac{i}{\pi}\right)^{n/2} \det{}^{-1/2}\left[\left(\frac{1}{2}\mathcal{R}\right)^{-1} \tanh \frac{1}{2}\mathcal{R}\right] \,. \tag{2.25}$$

These correspond respectively to the Euler number, as defined in 2.7 and the Hirzebruch signature. This will be explained in more detailed in Section 6 where their infinite dimensional counterpart will be analyzed.

We started from the definition of the analytical index of an elliptic operator acting on the space of sections of some fiber bundle and expressed it as the trace over the states of a quantum mechanical system —with the appropriate insertion of $(-1)^F$. Then through very simple considerations involving the equivalence between the path integral and the canonical formulation of quantum mechanics, we proved the remarkable fact that this index is equal to a topological invariant integral. We are going to devote the rest of this section to a very brief analysis of the structure of these invariants. All the expressions obtained so far for the topological indices involve trigonometric functions in the curvature two form of the bundle involved (for example in the absence of gauge fields, the tangent bundle of M .) These functions can be expanded in formal power series and give polynomials of finite degree: after a certain power all terms are zero since differential forms of degree higher than the dimension of M automatically vanish. These polynomials fall in the general class of invariants polynomials. A polynomial in the components of a $k \times k$ matrix A is called an invariant polynomial if:

$$P(A) = P(g_1 A g) \tag{2.26}$$

for all $g \in GL(k, C)$. It is a symmetric function of the eigenvalues λ of A and its general form is:

$$P(A) = a_0 + a_1 s_1(\lambda) + a_2 s_2(\lambda) + a_3 s_1(\lambda)^2 + ..., \tag{2.27}$$

where the $s_n(\lambda)$ are symmetric polynomials of degree n in λ. It follows from the Bianchi identity that P is closed when A is a matrix valued curvature two form. The polynomial P has another important property, namely topologically invariant integrals. It is proved by showing that $P(\Omega) - P(\Omega')$ is an exact form when Ω and Ω' correspond respectively to the curvature matrix of two different connections on the same bundle. These closed forms will in general represent some class of the cohomology groups $H_p(M, R)$. So given a connection on a bundle E, the polynomials give a map from the bundle to the de Rham cohomology groups of M. The numbers obtained by integrating these polynomials over closed submanifolds of M are not only independent of the connection chosen but are also invariant under deformations of the submanifolds. This follows from Stokes theorem and the fact that the polynomials are closed forms. We simply list some of the most important characteristic polynomials. For complex bundles E over M with $GL(k, C)$ transition functions, one define the Chern form as

$$c(\Omega) = Det(I + \frac{i}{2\pi}\Omega) = 1 + c_1(\Omega) + c_2(\Omega) + ... \tag{2.28}$$

where the subscript corresponds to the degree.

$$c_1 = \frac{i}{2\pi} \operatorname{Tr} \Omega \tag{2.29}$$

$$c_2 = \frac{1}{8\pi^2}(\operatorname{Tr}(\Omega \wedge \Omega) - \operatorname{Tr} \Omega \wedge \operatorname{Tr} \Omega) \tag{2.30}$$

$$c(\Omega) = \prod_k (1 + \frac{i}{2\pi}\lambda_k) = \prod_i (1 + x_k) \tag{2.31}$$

So the Chern form c_k is simply the symmetric polynomial of degree k in the formal eigenvalues of Ω: $x_k = \frac{i}{2\pi}\lambda_k$. Any invariant polynomial can thus be expressed in term of the Chern forms. Obviously in the expression above we can consider that each x_i is the curvature two form of a complex line bundle. So as far as characteristic polynomials are concerned every bundle is seen as a collection of line bundles, although it may not be a direct product of them. This is called the splitting principle. The principal properties of $c(\Omega)$, known as the total Chern class, are

$$c(E \oplus F) = c(E) \wedge c(F) \tag{2.32}$$

Another characteristic polynomial which appeared in our previous computations is the Chern character: $ch(\Omega) = \mathrm{Tr}\, e^{i\Omega/2\pi}$. It is additive with respect to the direct sum of vector bundles. There are similar constructions when the structure group of the bundle is $SO(n)$, which is the case for the tangent bundle in Riemannian geometry. Just as above we can introduce the formal eigenvalues x_i of the skew curvature matrix $R/2\pi$ and define the total Pontrjagin class by:

$$
\begin{aligned}
p(E) &= Det(I + \frac{i}{2\pi}\Omega) = 1 + p_1(\Omega) + p_2(\Omega) + ... \tag{2.33} \\
&= \prod_i (1 + x_i^2) \,. \tag{2.34}
\end{aligned}
$$

Finally for the tangent bundle of a Riemannian manifold M one can define two characteristic polynomials of primary interest for us, the $A-$roof genus:

$$\hat{A}(M) = \prod_i \frac{x_i/2}{\sinh x_i/2} \,, \tag{2.35}$$

and the Hirzebruch $L-$polynomial:

$$L(M) = \prod_i 2\frac{x_i/2}{\tanh x_i/2} \,. \tag{2.36}$$

Both of them can be expanded in terms of the Pontrjagin classes defined above. We can now summarize the results we have obtained so far. In each of the cases analyzed we computed the index of a Dirac operator D_V coupled to external gauge and gravitational fields. More precisely:

$$D_V, : \Gamma(S_+ \otimes V) \to \Gamma(S_- \otimes V) \,, \tag{2.37}$$

where $\Gamma(E_\pm)$ is the space of sections of the vector bundle $E = S_\pm \otimes V$. Here S_+ and S_- denote the positive and negative chirality components of the spinor bundle and V is the vector bundle corresponding to the appropriate representation of the gauge group. The Atiyah-Singer index theorem says that

$$\mathrm{Ind}\,(D_V) = \int_M \hat{A}(M).ch(V)\,. \tag{2.38}$$

Since M is finite dimensional, the expansion of the $A-$roof genus and of the Chern character gives a finite polynomial in the curvature two forms. All the different cases analyzed above are obtained by replacing V by the appropriate vector bundle. For example the Hirzebruch signature corresponds to $V = S_+ \oplus S_-$.

3 The Character Valued Index

There is a possible modification of the index theorem described above when a group G acts on the manifold M. It leads to what are called character valued index theorems whose simplest example is the Lefschetz fixed point theorem. These extensions are extremely simple to deal with in the framework we have developed above. We present them not only for their intrinsic interest but also because they shed light on what will follow concerning the Dirac-Ramond index. Our treatment follows closely the one given by Goodman and Witten [22]. We will see how these character index theorems provide a simple intuitive understanding of the extension of the Atiyah-Singer index theorem to loop space, but before doing that we should study them in their more traditional finite dimensional setting. Suppose there is a group G acting on M (for example an isometry) such that an element $g \in G$ commutes with Q and $(-1)^F$. The elements of the Hilbert space will transform as representations of G and once again for non-zero eigenvalues these representations come in pairs with opposite eigenvalues of $(-1)^F$. We can define a modified index as:

$$I_g(Q) = \text{Tr}\, g\,|_{Ker\dot{Q}_+} - \text{Tr}\, g\,|_{KerQ_-} = \text{Tr}(-1)^F g \,. \tag{3.1}$$

This is called the character valued index and for $g = 1$ it coincides with the ordinary index. It is quite clear that we are now computing the matrix elements of $(-1)^F$ between states which differ from each other by a symmetry operation g. It is easy to implement this prescription in the path integral formalism by changing the boundary conditions of the fields to be

$$x(\tau) = g.x(0) \tag{3.2}$$

$$\psi(\tau) = g.\psi(0) \tag{3.3}$$

We notice that a certain subset of fields will play a special role, namely the one associated with the fixed point set M_g of M under g —in particular M_g may be a set of isolated fixed points or the empty set. The boundary conditions of the fields describing M_g and its tangent space are unchanged. Their contribution to the path integral will identical to the one computed in the previous section with M replaced by M_g. Moreover one can readily see that they provide the dominant contribution and that the full result will be localized on this fixed point set. Indeed any path in the functional integral contributes a term proportional to $\exp -(g.x, x)/\tau$, where $(g.x, x)$ is the square of the distance between the point x and its image under the action of the group $g.x$. If $(g.x, x)$ is different from zero, the contribution of this path will be suppressed in the limit $\tau \to 0$. A slightly different argument makes this obvious. We have seen that the integral over the manifold M comes from the bosonic zero modes. Since in the present case there are zero modes associated only with the fixed point set, there will be no volume element for the normal bundle. This is the reason why the integral will be localized on M_g. We can then restrict the analysis to neighborhoods of M_g, parameterized by coordinates on M_g and its normal bundle N_g in M. Since we saw that all the different forms of the index theorem can be derived from the spinor case, we will restrict our analysis to this case. Consider the action of a compact Lie group G on a spin manifold by orientation preserving isometries compatible with the spin structure (i.e which lifts to the spin bundle). We can without loss of generality restrict our attention to an S^1 action. For every connected component of M_g the normal bundle has a canonical decomposition in M, invariant under g:

$$N_g = \oplus_\nu N_g(\exp i\nu) \,, \tag{3.4}$$

where N_g has a complex structure and g acts by multiplication by $\exp i\nu$. We are ignoring here the possible complication arising from a -1 eigenvalue. Notice that only a finite number of different eigenvalues can occur. As we will see, this last fact governs in a crucial way the Atiyah-Hirzebruch theorem and is in striking contrast with the Dirac-Ramond case to be considered later. First let us define the complex coordinates $X^\mu = x^\mu + ix^{\bar\mu}$ and the corresponding tangent bundle coordinates described by the fermionic fields $\Psi^\mu = \psi^\mu + \psi^{\bar\mu}$. The action of $G = S^1$ on the coordinates of the normal bundle is given by:

$$g.X = \exp i\nu X \tag{3.5}$$

$$g.\Psi = \exp i\nu \Psi, \tag{3.6}$$

where $\nu = \alpha n$, n is an integer and α parameterizes the transformation. If we look back at the derivation of $\hat A$-genus we see that no modification is necessary for the coordinates of the fixed point set since their boundary conditions are unchanged. However for the normal bundle (seen again as a collection of complex line bundles under the splitting principle) we have for each factor in the expression of the $\hat A$-genus 2.35 the following substitution:

$$\frac{x_i}{\sinh x_i} \rightarrow \frac{1}{2i\sin(\nu/2 + ix_i/2)} \tag{3.7}$$

The modification of the boundary conditions shifts $2\pi n \rightarrow 2\pi n + \nu$ in the infinite product representation of the sine function which occurs in the computation of the determinants, while the absence of zero mode removes the factor x_i in the numerator. The general expression for the character index on a manifold M is:

$$I_g = \int_{M_g} \prod_j \frac{ix_j/2}{\sin ix_j/2} \prod_r \frac{1}{2i\sin(\bar\theta_r/2)} \tag{3.8}$$

where $\bar\theta_r = \alpha\theta_r + ix_r$. The θ_r are the 'eigenvalues' of the group action on TM, x_r are the 'eigenvalues' of the curvature two form at the fixed points, and α is the parameter of the infinitesimal transformation. A sum over the disconnected parts of the fixed point set of M under the action of G is implicit in the above formula. We leave as an simple exercise the proof of the Lefschetz fixed point theorem which states that the Euler characteristic of a manifold with group action is the Euler characteristic of the fixed point set.

One of the most remarkable conclusion to be drawn from the character valued form of the Atiyah-Singer index theorem is the Atiyah-Hirzebruch theorem [23] which asserts that $\hat A(M) = 0$ for any spin manifold with a group action. This result, which was used extensively by Witten to study the absence of chiral fermions in large classes of Kaluza-Klein theories [15], is very easy to obtain. For simplicity we consider only the case of a set of isolated fixed points, labelled by an integer k. The extension to the general case is immediate. With the notation $z = \exp i\alpha$ we can rewrite (3.8) as:

$$I(z, M) = \sum_n z^n I_n = \sum_k \prod_i \frac{\exp(i\alpha n_k^i/2)}{1 - \exp(i\alpha n_k^i)}. \tag{3.9}$$

The sum over the characters of S^1 has only finitely many terms since the elliptic operator has only a finite number of zero modes. So for real values of α, $I(z, M)$ is non singular. From the right-hand side we learn that $I(z)$ is a rational function of z and is zero at $z = 0$ and $z = \infty$. In each term we can have poles at $|z| = 1$ but by the previous argument we learn that the sum of these poles must cancel. This simply corresponds to the fact that the fixed points effectively contribute with different signs depending on the respective orientation of the normal bundles (think of the rotation of a sphere leaving the north and the south poles fixed). The argument of Atiyah-Hirzebruch is simply that a rational function of z which vanishes at infinity and has no poles is zero. So $I(z) = 0$. This generalizes to arbitrary compact groups by looking at the action of elements of the maximal torus.

At this point one may wonder whether there is a theorem, similar to the one of Atiyah and Hirzebruch, for other representations than the spinorial one. It was conjectured by Witten [15] that in the Rarita-Schwinger case the result would be a constant. As was mentioned in the introduction, this conjecture was the starting point of new researches in topology. We will briefly come back to this problem in Section 7.

4   The Analytical Index of the Dirac-Ramond operator

Up to now we have concentrated our attention on the properties of the Dirac operator and found a simple way to derive the Atiyah-Singer index theorem using supersymmetric quantum mechanical methods. We would like to extend this analysis to the equivalent operator which appears in fermionic string theory *i.e.* the Dirac-Ramond operator. The goal is to uncover new topological invariants relevant to string theory rather than to field theory. They should reveal the properties of the loop space $\mathcal{L}M$ of the manifold M in which the string is embedded. The main step in the computation of the Dirac index was to identify a system whose supersymmetric charge was precisely the Dirac operator. The proof of the theorem followed from the simple computation of a partition function by path integral methods. For that reason we will concentrate in this section on the extension of partition function $\mathrm{Tr}(-1)^F e^{-\tau H}$ from the point particle case to the string case. As in before most of the relevant features are already present in the absence of interaction with external fields, i.e. when the string embedding space M is flat. This will permit us to use freely the language of conformal field theory to develop our intuition and simplify some arguments. Most of this analysis does not extend to the interacting case, but we will see in the next section that what survives is enough to compute the index of the Dirac-Ramond operator, even though conformal invariance is apparently lost. Our starting point will be the simplest generalization to two dimensions of the $N = \frac{1}{2}$ supersymmetric Lagrangian which gave us the $\hat{A}$-genus :

$$S = \int d^2 z \left[\partial_z x^\mu \partial_{\bar{z}} x^\mu - \psi^\mu \partial_{\bar{z}} \psi^\mu \right], \tag{4.1}$$

where we have used the usual complex variable notation. The integration is over the torus (taken to be a parallelogram with sides $(1, \tau)$ in the complex plane) and $x^\mu(z, \bar{z})$ and $\psi^\mu(z)$ are respectively two dimensional scalars and right-moving Majorana-Weyl spinors. The rest of the notation follows the one adopted in Section 2. We have to specify a spin structure, that is choose the boundary conditions for the ψ fields. We take them to be periodic along the two sides of the parallelogram for reasons which will be clear later. We also have to understand what operator plays the role of the Dirac operator. We expect it to be the generator of supersymmetry of (4.1) expressed in terms of second quantized fields. It is convenient to use superfield notation. Let $X^\mu = x^\mu + \theta\psi^\mu$. Defining $D = \partial_\theta + \theta\partial_z$ we have:

$$S = \int d^2z \, d\theta \, \partial_{\bar{z}} X^\mu D X^\nu \tag{4.2}$$

In the superfield formulation the Lagrangian is explicitly supersymmetric. The holomorphic supersymmetry transformations of the right-moving sector are $\delta x^\mu = \epsilon\psi^\mu$ and $\delta\psi^\mu = \epsilon\partial_z x^\mu$ with $\epsilon = \epsilon(z)$. They can be derived as usual by computing $[\epsilon Q, X] = \epsilon\delta X$ with $Q = \partial_\theta - \theta\partial_z$. The generator of supersymmetry is

$$Q = \int dz\,\psi^\mu \partial_z x^\mu \tag{4.3}$$

as can easily be checked using the obvious canonical commutation relations for the x and ψ fields. There is another way to look at the operator Q. The energy-momentum superfield is given by

$$T(z,\theta) = T_F + \theta T_B\,, \tag{4.4}$$

with $\qquad T_F = \psi^\mu \partial_z x^\mu \qquad$ and $\qquad T_B = \partial_z x^\mu \partial_z x^\mu + \psi^\mu \partial_z \psi^\mu\,. \tag{4.5}$

Considering for now the theory as defined on the complex plane rather than on a torus we have:

$$T_F = \sum_n z^{-n-\frac{3}{2}} G_n \qquad T_B = \sum_n z^{-n-2} L_n\,; \tag{4.6}$$

where the G_n and L_n are respectively the odd and even generators of the Neveu-Schwarz-Ramond algebra of superconformal holomorphic transformations. The powers of z are such that when one goes from the plane to the cylinder by the transformation $z \to \log z$ one recovers the correct Fourier coefficients. They satisfy well known graded commutation relations with a central charge $\hat{c} = \frac{2}{3}c$ whose value here is equal to the dimension of the embedding space-time:

$$[L_m, L_n] = (m-n)L_{m+n} + \frac{\hat{c}}{8}(m^3 - m)\delta_{m+n,0} \tag{4.7}$$

$$[L_m, G_n] = (\tfrac{1}{2}m - n)G_{m+n} \tag{4.8}$$

$$\{G_m, G_n\} = 2L_{m+n} + \frac{\hat{c}}{2}(m^2 - \tfrac{1}{4})\delta_{m+n,0} \tag{4.9}$$

Since we are only interested in the Ramond sector of the theory, all the indices m and n will be integers. For the left-moving sector which is not supersymmetric, we just have $\bar{T} = \partial_{\bar{z}} x^\mu \partial_{\bar{z}} x^\mu$ and $\bar{T} = \sum_n \bar{z}^{-n-2}\bar{L}_n$ and each scalar field contributes $c = 1$. The generators of the SL_2 algebra are L_{-1}, L_0, L_1 and in particular the radial quantization Hamiltonian of the theory is given by $H = L_0 + \bar{L}_0$ while the rotation operator is $P = L_0 - \bar{L}_0$. We have

$$[H, P] = 0 \quad [H, G_0] = 0 \quad [P, G_0] = 0 \quad \text{and} \tag{4.10}$$

$$[\bar{L}_0, H] = 0 \quad [\bar{L}_0, P] = 0 \quad [\bar{L}_0, G_0] = 0 \tag{4.11}$$

This part of the Virasoro algebra is all what we will really need. It is of course satisfied whether or not conformal invariance is maintained at the quantum level. It is easy to verify that $G_0 = Q$. This operator is the well known Ramond operator. It is the generalization to string theory of the Dirac operator. In fact one has $G_0 = \not{D} + \cdots$ as one may see by using the expansions $\partial_z x = \sum nz^{-n-1}x_n$ and $\psi = \sum z^{-n-\frac{1}{2}}\psi_n$ and the canonical commutation relations for the x^μ and ψ^μ fields. So G_0 is to the loop space of M what $\not{D}$ is to M. The Ramond operator is precisely the operator we were looking for and we have now to compute its index. The Ramond operator shares many properties with the Dirac operator. It commutes with the Hamiltonian and anticommutes with the world-sheet fermion parity $(-1)^F$:

$$(-1)^F G_0 + G_0 (-1)^F = 0 \; . \tag{4.12}$$

However the relation between G_0^2 and the Hamiltonian is slightly complicated by the presence of the central charge:

$$G_0^2 = L_0 - c/24 \; . \tag{4.13}$$

All the supersymmetric invariant states $|\psi_0\rangle$ satisfy

$$G_0 |\psi_0\rangle = 0 \qquad \text{and} \qquad L_0 |\psi_0\rangle = c/24 \, |\psi_0\rangle \tag{4.14}$$

while the non-supersymmetric states $|\psi_h\rangle$ with non-zero L_0 eigenvalues h and ± 1 fermion parity are paired by G_0:

$$L_0 |\psi_h\rangle_+ = h \, |\psi_h\rangle_+ \tag{4.15}$$

$$(-1)^F |\psi_h\rangle_+ = |\psi_h\rangle_+ \tag{4.16}$$

$$G_0 |\psi_h\rangle_+ = |\psi_h\rangle_- \tag{4.17}$$

$$L_0 |\psi_h\rangle_- = h \, |\psi_h\rangle_- \tag{4.18}$$

$$(-1)^F |\psi_h\rangle_- = - |\psi_h\rangle_- \tag{4.19}$$

Since the left moving sector is not supersymmetric the states annihilated by G_0 will be the direct product of the Ramond vacuum defined above with symmetric tensors of any rank created by the repeated application of the creation operators $\bar{x}_{-n}$. Finally notice that the algebra 4.7 is equivalent to the transformation law for the energy-momentum tensor under a change of coordinates $z_i \to z_j$:

$$T_i = \left(\frac{\partial z_j}{\partial z_i} \right)^2 T_j + \frac{c}{12} S_{ij} \; , \tag{4.20}$$

where S_{ij} is the usual Schwarzian derivative given by:

$$S_{ij} = S(z_i, z_j) = \frac{\dfrac{\partial^3 z_j}{\partial z_i^3}}{\dfrac{\partial z_j}{\partial z_i}} - \frac{3}{2} \left(\frac{\dfrac{\partial^2 z_j}{\partial z_i^2}}{\dfrac{\partial z_j}{\partial z_i}} \right)^2 \tag{4.21}$$

In particular the Schwarzian provokes a universal shift in energy when one goes from coordinates on the plane to coordinates on the torus. Since the L_0 and $\bar{L}_0$ modes were defined for the complex plane they will be shifted by going to the torus coordinates $(z \to e^{2\pi i z})$. The Hamiltonian and the rotation operator become:

$$H = (L_0 + \epsilon) + (\bar{L}_0 + \bar{\epsilon}) \tag{4.22}$$

$$P = (L_0 + \epsilon) - (\bar{L}_0 + \bar{\epsilon}) \tag{4.23}$$

It is now easy to write down the partition function for our model on a torus specified by a given value of the moduli $\tau = \tau_1 + i\tau_2$. Inserting a $(-1)^F$ to take into account the periodic boundary conditions of the fermion field we have:

$$\text{Tr}(-1)^F \exp(-2\pi\tau_2 H + 2\pi i\tau_1 P) = \text{Tr}\,(-1)^F q^{L_0 + \epsilon} \bar{q}^{\bar{L}_0 + \bar{\epsilon}}, \qquad (4.24)$$

where we have used $q = e^{2\pi i\tau}$. Notice the presence of the $e^{2\pi i\tau_1 P}$ factor which corresponds to a relative constant rotation between the initial and final closed strings and insures the proper boundary conditions compatible with our choice of complex structure. This rotation simply corresponds to the existence of an S^1 action acting on the configuration space of the string i.e. the loop space of the embedding space. It will play a crucial role in what follows. Using the relations (4.13) and (4.12) it is clear that the above partition function is in fact the index of the operator G_0. We can write:

$$\text{Ind}\,G_0 = \text{Ind}\,Q = \text{Tr}\,(-1)^F q^{L_0 + \epsilon} \bar{q}^{\bar{L}_0 + \bar{\epsilon}} \qquad (4.25)$$

Several comments are in order about the precise form of this formula before we can understand why it gives the index of the Ramond operator. Let us examine (4.25) and use (4.13). One gets:

$$\begin{aligned}
\text{Ind}\,(G_0) &= \text{Tr}(-1)^F q^{G_0^2}(\bar{q})^{\bar{L}_0 + \bar{\epsilon}} \qquad &(4.26) \\
&= \bar{q}^{\bar{\epsilon} + \bar{h}} \sum_{k \in \mathbf{N}} I_k \bar{q}^k \qquad &(4.27)
\end{aligned}$$

The first line is just a rewriting of (4.25) while the second equation obtains from the following observations:

- the pairing and the $(-1)^F$ factor in the trace imply that only the supersymmetric states contribute. From (4.13) their L_0 eigenvalue is $-\epsilon = c/24$;

- since G_0 and $\bar{L}_0$ commute, each of these states contributes with a weight $\bar{q}^{\bar{L}_0 + \bar{\epsilon}}$;

- since the energy is bounded from below, there is a state of lowest energy with $H = L_0 + \bar{L}_0 = -\epsilon + \bar{h}$. For the $N = \frac{1}{2}$ system considered above it will of course have the quantum numbers of a spinor on M as is usual for the ground state of the Ramond sector. On this ground state the index of G_0 will simply be the ordinary $\hat{A}$-genus . All the higher states will be obtained by repeated application of the creation operators and form the traditional spectrum of the corresponding string theory. This gives the summation over the positive integers in (4.26). For each level we will get the index of the Dirac operator coupled to the corresponding vector bundle.

Since P is the generator of an isometry, we can view (4.26) as a sort of G-index formula and each I_k is an integer given by the index of G_0 on the corresponding representation of P. In other words it is a character valued $U(1)$ index formula (very much like the one obtained in Section 2.)

5 Index Theorem in Loop Space

We now turn to the determination of the topological index, that is the evaluation of the functional integral which will give us an expression of (4.26) as a function of $\bar{q}$ and characteristic classes of M. This section is mainly devoted to the computation of some infinite dimensional determinants on a torus. The reader who is not interested in the details can go directly to the main result which is given in equation (5.26).

As in the point particle case we will expand the fields around the constant solutions of the equations of motion and keep only the terms of lowest order in the background metric. Let $x^\mu = x_0^\mu + \xi^\mu$ and $\psi^\mu = \psi_0^\mu + \zeta^\mu$. Using Riemann normal coordinate expansion around x_0 we have

$$g_{\mu\nu}(x_0) = \delta_{\mu\nu} \tag{5.1}$$

$$\partial_\lambda g_{\mu\nu}(x_0) = 0 \tag{5.2}$$

$$\partial_\rho \partial_\lambda g_{\mu\nu}(x_0) = -\frac{1}{3}(R_{\mu\rho\nu\lambda} + R_{\mu\lambda\nu\rho}) . \tag{5.3}$$

The component lagrangian becomes:

$$\mathcal{L} = \partial_{\bar{z}}\xi^\mu \partial_z \xi^\mu + \mathcal{R}_{\mu\nu}\partial_{\bar{z}}\xi^\mu \xi^\nu + \partial_{\bar{z}}\zeta^\mu \zeta^\nu \tag{5.4}$$

where $\mathcal{R}_{\alpha\beta} = \frac{1}{2}R_{\mu\nu\alpha\beta}\psi^\mu\psi^\nu$. We have [1]

$$\text{Ind } G_0 = (2\pi)^{-d/2} \int_M d^d x d^d \psi \left[\frac{\det'(\partial_{\bar{z}})}{\det'(-\partial_{\bar{z}}\partial_z + \mathcal{R}\partial_{\bar{z}})} \right]^{\frac{1}{2}} \tag{5.5}$$

We have used an obvious matrix notation and dropped the subscript 0 for the zero modes. The prime over the determinant means that the zero modes have to be excluded.

For later convenience the following notations are introduced. A complete set of normal modes with the appropriate periodicity in $z \to z+1$ and $z \to z+\tau$ is given by $\exp(\alpha z + \beta \bar{z})$ with:

$$\alpha = \frac{2\pi i}{\bar{\tau} - \tau}(m + n\bar{\tau}) \qquad \text{and} \qquad \beta = -\frac{2\pi i}{\bar{\tau} - \tau}(m + n\tau) \tag{5.6}$$

Also let:

$$\Gamma = \{\omega = m + n\tau \,|\, m, n \in \mathbf{Z}\} . \tag{5.7}$$

Equivalently, $\bar{\Gamma}$ and $\bar{\omega}$ denote the replacement of τ by $\bar{\tau}$ in (5.7), while the subscript in Γ_0 indicates the omission of the point $(m, n) = (0, 0)$ from the lattice Γ. We will often denote the product or the summation over all the sites of Γ_0 by a simple $\prod'$ or $\sum'$.

It is clear at this point that if we keep only the $n = 0$ set of normal modes in the computation of the Ramond index one should recover the particle case result *i.e.* the index of the Dirac operator. This fact can be used to check easily some of the determinant normalizations. As in the point particle case we could, paying attention

---

[1]We will not consider here the possible global anomalies related to global sign ambiguities in the definition of the square roots of the determinants. An analysis of this problem and its relation to Stieffel-Whitney classes in the case of the point particle was done in [17].

to the normalization of the zero modes, rescale the fields and the coordinates to get rid of the explicit τ_2 dependence in the subsequent calculations. We prefer not to do so.

The determination of the index in (5.5) involves the evaluation of the product of determinants:

$$\left[\frac{\det'(-\partial_{\bar{z}}\partial_z + \mathcal{R}\partial_{\bar{z}})}{\det'(\partial_{\bar{z}})}\right]^{-\frac{1}{2}} = \left[\frac{\det'(-\partial_{\bar{z}}\partial_z + \mathcal{R}\partial_{\bar{z}})}{\det'(-\partial_{\bar{z}}\partial_z)}\right]^{-\frac{1}{2}}\left[\frac{\det'(\partial_{\bar{z}})}{\det'(-\partial_{\bar{z}}\partial_z)}\right]^{\frac{1}{2}}$$

$$= \det'(-\partial_z + \mathcal{R})^{-\frac{1}{2}}. \tag{5.8}$$

To go from the first to the second equation we have used the known values of the free boson and fermion determinants. They can be determined by writing the product of eigenvalues α and β of $(-\partial_{\bar{z}}\partial_z)$ as:

$$\prod_{\bar{\omega}\in\Gamma_0}\left|\frac{\pi}{\tau_2}\bar{\omega}\right|^2 = \exp G'(0); \tag{5.9}$$

with

$$G'(0) = -\frac{\partial}{\partial s}\sum_{\bar{\omega}\in\bar{\Gamma}_0}\left|\frac{\pi}{\tau_2}\bar{\omega}\right|^{-2s} \quad \text{at s=0} \tag{5.10}$$

$$= -2\ln|2\tau_2\eta(q)\eta(\bar{q})|. \tag{5.11}$$

Here $\eta(q) = q^{\frac{1}{24}}\prod_{n=1}^{\infty}(1-q^n)$ is the Dedekind function We then obtain for the partition function of a periodic right moving free fermion whose ground state energy (Fermi level) in the Ramond sector is known to be $1/16$:

$$\text{Tr}\,(-1)^F q^{L_0+\epsilon} = q^{-\frac{1}{48}}q^{\frac{1}{16}}\prod_{n=1}^{\infty}(1-q^n) = \eta(q) = (2\tau_2)^{-\frac{1}{2}}\det'(\partial_{\bar{z}})^{\frac{1}{2}} \tag{5.12}$$

Here L_0 is the Hamiltonian of a $c = 1/2$ free right moving fermion field. Similarly the Bose-Einstein distribution gives:

$$\text{Tr}\,q^{L_0-\frac{1}{24}}\bar{q}^{\bar{L}_0-\frac{1}{24}} = [\eta(q)\eta(\bar{q})]^{-1} = |2\tau_2|\det'(-\partial_{\bar{z}}\partial_z)^{-\frac{1}{2}} \tag{5.13}$$

In d dimensions the properly normalized result reads:

$$\left[\frac{\det'(\partial_{\bar{z}})}{\det'(-\partial_{\bar{z}}\partial_z)}\right]^{\frac{1}{2}} = \left[\frac{1}{\sqrt{2\tau_2}\eta(\bar{q})}\right]^d \tag{5.14}$$

We have to evaluate a determinant of the type:

$$\det'(\partial_{\bar{z}} - F) = \prod_{\Gamma_0}\left[\frac{\pi}{\tau_2}(m+n\tau) - F\right] \tag{5.15}$$

$$= \prod{}'\left(\frac{\pi}{\tau_2}\omega\right)\prod{}'\left(1 - \frac{\tau_2 F}{\pi\omega}\right). \tag{5.16}$$

Now let us recall the definition of the Weierstrass σ–function [24]:

$$\sigma(z) = z\prod{}'(1 - \frac{z}{\omega})\exp\left(\frac{z}{\omega} + \frac{z^2}{2\omega^2}\right). \tag{5.17}$$

Notice the similarity between the σ function and the sine function —in fact the subsequent results will merely follow from a substitution of the sine by σ in the formulae of Section 2. This is most easily seen by writing down another representation for the σ function[2]:

$$\sigma(z) = e^{\eta_1 z^2} \cdot \frac{\sin(\pi z)}{\pi} \cdot \prod_1^\infty \frac{(1 - q^n e^{2\pi i z})(1 - q^n e^{-2\pi i z})}{(1 - q^n)^2} \tag{5.18}$$

$$= \frac{\theta(z, \tau)}{\theta'(0, \tau)} e^{\eta_1 z^2}, \tag{5.19}$$

where $\eta_1 = \zeta(1/2)$ is the value of the logarithmic derivative $\zeta(z)$ of $\sigma(z)$ at the half period [24]. Comparing (5.17) and (5.15) we have:

$$\det{}'(-\partial_{\bar{z}} - F) = -2\Lambda\eta^2(q) \cdot \frac{1}{\pi F} \cdot \sigma\left(\frac{\tau_2 F}{\pi}\right), \tag{5.20}$$

where $\log \Lambda = -\sum'\left(\frac{z}{\omega} + \frac{z^2}{2\omega^2}\right)$. The lack of absolute convergence of this factor will play an important role in what follows. We can now repeat the same type of manipulation on eq. (5.8). Let the formal eigenvalues of $\mathcal{R}$ be: $\mathcal{R}_{2i+1,2i+2} = -\mathcal{R}_{2i+2,2i+1} = 2\pi x_i$ ($i = 1, \cdots, d/2$). Then the second bracket in (5.8) reads:

$$\left[\frac{\det{}'(-\partial_{\bar{z}}\partial_z + \mathcal{R}\partial_{\bar{z}})}{\det{}'(-\partial_{\bar{z}}\partial_z)}\right]^{-\frac{1}{2}} = \prod_{i=1}^{\frac{d}{2}} \left[\prod_{\bar{\Gamma}_0}\left(1 - \frac{2i\tau_2 x_i}{m + n\bar{\tau}}\right)\right]^{-1} \tag{5.21}$$

$$= \left(\prod_{i=1}^{d/2} \frac{2i\tau_2 x_i}{\sigma(2i\tau_2 x_i)}\right) \exp\left[\frac{1}{2} G_1(\bar{\omega}) \sum_i z_i^2\right].$$

In the last equation Λ has be rewritten in terms of the Eisenstein series G_1 (see below). It is at this point that the role of Λ is important. Indeed since we are performing an index computation, no arbitrary process of regularization should be involved. There are many possible ways to deal with the problem of convergence of the series G_1, but the only unambiguous one is simply to require that its multiplying coefficient in Λ vanishes. This introduces the most remarkable constraint: $p_1(M) = \sum x_i^2 = \operatorname{Tr} R^2 = 0$. So the first conclusion is that for the index to have a well defined meaning the first Pontrjagin class of the manifold must vanish. Also it should be rather obvious at this stage that had one introduced extra gauge fields this condition would have been replaced by $\operatorname{Tr}(R^2 - F^2) = 0$, which is nothing but the Green-Schwarz anomaly cancellation condition in string theory [19]. Moreover we will see in a moment that once this condition is satisfied, the index will have very simple modular properties. This is of course the result which was expected in the introduction. The connection between the field theory anomaly and the modular properties was studied in full detail in [10,11,25]. From now on we will suppose that the above condition is satisfied. Let us now rearrange the above result. With the help of the Eisenstein series defined by:

$$G_k(\omega) = \sum_{\omega \in \Gamma_0} \frac{1}{\omega^{2k}} \tag{5.22}$$

---

[2]Most definitions and conventions can be found in the appendix.

the function $\sigma(z)/z$ has the expansion:

$$\frac{\sigma(z)}{z} = \exp \sum_{\omega \in \Gamma_0} \left[\ln(1 - \frac{z}{\omega}) + \frac{z}{\omega} + \frac{1}{2}(\frac{z}{\omega})^2 \right] \tag{5.23}$$

$$= \exp \left[-\sum_{k=3}^{\infty} \sum_{\omega \in \Gamma_0} \frac{1}{k} z^k \frac{1}{\omega^k} \right] \tag{5.24}$$

$$= \exp \left[-\sum_{k=2}^{\infty} \frac{1}{2k} z^{2k} G_k(\omega) \right] . \tag{5.25}$$

Inserting (5.21) and (5.14) in (5.5) we have the final expression for the index of the Dirac-Ramond operator:

$$\text{Ind } (G_0) = \bar{q}^{\bar{\epsilon}+\bar{h}} \sum_{k \in \mathbf{N}} I_k \bar{q}^k = \int_M \prod_{i=1}^{d/2} \frac{ix_i/2\pi}{\sigma(ix_i/2\pi|\bar{\omega})} \cdot \frac{1}{\eta(\bar{q})^d} \tag{5.26}$$

$$= \int_M \prod_{i=1}^{d/2} \exp \left[\sum_{2}^{\infty} \frac{1}{2k} (\frac{ix_i}{2\pi})^{2k} G_k(\bar{\omega}) \right] \cdot \frac{1}{\eta(\bar{q})^d} \tag{5.27}$$

$$= \int_M \prod_{i=1}^{d/2} \left[\frac{ix_i}{\theta\prime(ix_i/2\pi, \bar{\tau})} \cdot \eta(\bar{q}) \right] . \tag{5.28}$$

Several remarks are in order. First the integration over the fermionic zero modes $d^d\psi$ has been performed. It results in replacing $\mathcal{R}$ by $\underline{R} \equiv R_{\mu\nu} dx^\mu dx^\nu$ and interpreting the x_i accordingly as 2-forms. Secondly the explicit τ_2 dependence has disappeared. This is because in the above formula only the terms which are forms of order d will contribute to the integral $i.e.$ the terms homogeneous in x_i of degree $d/2$. It is easy to check that such terms do not contain any residual τ_2 factor. It is important to notice the modular properties of the result. They are best seen in (5.27): if the dimension is a multiple of 24 we see that the index is modular invariant under the full modular group. Its behaviour at the cusp ($q = 0$) is easily seen to be $q^{-d/24}$ which gives us $\bar{\epsilon}+\bar{h}+d/24 = 0$. Notice that the behaviour at the cusp corresponds to the $\hat{A}$-genus of M. Finally we point out that (5.27) and (5.28) have been written using the condition $p_1(M) = 0$. In the form (5.27) the result is readily seen to be a generating function for the topological index of the Dirac operator coupled to tensors of rank n where n is the level considered. For example as we just saw, the zero mode corresponds to the $\hat{A}$ genus, $i.e.$ the index of the Dirac operator for the spinors in the Ramond vacuum, while for higher levels the result is simply multiplied by the appropriate Chern character. In that sense we have a generating functional for the field theory anomalies [10,11]. In the following we will loosely refer to this generating functional as $\hat{A}(\tau)$.

Before proceeding to other indices corresponding to the Euler number and Hirzebruch signature of loop space, just as we did in the point particle case, it is worth trying to understand better what is the structure of the formula we just derived. We first realize that we really got a fixed point index theorem. We already saw in detail that the analytic part of the index was expressed as a sum of characters of the S^1 action which naturally acts on the loop $X(\sigma) \rightarrow X(\sigma + 2\pi\tau_1)$ and whose generator is P. As we saw these characters are just the integral powers of q (after we have analytically continued to the whole complex τ plane and extracted the prefactor $q^{-d/24}$

coming from the conformal anomaly). The integer coefficient of q^n in this expansion is clearly nothing but the index of the ordinary Dirac operator coupled to the linear combination of symmetric tensor products of the tangent bundle corresponding to the level n mass state of the string (eigenvalue of the L_0 operator). At level one the Fock space corresponds to the oscillator x_{-1}^μ acting on the ground state, at level two we have either x_{-2}^μ or the symmetric combination $x_{-1}^\mu x_{-1}^\nu$ and so on for the higher levels. So one could find this integer coefficient just by applying the formula for the ordinary index of the Dirac operator which involves the $\hat{A}$ genus and the Chern character of the corresponding bundle (remember that the Chern character of a direct sum is the sum of Chern character, while it is the product of these character for a tensor product). We still have to find a character index interpretation for the right hand side of the Dirac-Ramond index formula, i.e. for its topological expression. Remember that we started from an infinite dimensional space $\mathcal{L}M$, the free loop space of maps from $S^1 \to M$ and built a Dirac like operator on it $Q = \int d\sigma \psi^\mu(\sigma) \left(-i \frac{D}{DX^\mu(\sigma)} + g_{\mu\nu}(x) \frac{dX^\nu}{d\sigma} \right)$. The first term is simply the most naive extension of the Dirac operator (think of σ as a continuous index) while the second term contains the generator of the $U(1)$ rotation $\frac{dX^\nu}{d\sigma}$. In that sense the operator we are considering is an equivariant operator and its index must be localized on the fixed point set of the S^1 action. This fixed point set is the set of constant loops which are the only one for which $X(\sigma + \alpha) = X(\sigma)$. These constant loops are just points on the manifold M itself. So according to our previous analysis of the character index we should look at the normal bundle to M in $\mathcal{L}M$. If we look at the mode expansion of the coordinates $X(\sigma) = X_0 + \sum_{n \neq 0} e^{2\pi i n\sigma} X_n$ one readily sees that all the non zero modes give the coordinates of the normal bundle which is simply the (infinite) direct sum of the tangent bundle of M: $N_M = \oplus_{n \neq 0} T_n$, where the subscript n is just a bookkeeping device. We may now show the equivalence between (2.19) and the formal expression obtained by applying the character valued index theorem to $\mathcal{L}M$[7,8]. We saw in Section 3 that the general expression for this index on a manifold N is:

$$I_g = \int_{N_b} \prod_j \frac{ix_j/2}{\sin ix_j/2} \prod_r \frac{1}{2i\sin(\bar{\theta}_r/2)} \tag{5.29}$$

where M_b is the fixed point set of the group action and $\bar{\theta}_r = \alpha\theta_r + ix_r$. The θ_r are the 'eigenvalues' of the group action on TN, x_r are the 'eigenvalues' of the curvature two form at the fixed points, and α is the parameter of the infinitesimal transformation. In our case $N_b = \mathcal{L}M_b = M$ and $T(\mathcal{L}M_b)$ is the ∞-dimensional vector space generated by the string Fourier modes, X_n^μ. By definition θ_r is $\theta_n^\mu = 2\pi n$ if $\alpha = \tau_1 - i\tau_2 = \bar{\tau}$ corresponds to the simultaneous H and P actions. Applying the character valued theorem we find

$$\prod_r \frac{1}{2i\sin(\bar{\theta}_r/2)} = \prod_j \prod_{n=1}^\infty \frac{\exp(-\pi i\bar{\tau}n + x_j/2)}{1 - \exp(-2\pi i\bar{\tau}n + x_j)} \cdot \frac{\exp(-\pi i\bar{\tau}n - x_j/2)}{1 - \exp(-2\pi i\bar{\tau}n - x_j)}$$

$$= \prod_j \prod_{n=1}^\infty \frac{\bar{q}^n}{(1 - \bar{q}^n e^{x_j})(1 - \bar{q}^n e^{-x_j})}. \tag{5.30}$$

Defining $\prod_{n=1}^\infty \bar{q}^n$ by zeta function regularization ($\sum_1^\infty n = -1/12$) yields

$$I_g = \int_M \prod_j \frac{ix_j/2}{\sin(ix_j/2)} \frac{\bar{q}^{-1/12}}{\prod_{n=1}^\infty (1 - \bar{q}^n e^{x_j}) \prod_{n=1}^\infty (1 - \bar{q}^n e^{-x_j})} \tag{5.31}$$

If $p_1(M) = 0$ then the identity between (5.31) and (2.19) is guaranteed by the product representation of the σ-function given in equation (5.18). We conclude that we have in fact derived the character valued index associated with the natural S^1 action on $\mathcal{L}M$.

We would like to point out that the result of interest is really $\hat{A}(\tau)$ itself and not any of its truncations. The special form of the result comes from the product over all the modes of the lattice Γ. It should really be seen as an extension of the Atiyah-Singer index theorem to the loop space and the link between the analytical index of the Dirac-Ramond operator and topological and cohomological properties of the loop space. Any new insight will come from considering this result in that light. In particular the modular properties of the result are best seen in the approach presented above and follow from the simple fact that we are considering an operator on the loop space. It is in striking contrast with the cobordism approach to elliptic cohomology where the modular properties are quite mysterious [26,27,28] although the results are of course equivalent.

6 Euler number, signature, etc...

We are in position to analyze the different topological invariants which arise from coupling the $N = 1/2$ "right" moving system with a left moving sector. This corresponds to introducing a twisting of the original Ramond-Dirac operator and in the point particle case to the introduction of the $\eta, \bar{\eta}$ system of Section 2. The boundary conditions for the left sector can be chosen to be Ramond or Neveu-Schwarz in space and periodic or anti-periodic in time corresponding to the four spin structures on the torus. None of these choices will spoil the topological character of the result since as we saw in Section 4 the topological invariance is entirely governed by the supersymmetry of the right sector. Of course if the new sector possesses its own $N = 1/2$ supersymmetric charge $\bar{G}_0$, the states with $\bar{L}_0 \neq \bar{\epsilon}$ will be projected out of the trace. This will happen in the case of periodic Ramond boundary conditions and will in a particular case give the Euler number of M.

We will by the subscript 0 (respectively 1, 2, 3) denote the Ramond periodic boundary conditions (Ramond anti-periodic, Neveu-Schwarz periodic, Neveu-Schwarz anti-periodic). In this notation we will obtain unified expression for the four indices. Gauge or gravitational background fields can be coupled to this sector. In the latter case, the only one treated explicitly here, the condition $p_1(M) = 0$ is not required anymore. The treatment of the external gauge field coupling could be done as easily and would require $\mathrm{Tr}\, R^2 - \mathrm{Tr}\, F^2 = 0$. Of the four indices constructed, two of them have a simple geometrical interpretation. They correspond respectively to the extension to loop space of the Euler number and the Hirzebruch signature that we treated briefly in Section 2. To construct the Euler and the Hirzebruch indices we use the isomorphism between the exterior algebra and the tensor product of the spinor algebra with itself. This correspondence is well understood in finite dimensions and there seems to be a formal infinite dimensional analogue. Let ϕ be a spinor and let ρ be a dual spinor. The tensor product $\phi\rho$ is a bispinor which may be decomposed into a complete series of gamma matrices and their antisymmetric products:

$$\phi\rho \;\propto\; \mathrm{tr}(\rho\phi) + \sum \gamma^\mu \, \mathrm{tr}(\rho\gamma^\mu\phi)$$

$$+ \sum_{\mu < \nu} \gamma^\mu \gamma^\nu \, \mathrm{tr}(\rho \gamma^\mu \gamma^\nu \phi) + \cdots \tag{6.1}$$

This is the standard correspondence between spinors and the exterior algebra via the association of the wedge products $dx^1 \wedge dx^2 \wedge \cdots \wedge dx^k$ with the antisymmetric products of gamma matrices.

This observation is used in the following way. The canonical anticommutation relations for the fermions in the $N = 1/2$ right moving sector are $\{\psi^\mu(\sigma), \psi^\nu(\sigma')\} \propto g^{\mu\nu} \delta(\sigma - \sigma')$. This means that the ψ's are the gamma matrices in loop space. The carrier of the representation space for the gamma matrices are the spinors on loop space. We now introduce a right moving fermionic Ramond superfield which transforms as a covector

$$N_\mu = \eta_\mu + \theta \varphi_\mu \tag{6.2}$$

The canonical commutation relations will tell us that the η's are dual to the 'gamma' matrices and therefore the carrier space for the representations are the dual spinors. The state created by the combination of ψ's and η' will be spinors tensored with dual spinors. These are just the exterior algebra. One can make this identification more explicit by writing down the tangent space in terms of oscillators and comparing it with the Hilbert space generated by looking at the oscillator content of the fermions ψ and η. The fermion parity in the right moving sector $(-1)^{F_R}$ may be interpreted as multiplication of the bispinor by γ_5 on the left. Likewise, the fermion parity in the left moving sector $(-1)^{F_L}$ may be interpreted as multiplication of the bispinor by γ_5 on the right.

The Euler characteristic compares forms of even degree with odd degree, therefore it may be computed as

$$\chi = \mathrm{Tr}(-1)^{F_L + F_R} \tag{6.3}$$

In the supersymmetry approach one calculates the above by having both the left and right moving sectors satisfying periodic boundary condition in the temporal direction.

In finite dimensions the Hirzebruch signature compares self dual forms with anti-self dual forms. In Dirac language, the duality transformation is implemented by multiplication by γ_5 on the left. The associated index is formally

$$\tau = \mathrm{Tr}(-1)^{F_R} \tag{6.4}$$

What we are then led to calculate is the path integral with periodic temporal boundary conditions for both ψ and η in the case of the Euler number, and with periodic ψ and anti-periodic η in the case of the Hirzebruch signature [17,22]. In analogy with (4.25) these path integrals compute respectively:

$$I_\chi = \mathrm{Tr}(-1)^{F_L + F_R} q^{L_0 + \epsilon} \bar{q}^{\bar{L}_0 + \bar{\epsilon}} \,, \tag{6.5}$$

$$I_\tau = \mathrm{Tr}(-1)^{F_R} q^{L_0 + \epsilon} \bar{q}^{\bar{L}_0 + \bar{\epsilon}} \,, \tag{6.6}$$

where now $L_0 + \epsilon$ and $\bar{L}_0 + \bar{\epsilon}$ contain contributions from the full system of two dimensional fields X, ψ and η [3]. As in the supersymmetric quantum mechanical case, the particular boundary conditions chosen to calculate the Euler number give rise to a left handed $N =$

---

[3]We have eliminated the auxiliary fields φ_μ

1/2 supersymmetry [22]. By expressing $L_0 + \bar{\epsilon}$ as the square of this new supersymmetry, we can infer as before that only states with $\bar{L}_0 + \bar{\epsilon} = 0$ contribute to the trace. So altogether only the zero modes should enter the expression for I_χ. By explicitly carrying out the functional integration we obtain:

$$I_\chi = \frac{1}{(2\pi)^{d/2}} \int d^d X_0 \, d^d \psi_0 \, [\det{}'(-\partial\bar{\partial} + \mathcal{R}\bar{\partial})]^{-1/2} [\det{}'(\bar{\partial})]^{1/2} [\det(-\partial - \mathcal{R})]^{1/2} . \quad (6.7)$$

This expression directly simplifies into

$$I_\chi = \frac{1}{(2\pi)^{d/2}} \int d^d X_0 \, d^d \psi_0 \, [\det(-\mathcal{R})]^{1/2} , \quad (6.8)$$

which as we promised only contains the contribution from the zero modes. Integrating over the fermionic zero modes ψ_0 leaves us with the Euler characteristic of $\mathcal{L}M$, which coincides with the Euler number of M itself, according to the Lefschetz fixed-point theorem:

$$I_\chi = \chi(M) = \int_M \prod_j (-x_j). \quad (6.9)$$

We remark that in this case we do not need to impose $p_1(M) = 0$ in order to have a well defined index: the Euler number of a manifold is defined even if the manifold does not admit a spin structure (or a *string structure* in the case of the loop space). Possible inconsistencies due to the use of spinors to obtain the result should cancel out in the final expression, as they do.

Performing the path integral for the Hirzebruch signature gives rise to the same formula as in (6.7), but the determinant arising from the left handed fermions is to be evaluated over states with anti-periodic boundary conditions in τ. The details of the computation for this case as well as the next two can be found in the Appendix. One obtains for the Hirzebruch signature of $\mathcal{L}M$:

$$I_\tau = \int_M \prod_k ix_k \cdot \frac{\theta_1(ix_k/2\pi, \bar{\tau})}{\theta(ix_k/2\pi, \bar{\tau})} \quad (6.10)$$

There is a simple interpretation of this result in terms of the Atiyah-Bott character index theorem [29][30] along the same lines as what we did in the previous section —more details can be found in [7,8]. Once again the manifold M is not required to satisfy $p_1(M) = 0$ in order for this index to be defined (*cf.* appendix). Using the notations introduced in the appendix one can see that the two topological invariants corresponding to the boundary conditions (NS,P) and (NS,A) are given by:

$$I_a = \int_M \prod_k ix_k \cdot \frac{\theta_a(ix_k/2\pi, \bar{\tau})}{\theta(ix_k/2\pi, \bar{\tau})} \quad \text{a=(2,3)} \quad (6.11)$$

As with the generalization of the $\hat{A}$-genus there is a simple interpretation of these results in terms of a generating function for the the indices of the ordinary Dirac operator twisted by the different bundles corresponding to spectrum of the appropriate sector.

As for the case of $\hat{A}(\tau)$ the topological invariants I_a have simple transformations under the modular group or more precisely under some subgroup of level two. This is easily understood since the different boundary conditions are preserved by some subgroup of Γ. Only some phase anomalies appear under such a modular transformation.

Then one has a modular invariant result only if the dimension is a multiple of eight. Note that it is possible to rewrite the four indices in terms of Jacobi elliptic functions. The necessary definitions are listed in the Appendix. It provides an immediate connection with the definition of the elliptic genus given in the work of Peter Landweber and Robert Stong [13] and Serge Ochanine [12,31]. Explicitely we have:

$$I_1 = \int_M \prod_k i x_k \cdot \frac{cn(\omega_1 i x_k / 2)}{\sqrt{\kappa'} sn(\omega_1 i x_k / 2)} \tag{6.12}$$

$$I_2 = \int_M \prod_k i x_k \cdot \frac{1}{\sqrt{\kappa} sn(\omega_1 i x_k / 2)} \tag{6.13}$$

$$I_3 = \int_M \prod_k i x_k \cdot \frac{dn(\omega_1 i x_k)}{\sqrt{\kappa \kappa'} sn(\omega_1 i x_k)} \tag{6.14}$$

$$\tag{6.15}$$

where $\omega_1 = \theta_3^2(0, \bar{\tau})$, $\sqrt{\kappa} = \frac{\theta_1(0, \bar{\tau})}{\theta_3(0, \bar{\tau})}$ and $\sqrt{\kappa'} = \frac{\theta_2(0, \bar{\tau})}{\theta_3(0, \bar{\tau})}$. It is then very easy to find the expression for the parameters ϵ and δ defined in [12] in terms of $\lambda(\bar{\tau}) = \kappa^2 = 1 - (\kappa')^2$. In particular it is important to notice that $\lambda(\bar{\tau})$ is invariant under the subgroup of the modular group $\Gamma(2)$ and vanishes at $q = 0$.

In conclusion, we have shown how to use two-dimensional $N = 1/2$ supersymmetric theories to probe the topology of the loop space $\mathcal{L}M$ of compact Riemannian manifolds M without boundaries. The existence of a group action on $\mathcal{L}M$ makes it possible to split the infinite dimensional kernels that appear in the definitions of various indices into finite dimensional representations of the group action. The results that we obtain suggest that the finite dimensional character-valued index theorems hold true when naively applied to the infinite dimensional manifold $\mathcal{L}M$, provided certain topological restrictions are satisfied by M in the case of certain indices. Finally, let us point out that we could have coupled the Dirac-Ramond operator to other bundles like gauge fields by simply modifying the left-moving sector, introducing additional fermi fields and coupling them to external gauge fields, just as we did in the point particle case in section 2. As in all the cases studied here, the structure of the final result would be $\hat{A}(\tau)$ multiplied by the corresponding generalized Chern character. It is clear that for us the fundamental result is the expression of $\hat{A}(\tau)$ obtained in Section 5, all the other topological invariants being specializations of it.

7   The Atiyah-Hirzebruch Theorem Revisited

It is time now to make the connection with the other line of approach to elliptic cohomology which was mentioned in the introduction, namely the one concerned with the generalization of the Atiyah-Hirzebruch theorem to the Rarita-Schwinger operator. Since it is not what motivated our study of loop space index theorems, we will be very brief; but in view of the historical role played by Witten's conjecture [15] in the mathematical development of elliptic genera, this new theory, we thought that these lectures would have been incomplete without what follows. We will closely follow Witten's exposition of his proof [5][8]. We remember that the conjecture claimed that the character index of the Rarita-Schwinger operator is constant on a manifold with an S^1 action. In fact we will see that his proof goes further and applies to the all the representations of the *Spin* group which occur in the loop space index theorems we just derive. Since we want to avoid the complication related to $p_1(M) = 0$ we will study the loop space signature. We first have to modify our previous analysis to take into account the extra S^1. We have now two S^1 actions: one is the natural action on $\mathcal{L}M$, while the other is the lift to $\mathcal{L}M$ of the group action on M. The character index can easily computed as in Section 3 by inserting the operator $\exp i\theta K$ in the trace over the states of the sigma model, where K is the generator of the S^1 action over M. This will produce a refinement of the index over $\mathcal{L}M$: the character expansion is now indexed by the eigenvalues of K as well as by the eigenvalues of P. We can write

$$I(\theta) = \sum_{n,k} I_{n,k} q^n e^{ik\theta} , \tag{7.1}$$

where $I_{n,k}$ is the index of the Dirac operator acting on the representation R occurring at level n restricted to the eigenspace k of K. More precisely:

$$D_R, : \Gamma(S_+ \otimes T_R) \to \Gamma(S_- \otimes T_R) \tag{7.2}$$

where T_R is the bundle associated to R. What Witten observed is that one can consider a slightly more general problem by twisting the boundary conditions of the fields $X(\sigma)$ as follows:

$$X(\sigma + 2\pi) = \exp(2\pi\alpha K)X(\sigma) \tag{7.3}$$

This gives a one parameter family where the eigenvalues of P are of the form $n + \alpha k$ instead of simply n. Notice that supersymmetry is not affected. The new index is

$$I(\theta, \alpha) = \sum I_{n,k}(\alpha) \, q^{n+\alpha k} e^{i\theta k} . \tag{7.4}$$

However the $I_{n,k}(\alpha)$ must be independent of α since a smooth deformation of the spectrum cannot change the index. This implies that $I(\theta, 1) = I(\theta, 0)$ or that

$$I_{n,k} = I_{n+k,k} . \tag{7.5}$$

But now it is enough to remember that there can be no negative energy states in the spectrum so $I_{-n,k} = 0$ with $n > 0$. If we iterate m times the above relation we find the desired result:

$$I_{n,k} = 0 \qquad \text{for} \qquad k \neq 0 , \tag{7.6}$$

which proves the constancy of the index. The reader should consult the detailed work of Taubes for more rigorous arguments [32].

Acknowledgements

I would like to thank my collaborators Orlando Alvarez, Tim Killingback and Michelangelo Mangano with whom the material of these lectures was developed. I am specially thankful to Orlando Alvarez for the countless exciting hours we spent in Berkeley discussing the various aspects of index theorems and string theory. I also benefited from discussions with M. Dugan, C. Kounnas, N. Marcus, I. Singer, I. Sonoda and B. Zumino. Finally I would like to thank N. Zichichi for giving me the pleasant opportunity to present these lectures in Erice.

A   Appendix

We collect in this appendix most of the computations needed to obtain the results of the preceeding sections. For the sake of completeness there is some overlap with Section 4. Two good general references for this appendix are Whittaker and Watson [24] and *Elliptic functions* by K. Chandrasekharan [33]. We have to evaluate a series of determinants which are all of the same type, the only difference being the boundary conditions. The four boundary conditions are Ramond or Neveu-Schwarz, *i.e.* periodic or anti-periodic in space, and periodic or anti-periodic in time. They are symbolically noted (R,P), (R,A), (NS,P) and (NS,A). In the following computations the last three will be labelled by a single index $a = (1,2,3)$ while occasionally the first one will be denoted by the subscript 0. We introduce the following notations: $\omega = m + n\tau$ with $\mathrm{Im}\,\tau > 0$ and m, n integers. A shift by one of the half-period of the lattice $\Gamma = \{\omega\}$ will be denoted by $\omega_a = (1/2, \tau/2, 1/2 + \tau/2)$. We give below a list of the definitions to be used later, where the products and sums run over the entire lattice Γ except when a prime denotes the omission of the origin:

$$\wp(z) = \frac{1}{z^2} + {\sum}' \left(\frac{1}{(z-\omega)^2} - \frac{1}{\omega^2} \right) \tag{A.1}$$

$$\zeta(z) = \frac{1}{z} + {\sum}' \left(\frac{1}{z-\omega} + \frac{1}{\omega} + \frac{z}{\omega^2} \right) \tag{A.2}$$

$$\sigma(z) = z {\prod}'(1 - \frac{z}{\omega}) \exp \left(\frac{z}{\omega} + \frac{z^2}{2\omega^2} \right) \tag{A.3}$$

$$\sigma_a(z) = \frac{\sigma(z + \omega_a)}{\sigma(\omega_a)} \exp\left(-\eta_a z\right) \tag{A.4}$$

$$= \prod \left(1 - \frac{z}{\omega - \omega_a} \right) \exp \left(\Lambda_a - \frac{z^2}{2} e_a \right), \tag{A.5}$$

where $\wp(\omega_a) = e_a$ and $\zeta(\omega_a) = \eta_a$ and $\Lambda_a = \dfrac{z}{(\omega - \omega_a)} + \dfrac{1}{2}\left(\dfrac{z}{\omega - \omega_a}\right)^2$. These are respectively the Weierstrass $\wp$ and ζ functions and the four sigma functions which are in one to one correspondence with the four theta functions through

$$\sigma(z) = \frac{\theta(z, \tau)}{\theta'(0, \tau)} e^{\eta_1 z^2} \tag{A.6}$$

$$\sigma_a(z) = \frac{\theta_a(z, \tau)}{\theta_a(0, \tau)} e^{\eta_1 z^2} \tag{A.7}$$

The four theta functions are respectively denoted $-\theta_{11}$, θ_{10}, θ_{01} and θ_{00} in Mumford's book [34]. For the convenience of the reader we also give their standard product expansions:

$$\theta(z, \tau)_0 \equiv \theta(z, \tau) = c' q^{1/8} 2 \sin \pi z \cdot \prod_{n=1}^{\infty} (1 - q^n e^{2\pi i z})(1 - q^n e^{-2\pi i z}) \tag{A.8}$$

$$\theta_1(z, \tau) = c' q^{1/8} e^{\pi i z} \cdot \prod_{n=1}^{\infty} (1 + q^n e^{2\pi i z})(1 + q^{n-1} e^{-2\pi i z}), \tag{A.9}$$

$$\theta_2(z,\tau) = c' \prod_{n=1}^{\infty} (1 - q^{n-1/2}e^{2\pi i z})(1 - q^{n-1/2}e^{-2\pi i z}) , \tag{A.10}$$

$$\theta_3(z,\tau) = c' \prod_{n=1}^{\infty} (1 + q^{n-1/2}e^{2\pi i z})(1 + q^{n-1/2}e^{-2\pi i z}) , \tag{A.11}$$

$$\text{with } c' = \prod_{n=1}^{\infty} (1 - q^n) . \tag{A.12}$$

Using the above one easily evaluates the general class of determinants [35]

$$\left[\frac{\det(\partial + R)}{\det(\partial)}\right]_a^{\frac{1}{2}} = \prod \prod_k \left(1 - \frac{z_k}{\omega - \omega_a}\right) \tag{A.13}$$

$$= \prod_k \left[\sigma_a(z_k) \exp \frac{z_k^2}{2} e_a \cdot \exp\left(-\sum \Lambda_a(z_k)\right)\right] , \tag{A.14}$$

where $z_k = 2i\tau_2 x_k$ and x_k are the formal eigenvalues of the curvature matrix $R/2\pi$. Note the presence of the regularization factor $N_{a,k} = \exp - (\sum \Lambda_a(z_k))$ which in Section 5 resulted in the condition $p_1(M) = 0$. The determinants in flat background can be evaluated as in Section 5 and give

$$(\det \partial)_a^{\frac{1}{2}} = \left(\frac{\theta_a(0,\tau)}{\eta(q)}\right)^{\frac{d}{2}} \tag{A.15}$$

$$\text{with} \quad \eta(q) = q^{\frac{1}{24}} \prod_1^{\infty} (1 - q^n) . \tag{A.16}$$

We then have

$$ch_a(\tau) \equiv \det(\partial + R)^{\frac{1}{2}} = \prod_k N_{a,k} \frac{\theta_a(z_k,\tau)}{\eta(q)} \cdot \exp\left(\eta_1 z_k^2 + \frac{z_k^2}{2} e_a\right). \tag{A.17}$$

Recalling the contribution coming from the boson fields and the right moving fermions:

$$\hat{A}(\tau) \sim [\det'(\partial + R)]^{-\frac{1}{2}} = \left[\prod_k \frac{\tau_2 \sigma_k(z_k)}{\pi z_k} \cdot \frac{\theta'(z_k,\tau)}{\eta(q)}\right]^{-1} \tag{A.18}$$

$$= \prod_k \left[\frac{z_k \pi}{\tau_2} \cdot \frac{\eta(q)}{\theta(z_k,\tau)} \cdot \exp \sum' \Lambda(z_k) \cdot \exp \eta_1 z_k^2\right] .$$

We have defined $\Lambda = \frac{z}{\omega} + \frac{1}{2}\left(\frac{z}{\omega}\right)^2$. We can now combine $\hat{A}(\tau)$ with any of the $ch_a(\tau)$. The regularization factors of both terms will cancel using $-\sum \Lambda_a + \sum' \Lambda = -x^2 e_a/2$. This results in

$$\hat{A}(\tau) \cdot ch_a(\tau) = \prod_k \frac{\pi z_k}{\tau_2} \cdot \frac{\theta_a(z_k,\tau)}{\theta(z_k,\tau)} \tag{A.19}$$

If the left moving fermions have (R,P) boundary conditions θ_a is simply replaced by θ in the above, resulting in the Euler number. Finally we give the relation with Jacobi elliptic functions. They are obtained by solving the equation

$$(y'(u))^2 = (1 - 2\delta y^2(u) + \epsilon y^4(u)) \tag{A.20}$$

Introducing the modulus κ, the simplest of them is given by $\epsilon = \kappa^2$ and $2\delta = \kappa^2 + 1$. It corresponds to the Jacobi $sn(u)$ function which has the following expression in terms of theta functions

$$sn(u) = \frac{1}{\sqrt{\kappa}} \cdot \frac{\theta(z,\tau)}{\theta_2(z,\tau)} \, , \qquad (A.21)$$

with $\sqrt{\kappa} = \theta_1(0,\tau)/\theta_3(0,\tau)$ and $z = u/(\pi\theta_3^2(0,\tau))$. The other basic functions are

$$cn(u) = \sqrt{\frac{\kappa'}{\kappa}} \cdot \frac{\theta_1(z,\tau)}{\theta_2(z,\tau)} \qquad (A.22)$$

$$dn(u) = \sqrt{\kappa'} \cdot \frac{\theta_3(z,\tau)}{\theta_2(z,\tau)} \, . \qquad (A.23)$$

The following relations are satisfied

$$cn^2 + sn^2 = 1 \qquad (A.24)$$

$$dn^2 + \kappa^2 sn^2 = 1 \qquad (A.25)$$

$$\kappa^2 + \kappa'^2 = 1 \, , \qquad (A.26)$$

with $\sqrt{\kappa'} = \theta_2(0,\tau)/\theta_3(0,\tau)$. It is easy to see that $(2\delta, \epsilon) = (\kappa^2 - 2, \kappa'^2)$ corresponds to sn/cn and $(2\delta, \epsilon) = (1 - 2\kappa^2, -\kappa^2\kappa'^2)$ to sn/dn. It is important to notice that the modular properties of all these functions are governed by $\lambda(\tau) = \kappa^2$ as can be seen from their defining equation. It is well known that $\lambda(\tau)$ is a modular function invariant under the subgroup of the modular group generated by $\tau \to \tau + 2$ and $\tau \to \tau/(1 - 2\tau)$ or equivalently, if a, b, c, d are integers such that $ad - bc = 1$, that $\lambda(\frac{a\tau + b}{c\tau + d}) = \lambda(\tau)$ if and only if b and c are even, so that a and d are odd.

References

[1] M. Atiyah and I. Singer. Index of elliptic operators, I. *Ann. of Math.* **87**, 484 (1968).

[2] M. Atiyah and I. Singer. Index of elliptic operators, III. *Ann. of Math.* **87**, 546 (1968).

[3] Schwarz Green and Witten. *Superstring Theory I, II.* Cambridge University Press, 1987.

[4] O. Alvarez, T. Killingback, M. Mangano, and P. Windey. The index of the Ramond operator. Unpublished, Santa Barbara, August 1985.

[5] E. Witten. Elliptic genera and quantum field theory. *Comm. Math. Phys.* **109**, 525 (1987).

[6] O. Alvarez, T. Killingback, M. Mangano, and P. Windey. The Dirac-Ramond operator in string theory and loop space index theorems. In *Proceedings of the Irvine Conference on Nonlinear Problems in Field Theory*, Nuclear Physics B (Proc. Suppl.) 1A, 1987.

[7] O. Alvarez, T. Killingback, M. Mangano, and P. Windey. String theory and loop space index theorems. *Comm. Math. Phys.* **111**, 1 (1987).

[8] E. Witten. The Dirac operator in loop space. In P.S. Landweber, editor, *Elliptic Curves and Modular Forms in Algebraic Topology*, Springer-Verlag, 1988. Lecture Notes in Mathematics.

[9] N. Warner K. Pilch, A. Schellekens. Path integral calculation of string anomalies. *Nucl. Phys.* **B287**, 362 (1987).

[10] A. Schellekens and N. Warner. Anomalies and modular invariance in string theory. *Phys. Lett.* **177B**, 317 (1986).

[11] A. Schellekens and N. Warner. Anomaly cancellation and self-dual lattices. *Phys. Lett.* **181B**, 339 (1986).

[12] S. Ochanine. Sur les genres multiplicatifs définis par des intégrales elliptiques. *Topology* , (1987).

[13] P.S. Landweber and R. Stong. Circle actions on spin manifolds and characteristic numbers. *Topology* , (1988). to appear.

[14] P.S. Landweber, editor. *Elliptic Curves and Modular Forms in Algebraic Topology.* Springer-Verlag, 1988.

[15] E. Witten. Fermion quantum numbers in Kaluza-Klein theory. In N. Khuri *et al.* , editor, *Proceedings of the 1983 Shelter Island Conference on Quantum Field Theory and the Foundations of Physics*, MIT Press, 1986.

[16] L. Alvarez-Gaumé. Supersymmetry and the Atiyah-Singer index theorem. *Comm. Math. Phys.* **90**, 161 (1983).

[17] D. Friedan and P. Windey. Supersymmetric derivation of the Atiyah-Singer index theorem and the chiral anomaly. *Nucl. Phys.* **B235**, 395 (1984).

[18] E. Witten. unpublished, 1983.

[19] M.B. Green and J.H. Schwarz. Anomaly cancellation in supersymmetric $D = 10$ gauge theory require $SO(32)$. *Phys. Lett.* **149B**, 117 (1984).

[20] E. Witten. Constraints on supersymmetry breaking. *Nucl. Phys.* **B202**, 253 (1982).

[21] S. Cecotti and L. Girardello. Functional measure, topology and dynamical supersymmetry breaking. *Phys. Lett.* **110B**, 39 (1982).

[22] M. Goodman. Proof of the character-valued index theorems. *Comm. Math. Phys.* **107**, 391 (1986).

[23] M.F. Atiyah and F. Hirzebruch. *Essays in Topology and Related Subjects*, page 18. Springer-Verlag, 1970.

[24] E.T. Whittaker and G.N. Watson. *A Course of Modern Analysis.* Cambridge University Press, 1980.

[25] A. Schellekens and N. Warner. Anomalies, characters and strings. Submitted to *Nucl. Phys. B*.

[26] P.S. Landweber. Elliptic cohomology and modular forms. In P.S. Landweber, editor, *Elliptic Curves and Modular Forms in Algebraic Topology*, 1988. to be published.

[27] D. Zagier. A note on the Landweber-Stong elliptic genus. In P.S. Landweber, editor, *Elliptic Curves and Modular Forms in Algebraic Topology*, Springer-Verlag, 1988. Lecture Notes in Mathematics.

[28] D.V. and G.V. Chudnovsky. Elliptic modular forms and elliptic genera. Columbia University preprint (1985), to appear in *Topology*.

[29] M. Atiyah and R. Bott. The Lefschetz fixed-point theorem for elliptic complexes: I. *Ann. of Math.* **86**, 374 (1967).

[30] M. Atiyah and R. Bott. The Lefschetz fixed-point theorem for elliptic complexes: II. *Ann. of Math.* **88**, 451 (1968).

[31] S. Ochanine. Genres elliptiques equivariants. In P.S. Landweber, editor, *Elliptic Curves and Modular Forms in Algebraic Topology*, Springer-Verlag, 1988. Lecture Notes in Mathematics.

[32] C. Taubes. Harvard preprint 1987. to be published.

[33] K. Chandrasekharan. *Elliptic Functions*. Volume 281 of *Grundlehren der matematischen Wisenschaften*, Springer-Verlag, 1985.

[34] D. Mumford. *Tata Lectures on Theta I. Progress in Mathematics*, Birkhäuser, 1983.

[35] L. Alvarez-Gaumé, G. Moore, and C. Vafa. Theta functions, modular invariance and strings. *Comm. Math. Phys.* **106**, 40 (1986).

DISCUSSION I

– *Liu:*

Why did you choose the particular regularization you have used?

– *Windey:*

The index surely does not depend on any particular regularization scheme one chooses. However it is important to preserve the pairing. What I wanted to do is to use the full power of supersymmetry.

– *Liu:*

You mentioned earlier that from the index for the Dirac operator one can obtain the index for more general elliptic operators. Could you please explain?

– *Windey:*

The Dirac operator acts on a Hilbert space given by the section of the spin bundle $\Gamma(S_+)$. By choosing a particular vector bundle V and having the Dirac operator act on $\Gamma(S_+ \otimes V)$ you can get more general results, in particular you will see in the written notes how to get the Euler characteristic and the Hirzebruch signature. The simple reason for all this is that you change the Hilbert space from fermions to differential forms.

– *Kiritsis:*

I would like to comment on something you implied but you did not said explicitly. In the case where the spectrum is not discrete but continuous you do get an index that does depend on r.

– *Windey:*

This is true.

– *Miele:*

Is there any theorem which ensures us that the dimensions of the kernel and co–kernel of the operators in question better be finite?

– *Windey:*

Yes. One assumes the operators one is working with are Fredholm operators, which are precisely defined by the condition of having finite dimensional kernel and co–kernel.

– Schuler:

Could you explain the role of isolated fixed points in the computation of the character valued index?

– Windey:

I did not make any assumption on the fact that the fixed points were isolated or not. I just showed you that if the manifold admits a group action, the character index is concentrated on the fixed point manifold Mg, whatever that manifold is. In the particular case of a sphere the S' action has two obvious isolated fixed points.

– Quackenbush:

I know there is a relation between the abelian anomaly in $(2n + 2)$ dimensions and the non-abelian anomaly in 2n dimensions. Is there a way to understand this in terms of the index theorem?

– Windey:

Yes, but you have to use the family index theorem which I didn't talk about. I will try to give you a brief description of how it works. First you must remember that gauge anomalies appear in the variation of the log $det \not{D} = \Gamma$ under a gauge transformation. Take a one parameter family of such transformation, the change in the determinant will be given by a phase depending on this parameter *theta* and the anomaly measures the local winding number of this function. What you then show is that there is a Dirac operator in $2n + 2$ dimension whose index (the one I computed for you) equals this winding number $\int_0^{2n} d\theta \, \frac{d\Gamma(A^\theta)}{d\theta}$.

Another way to say it is that what appeared to be a function on the space of gauge fields is in fact a section of the determinant line bundle and the anomaly appears if this bundle is non trivial. The index theorem in $2n + 2$ dimension provides a representative for this first Chern class. The deep meaning of all this is made clear (or obscure depending on your taste) by the family index theorem. What family means is simply the following. Instead of considering the Dirac operator in a fixed background gauge field, you consider a family of Dirac operators parametrized by all possible non gauge equivalent gauge fields. At each point in this parameter space you have a kernel and a cokernel. The patching (a bit mysterious since the dimension of the kernel can jump) of this vector space over the parameter space is the index bundle. The family index is then used to compute the Chern character of this bundle. On all this you may look at the lectures of Alvarez–Gaumè at this school and references therein.

– Szczerba:

Where did you use the fact that the manifold is a compact?

– *Windey:*

By making assumptions about the spectrum.

– *Kiritsis:*

I know there is a generalization of index theorem in the non–compact case due to Callias.

DISCUSSION II

– *Quanckenbush:*

On the string world sheet you have conformal invariance which requires the introduction of ghosts to preserve the symmetry on the quantum level. Do these ghosts contribute to the computation of the index?

– *Windey:*

The way we have done the computation is to restrict our attention to the sigma–model. One way you can think about it is to consider the theory in the transverse gauge where the ghosts have been taken care of, two degrees of freedom have disappeared, and you are just left with a sigma–model. The other point of view is that you are simply examining some property of the sigma-model and you want to see if you can get some classification of them from the index theorems. Remember that we don't integrate over the moduli space.

For example, in the simple computation I presented, there were terms missing from the Lagrangian which can give you a vanishing central charge, which can make the left and right central charges equal, etc. I have just concentrated on the simplest model. It is important to remember that we are really doing a sigma-model computation to get an index of the Dirac operator in loop space.

– *Liu:*

Does the index of the Dirac–Ramond operator represent an obstruction to conformal invariance?

– *Windey:*

No. As I said we don't need conformal invariance. The result that you get is modular invariant, even though you started with a theory which was not conformally invariant. In a way you have recovered an invariance which was not there in the beginning. This is because the only contribution to this quantum mechanical calculation comes from the quadratic fluctuation of the fields, you have removed the higher order terms, which do not contribute. The computation is exact and the result is a modular function.

– *Liu:*

Could you explain how the vacuum expectation value of the current contributes to the anomaly?

– *Windey:*

Suppose you want to compute the divergence of the axial current. If you take the equations of motion into account, it should be equal to zero. However, you are computing the vacuum fluctuations and their contribution to the trace. If you have no vacuum fluctuations, you would have no anomaly because the result would be the classical result, and classically you know this current is conserved.

Another way to see that it is something quantum mechanical is to use the Fujikawa method where you make a chiral transformation of the fields and the anomaly appears as a noninvariance of the measure which is related to the quantum fluctuations.

– *Diaz:*

Witten has looked at a problem similar to the one you considered today. He uses N=1 supersymmetry and he does not get as a condition the vanishing of the first Pontjagin class. Could you please explain why?

– *Windey:*

You will find in the written notes a complete list of references. Witten has looked at the same problem and his analysis covers all the cases we covered, that means the $N = \frac{1}{2}$ as well as N=1 case. You will also find the analysis of the N=1 case in the written version of the lectures. But let me try explain.

For the $N = \frac{1}{2}$ system the field content of the left and right moving sectors are different. You have a chiral asymmetry at the level of the Lagrangian, so the left and right central charges are not equal. This system is if you want, the pure Dirac operator on loop space. For its index to be well defined you need $Tr R^2 = 0$.

You can couple this minimal operator to left handed fermions. The simplest way to do it is to have N=1 supersymmetry. Depending on the boundary conditions you choose, you get the signature or the Euler number on loop space. The system is now left-right symmetric and the central charges are equal. This corresponds in fact the type II string. We have two $N = \frac{1}{2}$ supersymmetries which get promoted to an N=1 supersymmetry. The disappearance of the left-right asymmetry removes the condition $Tr R^2 = 0$. Other left sector would require $Tr R^2 - Tr F^2 = 0$.

– *Diaz:*

If the sigma-model propagates on an orbifold, would the different indices obtained provide us with a classification of the orbifolds?

– *Windey:*

In principle you should be able to redo the calculation on an orbifold. I have not done the computation.

– *Diaz:*

Can you complete this sigma-model to Chern-Simons forms and compute the indices?

– *Windey:*

Yes, that is the comment I made at the end of my lecture. If you take two different sigma–models on two different manifolds, M and M', there is no reason why the two sigma-models cannot describe the same quantum field theory. As a simple example, you could take the Wess–Zumino–Witten model coupled to a Chern–Simons term, which is known to be equivalent to free fermions. The fact that you have two different manifolds does not mean that you have different quantum field theories. This observation is originally due to Witten.

The index provides a new tool which you can use to distinguish between two inequivalent quantum field theories. If the indices are different, the two quantum field theories are different.

– *Miele:*

Earlier, you computed energy momentum tensor and you found that energy at the vacuum depends on the geometry of the manifold. What is the physical meaning of this and is it related to the curvature of the manifold or is it a quantum effect?

– *Windey:*

I guess you are talking about the shift ε . The universal shift, ε , which you get when you go from the plane to the cylinder has to do with the fact that the energy momentum tensor is not a conformal field.

Suppose you take a statistical model defined on a finite width strip. As you go to the critical point, you want to know if you get something like scaling when you change the size of the sample. This is known as "finite-size scaling". The shift in the vacuum energy is proportional to the central charge. The shift will be different for models with different values of C. This was analysed in a beautiful paper by J. Cardy.

– *Lewellen:*

Are you saying that if string vacua have different indices that they are not vacua of the same theory; that they are not dynamically connected?

– *Windey:*

If you restrict your attention to the level of the sigma-model, that statement is correct. If you have two quantum field theories with different elliptic indices they cannot be equivalent theories.

- Lewellen:

Certainly at the two-dimensional level, but at the string level?

- Windey:

As long as you look at first quantized string theory they are different if they have different indices.

-- Lewellen:

Presumably, if you have a second quantized theory, they might be connected.

- Schuler:

Could you comment more on the connection with statistical mechanics?

- Windey:

Consider statistical mechanics on a torus. You can classify the models by the different values of c. For $c < 1$ you have a discrete series starting with $c = 1/2$, this is the Ising model, and so on. For this discrete series you have a discrete set of highest weight vectors.

For $c < 1$ you have no continual symmetry, no current. For the discrete series you have no system which seems to have chiral properties and no interesting result can be derived from the index. However if $c \neq \bar{c}$ more interesting things could happen. From the point of view of statistical mechanics I don't see why these models should not be considered.

WEAK HAMILTONIAN AMPLITUDES ON THE LATTICE

Guido Martinelli

Theoretical Physics Division
CERN
1211 Geneva 23, Switzerland

1.   INTRODUCTION

The understanding of non-leptonic kaon decays has proved to be a rather difficult problem in particle physics. Although the overall theoretical picture seems rather sound, the explanation of the $\Delta I = \frac{1}{2}$ rule and an accurate quantitative evaluation of CP violation in kaon systems are still lacking. In particular, the prediction (postdiction) of the CP violation parameters (ε and ε') would allow us to probe still-hidden low-energy sectors of the theory, the top quark for example, and the physics beyond the Standard Model.

Monte Carlo simulation of lattice QCD offers the possibility of a first-principle, non-perturbative computation of the matrix elements of the weak Hamiltonian in kaon decays.

In these lectures I will report the most recent developments in this field. The plan of the lectures is the following. In Section 2 I will recall some facts about kaon decays. In Section 3 I will give the basic elements of the short-distance expansion of the effective weak Hamiltonian in the Standard Model. I will describe in Section 4 the general theoretical framework of lattice QCD, with particular emphasis on the chiral properties of the theory. Finally, in Section 5, I will present the technique and give the most recent results of the computation of the matrix elements of the weak Hamiltonian by Monte Carlo simulation.

2.   NON-LEPTONIC DECAYS OF KAONS

2.1   The $\Delta I = \frac{1}{2}$ Rule

Kaons may decay in the following two-pion final states:

$$|\pi^+ \pi^0> = |I = 0, I_Z = 1>$$

$$|\pi^+ \pi^-> = \sqrt{\frac{2}{3}}|I = 0, I_Z = 0> + \sqrt{\frac{1}{3}}|I = 2, I_Z = 0>$$

$$|\pi^0 \pi^0> = \sqrt{\frac{1}{3}}|I = 0, I_Z = 0> - \sqrt{\frac{2}{3}}|I = 2, I_Z = 0>$$

$$(2.1)$$

The Superworld II
Edited by A. Zichichi
Plenum Press, New York, 1990

Experimentally the amplitude $K \to 2\pi$ with the two pions in an $I = 2$ state, $a_{\Delta I = 3/2}$, is much smaller than the amplitude $K \to 2\pi$ with the pions in an $I = 0$ state, $a_{\Delta I = \frac{1}{2}}$:

$$\frac{a(K^+ \to \pi^+ \pi^0)}{a(K_S \to \pi\pi)_{\Delta I = \frac{1}{2}}} \sim \frac{1}{20} \tag{2.2}$$

indicating that the weak Hamiltonian is dominated by its $\Delta I = \frac{1}{2}$ part.

2.2 CP Violation in the K°-$\overline{K}^\circ$ System

Let us write the neutral kaon wave function as:

$$|K(t)> = a(t)|K^0> + b(t)|\overline{K}^0> \tag{2.3}$$

where $|K(t)>$ obeys the equation

$$i\frac{d}{dt}|K(t)> = H|K(t)> = (M - i\Gamma)|K(t)> \tag{2.4}$$

H is the weak Hamiltonian and M and Γ are Hermitian matrices. CP conservation would imply that M_{12} and Γ_{12} are real.

Let us define the K° and $\overline{K}^\circ$ states according to the following convention:

$$CP|K^0> = |\overline{K}^0>$$
$$CP|\overline{K}^0> = |K^0> \tag{2.5}$$

The eigenstates of the weak Hamiltonian H will have the form

$$|K_S> = \frac{1}{2[1 + |\sigma|^2]^{\frac{1}{2}}}\left[(1 + \sigma)|K^0> + (1 - \sigma)|\overline{K}^0>\right]$$

$$|K_L> = \frac{1}{2[1 + |\sigma|^2]^{\frac{1}{2}}}\left[(1 + \sigma)|K^0> - (1 - \sigma)|\overline{K}^0>\right] \tag{2.6}$$

with eigenvalues $\lambda_{S,L} = im_{S,L} + \gamma_{S,L}/2$ ($|K_{S,L}(t)> = e^{-\lambda_{S,L} t}|K_{S,L}(0)>$).

In the absence of CP violation σ is equal to zero and K_S and K_L are eigenstates of CP [$CP|K_{S,L}> = \pm |K_{S,L}>$]. The experimental observation in 1964 of the decay of the $K_L \to \pi^+\pi^-$ [1] implies that the weak effective Hamiltonian has a CP-violating part, a two-pion system being even under CP. This means that σ is different from zero, i.e., that M_{12} and Γ_{12} have an imaginary part:

$$\sigma = \frac{-Im\, M_{12} + iIm\, \Gamma_{12}}{i(m_S - m_L) + (\gamma_S - \gamma_L)/2} \tag{2.7}$$

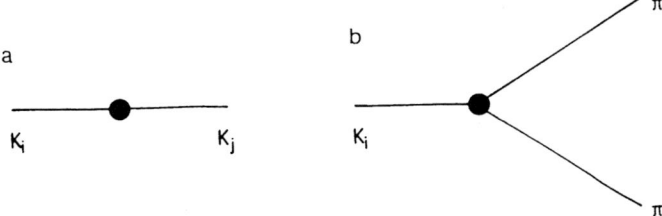

Fig. 1.   The two diagrams which contribute to CP violation are shown:
CP violation in the K°-$\bar{K}^\circ$ mass matrix (Fig. 1a) and in the
$K \to 2\pi$ decay (Fig. 1b).

The general form of the Hamiltonian matrix elements for kaon decays is
the following:

$$H_{ij} = m_K \delta_{ij} + <K_{i,}^0|H_W^{\Delta S=2}|K_j^0> +$$
$$\sum_n \frac{<K_i^0|H_W^{\Delta S=1}|n><n|H_W^{\Delta S=1}|K_j^0>}{m_K - m_n + i\eta} \qquad (2.8)$$
$$i = 1, 2 \to K^0, \bar{K}^0$$

From Eq. (2.8) we note that there are two possible sources of CP viola-
tion: one in the $\Delta S = 2$ K°-$\bar{K}^\circ$ mass matrix and one coming from direct
$\Delta S = 1$ $K \to 2\pi$ (3π) transitions, as illustrated in Fig. 1. $<K_i^\circ|H_w^{\Delta S=2}|K_j^\circ>$
and virtual intermediate states in the last term of Eq. (2.8) contribute
to M_{12}; the imaginary part of Γ_{12} is given by physical intermediate
states:

$$\Gamma_{ij} = 2\pi \sum_n \rho_n < K_i^0|H_W^{\Delta S=1}|n > <n|H_W^{\Delta S=1}|K_j^0 > \qquad (2.9)$$

where ρ_n is the density of states n.

In the next section we discuss the derivation of the Hamiltonian H
of Eq. (2.8) in the Standard Model.

3.   THE EFFECTIVE WEAK HAMILTONIAN[2]

In the Standard Model, the basic Lagrangian responsible for non-
leptonic decays is given by:

$$\mathcal{L} = \frac{g}{2\sqrt{2}} \left[J_\mu^+ W^\mu + h.c. \right] \qquad (3.1)$$

where W^μ is the field corresponding to the W boson and J_μ^+ is the quark
current:

$$J_\mu^+ = (\bar{u}, \bar{c}, \bar{t}) \, \gamma_\mu^L \, \cup_{C.K.M.} \begin{pmatrix} d \\ s \\ b \end{pmatrix} \qquad (3.2)$$

$\gamma_\mu^L = \gamma_\mu(1-\gamma_5)/2$ and U_{CKM} is the Cabibbo-Kobayashi-Maskawa mixing matrix

$$\begin{pmatrix} c_1 & s_1 c_3 & s_1 s_3 \\ -s_1 c_2 & c_1 c_2 c_3 + s_2 s_3 e^{i\delta} & c_1 c_2 c_3 - s_2 s_3 e^{i\delta} \\ -s_1 s_2 & c_1 s_2 c_3 - c_2 s_3 e^{i\delta} & c_1 s_2 s_3 + c_2 c_3 e^{i\delta} \end{pmatrix} \tag{3.3}$$

With only four flavours it would be possible to rotate the quark fields in order to remove any CP-violating phase. With six flavours, any CP-violation effect must be proportional to the only rephasing invariant that can be constructed out of the quark mass matrix[3]*:

$$s_1^2 s_2 s_3 \, sin\delta \tag{3.4}$$

The fact that only in the three-family case is it not possible to rotate away the CP-violating phase implies that at least one quark for each family must participate in any CP-violating process.

From the Lagrangian in Eq. (3.1), it is easy to derive the flavour-changing, $\Delta S = 1$ amplitudes between hadronic states h_2 and h_1:

$$< h_2|H|h_1 > = \frac{g^2}{2} \int d^4 x D(x, M_W^2) g^{\mu\nu} < h_2|T\{J_\mu^+(x) J_\nu(0)\}|h_1 > \tag{3.5}$$

The typical distance involved in the decay of a kaon is much larger than the inverse W mass, M_W. This means that we can replace the renormalized current × current expression in Eq. (3.5) by its Wilson short-distance expansion[4], as illustrated in Fig. 2. Thus we end up with an effective

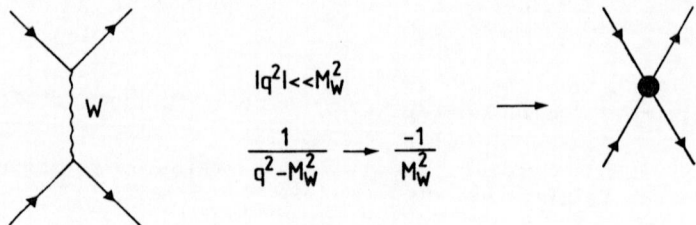

Fig. 2. Feynman diagram for a non-leptonic decay. Since $|q^2| \ll M_W^2$, we can shrink the W propagator to a point, thus obtaining the Fermi Hamiltonian.

---

* The combination in Eq. (3.4) is experimentally bounded to be less than $4 \cdot 10^{-4}$.

four-fermion Hamiltonian* :

$$H_{eff} = \frac{G_F}{\sqrt{2}} \sum_n C_n(\mu, M_W) O^n(\mu) \tag{3.6}$$

The Wilson coefficients $C_n(\mu, M_W)$ are specified once the renormalization conditions of the composite, local operators O^n are given. The latter involve the normalization scale μ, and the μ-independence of the right-hand side of Eq. (3.6) gives the renormalization group equations for the evolution of the coefficients C_n. In an asymptotically free theory like QCD, it is possible to calculate $C_n(\mu, M_W)$ in perturbation theory (see Fig. 3), provided $\mu \gg \Lambda_{QCD}$. The evaluation of the hadronic matrix elements of the operators O^n, $\langle h_2 | O^n | h_1 \rangle$, is on the other hand a completely non-perturbative problem.

The Hamiltonian in Eq. (3.6) can be decomposed into two parts which transform in a different way under the chiral group $SU(3)_L \times SU(3)_R$:

$$H_{eff} = H^{(8,1)} + H^{(27,1)} \tag{3.7}$$

The octet part of the Hamiltonian, $H^{(8,1)}$ contributes only to $\Delta I = \frac{1}{2}$ transitions, while $H^{(27,1)}$ contributes both to the $\Delta I = \frac{1}{2}$ and $\Delta I = 3/2$ decay amplitudes. In perturbation theory, the renormalization of the operator induced by the interaction between quarks and gluons (see Fig. 3) enhances the coefficient of the octet part with respect to the 27-plet part[5]:

$$C_8/C_{27} \sim 2 \div 3 \tag{3.8}$$

but it is not sufficient to explain the experimentally observed ratio [cf. Eq. (2.2)].

In the past, Shifman et al.[6] proposed that the enhancement was due to the presence of new operators which come into play only at scales $\mu < m_c$, the so-called "penguin" operators (see Fig. 4). According to Ref. 6), the coefficients of these operators are small $[O(\alpha_s \ln m_c/\mu)]$, but their matrix elements should be large, due to their left-right structure. The analysis of Ref. 6) is unfortunately not justified, since by general arguments the chiral behaviour of the matrix elements of the "penguin" operators is the same as that of the usual left-left operators (they vanish in the chiral limit). Thus there is no a priori reason why matrix elements of the "penguin" operators should be so large to account for the $\Delta I = \frac{1}{2}$ rule.

This means that a real explanation of the $\Delta I = \frac{1}{2}$ rule can be found only through an explicit, non-perturbative computation of the hadronic matrix elements of the relevant operators in H_{eff}.

The "penguin" diagram of Fig. 4 is also responsible for the $\Delta I = \frac{1}{2}$ CP-violating part of the weak Hamiltonian, because a virtual up, charm and top quark contribute as intermediate states. "Penguin" operators contribute to the last term on the right-hand side in Eq. (1.7). The $\Delta S = 2$ part of the effective weak Hamiltonian responsible for CP violation [second term on the right-hand side in Eq. (2.8)] can be derived in the Standard Model from the box diagram in Fig. 5. This diagram gives a contribution both to the real and the imaginary (CP-violating) parts of M. We think that the imaginary part is dominated by short-distance effects[7].

---

* Strictly speaking, upon renormalization, two fermion operators can also appear in H_{eff}.

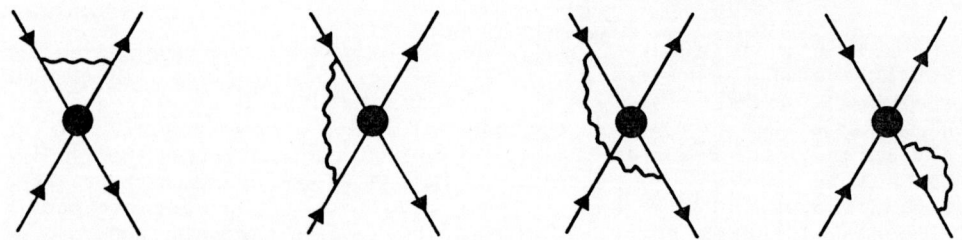

Fig. 3.   Diagrams contributing to the renormalization of a four-fermion operator.   The wavy lines are gluons.

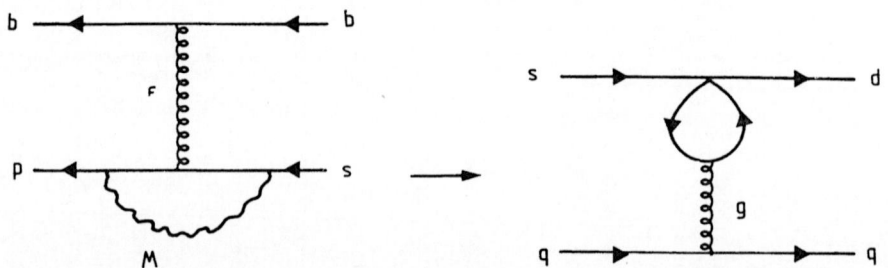

Fig. 4.   "Penguin diagram": the $1/q^2$ pole of the gluon propagator is killed by the term $\sim g^{\mu\nu} q^2$ which comes from the quark loop. The net result is that this diagram generates a local, left-right, four-fermion operator.

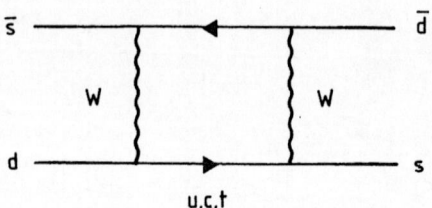

Fig. 5.   "Box" diagram responsible for K°-$\bar{K}^\circ$ (B°-$\bar{B}^\circ$) mixing.

138

In this case, one can approximate the "box" with the insertion of a local, four-fermion operator

$$O^{\Delta S=2} = \bar{s}\gamma^\mu(1-\gamma_5)d\,\bar{s}\gamma_\mu(1-\gamma_5)d \tag{3.9}$$

One finds[8]:

$$Im M_{12} = \frac{-G_F^2}{6\pi^2}\,f_K^2 m_K M_W^2 s_1^2 s_2 s_3 s_\delta S\left(\frac{m_c^2}{M_W^2},\frac{m_t^2}{M_W^2}\right)B \tag{3.10}$$

$Im M_{12}$ contains information on the top quark (and eventually on new physics) through the dependence of S on m_t^2/M_W^2. The so-called B parameter in Eq. (2.8) is proportional to the K°-$\bar{K}^\circ$ matrix element of $O^{\Delta S=2}$:

$$B = \frac{3}{8\,f_K^2 m_K^2} < K^0|\bar{s}\gamma_\mu(1-\gamma_5)d\bar{s}\gamma(1-\gamma_5)d|\bar{K}^0> \tag{3.11}$$

where the factor $8f_K^2 m_K^2/3$ is the value of the matrix element in the vacuum insertion approximation. A precise knowledge of B and of the mixing angles would allow us to put limits on the mass of the top quark[*].

So we see that, as was the case for the $\Delta I = \frac{1}{2}$ rule, a quantitative understanding of the physics of K°-$\bar{K}^\circ$ (B°-$\bar{B}^\circ$) mixing and of CP violations requires a precise non-perturbative evaluation of the matrix elements of local operators between hadronic states.

In spite of some present practical difficulties to be discussed below, lattice QCD is the only approximation which allows the computation of weak amplitudes in a well-justified theoretical framework, based on asymptotic freedom and renormalization group . Moreover, unlike other approaches, such as the bag model, the accuracy of the results can be systematically improved in time.

4.   FERMIONS ON THE LATTICE

From the previous discussion we have learned that the effective Hamiltonian has the form

$$H_{eff} = \sum_n C_n(\mu, M_W)\hat{O}^n(\mu) \tag{4.1}$$

We compute the Wilson coefficients in QCD perturbation theory and the hadronic matrix elements of the operators by Monte Carlo simulation of QCD on a lattice. However, the construction of renormalized, finite operators on the lattice suffers from two different but related problems: severe ultra-violet divergences in the limit of vanishing lattice spacing and bad chiral behaviour. Both problems arise because of the hard chiral symmetry breaking entailed by the Wilson term[10][11] in the lattice fermionic action, a term which is necessary to avoid quark species doubling and to control flavour symmetry. The way to circumvent this problem has

---

[*]Similar diagrams occur in B°-$\bar{B}^\circ$ mixing; for a recent review see Ref. 9).

been shown in Refs. 11) and 12) *. I will report the main results of that theoretical analysis in this section.

4.1 Basic Definitions and Ward Identities

A quite general form of the fermion action on the lattice is

$$S_\psi = \sum_{x,f} \left\{ \frac{-1}{2a} \sum_\mu \left[\bar\psi_f(x)(r - \gamma_\mu)U_\mu(x)\psi_f(x + \hat\mu) + \right. \right.$$

$$\bar\psi_f(x + \hat\mu)(r + \gamma_\mu)U_\mu^+(x)\psi_f(x) \Big] + \bar\psi_f(x)\left[M_o + \frac{4r}{a} \right]\psi_f(x) \Bigg\} =$$

$$\sum_f \bar\psi_f \Delta_f(U)\psi_f \tag{4.2}$$

ψ ($\bar\psi$) and U_μ are the quark and gluon fields respectively, and f denotes the flavour. The terms proportional to the Dirac matrices γ^μ are the naive lattice transcription of the covariant derivative $\bar\psi\gamma^\mu D_\mu\psi$ and an "irrelevant" operator, proportional to the parameter r, has been introduced to avoid the fermion species doubling (Wilson term[10]). The Wilson term acts as a mass term and breaks chiral symmetry explicitly, even in the limit of the vanishing quark mass M_o. As a consequence, the Ward identities for the vector and axial vector currents contain anomalous pieces, and tree-level chiral properties of composite operators are spoiled by the interaction between quarks and gluons. The Ward identities for the vector and axial vector currents can be easily derived by a chiral rotation of the fermion fields in the action of Eq. (4.2):

$$\psi \to \psi + i(\alpha_V^a + \gamma_5\alpha_A^a)\frac{\lambda^a\psi}{2}$$
$$\bar\psi \to \bar\psi - i\bar\psi(\alpha_V^a - \gamma_5\alpha_A^a)\frac{\lambda^a}{2} \tag{4.3}$$

One obtains

$$< \alpha|\nabla_\mu V_\mu^a(x)|\beta > = < \alpha| - \bar\psi(x)|\left[\frac{\lambda^a}{2}, M_0 \right]\psi(x)|\beta > +$$

$$< \alpha|X_V^a(x)|\beta >$$

$$< \alpha|\nabla_\mu A_\mu^a(x)|\beta > = < \alpha|\bar\psi(x)\left\{ \frac{\lambda^a}{2}, M_0 \right\}\psi(x)|\beta > +$$

$$< \alpha|X_A^a(x)|\beta > \tag{4.4}$$

where

$$\nabla_\mu f(x) = f(x) - f(x - \hat\mu)$$

---

* It is not clear whether the alternative lattice formulations of QCD (e.g., staggered fermions[13]) can be used in this context because of the complicated mixing between flavour and spin quantum numbers.

Fig. 6. The operator X_A^a is of order a (Fig. 6a). One-loop corrections, because of the bad ultra-violet behaviour of X_A^a $(\sim p^2)$ give a linearly divergent contribution as a → 0 (Fig. 6b).

$$V_\mu^a(x) \;=\; \frac{1}{2}\left[\bar\psi(x)U_n(x)\gamma_\mu\frac{\lambda^a}{2}\psi(x+\hat\mu)\;+\;h.c.\right]$$

$$A_\mu^a(x) \;=\; \frac{1}{2}\left[\bar\psi(x)U_n(x)\gamma_\mu\gamma_5\frac{\lambda^a}{2}\psi(x+\hat\mu)\;+\;h.c.\right]$$

(4.5)

The Ward identities in Eq. (4.4) look very similar to the corresponding identities in the continuum limit but for the last term on the right-hand side. In the free theory, this term is of order O(a), a being the lattice spacing, and disappears as a → 0. Unfortunately, in the real case, the interaction promotes these anomalous pieces, so that not only do they not vanish as a → 0, but become linearly divergent ($\sim 1/a$), as illustrated in Fig. 6. For the vector current this is not a problem, since one can show that X_V^a is itself the divergence of a current:

$$X_V^a \;=\; \nabla_\mu^a J_\mu^a \tag{4.6}$$

X_V^a can be moved on the left-hand side of the Ward identity. The current $V_\mu^a = V_\mu^a - J_\mu^a$ now obeys the same Ward identity of the continuum theory. This happens because the Wilson term respects vector symmetries (e.g., baryon number), which is not the case for the axial current. In the latter case, however, one can show that the matrix elements of X_A^a between on-shell states α and β can be written as

$$<\alpha|X_A^a|\beta> \;=\; -\;<\alpha|\bar\psi\left\{\frac{\lambda^a}{2}, M_0 - \overline{M}(M_0)\right\}\psi|\beta> \;+$$

$$(Z_A\,-1)<\alpha|\nabla_\mu A_\mu^a|\beta> +O(a) \tag{4.7}$$

The relation (4.7) is true at all orders in the coupling constant and is the more general one which is compatible with the symmetries of the action. O(a) indicates matrix elements of operators of dimension larger than four, which vanish as a → 0. One can also show that the coefficient $\overline{M}(M_0)$ is linearly divergent in $1/a$ and that $Z_A = Z_A(g_0)$ is a function only of the bare lattice coupling constant g_0 (and r). Let us rewrite the last equation (4.4) as follows

$$Z_a<\alpha|\nabla_\mu A_\mu^a|\beta> \;=\; <\alpha|\left\{\frac{\lambda^a}{2}, M_0 - \overline{M}(M_0)\right\}\psi|\beta> \;+$$

$$<\alpha|\overline{X}_A^a|\beta> \tag{4.8}$$

where

$$\overline{X}_A^a = X_A^a + \bar{\psi}\left\{\frac{\lambda^a}{2}, \overline{M}\right\}\psi + (Z_A - 1)\,\nabla_\mu A_A^a$$

has vanishing matrix elements between on-shell states in the continuum limit $[<\alpha|\overline{X}_A^a|\beta> \sim O(a)]$. From Eq. (4.8) we see that we have recovered the usual continuum Ward identity, provided we identify the good axial current A_μ^a and the bare quark mass m with:

$$\hat{A}_A^a = Z_A A_A^a \qquad m = M_0 - \overline{M}(M_0) \tag{4.9}$$

We also note that Eq. (4.8) is not sufficient alone to fix Z_A and m separately, but only the ratio Z_A/m. We need at least another Ward identity to disentangle the two constants. The simplest Ward identity we can think of involves two- and three-point correlation functions of vector and axial vector currents:

$$Z_A^{-1} < \nabla_\mu \hat{A}_\mu^a(x)\,\hat{A}_\nu^b(y)\hat{V}_\rho^c(0) > = < \bar{\psi}(x)\left\{\frac{\lambda^a}{2}, m\right\}\gamma_5\psi(x)\hat{A}_\nu^b(y)\hat{V}_\rho^c(0) >$$

$$+\, i\, f^{abd}\, \frac{Z_A}{Z_V}\, \delta(x-y) < \hat{V}_\nu^d(y)\hat{V}_\rho^c(0) > \; +\, i\, f^{acd}\, \frac{Z_V}{Z_A}\, \delta(x) < \hat{A}_\rho^d(y)\hat{A}_\rho^d(0) > \; +$$

$$< \overline{X}_A^a(x)\hat{A}_\nu^b(y)\hat{V}_\rho^c(0) > \; + \cdots \tag{4.10}$$

The dots indicate possible Schwinger terms. Z_V is the renormalization constant analogous to Z_A for the vector current V_ρ, [Eq. (4.5)], $(V_\rho = Z_V V_\rho)$, and m and Z_A have been defined before. The last term on the right-hand side of Eq. (4.10) is non-zero even in the continuum limit. The matrix elements of $\overline{X}_A^a$ vanish between on-shell states. $\overline{X}_A^a$ can, however, still give rise to contact terms [i.e., in Eq. (4.10) terms proportional to $\delta(x-y)$ or $\delta(x)$] in more complicated correlation functions because of extra divergences present when more fields are at the same point. These terms transform the anomalous Ward identity in Eq. (4.10) into the corresponding identity of the continuum:

$$< \nabla_\mu \hat{A}_\mu^a(x)\hat{A}_\nu^b(y)\hat{V}_\rho^c(0) > = < \bar{\psi}(x)\left\{\frac{\lambda^a}{2}, m\right\}\gamma_5\psi(x)\hat{A}_\nu^b(y)\hat{V}_\rho^c(0) >$$

$$+\, i\, f^{abd}\, \delta(x-y) < \hat{V}_\nu^d(y)\hat{V}_\rho^c(0) > \; +\, i\, f^{acd}\, \delta(x) < \hat{A}_\nu^b(y)\hat{A}_\rho^d(0) > \; + \cdots \tag{4.11}$$

as can be proved in perturbation theory[11]. This implies the following relation:

$$\sum_x < \left[\nabla_\mu A_\mu^a(x) - \bar{\psi}(x)\left\{\frac{\lambda^a}{2}, \frac{m}{Z_A}\right\}\gamma_5\psi(x)\right] A_\nu^b(y)V_\rho^c(0) > =$$

$$i\, f^{abd}\, \left(\frac{Z_V}{Z_A^2}\right) < V_\nu^d(y)V_\rho^c(0) > \; +\, i\, f^{acd}\, \frac{1}{Z_V} < A_\nu^b(y)A_\rho^d(0) > \tag{4.12}$$

Equations (4.8) and (4.12) allow us to separate Z_A and m in order to find the axial current and the quark mass of the continuum theory.

The strategy based on continuum Ward identities as a means of identifying the "good" parameters and renormalized operators (m and A_μ^a in this case) is quite general, as will also be clear from the discussion of the renormalization of four-fermion operators given in the next subsection. Current algebra relations of the type discussed above have been verified by numerical simulation on the lattice in Ref. 14.

4.2 Renormalization of Four-Fermion Operators

The problem of the construction of renormalized, four-fermion operators is quite intriguing because of the presence of severe power divergences and of the bad chiral behaviour of the bare operators on the lattice, induced by the Wilson term.

It is simple to start by giving a specific example. Let us consider the (8,1) operator:

$$O_\alpha = (\bar\psi_L \frac{\lambda\alpha}{2} \gamma_\mu \psi_L)(\bar\psi_R \gamma^\mu \psi_R) \qquad (4.13)$$

where $\psi_{L,R} = (1 \mp \gamma_5)\psi/2$, and λ_α is one of the Gell-Mann matrices acting in flavour space. Even at zero order in the bare strong coupling constant and in the chiral limit, the operator in Eq. (4.13) can mix an operator of dimension three with a power-divergent coefficient through the diagram in Fig. 7:

$$\delta O_\alpha = \frac{1}{a^3}\left(\bar\psi_L \frac{\lambda^\alpha}{2} \psi_R + \bar\psi_R \frac{\lambda^\alpha}{2}\psi_L\right) \qquad (4.14)$$

δO_α is a (3,3̄) operator. The factor $1/a^3$ is there because of dimensional reasons. Usually, for massless quarks, this mixing is not possible because the original operator cannot flip the helicity; on the lattice the mixing is given by the Wilson term which acts as a mass term when inserted into the loop of the diagram in Fig. 7. This example shows that a bare operator, which one would naively expect to be an (8,1) operator, is really a mixture of (8,1) and (3,3̄) operators. A definite chiral behaviour and the removal of lattice artifact divergences can be obtained at the same time, using the Ward identities along a line similar to the one followed in the previous subsection for the axial current.

First, one subtracts from the naive operator a combination of operators with equal or lower dimensions and all possible naive chiralities, in such a way as to obtain an operator with the correct chiral properties with respect to the "good" axial and vector charges. Generally, at this point not all divergences have been eliminated, but one is left with an overall multiplicative renormalization which is determined by requiring the lattice operators to be normalized equally to the continuum ones.

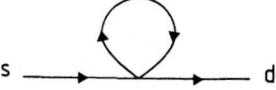

Fig. 7.   The insertion of the Wilson term in the loop of this diagram mixes the four-fermion operator with a two-fermion operator by flipping the helicity of the quarks.

143

In the chiral limit, this procedure is straightforward. Let us consider a particular operator O_α which transforms, under naive axial transformation, according to some representation R:

$$\frac{\delta O_\alpha}{\delta \varepsilon^a} = i\, R^a_{\alpha\beta} O_\beta \qquad (4.15)$$

R^a are the axial generators of the particular representation to which O belongs; α and β are flavour labels appropriate to R. The operator which truly transforms according to R is

$$\tilde{O}_\alpha = O_\alpha + \sum_{n,\sigma} d^n_{\alpha\sigma} O^{(n)}_\sigma \qquad (4.16)$$

with $O^{(n)}$ transforming according to $R^{(n)} \neq R$. The coefficients $d^{(n)}$ are restricted by the condition that they must respect the conserved vector symmetry.

In the chiral limit, the integrated axial current Ward identity for $\tilde{O}$ reads:

$$\sum_x \nabla_\mu \, <\alpha|T(\hat{A}^a_\mu(x)\tilde{O}(0))|\beta> =$$

$$\sum_x <\alpha|T(\bar{X}^a(x)\tilde{O}(0))|\beta> \; + \; <\alpha|i\,\frac{\delta\tilde{O}}{\delta\varepsilon^a}|\beta> \qquad (4.17)$$

where

$$\frac{\delta\tilde{O}}{\delta\varepsilon} = i\left(RO + \sum_n d^n R^n O^{(n)}\right) \qquad (4.18)$$

and we have omitted flavour indices. For the appropriate choice of the coefficients $d^{(n)}$, the contact terms arising in the correlation function of $\bar{X}$ with $\tilde{O}$ will correct the complicated naive chiral variation, Eq. (4.18), in such a way as to reproduce the desired transformation property. The condition for this is:

$$\sum_x <\alpha|T(\overline{X}^a(x)\tilde{O}_\alpha(0)|\beta> \; + \; <\alpha|i\,\frac{\delta\tilde{O}}{\delta\varepsilon^a}|\beta> =$$

$$-R^a_{\alpha\beta}<\alpha|\tilde{O}_\beta|\beta> \qquad (4.19)$$

Equations (4.19) are a set of linear, inhomogeneous equations in $d^{(n)}$ which has a unique solution. The $\tilde{O}$ thus constructed is not yet finite, for $a \to 0$. Equations (4.17) and (4.19) show that the only freedom left is to take linear superpositions of operators transforming in the same way under chiral rotations. We thus define

$$\hat{O} = Z_{LATT}\tilde{O} \qquad (4.20)$$

Z_{LATT} being a matrix which mixes equivalent representations only, and such that:

144

$$< \alpha | Z_{LATT} \tilde{O} | \beta > = \ \text{finite} \qquad (4.21)$$

This condition is consistent with the left-hand side of Eq. (4.17). In fact, if both A_μ and $Z_{LATT}\tilde{O}$ have finite matrix elements, possible divergent terms in their product are localized operators whose contribution to the left-hand side of Eq. (4.17) vanishes upon taking the four-divergence and summing over x.

To complete the programme, we must specify more precisely the normalization condition (4.21). In the chiral limit, this requires the introduction of a subtraction point μ, so that:

$$Z_{LATT} = Z_{LATT}(\mu a, g_0)$$
$$\hat{O}(\mu) = Z_{LATT}\tilde{O} \qquad (4.22)$$

Z_{LATT} $(\mu a, g_0)$ is completely determined, including the finite terms, by the requirement that $O(\mu)$, for some value $\mu \ll a^{-1}$, obeys the same normalization condition as the continuum operator $O(\mu)$.

The lattice definition of the weak Hamiltonian is then:

$$H_{eff} = \frac{G}{\sqrt{2}} \, sin\theta_c \, cos\theta_c \, \sum_n C^n(\mu, M_W) \hat{O}^n(\mu) \qquad (4.23)$$

where the Wilson coefficients are those computed in the continuum.

By construction, the operator $\tilde{O}$ obeys the continuum Ward identity

$$\int dx < \alpha | \nabla_\mu \hat{A}_\mu^a(x) Z_{LATT}\tilde{O} | \beta > = -R^a < \alpha | Z_{LATT}\tilde{O} | \beta > \qquad (4.24)$$

where all the matrix elements in Eq. (4.24) are finite. Equation (4.24) is equivalent to the statement that the operator $\tilde{O}$ obeys soft pion theorems. In fact, upon integration on x, only the pion pole can give a contribution to the correlation function on the left-hand side of Eq. (4.24), as illustrated in Fig. 8. So we obtain

$$f_\pi < \alpha + \pi^a | \tilde{O} | \beta > = R^a < \alpha | \tilde{O} | \beta > = < \alpha | \left[Q_5^a, \tilde{O} \right] | \beta > \qquad (4.25)$$

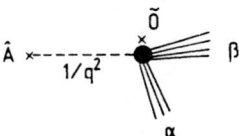

Fig. 8.   Pion pole contribution to the matrix element of the renormalized operator.

i.e., the usual soft pion theorem of the continuum theory. This also implies that we can compute the physical $k \to \pi\pi$ amplitude from the $k \to \pi$ matrix element alone. At this point, we may turn the argument around. We can forget the Ward identity in Eq. (4.24) and determine the coefficients $d^{(n)}$ in the expansion (4.16) simply by requiring that the matrix elements of $\hat{0}$ on the lattice obey the standard soft pion theorems. The latter conditions determine the operator apart from the overall normalization matrix Z_{LATT}.

To complete the analysis, we have to accomplish two further steps:

   i)   identify in the various cases the operators we need to subtract from the bare operators relevant to the weak Hamiltonian;

   ii) give a practical set of rules to determine the coefficients $d^{(n)}$ from the matrix elements on pseudoscalar meson states of the previous operators.

A final remark is necessary. The use of perturbation theory to fix coefficients d^n which are finite (i.e., the coefficients of operators of dimension six) is completely justified. On the other hand, the use of perturbation theory to fix power-divergent coefficients leads in general to wrong results, and it is much safer to use a non-perturbative technique such as the one suggested here.

4.3   Renormalized $\Delta S = 2$, $\Delta I = \frac{1}{2}$ and Penguin Operators

Let us consider the $\Delta S = 2$ operator:

$$O_{LL}^{\Delta s=2} = (\bar{s}_L \gamma^\mu d_L)(\bar{s}_L \gamma^\mu d_L) \qquad (4.26)$$

which is relevant for the K°-$\bar{K}^\circ$ transition. This operator cannot mix with lower-dimension operators, since it is not possible to write down two quark operators with $\Delta S = 2$. As a consequence, the renormalized operator has the form

$$\hat{O}_{LL}^{\Delta S=2} = Z_{LATT}^{\Delta S=2}(\mu a, g_0)\left[O_{LL}^{\Delta S=2} + \sum_i c_i(\bar{s}\Gamma_i d)(\bar{s}\Gamma_i d)\right] \qquad (4.27)$$

where the coefficients $c_i = c_i(g_0)$ are finite, depend only on g_0 and r and can be safely computed in perturbation theory[15]. The operators $(\bar{s}\Gamma_i d)(\bar{s}\Gamma_i d)$ (Γ_i are Dirac matrices) respect the vector symmetry but they belong to a chiral representation different from (27,1).

Let us now consider the operators which are relevant for $\Delta I = \frac{1}{2}$:

$$O^- = \left[(\bar{s}_L\gamma^\mu d_L)(\bar{u}_L\gamma^\mu u_L)(u_L\gamma^\mu d_L)\right] - (u \to c)$$

$$O^+ = \frac{1}{5}\left[(\bar{s}_L\gamma^\mu d_L)(\bar{u}_L\gamma^\mu u_L) + (\bar{s}_L\gamma^\mu d_L) + (\bar{s}_L\gamma^\mu u_L)(\bar{u}_L\gamma^\mu d_L)\right.$$

$$\left. + 2\,(\bar{s}_L\gamma^\mu d_L)(\bar{d}_L\gamma^\mu d_L) + 2(\bar{s}_L\gamma^\mu d_L)(\bar{s}_L\gamma^\mu s_L)\right]$$

$$\left[(\bar{s}_L\gamma^\mu d_L)(\bar{c}_L\gamma^\mu c_L) + (\bar{s}_L\gamma^\mu c_L)(\bar{c}_L\gamma^\mu d_L)\right]$$

(4.28)

Fig. 9.   Example of diagrams by which the operators $O^{\pm}$ defined in Eqs. (4.28) of the text can mix with operators of lower dimensions.

The operators in Eqs. (4.28) can mix not only with operators of dimension six with finite coefficients as before, but also with the following five- and three-dimension operators:

$$\bar{s}\sigma_{\mu\nu}F_{\mu\nu}d \qquad \bar{s}\sigma_{\mu\nu}\tilde{F}_{\mu\nu}d$$

$$\bar{s}d \qquad \bar{s}\gamma_5 d \qquad\qquad (4.29)$$

The coefficients of the operators in Eq. (4.29) will in general be divergent as powers in $1/a$. Examples of diagrams responsible for the mixing are shown in Fig. 9. As in the case we considered in the previous subsection [Eq. (4.13)], the mixing arises because of the insertion of the Wilson term in the loops of the diagrams of Fig. 9. The coefficient of $\bar{s}\sigma_{\mu\nu}\tilde{F}_{\mu\nu}d$ is always finite because of CPS symmetry[12]. The coefficient of the operator $\bar{s}\sigma_{\mu\nu}F_{\mu\nu}d$ which, for dimensional reasons, is linearly divergent in $1/a$, can be made finite by the GIM cancellation when we allow the charm to propagate in the loop. For this, in our numerical simulation, the lattice coupling must be chosen in such as way as to satisfy the condition (see also the next section):

$$m_c a \ll 1 \qquad\qquad (4.30)$$

In this case, the GIM mechanism cancels the most divergent term, which is independent of the quark masses, and makes finite the coefficient of $\bar{s}\sigma_{\mu\nu}F_{\mu\nu}d$:

$$\left(\frac{1}{a} + \cdots\right)\bar{s}\sigma_{\mu\nu}F_{\mu\nu}d \to (m_c - m_u)\bar{s}\sigma_{\mu\nu}F_{\mu\nu}d \qquad\qquad (4.31)$$

The coefficients of the scalar and pseudoscalar densities are always power-divergent and must be evaluated non-perturbatively.

The "penguin" operators, relevant for the calculation of ε'/ε, short of putting the top quark on the lattice, always require the subtraction of three operators ($\bar{s}\sigma_{\mu\nu}F_{\mu\nu}d$, $\bar{s}d$ and $\bar{s}\gamma_5 d$) with power-divergent coefficients.

We are now ready to give the practical recipe for fixing the coefficients of $\bar{s}d$ and $\bar{s}\gamma_5 d$ (the other finite coefficients being taken from perturbation theory) for the operators $O^{\pm}$ given in Eqs. (4.28). For small momentum transfer we have:

$$< 0|O^{\pm} + \delta_6 O^{\pm} + \delta_5 O^{\pm}|K^0 > Z = i\,\delta_1^{\pm}$$

$$< \pi^+(p)|O^{\pm} + \delta_6 O^{\pm} + \delta_5 O^{\pm}|K^+(k) > = \delta_2^{\pm} + \gamma_2^{\pm}\,(p \cdot k) \qquad\qquad (4.32)$$

where $0^{\pm} + \delta_6 0^{\pm} + \delta_5 0^{\pm}$ are the bare operators $0^{\pm}$ added to the corrections due to operators of dimension six ($\delta_6 0^{\pm}$) and five ($\delta_5 0^{\pm}$), (the coefficients of the mixing being taken from a perturbative calculation). Similarly:

$$< 0|\bar{s}\gamma_5 d|K^0 > = i \, \delta_p$$

$$< \pi^+(p)|\bar{s}d|K^+(k) > = \delta_S + \gamma_S \, (p \cdot k) \qquad (4.33)$$

On the other hand, in the continuum, at lowest order in chiral symmetry breaking, soft pion theorems imply the following relations[16]:

$$<0|O^{\pm}|K^0> = i \, (m_K^2 - m_\pi^2)\delta^{\pm}$$

$$<\pi^+(p)|O^{\pm}|K^+(k)> = \frac{-m_K^2}{f_\pi} \, \delta^{\pm} + \gamma^{\pm} \, (p \cdot k)$$

$$<\pi^+\pi^-|O^{\pm}|K^0> = i \, \frac{m_K^2 - m_\pi^2)}{f_\pi} \, \gamma^{\pm} \qquad (4.34)$$

(All $K \rightarrow 2\pi$ matrix elements are taken for vanishing four-momentum transfer.)

We see that $\gamma^{\pm}$ is the number we need to predict the amplitude at lowest order in the chiral expansion.

By a suitable combination of $0^{\pm} + \delta_6 0^{\pm} + \delta_5 0^{\pm}$ with $\bar{s}d$ and $\bar{s}\gamma_5 d$, we can enforce the relations in Eq. (4.34). Defining the renormalized $0^{\pm}$ as $0^{\pm} = Z_{LATT}^{\pm}(\mu a)(0^{\pm}+\delta_6 0^{\pm}+\delta_5 0^{\pm}+d_s^{\pm}\bar{s}d+d_p^{\pm}\bar{s}\gamma_5 d)$, one finds:

$$\gamma^{\pm} = Z_{LATT}^{\pm}(\mu a) \, (\gamma_2^{\pm} - \delta_2^{\pm}\gamma_s/\delta_s) + 0 \, (m_K^2) \qquad (4.35)$$

The factor $Z_{LATT}^{\pm}$ has been added to connect the lattice operators to the corresponding ones in the continuum [Eq. (4.20)]. The $0(m_K^2)$ terms in Eq. (4.35) depend on $\delta^{\pm}$, but are of highest order in the chiral expansion. The procedure described above for $0^{\pm}$ cannot be used for "penguin" operators. In fact, the parity-conserving part of "penguin" operators requires, as we have seen above, the subtraction of two operators with power-divergent coefficients ($\bar{s}\sigma_{\mu\nu}F_{\mu\nu}d$ and $\bar{s}d$). It still requires however only one subtraction for the parity-violating part. So in principle, we can use the $k \rightarrow 0$ amplitude to fix the coefficient of $\bar{s}\gamma_5 d$ and compute directly $k \rightarrow 2\pi$ on the lattice. In practice this is rather hard to achieve, because it would require the use of lattices too big for presently-available computer resources.

5. NUMERICAL COMPUTATION OF THE WEAK HAMILTONIAN MATRIX ELEMENTS

5.1 Basic Notions on Monte Carlo Techniques

We start by defining operators carrying the same quantum number of the particles under study. For example, in the case of the pion and of the rho, we may use

$$
\begin{aligned}
\pi^+(x) &= \bar{u}^A(x)\gamma_5 d^A(x) \\
\rho_\mu^+(x) &= \bar{u}^A(x)\gamma_\mu d^A(x)
\end{aligned}
\tag{5.1}
$$

From the action in Eq. (4.2), it is straightforward to compute the correlation function for these operators. In the pion case:

$$
G(\vec{x},t) = <\pi(\vec{x},t)\pi^+(0,0)> =
$$

$$
\frac{\int d[U] \prod_f [det\Delta_f(U)]\, t_r\, [S^u(x,0)\gamma_5 S^d(0,x)\gamma_5]\, e^{-S_G(U)}}{\int d[U] \prod_f [det\Delta_f(U)]\, e^{-S_G(U)}}
\tag{5.2}
$$

$S_G(u)$ is the gluon action; Δ_f has been defined in Eq. (4.2). $S^{u,d}(x,0)$ is the up, down quark propagator between 0 and x:

$$
S^{u,d}(x,0) = \left[\Delta_{u,d}(x,0)\right]^{-1}
\tag{5.3}
$$

The right-hand side of Eq. (5.2) is represented by the diagram in Fig. 10. In a Monte Carlo simulation the integral over the gluon fields U_μ is replaced by the sum over gluon field configurations generated by some numerical algorithm (Metropolis, Langevin,...[17]) on a finite, generally hypercubic lattice. $\Delta(x,0)$ is inverted [see Eq. (5.3)] by some numerical technique, such as the Gauss-Seidel method, for example. All the numerical results listed below have been obtained in the so-called "quenched" approximation[*].

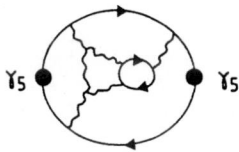

Fig. 10.   Typical diagram for the pion propagator.

---

[*]For further information on the validity of this approximation in which one neglects the effects of the fermion determinant, see, for example, Ref . 17.

We may fix the momentum by summing $G(\vec{x}, t)$ over the space components:

$$G(\vec{q}, t) = \sum_{\vec{x}} G(\vec{x}, t) e^{i\vec{q} \cdot \vec{x}} \tag{5.4}$$

$G(\vec{q}, t)$ will propagate all possible intermediate states of momentum $\vec{q}$ carrying the same quantum numbers of the pion. At zero momentum, considering only one-particle intermediate states, one has:

$$G(t) = \sum_{\vec{x}} G(\vec{x}, t) = \sum_{n} \frac{Z_n}{2m_n} e^{-m_n t} \tag{5.5}$$

$m_n = m_\pi, m_\pi^*, \ldots$ For large time distances, the pole corresponding to the lowest-lying state, the pion, will stem out since the heavier states are exponentially suppressed as $\exp[-(m_n - m_\pi)t]$:

$$G(t) \underset{t \to \infty}{\longrightarrow} \frac{Z_\pi}{2m_\pi} e^{-m_\pi t} \tag{5.6}$$

When t is too small, one has a systematic overestimate of the mass of the lowest-lying state. The logarithm of $G(t)$, as we observe it in a Monte Carlo simulation, is plotted in Fig. 11 as a function of t. At large t, $\ln[G(t)]$ goes like a straight line because only one particle is propagating [Eq. (5.6)]. The slope of the straight line corresponds to the mass of the particle in lattice units ($m_\pi a$ in this case), and the intercept in t = 0 allows the determination of Z_π and consequently of the meson decay constants (f_π, f_ρ, ...; see also below).

A check of consistency of the results can be obtained by using some other operator with the same quantum numbers: different operators should lead to the same value of m_π. A possible alternative for the pion is the use of the fourth component of the axial current

$$A_0^+(x) = \bar{u}^A(x)\gamma^0\gamma^5 d^A(x) \tag{5.7}$$

We may use the operators in Eqs. (5.1) and (5.7) to compute the pseudo-scalar meson decay constants, which are of particular interest in the physics of D and B mesons.

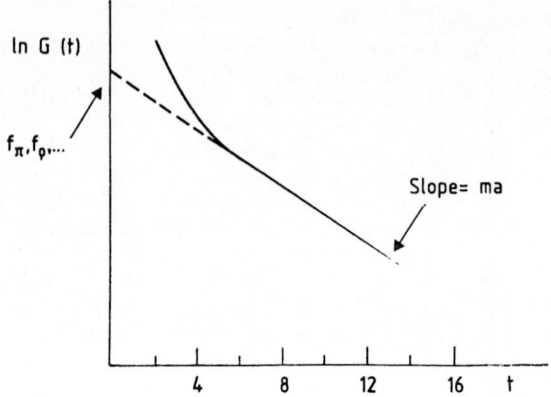

Fig. 11. $\ln[G(t)]$ plotted versus t.

Consider the following correlation functions:

$$\sum_{\vec{x}} <0|A_0(\vec{x},t)\pi^+(0,0)|0> \xrightarrow[t\ a>>1]{} \frac{<0|A_0|\pi><\pi|\pi^+|0>}{2m_\pi} e^{-m_\pi t}$$

$$\sum_{\vec{x}} <0|\pi(\vec{x},t)\pi^+(0,0)|0> \xrightarrow[t\ a>>1]{} \frac{<0|\pi|\pi><\pi|\pi^+|0>}{2m_\pi} e^{-m_\pi t} \qquad (5.8)$$

$$\frac{<A_0\pi^+>}{<\pi\pi^+>} \xrightarrow[t\ a>>1]{} const = \frac{f_\pi m_\pi}{Z_A <0|\pi|\pi>} \qquad (5.9)$$

By fitting the propagators in Eq. (5.8), using Eq. (5.9) and Z_A computed on the lattice with the Ward identities (see Section 3), we can obtain f_π (f_k, ...). One finds[18]:

$$\frac{f_K}{f_\pi} \sim 1.2 \pm 0.10 \exp 1.25$$

$$\frac{f_D}{f_\pi} = 1.37 \pm 0.20 \qquad \frac{f_F}{f_\pi} = 1.60 \pm 0.25 \qquad (5.10)$$

Since f_D and f_F have not been measured, to appreciate the accuracy of the results we can compare the lattice J/ψ decay constant, evaluated with the same technique, to its experimental value:

$$1/f_{J/\psi} = 0.100 \qquad (1/f_{J,\psi}) \quad exp = 0.125 \qquad (5.11)$$

where the last figure is derived from the width $\Gamma_{J/\psi} \to e^+e^-$. Thus we expect $f_{D,F}$ to have a ~15% systematic error. By assuming the potential model for the two-quark system, we can also derive $f_{B_{s,d}}$:

$$f_{B_{s,d}} \simeq 140, 110 \ MeV \qquad (5.12)$$

This number is relevant for $B_0-\bar{B}_0$ mixing.

5.2 How to extract the matrix elements of the operators

We now consider more complicated three-point correlation functions involving two pseudoscalar operators as in Eq. (5.1) and the operator of which we want to compute the matrix elements ($0^\pm$, $\bar{s}d$ or $\bar{s}\sigma_{\mu\nu}F^{\mu\nu}d$ for example):

$$G(t_1,t_2) = \sum_{\vec{x}_1,\vec{x}_2} e^{i\vec{q}\vec{x}_1} <\bar{\psi}(\vec{x}_1,t_1)\gamma_5\psi(\vec{x}_1,t_1)O(0)\bar{\psi}(\vec{x}_2,t_2)\gamma_5\psi(\vec{x}_2,t_2)>$$

$$O = (\bar{\psi}\Gamma\psi)(\bar{\psi}\Gamma\psi) \text{ or } (\bar{\psi}\Gamma\psi) \qquad (5.13)$$

In most of the cases, for practical reasons the time position of the operator ($t = 0$) and of one of the two pseudoscalar operators (let us say t_2) are fixed, but we may vary t_1 at will. On a finite, periodic lattice of length T, we may have the two possibilities illustrated in Figs. 12a and 12b. In the first case (Fig. 12a) $T-t_1$ and t_2 are large, so that we isolate the lowest-lying one-particle states, but $|t_1-t_2| >> T-t_1$ and t_2. The correlation function in Eq. (5.13) is then dominated by the matrix

151

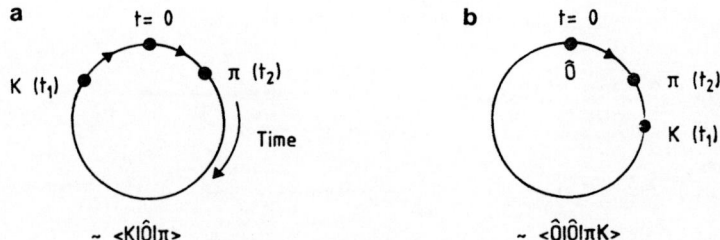

Fig. 12a.   The circle represents the time. The operator is at the ori-
             gin and with the time positions shown in the figure, the
             three-point correlation is dominated by the matrix elements
             of the operator sandwiched between a kaon and a pion.

Fig. 12b.   The three-point correlation is dominated by the operator
             matrix element between the vacuum and a two-pseudoscalar
             state.

element of the operator sandwiched between the pseudoscalar particles (a
kaon and a pion, for example):

$$G(t_1, t_2) \sim G_K(T - t_1) < K|O|\pi > G_\pi(t_2) \qquad (5.14)$$

The second possibility (Fig. 12b) is that t_2 and t_1 are both large but
$T-t_1$, $T-t_2 \gg t_1$, t_2. In this case, the correlation function is dominated
by the matrix element $\langle 0|O|\pi k \rangle$. Neglecting final state interactions,

$$G(t_1, t_2) \sim < 0|O|\pi K > G_\pi(t_2) G_K(t_1) \qquad (5.15)$$

All the time distances in this game must be large so as to isolate the
wanted signal. We may check the range of validity in t_1 of the expressions in Eqs. (5.14) and (5.15) by noticing that, in both cases:

$$R = \frac{G(t_1, t_2)}{G_K(t_1) G_\pi(t_2)} \sim \text{const} \qquad (5.16)$$

[on a periodic lattice $G_\pi(t_1) = G_\pi(T-t_1)$]. R, computed by a Monte Carlo
simulation on a $16^3 \times 48$ lattice as a function of t_1 is shown in Fig.
13 [18]. Notice how R depends on t_1 only in the transition region between
the two plateaux corresponding to $\langle 0|O|\pi k \rangle$ and $\langle k|O|\pi \rangle$.

5.3 $\Delta T = 3/2$ and the B Parameter[19][20][21]

By the technique of the subsection 5.2, one computes the matrix
elements of the operator:

$$\hat{O}_{LL}^{\Delta S=2} = Z_{LATT}\left(O_{LL}^{\Delta S=2} + C_1(\bar{s}\gamma_5 d\bar{s}\gamma_5 d - \bar{s}d\bar{s}d) + \cdots \right) \qquad (5.17)$$

between K° and $\bar{K}^\circ$. O_{LL} is the bare lattice operator; c_1, ... are finite
coefficients which depend only on g_0 and r and can be computed in perturbation theory. The corresponding operators ($\bar{s}\gamma_5 d\bar{s}\gamma_5 d$ for example) are
there because of the Wilson term. The matrix elements of O_{LL} between
kaon states should have the following dependence on the momenta:

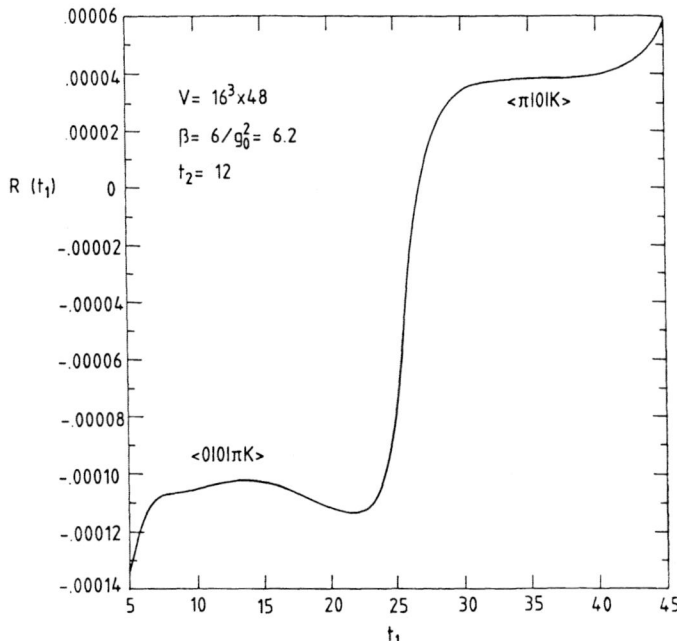

Fig. 13. $R = G(t_1,t_2)/G_K(t_1)G_\pi(t_2)$ as a function of t_1 on a $16^3 \times 48$ lattice.

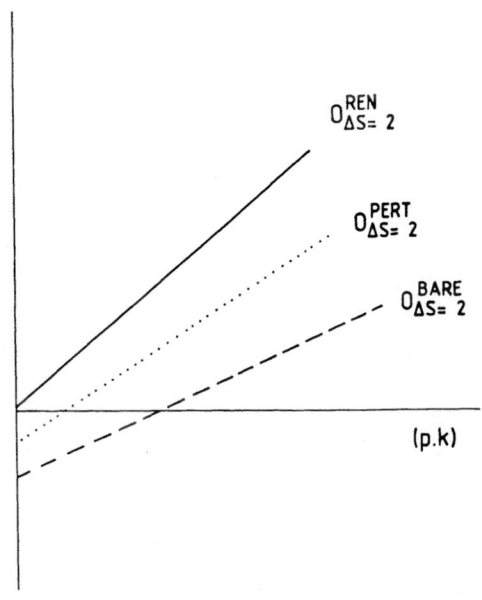

Fig. 14.   The matrix element of the $\Delta S = 2$ operator as a function of p·k.   The matrix element of the renormalized operator vanishes as p·k → 0 (solid line); the matrix element of the bare operator has a constant term (dashed line); the operator, renormalized by using lowest-order perturbation theory, has a matrix element with a smaller intercept in p·k = 0 than the bare one (dotted line).

$$< K^0(p))|\hat{O}_{LL}^{\Delta S=2}|\overline{K}^0(k)) \; = \; \gamma_{LL}(p \cdot k) \qquad\qquad (5.18)$$

However, in numerical simulations it was found that the perturbative sub-traction, which in principle is perfectly legitimate for this operator, does not work properly at $g_0 \sim 1$ [18*]. The situation is illustrated in Fig. 14. The matrix element of the exactly renormalized operator O_{LL} vanishes as $p \cdot k \to 0$ [Eq. (5.18)]. The bare operator certainly has a non-vanishing constant term

$$< K^0(p)|O_{LL}^{\Delta S=2}|\overline{K}^0(k) > \; = \; \delta_B \; + \; \gamma_B(p \cdot k) \qquad\qquad (5.19)$$

The operator with c_1, ... computed at lowest-order in perturbation theory, still has a constant term different from zero, but smaller than for the bare operator (lowest-order perturbation theory always goes in the right direction). In the absence of a better determination of the coefficients c_1, ..., we can only compute γ_{LL} by measuring the slope in $p \cdot k$, even if the matrix element of the operator is different from zero at $p \cdot k = 0$. This leaves a systematic error in the evaluation of γ_{LL} because of the imperfect renormalization of the operator, due to the approximate knowledge of the coefficients c_1, ... In Fig. 15 I report the results obtained at two different values of the bare coupling constant g_0: $\beta = 6/g_0^2 = 6.0$ on a $20 \times 10 \times 10 \times 40$ lattice and $\beta = 6.2$ on a $16^3 \times 48$ lattice[18]. The results are compatible with the expected renormalization group behaviour but, as discussed before, a constant term is still present in the matrix element. From the slope of the curve in Fig. 15, one can compute the B parameter and the $\Delta I = 3/2$ amplitude using the relation[22]:

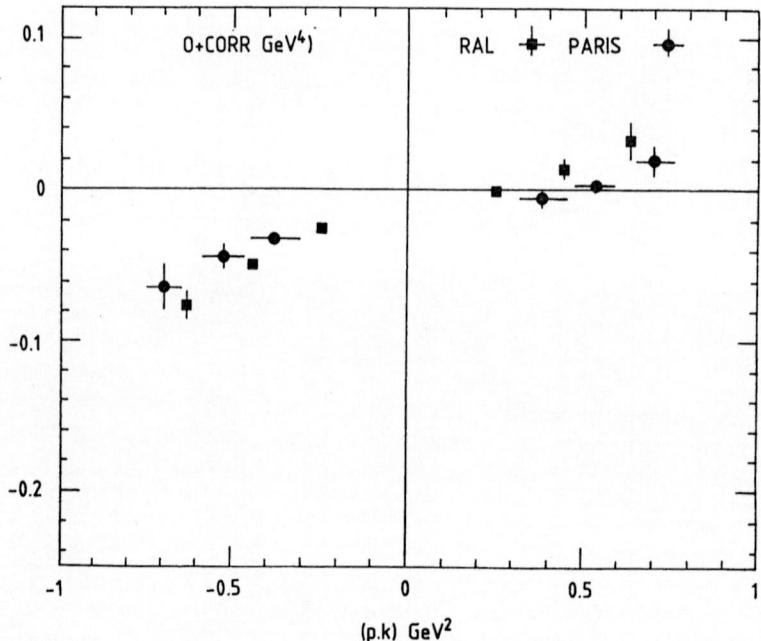

Fig. 15. O_{LL}, renormalized in lowest-order perturbation theory, is given as a function of (p·k) [Ref. 16)].

---

*We have independent evidence of that from other computations[21].

$$< \pi^+\pi^0|H_W|K^+ > = \left(\frac{G_F}{2\sqrt{2}}\right) sin\vartheta_c cos\vartheta_c C(\mu, M_W) \times$$

$$\frac{3}{4}\left(\frac{m_K^2 - m_\pi^2}{F_\pi}\right) \frac{< K^0|\hat{O}_{LL}^{\Delta S=2}|\overline{K}^0 >}{m_K^2} \tag{5.20}$$

From the combined results at β = 6 and 6.2, one finds[*]:

$$B = 0.6 \pm 0.2$$

$$\frac{< \pi^+\pi^0|H_W|K^+ >}{m_K} = (0.65 \pm 0.25)10^{-8} \quad \exp(3.7 \times 10^{-8}) \tag{5.21}$$

In spite of the large error, the result for the $\Delta I = 3/2$ amplitude is good, given the fact that this is a first principle computation with no free parameters. The value of B reported still has an error which is too large to make this result useful. There are two main sources of error besides the imperfect knowledge of c_1, ... The first is the fact that it is difficult to approach the chiral limit, i.e., terms of order $(p\cdot k)^2$ are still important in the range of $(p\cdot k)$ which we can explore on these lattices. The second is the statistical error. We are actually planning to use the APE computer in order to work on a $16^3 \times 40$ lattice at β = 6 so as to reduce the minimum $(p\cdot k)$ by a factor 2.5 and to decuplicate the statistics. In this way we expect to reduce the error on the B parameter to something of the order of 10%.

Many of the problems one encounters with light quarks disappear as we consider systems made up of a heavy and a light quark. The extra operators give a very small contribution (so that a large error for c_1, ... is also unimportant) and statistical fluctuations become small. The result for the D°-$\overline{D}^\circ$ parameter B is:

$$B_{D_0 - \bar{D}_0} = 1.00^{+0.20}_{-0.15} \tag{5.22}$$

supporting the prejudice that vacuum saturation works better for heavy quarks.

We finally observe that even with an error as large as those in Eqs. (5.21), the octet enhancement should be clearly visible. Moreover, one can compute the ratio

$$a_{\Delta I=\frac{1}{2}} \; / \; a_{\Delta I=\frac{3}{2}} \tag{5.23}$$

for which many statistic/systematic errors cancel out.

5.4 The $\Delta I = \frac{1}{2}$ Matrix Element

In this case, we have to compute by numerical simulation the matrix elements of the four-fermion operators $O^\pm + \delta_6 O^\pm + \delta_5 O^\pm$ [cf. Eq. (4.32)] and of the two quark operators $\bar{s}d$ and $\bar{s}\gamma_5 d$ in order to construct the renormalized operators $O^\pm$. The coefficient of $\bar{s}\sigma_{\mu\nu}F_{\mu\nu}d$, which enters into $\delta_5 O^\pm$, can be computed in two-loop perturbation theory, provided we can

---

[*]Preliminary results[18].

put the charm quark on the lattice ($m_c a \ll 1$). Such tedious computation is by now completed[23]. The coefficients of $\delta_6 O^\pm$ were computed in Refs. 15. The coefficients of the scalar, $\bar{s}d$, and pseudoscalar, $\bar{s}\gamma_5 d$, densities can be computed following the method explained in subsection 4.3. The plateaux which allow the evaluation of the matrix elements of $O^\pm$, $\bar{s}\sigma_{\mu\nu}F_{\mu\nu}d$, $\bar{s}d$ and $\bar{s}\gamma_5 d$ have been clearly seen in Monte Carlo simulations[18]. The signal persists even after the non-perturbative subtraction of ($\bar{s}d$) (subsection 4.3). Unfortunately, the analysis has not yet been completed and I do not have the results to present here. We expect to be able to present the first results for this $\Delta I = \frac{1}{2}$ amplitude within a couple of months.

6.   SUMMARY

I have discussed at great length the computation of the matrix elements of the weak Hamiltonian in lattice QCD. The theoretical framework for these computations is now established and the first results for the easiest matrix elements have already appeared. We expect to have the first results for the $\Delta I = \frac{1}{2}$ in a very short time.

REFERENCES

1. J. Christenson et al., Phys. Rev. Lett. 13:138 (1964).

2. For a recent review on the effective weak Hamiltonian, see:
   J.F. Donoghue et al., Phys. Reports 131:319 (1986).

3. D.D. Wu, Phys. Rev. D33:860 (1986);
   O.W. Greenberg, Phys. Rev. D32:1841 (1985);

4. K.G. Wilson, Phys. Rev. 179:1499 (1969).

5. M.K. Gaillard and B.W. Lee, Phys.Rev.Lett. 33:108 (1974);
   G. Altarelli and L. Maiani, Phys. Lett. 52B:351 (1974).

6. M.A. Shifman et al., Nucl. Phys. B120:316 (1977); Sov. Phys. JEPT
   45:670 (1977); see also:
   F. Gilman and M.B. Wise, Phys. Rev. D20:2372 (1979); Phys. Rev.
   D27:1128 (1983).

7. J.F. Donoghue and B.R. Holstein, Phys. Rev. D29:2088 (1984).

8. A.J. Buras et al., Phys. Lett. 134B:373 (1984).

9. G. Altarelli and P. Franzini, CERN preprint TH.4745/87 (May 1987).

10. K.G. Wilson, in "New Phenomena in Subnuclear Physics", A. Zichichi,
    ed., Plenum Press, New York (1977).

11. M. Bochicchio et al., Nucl. Phys. B262:331 (1985).

12. L. Maiani et al., Phys. Lett. 176B:445 (1986); CERN preprint
    TH.4517/86 (August 1986).

13. See, for example:
    S. Sharpe, Proceedings of the XXII Moriond Conference, March 1987 –
    SLAC-PUB-4316.

14. L. Maiani and G. Martinelli, Phys. Lett. 178B:265 (1986).

15. G. Martinelli, Phys. Lett. 141B:395 (1984);
    C. Bernard et al., UCLA/87/TEP/18 (May 1987).

16. For a recent reappraisal, see:
    C. Bernard et al., Phys. Rev. D32:2343 (1985).

17. See, for example, the introductory course on lattice QCD by
    G. Martinelli, Proceedings of the Nato Advanced Study Institute –
    "Heavy Ion Collisions", 2-15 September 1984, P. Bouche et al., eds.,
    Plenum Press, New York (1986).

18. M.B. Gavela et al., in preparation.

19. The first results for this quantity were obtained by:
    N. Cabibbo et al.,Nucl. Phys. B244:381 (1984), and
    R.C. Brower et al., Phys. Rev. Lett. 53:1318 (1984).

20. L. Maiani and G. Martinelli, Phys. Lett. 181B:344 (1986).

21. A. Soni, Proceedings of the XXII Moriond Conference, March 1987, and references therein.

22. J.F. Donoghue et al., Phys. Lett. 119B:412 (1982).

23. G. Curci et al., in preparation.

DISCUSSION I

–Schuler:

You mentioned index theorems. Does it mean that a topological charge can be defined on the lattice in a way that the index theorem equation is valid?

– Martinelli:

Yes. The works by M. Luscher and by M. Bochicchio, G.C. Rossi, G. Immirzi and K. Yoshida have shown how to define the topological charge on the lattice. These definitions have a topological meaning and they are not affected by perturbative corrections even when $a \neq 0$.

– Kiritsis:

How do you manage to have the $SU(3)_L \times SU(3)_R$ currents conserved and $U(1)_A$ anomaly on the lattice?

– Martinelli:

The lattice gives the correct $U(1)$–anomaly, while for $SU(3)_L \times SU(3)_R$, one obtains "good" conserved currents through the procedure explained in this lecture.

– Gourdin:

This morning you produced a formula where a function $S\left(\frac{m_c^2}{M_W^2} , \frac{m_t^2}{M_W^2}\right)$ comes in. Do you know with enough accuracy such a function in order to be able to deduce from experiments information on the top mass? Personally I cannot believe that.

– Martinelli:

The short distance contribution on the imaginary part of the amplitude $K^0 \rightarrow K^0$ is quite accurately known, while the long distance contribution is estimated to give a correction of the order of the 20% (Donoghue).

– Morelli:

Is the r–parameter of the Wilson action completely arbitrary or it is fixed by some conditions?

– Martinelli:

r is arbitrary. It exists a certain function of M_0 and r which corresponds to the bare quark mass $m = M_0$. $\overline{M}(M_0.r)M_0$ will depend on r, since m does not.

– *Morelli:*

This morning you referred to a three flavour model. How more flavours affect the calculations of the CP–violating processes?

– *Martinelli:*

I discussed only the representations of $SU(3)_L \times SU(3)_R$ for simplicity. The same arguments can be extended to any number of flavours.

– *Toppan:*

I would like to know what is the present status of maximal CP–violation and in particular if it is possible to give a meaning to this concept in a convention–independent way.

– *Martinelli:*

C. Jarlskog and H. Frampton have discussed this problem and they have shown that every CP–violating process, in the 3 family case, is proportional to the invariant $s_1^2 s_2 s_3 \; sin\delta$ for $(c_i \sim 1)$.

– *Miele:*

Do you make the assumption that $\varepsilon' = 0$?

– *Martinelli:*

I do not assume $\varepsilon' = 0$. The relations which I have used are based on the definition of η_{+-} and η_{00} in terms of ε and ε', using the fact that one expects $\varepsilon'/\varepsilon << 1$ on experimental basis. In the super–weak model, introduced by Wolfenstein, $\varepsilon' = 0$.

DISCUSSION II

– *Balog:*

May you comment on the 1/N approach by Bardeen, Buras et al.?

– *Martinelli:*

They start with $H_{eff} = C(\mu, M_W)O(\mu)$ and compute $< O >$ in the 1/N expansion using the chiral Lagrangian, which is given by $Tr(\partial_\mu \sum \partial^\mu \sum^+ + \ldots)$ with $\sum = exp(i\pi^a \lambda^a /8f\pi)$. There is no octet enhancement at lowest order in 1/N, but they claim it is there when they include the first non-trivial order. My main objection to their approach is that it is not clear how to match the operators in the chiral theory with the operators in the quark language, beyond the lowest order in 1/N. In fact they have not equal renormalization properties.

– *Liu:*

Can you show that on the lattice the coefficient in front of Wess–Zumino term is quantized?

– *Martinelli:*

If you can give a topological meaning to the Wess–Zumino term as it has been done for the topological charge, the answer is probably yes.

– *Liu:*

Can you prove that in the low energy effective action of n flavours mass QCD on the lattice $SU_V(n)$ remains unbroken, while $SU_A(n)$ is spontaneous broken?

– *Martinelli:*

D. Weingarten on the lattice and then E. Witten have shown that axial symmetry is spontaneously broken.

– *Balog:*

Why do you calculate the $K\pi$ amplitude and not $K\pi\pi$?

– *Martinelli:*

In principle we could calculate the $K \to \pi\pi$ amplitude but this is difficult in practice. We can use the $K \to \pi$ amplitude applying soft–pions theorems which are known to be accurate at the 20-30% level. For ε'/ε, since we cannot put the top quark on the lattice it will be necessary to calculate the amplitude $K \to \pi\pi$.

– Schuler:

I would like to know how far you are away from putting $SU(2)_L \times U(1)$ on the lattice?

– Martinelli:

We know how to restore global symmetries such as the chiral symmetry. For $SU(2) \times U(1)$ we have a local symmetry but in principle things go the same way. One must choose a complete set of Ward identities: for instance add counterterms like cW_μ^2 in order to ensure a massless W.

– Singh:

What is the mass of the "pion" entering in your calculations?

– Martinelli:

The smallest value is $m_\pi \sim 500$ MeV.

– Singh:

Could you explain how you calculate the structure function of pion on the lattice?

– Martinelli:

What we want to calculate is $< p|T(J_\mu(x)J_\nu(0))|p >$, i.e. a graph of the type

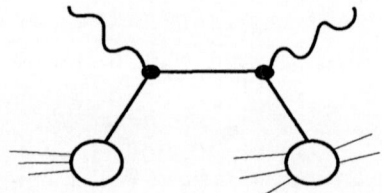

Using Wilson OPE this calculation reduces to the calculation of the insertion of operators of the form $O_{\mu_1}\ldots \mu_n = \overline{\psi}\gamma_{\mu_1}D_{\mu_2}\ldots D_{\mu_n}\psi$. The first non trivial operator is the energy–momentum tensor which corresponds to the average momentum carried by the quarks in the pion $< x >$. The second operator corresponds to $< x^2 >$. Both have been computed in a Monte Carlo simulation.

– Singh:

What are the motivations of SUSY QCD? Is there any sort of cancellation of unwanted terms?

162

– Martinelli:

At the beginning of the eighties there has been great interest in putting SUSY on the lattice. It turned out that SUSY must be broken explicitly on the lattice if D=4. The recent work of Curci e Veneziano has shown how to recover from this broken theory the supersymmetric currents which obey the correct SUSY algebra.

– Gourdin:

This morning you produced the result of your calculation for the $\Delta I = 3/2$ amplitude to be

$$(6 \pm 3)10^{-8}$$

Question 1)	What are the reasons for the 50% error?
Question 2)	What is the experimental error?
Question 3)	What is the result obtained with QCD sum rules?

– Martinelli:

The error on the lattice result is of statistical and systematic origin. Using larger lattices, smaller pion masses and higher statistics we will reduce the error. The present result for the B parameter is $B = 0.5 \pm 0.2$ but we plan to reduce the error to ~ 0.07 in the future. The experimental error for the $\Delta T = 3/2$ amplitude is negligible. For what QCD sum rules are concerned, the calculation of the $\Delta T = 3/2$ gives a result close to the experimental one while for the $\Delta T = 1/2$ amplitude no stringent result has been obtained.

– Morelli:

I saw the word "quenched", what does it mean?

– Martinelli

The main systematic error of our calculation is due to the quenched approximation. In the functional integral $\int d\psi d\bar{\psi} dU \, e^{-\bar{\psi}\Delta(U)\psi} O(\psi, \bar{\psi}, U)$ performing the Gaussian integral on ψ and $\bar{\psi}$ you get: $\int dU \, det \, (\Delta(U)) O(\Delta^{-1}(U), U)$. The quenched approximation corresponds to set $det \, \Delta(U) = 1$, i.e. to ignore diagrams which contain fermionic loops.

The error due to this approximation depends on the quantity you want to calculate. It is a good approximation for the structure function of π, since the sea quarks carry a small fraction of the momentum, but it is a bad approximation when you consider the decay $\rho \to \pi\pi$, since it gives $\Gamma_{\rho \to \pi\pi} = 0$. For H_{eff} it is difficult to know what is the effect of the quenching.

STRING FIELD THEORY

Taichiro Kugo

Department of Physics, Kyoto University

Kyoto 606, Japan

ABSTRACT

An introductory review of bosonic string field theory is given.

1. INTRODUCTION

As for the method for calculating string amplitudes in the perturbation regime, there is a beautiful and powerful formulation a la Polyakov based on the conformal field theory on 2-dimensional Riemann surfaces.[1] This is actually an arena where many physicists and mathematicians have been doing active investigations and made much developments recently.[2]

However we also would like to know the non-perturbative aspects and the underlying geometrical principle of string theory : i) How does the expected compactification of extra space dimensions occur ? Which background geometry is preferred ? These should be answered since the closed-string determines the space-time structure of its own. ii) Is superstring generated spontaneously from the d = 26 bosonic string theory as was conjectured first by Freund[3] ? iii) What is the most fundamental principle to determine the string dynamics ? Is it possible to formulate it as a geometry of string configuration space ?

Covariant string field theory would probably be the most promising approach to these questions. Actually a considerable progress has been made in constructing gauge string field theory recently since the pioneering work by Siegel.[4] As a highlight in the approach, for instance, we already have a pregeometrical string field theory[5] possessing purely cubic action Φ^3 which provides us with a very interesting viewpoint for the string dynamics.

I give here an introductory review of bosonic string field theory. As for the interacting string field theories, there are mainly two approaches, the one based on the joining-splitting type interaction vertex and the one based on the mid-point interaction vertex. The former approach first suggested by Siegel[4] is fully developed by Kyoto group[6] (Hata, Itoh, Kunitomo, Ogawa and myself) and also by Neveu and West[7] partly, and the latter is constructed by Witten.[8] I review here mainly the former type theory, which I call HIKKO's theory for brevity, and briefly Witten's one.

2. STRING FIELD

String field $\Phi = \Phi[Z]$ is a functional of string coordinate $X^\mu(\sigma)$ in d-dimensional space-time ($\mu = 0,1,\ldots,$ d-1) and Faddeev-Popov (FP) ghost and anti-ghost coordinates $c(\sigma)$ and $\bar{c}(\sigma)$:

$$Z = [\, X^\mu(\sigma),\ c(\sigma),\ \bar{c}(\sigma)\, ;\, \alpha\,] \tag{2.1}$$

Here we have put an additional parameter α, which will play the role of (unphysical) "string-length" in HIKKO's approach but is not present in Witten's open-string theory. We are assuming that the parameter σ, distinguishing positions on the string, runs over $0 \leq \sigma \leq \pi$ for open-string and $-\pi \leq \sigma \leq \pi$ for closed-string. The boundary conditions of the coordinates are

open : $\qquad X^\mu(\sigma) = c(\sigma) = \bar{c}'(\sigma) = 0$ at $\sigma = 0$ and π, $\qquad$ (2.2a)

closed : $\quad X$,c, $\bar{c}$; periodic functions of σ with period 2π, (2.2b)

where the prime denotes $\partial/\partial\sigma$.

The string field Φ is subject to the following conditions :

(i) Φ carries ghost number $N_{FP} = -1$ and hence <u>Grassmann-odd</u> quantity. The ghost number N_{FP} is defined such that ghost and anti-ghost coordinates $c(\sigma)$ and $\bar{c}(\sigma)$ carry $N_{FP} = +1$ and -1, respectively.

(ii) Φ is a <u>real</u> (hermitian) field in the sense that

$$\Phi^\dagger[Z] = \Phi[\tilde{Z}], \quad \tilde{Z} \equiv [\, X^\mu(\pi-\sigma),\ -c(\pi-\sigma),\ \bar{c}(\pi-\sigma);\ -\alpha\,] \tag{2.3}$$

For open-string case, Φ is generally matrix-valued and then $\Phi^\dagger$ implies also the matrix dagger. [For this association of matrix indices (Chan-Paton factor), only the groups U(N) (orientable) and O(N), USp(2N) (non-orientable) are known to be allowed.[9] For simplicity, however, our open-string field is assumed to be $\Phi_i^{\ j}$, carrying N x N* index of U(N).]

(iii) For closed-string case, Φ is further constrained by

$$\mathcal{P}\Phi = \Phi \qquad\qquad \text{or} \qquad (L_+ - L_-)\Phi = 0. \tag{2.4}$$

Here $L_+ - L_-$ is the rigid translation operator of σ-coordinate,

$$L_+ - L_- = \int_{-\pi}^{\pi} d\sigma\, (\, X^{\mu\prime} P_\mu + c'\pi_c + \bar{c}'\pi_{\bar{c}}\,) \tag{2.5}$$

with $P_\mu(\sigma) = -i\pi\delta/\delta X^\mu(\sigma)$, etc, and $\mathcal{P}$ is the projection operator onto the $L_+ - L_- = 0$ sector :

$$\mathcal{P} = \int_{-\pi}^{\pi} \frac{d\theta}{2\pi}\, \exp\, i\theta(L_+ - L_-). \tag{2.6}$$

Therefore the constraint (2.4) is a physical requirement that the closed-string field Φ should be invariant under an arbitrary shift of the origin of the σ-coordinate since there is no distinguished point in σ of closed string.

The coordinates and their conjugate momenta are expanded into the oscillator modes as follows :

$$A_\pm^\mu(\sigma) \equiv \eta^{\mu\nu} P_\nu(\sigma) \mp X^{\mu\prime}(\sigma) \equiv \sum_{n=-\infty}^{\infty} \alpha_n^{\mu(\pm)} e^{\pm in\sigma}, \tag{2.7a}$$

$$C_\pm(\sigma) \equiv i\pi_{\bar{c}}(\sigma) \mp c(\sigma) \equiv \sum_{n=-\infty}^{\infty} c_n^{(\pm)} e^{\pm in\sigma}, \tag{2.7b}$$

$$\bar{C}_\pm(\sigma) \equiv \bar{c}(\sigma) \mp i\pi_c(\sigma) \equiv \sum_{n=-\infty}^{\infty} \bar{c}_n^{(\pm)} e^{\pm in\sigma}. \tag{2.7c}$$

where $\eta^{\mu\nu}$ denotes the flat Minkowski metric diag$(-1,+1,+1,\ldots,+1)$. The indices $(\pm)$ attached to the oscillator modes $\alpha_n^{\mu(\pm)}$, $c_n^{(\pm)}$ and $\bar{c}_n^{(\pm)}$ distinguish the right-mover$(+)$ and the left-mover$(-)$ for the closed-string case but should be omitted in the open-string case for which the both modes are the same by the reflection effect from the end points. These oscillators satisfy

$$[\alpha_n^{\mu(\pm)}, \alpha_m^{\nu(\pm)}] = n\eta^{\mu\nu}\delta_{n+m,0} , \quad \{c_n^{(\pm)}, \bar{c}_m^{(\pm)}\} = \delta_{n+m,0} . \tag{2.8}$$

$$a_{-n} = a_n^\dagger \quad for \quad a_n = \alpha_n^{\mu(\pm)}, c_n^{(\pm)}, \bar{c}_n^{(\pm)} \ (n \geq 1) .$$

For the open-string case, zero-modes exist only in the $X^\mu(\sigma)$ and $\bar{c}(\sigma)$ coordinates but not in $c(\sigma)$ because of the boundary conditions (2.2a). They are the center-of-mass coordinate x^μ of string and $\bar{c}_0$ in (2.7c), to which the zero-modes α_0^μ in (2.7a) and c_0 in (2.7b) are the conjugates :*)

$$open: \quad \alpha_0^\mu = p^\mu = -i\frac{\partial}{\partial x_\mu} , \quad c_0 = \frac{\partial}{\partial \bar{c}_0} . \tag{2.9a}$$

On the other hand, in closed-string case, all the coordinates $X^\mu(\sigma)$, $c(\sigma)$ and $\bar{c}(\sigma)$ contain zero-modes, denoted by x^μ, $-\frac{i}{2}\partial/\partial\pi_c^0$ and $\frac{1}{2}\bar{c}_0$, respectively, and related with the zero-modes in (2.7) by

$$closed: \quad \alpha_0^{\mu(+)} = \alpha_0^{\mu(-)} = \frac{1}{2}p^\mu , \tag{2.9b}$$

$$\bar{c}_0^{(\pm)} = \frac{1}{2}\bar{c}_0 \mp i\pi_c^0 , \quad c_0^{(\pm)} = \frac{\partial}{\partial\bar{c}_0} \pm \frac{i}{2}\frac{\partial}{\partial\pi_c^0}$$

Note that there is only one ghost zero-mode $\bar{c}_0$ in open-string, while there are two in closed-string for which we are using $\bar{c}_0$- and π_c^0- representations.

As is well-known in a harmonic oscillator system, one can avoid the explicit use of functions to represent the state by using the Dirac's ket representation ; i.e.,

$$e^{-\frac{1}{2}x^2} \longrightarrow |0\rangle , \tag{2.10}$$

$$H_n(x) e^{-\frac{1}{2}x^2} \longrightarrow \frac{1}{\sqrt{n!}}(a^\dagger)^n|0\rangle .$$

In the same way, we can use the ket representation to express the dependence of our string field Φ on all the non-zero modes avoiding the (formidable) explicit functional representation ; then $|0\rangle$ is the vacuum for all the oscillators in (2.7),

$$(\alpha_n^\mu , c_n , \bar{c}_n)|0\rangle = 0 \quad for \quad n \geq 1 . \tag{2.11}$$

We still keep the coordinate representations for the zero-modes and express the stirng field Φ as follows by making explicit the dependence on the ghost zero-modes

$$|\Phi(x,\bar{c}_0,\pi_c^0,\alpha)\rangle = -\bar{c}_0|\phi(x,\alpha)\rangle + |\psi(x,\alpha)\rangle + \bar{c}_0\pi_c^0|\chi(x,\alpha)\rangle + i\pi_c^0|\eta(x,\alpha)\rangle. \tag{2.12}$$

The π_c^0 parts should of course be omitted in the open-string case. The physical components are contained in the $|\phi(x,\alpha)\rangle$ part, whose lower excitations are exemplified for the open-string case as

$$|\phi(x,\alpha)\rangle = \varphi(x,\alpha)|0\rangle + \left[A_\mu(x,\alpha)\alpha_{-1}^\mu|0\rangle + i\bar{C}(x,\alpha)c_{-1}|0\rangle + C(x,\alpha)\bar{c}_{-1}|0\rangle \right] + \cdots . \tag{2.13}$$

The above ghost number assignment $N_{FP} = -1$ for Φ corresponds to the assignment $N_{FP} = 0$ for this physical component ϕ , hence implying that

---

*) Do not confuse c_0 appearing in (2.7b) for the open-string case with the zero-mode c_0 in closed-string $c(\sigma)$. The former is not the zero-mode in $c(\sigma)$ but the zero-mode (momentum) $\partial/\partial\bar{c}_0$ contained in $i\pi_{\bar{c}}(\sigma)$.

the coefficient fields φ, A_μ, $\bar{C}$ and C are assigned the ghost numbers N_{FP} = 0, 0, -1 and +1, respectively. In this sense, N_{FP} is the net ghost number carried also by the coefficient fields. Therefore the condition N_{FP} = -1 for Φ is not actually a constraint on Φ but merely an assignment of ghost numbers on the coefficient fields. On the other hand, we can define an internal ghost number, denoted by n_{FP}, counting only the ghost numbers of oscillators c_n and $\bar{c}_n$ alone. We shall impose an actual constraint n_{FP} = -1 on Φ for the case of gauge-invariant action below, in which case the states $c_{-1}|0\rangle$ and $\bar{c}_{-1}|0\rangle$ in (2.13) carring n_{FP} = +1 and -1, for instance, are forbidden.

3. STRING FIELD THEORY: FREE CASE

As is now well-known, Lorentz covarint (first) quantization of one-string system was carried out by Kato and Ogawa[10] based on the BRS invariance.[11,12] For the orthonormal gauge $g^{ab} \propto \eta^{ab}$, the BRS operator Q_B for open-string is found to be given by

$$Q_B = \int_0^\pi \frac{d\sigma}{\pi} \left\{ i\pi_{\bar{c}}\left[-\frac{1}{2}(\eta^{\mu\nu}P_\mu P_\nu + \eta_{\mu\nu}X^{\mu\prime}X^{\nu\prime}) + i(c'\bar{c} - \pi'_{\bar{c}}\pi_c) \right] \right.$$
$$\left. - c(P_\mu X^{\mu\prime} + c'\pi_c + \pi'_{\bar{c}}\bar{c}) \right\} \tag{3.1a}$$

$$= \sum_{\pm} \int_0^\pi \frac{d\sigma}{2\pi} : C_\pm\left(-\frac{1}{2}A_\pm^2 - i\frac{dC_\pm}{d\sigma}\bar{C}_\pm + \alpha(0) \right): \tag{3.1b}$$

$$= -\sum_{n=-\infty}^{\infty} : c_{-n}\left(L_n^X + \frac{1}{2}L_n^{FP} - \alpha(0)\delta_{n,0} \right): \tag{3.1c}$$

where : : denotes the normal product, $A_\pm^\mu$, $C_\pm$ and $\bar{C}_\pm$ in (3.1b) are the variables defined in (2.7), and L_n^X and L_n^{FP} are the Virasoro opertors for usual string and ghost modes, respectively, given by

$$L_n^X = \frac{1}{2}\sum_{m=-\infty}^{\infty} : \alpha_{-m}\cdot\alpha_{n-m}: \quad , \qquad L_n^{FP} = \sum_{m=-\infty}^{\infty}(n+m) : \bar{c}_{n-m}c_m : \tag{3.2}$$

Kato and Ogawa have shown the following : i) The BRS operator Q_B satisfies the nilpotency

$$Q_B^2 = 0 , \tag{3.3}$$

if and only if the space-time dimension d equals 26 and the intercept parameter $\alpha(0)$ = 1. ii) Define physical states by imposing

physical state condition : $\quad Q_B|\Phi\rangle = 0 ,$ (3.4)

"gauge" condition : $\quad \bar{c}_0|\Phi\rangle = 0 .$ (3.5)

Then, the physical states obey the on-shell equation

$$L|\Phi\rangle = 0 \qquad \left(L \equiv \{Q_B, \bar{c}_0\} = -L_0^X - L_0^{FP} + \alpha(0) \right)$$
(3.6)

and have the form

$$|\Phi\rangle = |DDF\rangle + \tilde{Q}_B|*\rangle \qquad \left(\tilde{Q}_B = Q_B|_{c_0 = \bar{c}_0 = 0} \right). \tag{3.7}$$

Here the first part $|DDF\rangle$ stands for the state spanned by the 24 transverse modes $X^{\mu=1\sim24}$ alone and has positive norm, and the other dangerous non-positive metric modes $X^{\mu=0,25}$, c and $\bar{c}$ appear only in the second term of the form $\tilde{Q}_B|*\rangle$, i.e., only in zero-norm combinations. Thus this gives rise to an alternative proof of the well-known no-ghost theorem of string theory.[13]

168

The first quantization of <u>closed</u>-string goes quite similarly. Here we only note the BRS operator in that case, which is defined by the same equation as (3.1a and b) with $\int_o^\pi d\sigma$ now replaced by $2 \times \int_{-\pi}^\pi d\sigma$. [The factor 2 is multiplied merely for the convenience sake.] Then the BRS operator of closed string is just the sum of two "open-string BRS operators" of the right- and left- moving modes :

$$Q_B = Q_{B+} + Q_{B-} \ ,$$

$$Q_{B\pm} = 2 Q_B^{open} \left[(\alpha_n, c_n, \bar{c}_n) \to (\alpha_n^{(\pm)}, c_n^{(\pm)}, \bar{c}_n^{(\pm)}) \right] \ . \tag{3.8}$$

The second quantization is performed simply by changing the string state $|\Phi\rangle$ at the first quantization level into a field operator (or classical field) Φ as usual. The action of the string field Φ is constructed so that it yields the previous equation (3.4), $Q_B |\Phi\rangle = 0$, as a field equation now; thus we have [14]

$$S_o = \Phi \cdot Q_B \Phi \tag{3.9}$$

as the free level action, where the dot denotes the inner product $\Phi \cdot \Psi \in \mathbb{C}$ defiend by

$$\Phi \cdot \Psi \equiv \begin{cases} \int [d\tilde{z}] \, tr \left(\Phi[\tilde{z}] \, \Psi[\tilde{z}] \right) & \text{for open-string} \\[2ex] \int [d\tilde{z}] \, \pi_c^0 \, \Phi[\tilde{z}] \Psi[\tilde{z}] & \text{for closed-string} \end{cases} \tag{3.10}$$

Owing to the nilpotency (3.3) of Q_B at $d = 26$, this action (3.9) clearly possesses the local gauge invariance under

$$\delta\Phi = Q_B \Lambda \tag{3.11}$$

with $\Lambda[Z]$ being the gauge transformation <u>functional</u> parameter. The string field Φ is real and constraiend to be the one carrying the internal ghost number $n_{FP} = -1$ as well as the net one $N_{FP} = -1$ in this gauge invariant action, and so the transformation parameter Λ should be purely imaginary, $\Lambda^\dagger[z] = -\Lambda[\tilde{z}]$ and carry $N_{FP} = n_{FP} = -2$, accordingly. A careful reader may have noticed the presence of the "measure" π_c^0 (the extra ghost zero-mode of closed-string) in the definition of the inner product (3.10) for the closed-string case, and wondered whether it may violate the partial integration law

$$Q_B \Phi \cdot \Psi = - (-)^{|\Phi|} \Phi \cdot Q_B \Psi \ , \tag{3.12}$$

which is necessary for the gauge invariance. [$|\Phi|$ is 1 or 0 when Φ is Grassmann-odd or -even, respectively.] Actually π_c^0 does not anti-commutes with Q_B and yields

$$\{ Q_B , \pi_c^0 \} = i (L_+ - L_-) \qquad \left(L_\pm \equiv - L_o^{X(\pm)} - L_o^{FP(\pm)} + \alpha(0) \right). \tag{3.13}$$

But, since the closed-string field Φ is constrained by $\mathcal{P}\Phi = \Phi$ as explained in (2.4) and so should be Λ, $\mathcal{P}\Lambda = \Lambda$, accordingly, this operator $L_+ - L_-$ vanishes both on Φ and Λ and hence the partial integration law (3.12) is valid also in closed-string case.

This system (3.9) is indeed satisfactory : We have $Q_B \Phi = 0$ as the equation of motion and can impose

$$\bar{c}_o \Phi = 0 \tag{3.14}$$

as a <u>gauge-fixing condition</u> corresponding to the gauge-invariance under

(3.11). So the two conditions (3.4) and (3.5) of Kato-Ogawa's are reproduced.

[Digression : It is interesting to note that a rather natural form of free gauge-invariant action (3.1) has not been settled until late of 1985, although many authors have suspected it undoubtedly. After the pioneering Siegel's work[4] on gauge-fixed free action, Banks and Peskin, and Kaku[15] proposed a non-local gauge-invariant action, and subsequently its local version was found by Siegel and Zwiebach, Banks and Peskin and Itoh, Kugo, Kunitomo and Ooguri,[16] independently, in Summer '85 which is known now as the minimal form of gauge-invariant action. The action (3.1) itself was given by Witten[8] and Neveu, Nicolai and West[14] finally in fall '85.]

The gauge invariance (3.11) is in fact an enormously huge invariance since Λ is a functional parameter. It is instructive to see that it actually contains i) the Yang-Mills gauge transformation for the open-string case and ii) the general coordinate transformation for the closed-string case (of their free level form, of course).

Open-String: The string field Φ and the transformation parameter Λ are expanded in their "local field" components as follows :

$$|\Phi\rangle = -\bar{c}_0 \left(\underset{N=0}{\varphi(x)|0\rangle} + \underset{N=1}{A_\mu(x)\alpha_{-1}^\mu|0\rangle} + \underset{N\geq2}{\cdots} \right) + \left(\underset{N=1}{\tfrac{i}{2}\tilde{B}(x)\bar{c}_{-1}|0\rangle} + \underset{N\geq2}{\cdots} \right), \quad (3.15)$$

$$|\Lambda\rangle = -\bar{c}_0 \left(i\,\varepsilon(x)\,\bar{c}_{-1}|0\rangle + \cdots \right) + \cdots .$$

[Here and hereafter in this section we omit the dependence on α parameter.] Then, the gauge transformation (3.11) yields

$$\delta\Phi = Q_B\Lambda \implies \begin{cases} \delta\varphi(x) = 0 \\ \delta A_\mu(x) = \partial_\mu\varepsilon(x) \\ \delta B(x) = 0 \qquad (B(x) \equiv \tilde{B}(x) + \partial^\mu A_\mu(x)) \\ \vdots \end{cases} \quad (3.16)$$

and the action (3.9) produces

$$\Phi\cdot Q_B\Phi = \int dx\, tr\left\{ \tfrac{1}{2}\varphi(\Box+2)\varphi + \tfrac{1}{2}A^\mu(\Box\eta_{\mu\nu}-\partial_\mu\partial_\nu)A^\nu + \tfrac{1}{2}B^2 + \cdots \right\}. \quad (3.17)$$

From (3.16) and (3.17) we see that $\varphi(x)$ is a tachyon field, $A_\mu(x)$ is the Yang-Mills field and $B(x)$ is the Nakanishi-Lautrup field : (3.16) and (3.17) reproduce the free part of the YM gauge transformation $D_\mu\varepsilon(x)$ and the action $-\tfrac{1}{4}\,tr\,F_{\mu\nu}^2$, respectively.

Closed-String: The local field expansions of Φ and Λ for the closed-string case are given by

$$|\Phi\rangle = -\bar{c}_0\Big\{ \underset{N_\pm=0}{\varphi(x)|0\rangle} + \Big[\tfrac{1}{2}\hat{h}_{\mu\nu}(x)(\alpha_{-1}^\mu\alpha_{-1}^\nu)^{(+-)}|0\rangle + \tfrac{1}{2}A_{\mu\nu}(x)(\alpha_{-1}^\mu\alpha_{-1}^\nu)^{[+-]}|0\rangle $$
$$ + \hat{D}(x)(c_{-1}\bar{c}_{-1})^{(+-)}|0\rangle + \underset{N_\pm\geq2}{S(x)(c_{-1}\bar{c}_{-1})^{[+-]}|0\rangle} \Big] + \cdots \Big\}$$
$$ \underset{N_\pm=1}{}$$

$$ + \tfrac{i}{2}\Big\{ \big[b_\mu(x)(\bar{c}_{-1}\alpha_{-1}^\mu)^{(+-)}|0\rangle + \underset{N_\pm\geq2}{e_\mu(x)(\bar{c}_{-1}\alpha_{-1}^\mu)^{[+-]}|0\rangle} \big] + \cdots \Big\},$$
$$ \underset{N_\pm=1}{}$$
$$(3.18)$$
$$|\Lambda\rangle = -\bar{c}_0\Big\{ -i\big[\varepsilon_\mu(x)(\bar{c}_{-1}\alpha_{-1}^\mu)^{(+-)}|0\rangle + \zeta_\mu(x)(\bar{c}_{-1}\alpha_{-1}^\mu)^{[+-]}|0\rangle \big] + \cdots \Big\}$$
$$ - \tfrac{1}{2\sqrt{2}}\eta(x)\,\bar{c}_{-1}^{(+)}\bar{c}_{-1}^{(-)}|0\rangle + \cdots ,$$

where we have omitted the π_c^0-proportional parts which do not contribute to the action (3.9) owing to the presence of the "measure" π_c^0 in (3.10), and have used the abbreviation,

$$(ab)^{(+-)} \equiv (a^{(+)}b^{(-)} + a^{(-)}b^{(+)})/\sqrt{2} ,$$
$$(ab)^{[+-]} \equiv (a^{(+)}b^{(-)} - a^{(-)}b^{(+)})/\sqrt{2} . \tag{3.19}$$

Then the action (3.9) becomes

$$\Phi \cdot Q_B \Phi = \int dx \Big[\tfrac{1}{2} \varphi(\Box+8)\varphi + \tfrac{1}{4} \hat{h}_{\mu\nu} \Box \hat{h}^{\mu\nu} + \tfrac{1}{4} A_{\mu\nu} \Box A^{\mu\nu}$$

$$- \tfrac{1}{2} \hat{D} \Box \hat{D} + \tfrac{1}{2} S \Box S - \tfrac{1}{2}(b_\mu^2 + e_\mu^2) \tag{3.20}$$

$$+ b^\mu(\partial^\nu \hat{h}_{\mu\nu} - \partial_\mu \hat{D}) - e^\mu(\partial^\nu A_{\mu\nu} - \partial_\mu S)\Big] + \cdots$$

and the gauge-transformation (3.11) gives

$$\delta\Phi = Q_B\Lambda \quad \Rightarrow \quad \begin{cases} \delta\hat{h}_{\mu\nu} = \partial_\mu \varepsilon_\nu + \partial_\nu \varepsilon_\mu , & \delta A_{\mu\nu} = \partial_\mu \zeta_\nu - \partial_\nu \zeta_\mu , \\ \delta\hat{D} = \partial\cdot\varepsilon , & \delta S = \partial\cdot\zeta + \eta , \\ \delta b_\mu = \Box \varepsilon_\mu , & \delta e_\mu = \Box \zeta_\mu + \partial_\mu\eta . \end{cases} \tag{3.21}$$

We can eliminate $b_\mu(x)$ and $e_\mu(x)$ since they are auxiliary fields as clear in the action (3.20), and gauge away $S(x)$ by using the gauge freedom $\eta(x)$ in (3.21). Then, performing also the field redefinitions

$$\hat{h}_{\mu\nu}(x) = h_{\mu\nu}(x) + \eta_{\mu\nu} D(x) , \qquad \hat{D}(x) = D(x) + \tfrac{1}{2} h_\mu^{~\mu}(x) , \tag{3.22}$$

we find

$$\Phi \cdot Q_B \Phi = \int dx \Big\{ \tfrac{1}{2} \varphi(\Box+8)\varphi$$

$$+ \Big[-\tfrac{1}{2\kappa^2}(\sqrt{-g}R)_{lin.} + \tfrac{1}{2!3!} F_{\mu\nu\rho}F^{\mu\nu\rho} + \tfrac{d-2}{4} D\Box D\Big] + \cdots\Big\}, \tag{3.23}$$

$$\begin{cases} \delta h_{\mu\nu} = \partial_\mu \varepsilon_\nu + \partial_\nu \varepsilon_\mu , & \delta\varphi = 0 , \\ \delta A_{\mu\nu} = \partial_\mu \zeta_\nu - \partial_\nu \zeta_\mu , & \delta D = 0 . \end{cases} \tag{3.24}$$

From this, we now see that φ is a tachyon and $h_{\mu\nu}$, $A_{\mu\nu}$ and D are graviton, anti-symmetric tensor gauge field and dilaton, respectively : The first term of the square bracket in (3.23) gives the linearized Einstein action

$$-\tfrac{1}{2\kappa^2}(\sqrt{-g}R)_{lin.} = \tfrac{1}{4} h^{\mu\nu}(\Box h_{\mu\nu} - 2\partial_\nu\partial^\rho h_{\mu\rho} + 2\partial_\mu\partial_\nu h_\rho^{~\rho} - \eta_{\mu\nu}\Box h_\rho^{~\rho}) \tag{3.25}$$

with the usual identification of $h_{\mu\nu}(x)$ with the metric $g_{\mu\nu}(x)$

$$g_{\mu\nu}(x) = \eta_{\mu\nu} + \kappa h_{\mu\nu}(x) . \tag{3.26}$$

The next term in (3.23) is the free action of anti-symmetric tensor gauge field $A_{\mu\nu}$ whose field strength is given by

$$F_{\mu\nu\rho} = \partial_\mu A_{\nu\rho} + \partial_\nu A_{\rho\mu} + \partial_\rho A_{\mu\nu} . \tag{3.27}$$

The gauge transformation (3.24) also reproduces correctly the linearized version of general coordinate transformation and anti-symmetric tensor gauge transformation, while the tachyon φ and dilaton D are left invariant.

EXERCISE : i) Show that the open-string BRS operator (3.1) is rewritten by making explicit the dependence on the ghost zero-modes $\bar{c}_0$ and $c_0 = \partial/\partial\bar{c}_0$ as

$$Q_B = c_o L + \bar{c}_o M + \tilde{Q}_B \ , \qquad \tilde{Q}_B \equiv Q_B\big|_{\bar{c}_o = c_o = 0} \ ,$$

$$L = -L_o^X - L_o^{FP} + 1 = -\tfrac{1}{2}p^2 - N + 1 \ , \qquad (3.28)$$

$$N = \sum_{n=1}^{\infty} \left\{ \alpha_{-n}\cdot\alpha_n + n\left(c_{-n}\bar{c}_n + \bar{c}_{-n}c_n\right)\right\} \ , \qquad M = 2\sum_{n=1}^{\infty} n\, c_{-n}c_n \ .$$

N is the mode-number counting operator, whose values are already indicated for the states in (3.15). Also derive the same form of expression for the closed-string Q_B in $\pi_c^o = 0$ sector with L and M then reading

$$L = -\tfrac{1}{2}p^2 - 2(N_+ + N_-) + 4 \ , \qquad M = M_+ + M_- \ , \qquad (3.29)$$

where $N_\pm$ and $M_\pm$ are the same operators as N and M in (3.28) with the right- and left-moving modes $\alpha_n^{(\pm)}$, $c_n^{(\pm)}$ and $\bar{c}_n^{(\pm)}$ substituted, respectively.

ii) Confirm the above "local field" expressions of the action and gauge-transformation, eqs.(3.16), (3.17), (3.20) and (3.21), using the fact that the inner product (3.10) is given equivalently in terms of bra-ket notation by

$$\Phi \cdot \Psi = \int dx\, d\bar{c}_o \ (tr) \ \langle \Phi(x,\bar{c}_o)|\Psi(x,\bar{c}_o)\rangle\big|_{\pi_c^o = 0}$$

for hermitian Φ.

---

4. STRING FIELD THEORY : INTERACTING CASE

We now come to the interacting string field theory. In the case of point-particle field theory, the simplest local interaction term $g\,\phi^3$ was written in the form

$$g\phi^3 = g\int dx_1\, dx_2\, dx_3 \ \phi(x_1)\phi(x_2)\phi(x_3) \, V(x_1,x_2,x_3) \ ,$$

$$V(x_1,x_2,x_3) = \delta(x_1 - x_3)\delta(x_2 - x_3) \ . \qquad (4.1)$$

Just similarly to this, the string interaction term is given in the following form by using a suitable vertex functional $V[Z_1, Z_2, Z_3]$:

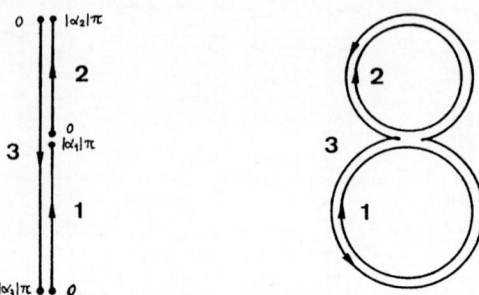

Fig.1.   The overlapping structure of the open and closed string vertixes in HIKKO's theory. These figures stand for the case in which the string 3 is the longest in σ-space.

$$g \Phi^3 = g \int [d Z_1 d Z_2 d Z_3] \, \Phi[Z_1] \Phi[Z_2] \Phi[Z_3] V[Z_1, Z_2, Z_3] . \qquad (4.2)$$

The vertex functional $V[Z_1, Z_2, Z_3]$ is essentially the δ-functional of string's coordinates Z. In HIKKO's theory[6], it has the overlapping structure as shown in Fig.1, which is familiar in the light-cone gauge string field theory[26]. Referring to this figure the vertex functional is given in the form

$$V[Z_1, Z_2, Z_3] = (\mathcal{P}_{123}) \, \pi_{\bar{c}}(\sigma_{int}) \, \delta\left(\sum_{r=1}^{3} \alpha_r\right)$$

$$\times \prod_{|\sigma| \leq \pi|\alpha_1|} \delta\left(X^{(1)}(\sigma_1) - X^{(3)}(\sigma_3)\right) \delta\left(\alpha_1 c^{(1)}(\sigma_1) - \alpha_3 c^{(3)}(\sigma_3)\right) \delta\left(\frac{\bar{c}^{(1)}(\sigma_1)}{\alpha_1^2} - \frac{\bar{c}^{(3)}(\sigma_3)}{\alpha_3^2}\right)$$

$$\times \prod_{\pi|\alpha_1| \leq \sigma \leq \pi|\alpha_3|} \delta\left(X^{(2)}(\sigma_2) - X^{(3)}(\sigma_3)\right) \delta\left(\alpha_2 c^{(2)}(\sigma_2) - \alpha_3 c^{(3)}(\sigma_3)\right) \delta\left(\frac{\bar{c}^{(2)}(\sigma_2)}{\alpha_2^2} - \frac{\bar{c}^{(3)}(\sigma_3)}{\alpha_3^2}\right),$$

$$\sigma_1(\sigma) = \frac{\sigma}{|\alpha_1|}, \quad \sigma_2(\sigma) = \frac{\sigma - \pi|\alpha_1| sgn(\sigma)}{|\alpha_2|}, \quad \sigma_3(\sigma) = \frac{\pi|\alpha_3| sgn(\sigma) - \sigma}{|\alpha_3|} . \qquad (4.3)$$

where this is the expression for the $|\alpha_3| = |\alpha_1| + |\alpha_2|$ case and $0 \leq \sigma \leq \pi|\alpha_3|$ for open string case and $-\pi|\alpha_3| \leq \sigma \leq \pi|\alpha_3|$ for closed string case. The ghost prefactor $\pi_{\bar{c}}(\sigma_{int})$ at the interaction point is necessary for the BRS invariance of the vertex as expalined below, and $\mathcal{P}_{123} = \mathcal{P}^{(1)} \mathcal{P}^{(2)} \mathcal{P}^{(3)}$ is the product of projection operators defiend in (2.6) which is necessary only in the closed-string case.

In Witten's theory[8], the vertex functional has the structure as shown in Fig.2. Since the strings interact at their mid-point $\sigma = \frac{\pi}{2}$, this type of vertex is called mid-point interaction vertex and was first considered by Goto.[17] Although the Witten's open string veretx has a natural extention to the closed string case as drawn in Fig.2, it is known[18] not to give an approprite vertex satisfying the gauge invariance as far as one imposes the constraint $\mathcal{P} \Phi = \Phi$ on the closed string field. So Witten's theory exists only for the open string at present.[19]

In HIKKO's theory, only for the open-string case, it is necessary further to introduce a 4-string interaction vertex $V^{(4)}$ in the form

$$V^{(4)}[Z_1, Z_2, Z_3, Z_4] = \int_{\sigma_-}^{\sigma_+} d\sigma_0 \, f(\sigma_0) \, \pi_{\bar{c}}(\sigma_{int}) \, V_0^{(4)}(\sigma_0) . \qquad (4.4)$$

Here $V_0^{(4)}(\sigma_0)$ stands for the overlapping δ-functional corresponding to the Fig.3, given similarly to the 3-string case, and σ_0 is a parameter specifying the interaction point (as shown in Fig.3) at which two strings cross over and rejoin. Since the interaction point is not fixed even if the string-lengths $\alpha_1 \sim \alpha_4$ are fixed, we integrated in (4.4) over σ_0 from its minimum σ_- to the maxium σ_+ with a suitable weight function $f(\sigma_0)$ determined later. The 4-string configurations at the end-points $\sigma_0 = \sigma_+$ and σ_- are shown in Figs.4 and 5, respectively.

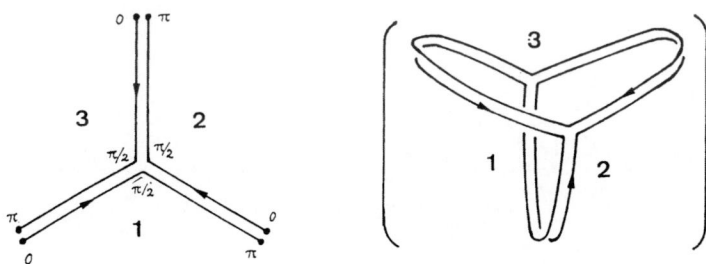

Fig.2.   The structure of Witten's open string vertex (and its straightforward extention to the closed string case).

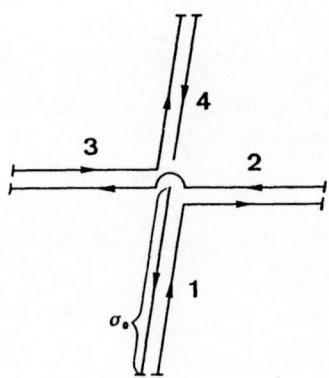

Fig.3.   Structure of the overlapping δ-functional $V_0^{(4)}(\sigma_0)$ in 4-string vertex.

a) $\sigma_0 = \sigma_+$

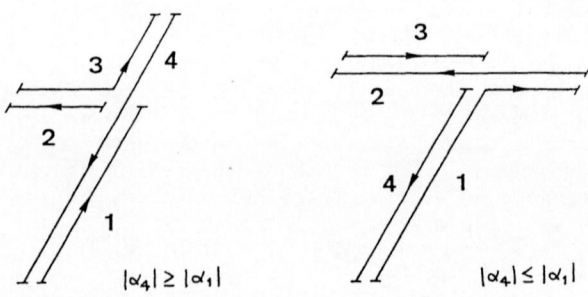

b) $\sigma_0 = \sigma_-$

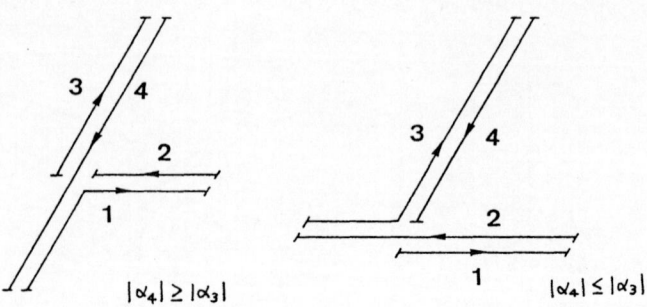

Fig.4.   4-string configurations at a) $\sigma_0 = \sigma_+$ and b) $\sigma_0 = \sigma_-$.

Commonly to HIKKO and Witten, the * product gives a string field $\Phi * \Psi$ from arbitrary two string field Φ and Ψ and is defined by using the 3-string vertex V as

$$\left(\Phi * \Psi\right)[Z_3] = \int [dZ_1 dZ_2]\, \Phi[Z_1]\Psi[Z_2]\, V[Z_1, Z_2, \tilde{Z}_3]. \tag{4.5}$$

Only in HIKKO's open-string case, another product $\Phi \circ \Psi \circ \Lambda$ of arbitrary three string fields Φ, Ψ and Λ is necessary to be introduced by using the 4-open-string vertex $V^{(4)}$ (4.4) :

$$\left(\Phi \circ \Psi \circ \Lambda\right)[Z_4] = \int \Phi[Z_1]\Psi[Z_2]\Lambda[Z_3] V^{(4)}[Z_1, Z_2, Z_3, \tilde{Z}_4]\, [dZ_1 dZ_2 dZ_3]. \tag{4.6}$$

With these definitions we can prove the following string field algebra[20-22] as will be shown later :

I) Open-String
distributive law $0(g^1)$ $\quad Q_B(\Phi * \Psi) = Q_B\Phi * \Psi + (-)^{|\Phi|}\Phi * Q_B\Psi$, $\quad$ (4.7a)

"associative" laws
$0(g^2)$
$$\left(\Phi * \Psi\right)*\Lambda - \Phi*\left(\Psi*\Lambda\right) \tag{4.7b}$$
$$= (-)^{|\Phi|+|\Psi|+|\Lambda|}\left\{ Q_B(\Phi \circ \Psi \circ \Lambda) - \left(Q_B\Phi \circ \Psi \circ \Lambda + (-)^{|\Phi|}\Phi \circ Q_B\Psi \circ \Lambda + (-)^{|\Phi|+|\Psi|}\Phi \circ \Psi \circ Q_B\Lambda\right)\right\},$$

$0(g^3)$ $\quad (-)^{|\Sigma|}\left(\Phi \circ \Psi \circ \Lambda\right)*\Sigma + \Phi*\left(\Psi \circ \Lambda \circ \Sigma\right)$
$$= (\Phi*\Psi)\circ\Lambda\circ\Sigma - \Phi\circ(\Psi*\Lambda)\circ\Sigma + \Phi\circ\Psi\circ(\Lambda*\Sigma), \tag{4.7c}$$

$0(g^4)$
$$(-)^{|\Sigma|+|\Xi|}\left(\Phi \circ \Psi \circ \Lambda\right)\circ\Sigma\circ\Xi + (-)^{|\Xi|}\Phi\circ\left(\Psi\circ\Lambda\circ\Sigma\right)\circ\Xi + \Phi\circ\Psi\circ\left(\Lambda\circ\Sigma\circ\Xi\right) = 0. \tag{4.7d}$$

II) Closed-String
(anti-) commutativity $\quad \Phi * \Psi = -(-)^{|\Phi||\Psi|}\Psi * \Phi$ $\quad$ (4.8a)

distributive law $0(g^1)$ $\quad Q_B(\Phi * \Psi) = Q_B\Phi * \Psi + (-)^{|\Phi|}\Phi * Q_B\Psi$, $\quad$ (4.8b)

Jacobi identity $0(g^2)$ $\quad \left(\Phi*\Psi\right)*\Lambda + (-)^{|\Phi|(|\Psi|+|\Lambda|)}(\Psi*\Lambda)*\Phi$
$$+ (-)^{|\Lambda|(|\Phi|+|\Psi|)}(\Lambda*\Phi)*\Psi = 0, \tag{4.8c}$$

III) Cyclic symmetries

open & closed $\quad \Phi \cdot \Psi = (-)^{|\Phi||\Psi|}\Psi \cdot \Phi$, $\quad$ (4.9a)

$$\Phi \cdot (\Psi*\Lambda) = (-)^{|\Phi|(|\Psi|+|\Lambda|)}\Psi \cdot (\Lambda * \Phi), \; etc, \tag{4.9b}$$

open $\quad \Phi \cdot (\Psi \circ \Lambda \circ \Sigma) = -(-)^{|\Phi|(|\Psi|+|\Lambda|+|\Sigma|)}\Psi \cdot (\Lambda \circ \Sigma \circ \Phi), etc. \tag{4.9c}$

These identities are for HIKKO's theory. The Witten's original * product is associative $(\Phi*\Psi)*\Lambda = \Phi*(\Psi*\Lambda)$ and non-commutative $\Phi*\Psi \not\propto \Psi*\Phi$. It is, however, not his * product itself but rather its graded commutator

$$[\Phi, \Psi\} = \frac{1}{2}\left(\Phi * \Psi - (-)^{|\Phi||\Psi|}\Psi * \Phi\right) \tag{4.10}$$

that actually appears in the action and the gauge transformation in his theory. Therefore if we understand the * product here in our closed-string identities (4.8) to stand for the graded commutator (4.10) in Witten's case, all our closed-string identities (4.8) including cyclic symmetries (4.9) are valid also in the Witten's theory; that is, with this understanding, the HIKKO's closed-string theory and Witten's

175

open-string one obey the same form of string field algebra. It should also be emphasized that any of the important identities (4.7) and (4.8) (except for (4.8a)) turns out to hold if and only if the space-time dimension d is 26 (and $\alpha(0) = 1$) just as $Q_B^2 = 0$.

Once these identities of string field algebra are established, it is an easy task to write down the gauge-invariant action as well as the gauge transformation.

The action of fully interacting string field theory is given by [6,8]

$$S = \Phi \cdot Q_B \Phi + \frac{2}{3} g \, \Phi^3 \left[+ \frac{2}{4} g^2 \Phi^4 \right],$$

$$\Phi^3 \equiv \Phi \cdot (\Phi * \Phi), \qquad \Phi^4 \equiv \Phi \cdot (\Phi \circ \Phi \circ \Phi). \tag{4.11}$$

[The Φ^4 term, or ($\circ\circ$) product, exists only in the HIKKO's open-string case.] This action is invariant under the gauge transformation

$$\delta \Phi = Q_B \Lambda + g(\Phi * \Lambda - \Lambda * \Phi) \left[+ g^2(\Phi \circ \Phi \circ \Lambda - \Phi \circ \Lambda \circ \Phi + \Lambda \circ \Phi \circ \Phi) \right]. \tag{4.12}$$

Writing the inner product as $\Phi \cdot \Psi = \int \Phi * \Psi$, Witten observed an analogy between his string field theory and Yang-Mills system :

$$\Phi \longleftrightarrow A = A_\mu \, dx^\mu, \qquad Q_B \longleftrightarrow d,$$

$$S_{Witten} = \int (\Phi * Q_B \Phi + \frac{2}{3} g \, \Phi * \Phi * \Phi) \longleftrightarrow Chern\text{-}Simons\ 3\text{-}form \int (AdA + \frac{2}{3} g A^3),$$

$$\delta \Phi = Q_B \Lambda + g(\Phi * \Lambda - \Lambda * \Phi) \longleftrightarrow \delta A = d\Lambda + g[A, \Lambda]. \tag{4.13}$$

As an analogous quantity to the field strength $F = dA + gA^2$, we have

$$\delta_B \Phi = Q_B \Phi + g \Phi * \Phi \left[+ g^2 \Phi \circ \Phi \circ \Phi \right], \tag{4.14}$$

which happens to be the full "BRS transformation" of HIKKO's obtained by generalizing the Siegel's linear one $\delta_B \Phi = Q_B \Phi$. The analogy proceeds further; corresponding to the Bianchi identity $0 = dF + g[A, F]$, we have (off-shell) nilpotency of the "BRS transformation" (4.14) :

$$0 = \delta_B(\delta_B \Phi) = -Q_B \, \delta_B \Phi + g(\delta_B \Phi * \Phi - \Phi * \delta_B \Phi)$$

$$\left[+ g^2(\delta_B \Phi \circ \Phi \circ \Phi - \Phi \circ \delta_B \Phi \circ \Phi + \Phi \circ \Phi \circ \delta_B \Phi) \right] \tag{4.15}$$

The commutator of two gauge transformations (4.12) with parameters Λ_1 and Λ_2 is found to be

$$\left[\delta(\Lambda_1), \delta(\Lambda_2) \right] = \delta(2g\Lambda_1 * \Lambda_2) \tag{4.16}$$

for the closed-string case (and Witten's case also with the * product understood as the graded commutator of his * product). This implies that the gauge-transformation algebra closes and the structure constant of this infinite dimensional "Lie algebra" is given by the 3-string vertex functional $2gV[z_1, z_2, \tilde{z}_3]$. For the open-string case, on the other hand, we find

$$\left[\delta(\Lambda_1), \delta(\Lambda_2) \right] \Phi = \delta(\Lambda_3) \Phi - g^2(S_{,\Phi} \circ \Lambda_1 \circ \Lambda_2 - \Lambda_1 \circ S_{,\Phi} \circ \Lambda_2 + \Lambda_1 \circ \Lambda_2 \circ S_{,\Phi}),$$

$$\Lambda_3 = g(\Lambda_1 * \Lambda_2) + g^2(\Phi \circ \Lambda_1 \circ \Lambda_2 - \Lambda_1 \circ \Phi \circ \Lambda_2 + \Lambda_1 \circ \Lambda_2 \circ \Phi) - (1 \leftrightarrow 2),$$

$$S_{,\Phi} = \frac{1}{2} \frac{\delta}{\delta \Phi} S = \delta_B \Phi, \tag{4.17}$$

implying that the gauge-transformation algebra of (HIKKO's) open-string closes only on-shell and the structure constant depends on the field Φ.

EXERCISE : i) Prove the nilpotency $\delta_B(\delta_B\Phi) = 0$, (4.15), of the "BRS transformation" (4.14) using the identities (4.7) and (4.8).
ii) The action S (4.11) is constructed so as to yield

$$\delta S = 2\,\delta\Phi \cdot \delta_B\Phi \qquad\qquad (4.18)$$

in the case of gauge transformation,

for arbitrary variation $\delta\Phi$ of Φ. Show that (4.18) is rewritten into the form $\delta S = 2\Lambda \cdot \delta_B(\delta_B\Phi)$ using (4.9) and (3.12), thus proving the gauge-invariance by the nilpotency $\delta_B(\delta_B\Phi) = 0$.
iii) Confirm the gauge-transformation algebra (4.16) and (4.17).
iv) Prove the following identity by using (4.7~9) :

$$\delta_B\Phi \cdot \delta_B\Phi = 0. \qquad\qquad (4.19)$$

Let us conclude this section by presenting the gauge-fixed BRS-invariant action. The gauge-fixed action is given from the gauge-invariant one (4.11) simply by setting the ψ-component of $\Phi = -\bar{c}_0\phi + \psi$ equal to zero*):

$$\hat{S}[\phi] = \left[\Phi \cdot Q_B\Phi + \tfrac{2}{3}g\Phi^3 \left(+\tfrac{2}{4}g^2\Phi^4\right)\right]_{\psi=0} = \phi \cdot L\phi + \tfrac{2}{3}g\phi^3 \left(+\tfrac{2}{4}g^2\phi^4\right). \qquad (4.20)$$

Note, however, that although the inside of the square bracket being the same form as (4.11) the Φ field here is a relaxed one subject to no constraint on its internal ghost number n_{FP}. This action $\hat{S}$ is invariant under the following BRS transformation $\hat{\delta}_B\phi$ obtained from the previous "BRS transformation" (4.14) of ϕ by setting $\psi = 0$ again :

$$\hat{\delta}_B\phi = \delta_B\phi\big|_{\psi=0} = \int d\bar{c}_o\,\delta_B\Phi\big|_{\psi=0}. \qquad (4.21)$$

Although no one has clarified the detailed gauge-fixing procedure how to derive this gauge-fixed action (4.20) and its BRS transformation (4.21) from the original gauge-invariant system (4.11), they are just the action and BRS transformation that HIKKO presented in their first paper prior to finding their gauge-invariatn action.**) So we can prove that i) the BRS transformation $\hat{\delta}_B\phi$ is on-shell nilpotent and ii) the action $\hat{S}$ is invariant under $\hat{\delta}_B\phi$. Similar gauge-fixed action and BRS transformation are proposed by Thorn[24] in the context of Witten's theory.

EXERCISE : i) Show that the equation of motion of the gauge-fixed system is given by $\delta\hat{S}/\delta\phi = 2\delta_B\psi\big|_{\psi=0} = 0$ by the help of (4.18). Show also that the off-shell nilpotency $(\delta_B)^2\phi = 0$ of the original "BRS transformation" (4.14) implies $(\hat{\delta}_B)^2\phi \propto \delta_B\psi\big|_{\psi=0}$, thus proving the on-shell nilpotency of $\hat{\delta}_B$.
ii) Derive an equation $\hat{\delta}_B\hat{S} = -[\delta_B\Phi \cdot \delta_B\Phi]_{\psi=0}$ by the help of (4.18) again, which proves the BRS invariance of $\hat{S}$ by the identity (4.19) shown previously.

---

*) Note that the χ and η components of closed-string $\Phi = -\bar{c}_o\phi + \psi + \bar{c}_o\pi_c^o\chi + i\pi_c^o\eta$ do not appear in the action and in $\delta_B\phi$ from the beginning owing to the "measure" π_c^o of the inner prduct and the constraint $P\Phi = \Phi$.
**) The gauge-fixing procedure for the free system is, however, now well understood. See e.g., Refs.23.

5. PROOF OF DISTRIBUTIVE LAW

It is easy to see that the distributive law (4.7a) or (4.8b) is equivalent to the identity

$$\sum_{r=1}^{3} Q_B^{(r)} |V\rangle = 0 \tag{5.1}$$

for the 3-string vertex functional $|V\rangle$ (in the ket notation). To prove this is the subject of this section. Let us start with the open-string case.

5.1. 3-String Vertex (Open-String)

Note that δ-fuction $\delta(x-y)$, for instance, generally implies not only the connection of the coordinates $(x-y)\,\delta(x-y) = 0$ but also the connection of the conjugate momenta $(\partial/\partial x + \partial/\partial y)\,\delta(x-y) = 0$. Therefore the δ-fuctional part of the vertex (4.3) i.e., the (continuum) product of infinite numbers of δ-fuctions which we denote by V_0 may be characterized more precisely by the following Goto-Naka[25] type connection conditions

$$\left(\Theta_1 \mathcal{O}^{(1)}(\sigma_1) + \Theta_2 \mathcal{O}^{(2)}(\sigma_2) - \mathcal{O}^{(3)}(\sigma_3) \right) |V_0\rangle = 0 , \tag{5.2}$$

$$\Theta_1(\sigma) = \theta(\pi\alpha_1 - \sigma) , \quad \Theta_2(\sigma) = \theta(\sigma - \pi\alpha_1) ,$$

$$\mathcal{O}^{(r)}(\sigma_r) = \left(X^{\mu(r)}(\sigma_r), \; \alpha_r^{-1} A_{\pm}^{\mu(r)}(\sigma_r), \; \alpha_r C_{\pm}^{(r)}(\sigma_r), \; \alpha_r^{-2} \bar{C}_{\pm}^{(r)}(\sigma_r) \right), \tag{5.3}$$

where $A_{\pm}^{\mu}$, $C_{\pm}$ and $\bar{C}_{\pm}$ are variables defined in (2.7) containing momenta P^{μ}, π_c and $\pi_{\bar{c}}$ also.

The solution to this eq.(5.2) expressed in terms of oscillator modes is well-known[26] and is given by

$$|V_0\rangle = e^{E(1,2,3)} |0\rangle \, \bar{\delta}(1,2,3) ,$$

$$E(1,2,3) = \sum_{r,s=1}^{3} \left\{ \sum_{m,n=0}^{\infty} \bar{N}_{nm}^{rs} \left(\tfrac{1}{2} \alpha_{-n}^{(r)} \cdot \alpha_{-m}^{(s)} + \alpha_r^{-1} \bar{C}_{-n}^{(r)} C_{-m}^{(s)} m\alpha_s \right) \right\}, \tag{5.4}$$

$$\bar{\delta}(1,2,3) = 2\pi \, \delta\left(\sum_r \alpha_r \right) (2\pi)^d \delta^d\left(\sum_r P_r \right) \delta\left(\sum_r \bar{c}_0^{(r)}/\alpha_r \right) ,$$

where $|0\rangle$ is the product $|0\rangle_1 |0\rangle_2 |0\rangle_3$ of Fock vacuua of 3 strings $r = 1,2,3$, and $\bar{N}_{nm}^{rs}$'s are Fourier components of the Neumann function explained shortly. Note that the exponent E, given by quadratic form of creation oscillators, may be rewritten into a manifestly OSp(d/2) symmetric form $\sum \bar{N}_{nm}^{rs} \tfrac{1}{2} a_{-n}^{N(r)} \eta_{NM} a_{-m}^{M(s)}$ if we use OSp(d/2) vector oscillator mode

$$a_n^N \equiv \left(\alpha_n^{\mu}, \; \gamma_n \equiv in\alpha C_n, \; \bar{\gamma}_n \equiv \alpha^{-1}\bar{C}_n \right) \tag{5.5}$$

satisfying OSp-covariant graded commutation relation

$$[a_n^N, a_m^M \} = n \eta^{NM} \delta_{n+m,0} , \quad \eta^{NM} = \begin{pmatrix} \eta^{\nu\mu} & 0 \\ \hline 0 & \begin{matrix} 0 & i \\ -i & 0 \end{matrix} \end{pmatrix} . \tag{5.6}$$

5.2. Neumann Function and Mandelstam Mapping

To explain how to obtain the expression (5.4) as well as for other purposes later, we first extend the operator $X^{\mu(r)}(\sigma_r)$ of each string r to $X^{\mu(r)}(\xi_r, \sigma_r)$ so as to include also the (imaginary) time variable ξ_r by letting $X^{\mu(r)}$ propagate freely on the 2-dimensional plane (ξ_r, σ_r)

$$\left[\left(\frac{\partial}{\partial \xi_r}\right)^2 + \left(\frac{\partial}{\partial \sigma_r}\right)^2 \right] X^{\mu(r)}(\xi_r, \sigma_r) = 0 \quad . \tag{5.7}$$

with $X^{\mu(r)}\big|_{\xi_r = 0} = X^{\mu(r)}(\sigma_r)$ and $\partial X^{\mu(r)}/\partial \xi_r \big|_{\xi_r = 0} = i P^{\mu(r)}(\sigma_r)$ as initial data.

Next, we glue these 3 strings' 2-dimensional planes (scaled by multiplying by the string-length parameter α_r) according to the overlapping structure of our vertex in Fig.1, and consider a complex ρ-plane as drawn in Fig.5:

$$\rho = \tau + i\sigma = \alpha_r(\xi_r + i\sigma_r) + i\beta_r + \tau_0 \quad .$$
$$(\beta_1 = 0 \ , \ \beta_2 = \alpha_1 \ , \ \beta_3 = \alpha_1 + \alpha_2) \tag{5.8}$$

We are assuming $\alpha_1 , \alpha_2 > 0$ and $\alpha_3 < 0$ so that each string strip corresponds to $\xi_r \leq 0$ and $0 \leq \sigma_r \leq \pi$. Then we can regard these 3 functions $X^{\mu(r)}(\xi_r, \sigma_r)$ as a function $X^\mu(\rho)$*) defined on the ρ-palne simply by identifying $X^\mu(\rho) = X^{\mu(r)}(\xi_r, \sigma_r)$ on each string strip.

Despite such a region-wise definition, $X^\mu(\rho)$ gives a smooth function if placed in front of the vertex δ-functional $|V_0\rangle$ owing to the connection condition (5.2). Therefore we can now define the Neumann function $N(\rho, \tilde{\rho})$ on the ρ-plane by

$$\langle \tilde{0} | T X^\mu(\rho) X^\nu(\tilde{\rho}) | V_0' \rangle = -\frac{1}{2} \eta^{\mu\nu} N(\rho, \tilde{\rho}) \ , \tag{5.9}$$

where $|V_0'\rangle$ is the δ-functional (5.4), $e^E|0\rangle$, with the total "momentum" conservation factor $\bar{\delta}(1,2,3)$ omitted. To treat the zero-modes on the same footing as the other non-zero modes, we take $|0\rangle$ in $|V_0'\rangle$ and $\langle \tilde{0}|$ in (5.9) satisfying

$$\langle \tilde{0} | p_r = x_r | 0 \rangle = 0 \quad \text{for} \quad r = 1, 2, 3, \tag{5.10}$$

and consider p_r and x_r as creation and annihilation operators, respectively, henceforth. T in (5.9) is the anti-time ordering with respect to the time variable ξ_r:

$$T \, \mathcal{O}(\rho) \mathcal{O}'(\tilde{\rho}) \equiv \theta(\tilde{\xi} - \xi) \mathcal{O}(\rho) \mathcal{O}'(\tilde{\rho}) + (-)^{|\sigma||\sigma'|} \theta(\xi - \tilde{\xi}) \mathcal{O}'(\tilde{\rho}) \mathcal{O}(\rho). \tag{5.11}$$

Note that the ordering has to be specified only when $\mathcal{O}$ and $\mathcal{O}'$ lie on the same string strip.

Now by the usual argument in field theory, the definition (5.9) together with the commutation relation $[\dot{X}^{\mu(r)}(\sigma), X^{\nu(r)}(\tilde{\sigma})] = \pi \eta_{\mu\nu} \delta(\sigma - \tilde{\sigma})$, the equation of motion $\Box_\rho X^\mu = 0$, the boundary conditions $X^{\mu(r)\prime}\big|_{\sigma_r = 0, \pi} = 0$ and $\langle \tilde{0} | V_0' \rangle = 1$ lead to

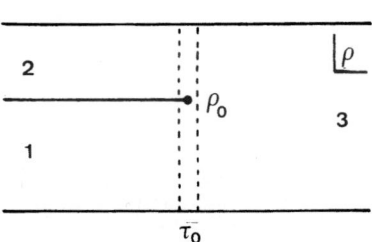

Fig.5 Complex ρ-plane

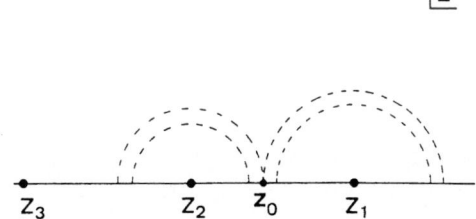

Fig.6 Complex z-plane

---

*) We should write it as $X^\mu(\rho, \rho*)$, more precisely.

$$\begin{cases} \Box_\rho \, N(\rho, \tilde{\rho}) = 2\pi \, \delta^2(\rho - \tilde{\rho}), \qquad \left(\Box_\rho = \left(\frac{\partial}{\partial \tau}\right)^2 + \left(\frac{\partial}{\partial \sigma}\right)^2 \right) \\ \frac{\partial}{\partial n} N = 0 \quad \text{at the boundary}, \end{cases} \tag{5.12}$$

as an equation determining the Neumann function. To solve this equation easily, let us perform a conformal mapping from the ρ-plane to the upper half-plane of z via the so-called Mandelstam mapping[27]

$$\rho(z) = \sum_{r=1}^{3} \alpha_r \, \ln(z - Z_r) \tag{5.13}$$

so that the complicated shape of string boundary on ρ-plane (Fig.5) is mapped simply to the real axis on z-plane as drawn in Fig.6. On the z-plane, eq.(5.12) becomes

$$\begin{cases} \Box_z \, N(\rho(z), \rho(\tilde{z})) = 2\pi \, \delta^2(z - \tilde{z}), \tag{5.14a} \\ \frac{\partial}{\partial y} N \Big|_{y=0} = 0, \qquad (z = x + iy). \tag{5.14b} \end{cases}$$

Eq.(5.14b) is the Poisson equation in 2 dimension which is known to have Coulomb potential solution $\ln r$ and the boundary condition (5.14b) is easily satisfied by the mirror image method; thus we find

$$N(\rho(z), \rho(\tilde{z})) = \ln|z - \tilde{z}| + \ln|z - \tilde{z}^*| \tag{5.15}$$

$$\equiv -\delta_{rs} \left\{ \sum_{n \geq 1} \frac{2}{n} e^{-n|\xi_r - \tilde{\xi}_s|} \cos(n\sigma_r) \cos(n\tilde{\sigma}_s) - 2 \max(\xi_r, \tilde{\xi}_s) \right\}$$

$$+ 2 \sum_{n,m \geq 0} \bar{N}_{nm}^{rs} e^{n\xi_r + m\tilde{\xi}_s} \cos(n\sigma_r) \cos(n\tilde{\sigma}_s). \tag{5.16}$$

Here in the second eq.(5.16) we have performed Fourier expansion assuming that ρ and $\tilde{\rho}$ lie on the strips of strings r and s, respectively, and defined the Fourier coefficients $\bar{N}_{nm}^{rs}$. The first terms proportional to δ_{rs} in (5.16) are extracted for convenience, which will turn out to be a "free propagation term" (giving a special solution to (5.12)).

It is now immediate to see that the defining equation (5.9) of the Neumann function actually reproduces the expression (5.16) if we take $|v_0'\rangle = e^E |0\rangle$ with the exponent E given by (5.4) : Indeed, in gneral, for such $|v_0'\rangle = e^E |0\rangle$ with E quadratic in creation operators and for any operator $\mathcal{O}(\rho)$ linear in the creation and annihilation operators, we have

$$\langle \tilde{0} | T \, \mathcal{O}(\rho) \, \mathcal{O}'(\tilde{\rho}) | v_0' \rangle = \overrightarrow{\mathcal{O}(\rho) \, \mathcal{O}'(\tilde{\rho})}, \tag{5.17a}$$

$$\overrightarrow{\mathcal{O}(\rho) \, \mathcal{O}'(\tilde{\rho})} \equiv \theta(\tilde{\xi}_s - \xi_r) \langle \tilde{0} | \mathcal{O}^{(+)}(\rho) \mathcal{O}'^{(-)}(\tilde{\rho}) |0\rangle + \theta(\xi_r - \tilde{\xi}_s) \langle \tilde{0} | \mathcal{O}'^{(+)}(\tilde{\rho}) \mathcal{O}^{(-)}(\rho) |0\rangle$$

$$+ \left[\mathcal{O}^{(+)}(\rho), \left[\mathcal{O}'^{(+)}(\tilde{\rho}), E \right] \right], \tag{5.17b}$$

where $\mathcal{O}^{(\pm)}(\rho)$ denote the annihilation and creation operator parts of $\mathcal{O}(\rho)$, respectively. Applying this formula (5.17b) to $X^\mu(\rho)$ and using

$$X^{\mu(+)}(\rho) = x_r^\mu + i \sum_{n=1}^{\infty} \frac{1}{n} \alpha_n^{\mu(r)} e^{n\xi_r} \cos(n\sigma_r),$$

$$X^{\mu(-)}(\rho) = i p_r^\mu \xi_r - i \sum_{n=1}^{\infty} \frac{1}{n} \alpha_{-n}^{\mu(r)} e^{-n\xi_r} \cos(n\sigma_r). \tag{5.18}$$

we see that the first two terms (free propagation term $\langle \tilde{0}| T \, \mathcal{O}(P) \mathcal{O}'(\tilde{P})|0\rangle$) and the last term in the RHS of (5.17b) just reproduce the first and second terms of (5.16), respectively.

By similar considerations to the FP ghost coordinate, we can confirm the validity of the ghost parts of E in (5.4). It is also possible to confirm the Goto-Naka connection conditions (5.2) directly by applying the operators (5.3) on the vertex $|V_o\rangle$ given in (5.4) if we use the expression (5.16) of the Neumann function and its smoothness across the interaction boundary. [The necessity of the total momentum conservation factor $\tilde{\delta}(1,2,3)$ in (5.4) becomes understandable there]

EXERCISE : i) Confirm that the Mandelstam mapping (5.13) maps the string boundaries of P-plane to the real axis on z-plane.
ii) All the Neumann coefficients $\bar{N}_{nm}^{rs}$ are calculable from the defining equation (5.15) = (5.16). Derive, for instance, the formula

$$\bar{N}_{nm}^{rs} = \frac{1}{nm} \oint_{Z_r} \frac{dz}{2\pi i} \oint_{Z_s} \frac{d\tilde{z}}{2\pi i} \frac{1}{(\bar{z}-\tilde{z})^2} \, e^{-n\zeta_r(z)-m\zeta_s(\tilde{z})}$$

valid for $n,m \geq 1$, where $P(z) = \alpha_r \zeta_r(z) + \tau_0 + i\beta_r$ ($\zeta_r \equiv \xi_r + i\sigma_r$).
iii) Confirm the formula (5.17). The quantity (5.17b) is called operator contraction.

5.3 Proof of $\sum_{r=1}^{3} Q_B^{(r)}|V\rangle = 0$

First note that the δ-functional $|V_o\rangle$ carries $N_{FP} = -1$ because of the factor $\delta(\sum \bar{c}_0^{(r)}/\alpha_r)$, but the desired vertex $|V\rangle$ should carry $N_{FP} = 0$ since $\delta_B \Phi = Q_B\Phi + g\Phi*\Phi + \ldots$ demands $\Phi*\Phi$ should have $N_{FP} = 0$ as $Q_B\Phi$. [Note that the functional integration measure $[dZ]$ in (4.5) carry $N_{FP} = +1$ as $d\bar{c}_0$ since the other non-zero mode parts appear in ghost-antighost pairs]. So a factor with $N_{FP} = +1$ is missing. We now show that the missing factor is the ghost operator at the interaction (splitting or joining) point and that the desired vertex is given by[6,7,4]

$$|V\rangle = C(\sigma_{int}) \, |V_o\rangle . \tag{5.19}$$

Recall the expression (3.1b) of $Q_B^{(r)}$ written in terms of the oeprators

$$\mathcal{O}_{\pm}^{(r)}(\sigma_r) \equiv (\alpha_r^{-1} A_{\pm}^{\mu(r)}(\sigma_r), \; \alpha_r C_{\pm}^{(r)}(\sigma_r), \; \alpha_r^{-2} \bar{C}_{\pm}^{(r)}(\sigma_r)). \tag{5.20}$$

Just for X^μ in the previous subsection, we can extend these oeprators to include the time argument ξ_r by free propagation, i.e., simply by the replacements $\mathcal{O}_{\pm}^{(r)}(\sigma_r) \to \mathcal{O}_{\pm}^{(r)}(\sigma_r \mp i\xi_r)$ and define the operators $\mathcal{O}(P) \equiv (A^\mu(P), C(P), \bar{C}(P))$ by

$$\mathcal{O}(P) \equiv \begin{cases} \mathcal{O}_+^{(r)}(\sigma_r - i\xi_r) & \text{for } \text{Im } P \geq 0 \\ \mathcal{O}_-^{(r)}(\sigma_r + i\xi_r) & \text{for } \text{Im } P \leq 0 \end{cases} \tag{5.21}$$

on the exteded P-plane [the original P-plane (Im$P \geq 0$) and its complex conjugate mirror plane (Im$P \leq 0$)] as drawn in Fig.7.
Then, since $Q_B^{(r)}$ is conserved, we can shift the time ξ_r of oeprators in the integrand (3.1b) of $Q_B^{(r)}$ and write

$$\left(\sum_{r=1}^{3} Q_B^{(r)}\right)|V\rangle = \Big[\frac{1}{2} \oint_{C_P} \frac{dP}{2\pi i} : C(P)\{-A^2(P) + 2\frac{dC(P)}{dP} \bar{C}(P)\} :$$

$$+ \sum_{r=1}^{3} \alpha(0) \int_{-\pi}^{\pi} \frac{d\sigma_r}{2\pi} C^{(r)}(\sigma_r) \Big] C(P_0)|V_o\rangle , \tag{5.22}$$

181

with the integration contour C_ρ depicted in Fig.7. Further, rewriting the normal products by Wick theorem as

$$: A(\rho) \cdot A(\rho) : \; = \lim_{\delta \to 0} \Big(T A(\rho) \cdot A(\rho+\delta) - \langle 0 | T A(\rho) \cdot A(\rho+\delta) | 0 \rangle \Big)$$

$$= \lim_{\delta \to 0} \Big(T A(\rho) \cdot A(\rho+\delta) - \frac{d}{\delta^2} \Big) + \frac{d}{12} \frac{1}{\alpha_r^2} \; ,$$

$$: \frac{dC(\rho)}{d\rho} \bar{C}(\rho) : \; = \lim_{\delta \to 0} \Big(T \frac{dC(\rho)}{d\rho} \bar{C}(\rho+\delta) - \frac{1}{\delta^2} \Big) + \frac{1}{12} \frac{1}{\alpha_r^2} \; , \tag{5.23}$$

and going onto the z-plane, we reach

$$\Big(\sum_{r=1}^{3} Q_B^{(r)} \Big) | V \rangle = T \Big[\frac{1}{2} \oint_{C_z} \frac{dz}{2\pi i} \lim_{\delta \to 0} \Big\{ \Big(\frac{d\rho(z')}{dz'} \Big)^{-1} C(z) \big[-A(z) \cdot A(z')$$

$$+ 2 \frac{dC(z)}{dz} \bar{C}(z') + \frac{d-2}{\delta^2} \Big(\frac{d\rho}{dz} \Big) \Big(\frac{d\rho'}{dz'} \Big) \big] \Big\}$$

$$+ \sum_{r=1}^{3} \Big(\alpha(0) - \frac{d-2}{24} \Big) \int_{-\pi}^{\pi} \frac{d\sigma_r}{2\pi} C^{(r)}(\sigma_r) \Big] C(z_0) | V_0 \rangle , \tag{5.24}$$

where $\rho(z') = \rho(z) + \delta$, the integration contour C_z is drawn in Fig.8 and

$$A^\mu(z) \equiv \Big(\frac{d\rho(z)}{dz} \Big) A^\mu(\rho(z)), \quad C(z) \equiv C(\rho(z)), \quad \bar{C}(z) \equiv \Big(\frac{d\rho(z)}{dz} \Big) \bar{C}(\rho(z)). \tag{5.25}$$

The operators $\mathcal{O} \equiv (A^\mu, C$ and $\bar{C})$ in (5.24) still contain annihilation operator parts $\mathcal{O}^{(+)}$, which can be eliminated by passing them through to the most right on the vacuum $|0\rangle$ in $|V_0\rangle = e^E |0\rangle$. Then the necessary commutator algebra is performed very systematically by the following Wick-like theorem :

EXERCISE : i) Show the formulas :

$$T \mathcal{O}_1 \mathcal{O}_2 \, e^E |0\rangle = \big(\hat{\mathcal{O}}_1 \hat{\mathcal{O}}_2 + \overrightarrow{\mathcal{O}_1 \mathcal{O}_2} \big) \, e^E |0\rangle ,$$

$$T \mathcal{O}_1 \mathcal{O}_2 \mathcal{O}_3 \, e^E |0\rangle = \Big[\hat{\mathcal{O}}_1 \hat{\mathcal{O}}_2 \hat{\mathcal{O}}_3 + \big(\hat{\mathcal{O}}_1 \overrightarrow{\mathcal{O}_2 \mathcal{O}_3} + \overrightarrow{\mathcal{O}_1 \mathcal{O}_2} \hat{\mathcal{O}}_3 + \overrightarrow{\mathcal{O}_1} \hat{\mathcal{O}}_2 \overrightarrow{\mathcal{O}_3} \big) \Big] e^E |0\rangle , \tag{5.26}$$

where $\overrightarrow{\mathcal{O}_1 \mathcal{O}_2}$ is the contraction (c-number) defined in (5.17a), and $\hat{\mathcal{O}}$ denotes

$$\hat{\mathcal{O}} \equiv \mathcal{O}^{(-)} + [\mathcal{O}^{(+)}, E] \tag{5.27}$$

giving an operator linear in creation operators alone.

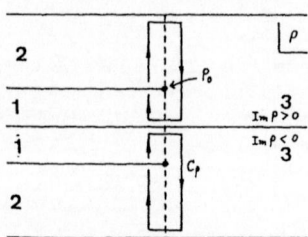

Fig.7. (Extended) ρ-plane and the integration contour C_ρ.

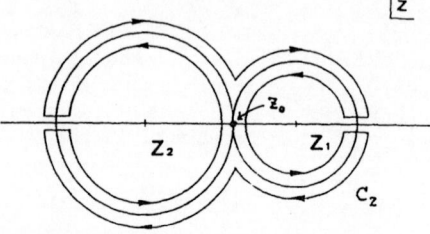

Fig.8. The contour C_z in (5.24).

ii) Derive the following contraction formulas using the Neumann fucntion (5.15):

$$\overbracket{A^\mu(z_1)\ A^\nu(z_2)} = \eta^{\mu\nu}\ \frac{1}{(z_1 - z_2)^2}\ ,$$

$$\overbracket{C(z_1)\ \bar{C}(z_2)} = -\ \frac{1}{z_1 - z_2}\ . \tag{5.28}$$

Using the following expansion of $P(z)$ around the interaction point $P_0 = P(z_0)$ determined by $dP/dz\big|_{z=z_0} = 0$,

$$P_0 - P(z) = a(z - z_0)^2 + b(z - z_0)^3 + \cdots \ ,$$

$$\left(\frac{dP}{dz}\right)^{-1} = -\frac{1}{2a}\ \frac{1}{z - z_0} + \frac{3b}{4a^2} + O(z - z_0)\ , \tag{5.29}$$

we can now calculate the contour integral (5.24) by evaluating the pole residue at $z = z_0$ since $A(z)$, $C(z)$ and $\bar{C}(z)$ are analytic functions regular even at $z = z_0$ (thanks to the dP/dz factor multiplied in (5.25)).

The no-contraction term in the Wick-like formula, in which all the operators $\mathcal{O}$ are replaced by $\hat{\mathcal{O}}$, yields

$$\frac{1}{2}\cdot\frac{1}{2a}\ \hat{C}(z_0)\left[-\hat{A}^2(z_0) + 2\frac{d\hat{C}(z_0)}{dz}\ \hat{\bar{C}}(z_0)\right]\hat{C}(z_0)|V_0\rangle\ , \tag{5.30}$$

which vanishes by $\hat{C}(z_0)\hat{C}(z_0) = 0$. This was the reason why we have chosen the ghost prefactor $C(\sigma_{int})$ in the vertex (5.19).

There are three kind of one-contraction terms :

(1) $\quad \hat{C}(z)\left\{-\overbracket{A(z)\cdot A(z')} + 2\frac{d\hat{C}}{dz}(z)\overbracket{\bar{C}(z')}\right\}\hat{C}(z_0)|V_0\rangle\ ,$

(2) $\quad \overbracket{C(z)\ 2\frac{d\hat{C}}{dz}(z)}\ \bar{C}(z')\ \hat{C}(z_0)|V_0\rangle\ , \tag{5.31}$

(3) $\quad \hat{C}(z)\ 2\frac{d\hat{C}}{dz}(z)\overbracket{\bar{C}(z')\ C(z_0)}|V_0\rangle\ .$

The contractions (1) and (2) contain $1/\delta^2$ singularities but are cancelled by the $(d-2)/\delta^2$ term in (5.24) (as they should be since we have started with normal ordered Q_B). With an elementary calculation, the contributions to the integral (5.24) of these one-contraction terms are found, respectively, to be [20]

(1) $\quad -\frac{d-2}{8}\left\{\frac{1}{4a}\frac{d^2\hat{C}}{dz^2}(z_0) - \frac{b}{4a^2}\frac{d\hat{C}}{dz}(z_0)\right\}\hat{C}(z_0)|V_0\rangle\ ,$

(2) $\quad \frac{1}{4a}\frac{d^2\hat{C}(z_0)}{dz^2}\ \hat{C}(z_0)|V_0\rangle\ , \tag{5.32}$

(3) $\quad \hat{C}(z_0)\left\{-\frac{1}{4a}\frac{d^2\hat{C}}{dz^2}(z_0) + \frac{3b}{4a^2}\frac{d\hat{C}}{dz}(z_0)\right\}|V_0\rangle\ .$

So, summing them up, we have

$$\left\{\frac{1}{4a}\frac{d^2\hat{C}}{dz^2}\hat{C}\left(-\frac{d-2}{8} + 1 + 2\right) + \frac{b}{4a^2}\frac{d\hat{C}}{dz}\hat{C}\left(\frac{d-2}{8} - 3\right)\right\}|V_0\rangle\ , \tag{5.33}$$

both the two terms of which vanish at $d = 26$! [Note that the factor $d-2$ in (1) of (5.32) reflects the $OSp(d/2)$ invariance of our vertex.] These are the all for the first integral term in (5.24), and its second term proportional to $\alpha(0) - (d-2)/24$ also vanishes by taking $\alpha(0) = 1$. Thus we have finished the proof of (5.1) for the open-string, i.e., have found that the distributive law $Q_B(\Phi*\Psi) = Q_B\Phi*\Psi + (-)^{|\Phi|}\Phi*Q_B\Psi$ holds if and only if $d = 26$ and $\alpha(0) = 1$.

The precise form of our open-string vertex is thus given by[20]

$$|V(1,2,3)\rangle = \mu(\alpha_1, \alpha_2, \alpha_3)\, G(\sigma_{int})\, |V_0(1,2,3)\rangle,\tag{5.34}$$

where the ghost prefactor $G(\sigma_{int}) = C(z_0) = i\pi_{\bar{c}}(z_0) - c(z_0)$ equals $i\alpha_r \pi_{\bar{c}}^{(r)}(\sigma_{int})$ since $c(z_0) = 0$, and the factor μ is a measure function of α_r-integration to be determined later.

5.4 Closed-String Vertex

The construction of the closed-string vertex $|V_{closed}\rangle$ is now very simple, since the closed-string is more or less a product of two "open-strings" consisting of the right- and left-moving modes, respectively. Indeed if we write the Goto-Naka type connection conditions for the closed-string vertex as drawn in Fig.1, we can split them into the right- and left-moving mode parts, each of which turns out to take exactly the same form as the conditions for the open-string. Therefore clearly we can take

$$|V_{closed}\rangle \simeq |V_{open}^{(+)}\rangle \otimes |V_{open}^{(-)}\rangle,\tag{5.35}$$

where the superscripts $(\pm)$ indicate the right- and left- moving modes substitued into the open-string vertex (5.34)*). Moreover, since the same splitting of $\pm$ modes occurs also in Q_B, as seen in (3.8), this vertex (5.35) satisfies the required BRS invariance (5.1) :

$$\left(\sum_{r=1}^{3} Q_{B\,closed}^{(r)}\right)|V_{open}^{(+)}\rangle \otimes |V_{open}^{(-)}\rangle = \sum_{r=1}^{3}(Q_{B+}^{(r)} + Q_{B-}^{(r)})|V_{open}^{(+)}\rangle \otimes |V_{open}^{(-)}\rangle = 0.\tag{5.36}$$

However (5.35) does not yet give the desired closed-string vertex $|V_{closed}\rangle$, which should carry ghost number $N_{FP} = -2$ in this case so as to compensate the ghost number number $N_{FP} = +2$ of $d\pi_c^{0(1)} d\pi_c^{0(2)}$ contained in $[dZ_1 dZ_2]$ of $\Phi_1 * \Phi_2$ additionally compared with the open-string case. So we need to find a $N_{FP} = -2$ factor by which to multiply (5.35) and it must be commutative with Q_B in order not to violate the already-satisfied BRS invariance (5.36). Taking account of the cyclic-symmetry requirement also, we take the following oeprators as such factor :

$$\mathcal{P}_{123} \sum_{r,s,t=1}^{3} \epsilon_{rst}\, \alpha_r\, \pi_c^{0(s)} \pi_c^{0(t)}.\tag{5.37}$$

Actually, although $\{Q_B, \pi_c^0\} = i(L_+ - L_-) \neq 0$, it vanishes in the presence of projection operator $\mathcal{P}_{123} = \mathcal{P}^{(1)}\mathcal{P}^{(2)}\mathcal{P}^{(3)}$. With this vertex, the distributive law is thus established also for the closed-string case.

Our closed-string vertex thus constructed can be shown to be written in the form [21]

$$|V_{closed}\rangle = \mathcal{P}_{123}\, \mu^2(\alpha_1,\alpha_2,\alpha_3)\, G(\sigma_{int}) \left(\frac{\alpha_1\alpha_2}{\alpha_3}\pi_c\right)|V_0(1,2,3)\rangle,$$

$$\pi_c = \alpha_1 \frac{\pi_c^{0(2)}}{\alpha_2} - \alpha_2 \frac{\pi_c^{0(1)}}{\alpha_1}, \qquad G(\sigma_{int}) = i\alpha_r \pi_{\bar{c}}^{(r)}(\sigma_{int}),\tag{5.38}$$

$$|V_0\rangle = e^{E_+ + E_-}|0\rangle\, \delta(\sum_r \alpha_r^{-1}\pi_c^{0(r)})\,\bar{\delta},$$

where $|V_0\rangle$ is the δ-functional of this case and $E_\pm$ is the same as the open-string's one E (5.4) with $\pm$ modes substituted, $\bar{\delta}$ being the same as

---

*) Of course, the total momentum conservation factor $\bar{\delta}$ in (5.4), which is common to the right- and left-movers, should be included only once.

in (5.4). The necessity of the measure μ^2 (square of the open-string case μ) will be seen later.

6. ASSOCIATIVE LAWS (OPEN-STRING)

Let us show that the "identities"

$$(\Phi_1 * \Phi_2) * \Phi_3 - \Phi_1 * (\Phi_2 * \Phi_3) \simeq 0 , \tag{6.1}$$

$$Q_B(\Phi_1 \circ \Phi_2 \circ \Phi_3) - \left[Q_B \Phi_1 \circ \Phi_2 \circ \Phi_3 + (-)^{|1|}\Phi_1 \circ Q_B \Phi_2 \circ \Phi_3 + (-)^{|1|+|2|}\Phi_1 \circ \Phi_2 \circ Q_B \Phi_3 \right] \simeq 0 \tag{6.2}$$

hold separately. As a matter of fact, we mean by $\simeq 0$ in these identities that they <u>almost</u> vanish although leaving "small" non-vanishing pieces. These non-vanishing pieces of (6.1) and (6.2), however, will turn out to have the same form and cancel with each other so that we will get the identity (4.7b).

6.1 Associative Law (6.1)

Recall the definition (4.5) of $\Phi_1 * \Phi_2$ with the vertex (5.34) used:

$$(\Phi_1 * \Phi_2)[Z_3] = \int [dZ_1 dZ_2] \, \Phi_1[Z_1] \Phi_2[Z_2] \mu(\alpha_1,\alpha_2,\alpha_3) G(\sigma_{int}^{123}) V_0[Z_1,Z_2,\tilde{Z}_3]. \tag{6.3}$$

Note that this product $\Phi_1 * \Phi_2$ (or equivalently, the overlapping δ-functional V_0 ,) has three distinct configurations as drawn in Fig.9 depending on the regions of α_1 and α_2 integrations. We call them A-, B- and C-type configurations, respectively. So, each to the terms $(\Phi_1 * \Phi_2)*\Phi_3$ and $\Phi_1 * (\Phi_2 * \Phi_3)$, 3^2 configurations contribute, as illustrated in Figs.10 and 11, respectively.

Using (6.3), the first term $(\Phi_1 * \Phi_2)*\Phi_3$ in (6.1) looks like

$$\left((\Phi_1 * \Phi_2) * \Phi_3 \right)[Z_4] = \int [dZ_1 dZ_2 dZ_3] \, \Phi_1[Z_1] \Phi_2[Z_2] \Phi_3[Z_3] \Delta[Z_1,Z_2,Z_3,\tilde{Z}_4], \tag{6.4}$$

where Δ is an effective 4-string vertex given by

$$\Delta[Z_1,Z_2,Z_3,Z_4] \equiv \int [dZ_5] \mu(\alpha_1,\alpha_2,-\alpha_5) \mu(\alpha_5,\alpha_3,\alpha_4) G(\sigma_{int}^{12\bar{5}}) G(\sigma_{int}^{534}) V_0[Z_1,Z_2,\tilde{Z}_5] V_0[Z_5,Z_3,Z_4]. \tag{6.5}$$

Since the main body V_0 of the 3-string vertex V is simply the δ-functional corresponding to the 3 configurations in Fig.9, this effective 4-string vertex Δ resultant from the RHS of (6.5) is also essentially the δ-functional, say $V_0^{(4)}$, corresponding to the 3^2 configurations shown in Fig.10. The secon term $\Phi_1 * (\Phi_2 * \Phi_3)$ has also similar expressions to (6.4) and (6.5) and the main body of the

A **B** **C**

Fig.9 Three configurations contributing to $\Phi_1 * \Phi_2$ corresponding to the cases (A) $|\alpha_3| = |\alpha_1| + |\alpha_2|$, (B) $|\alpha_1| = |\alpha_2| + |\alpha_3|$ and (C) $|\alpha_2| = |\alpha_3| + |\alpha_1|$, respectively.

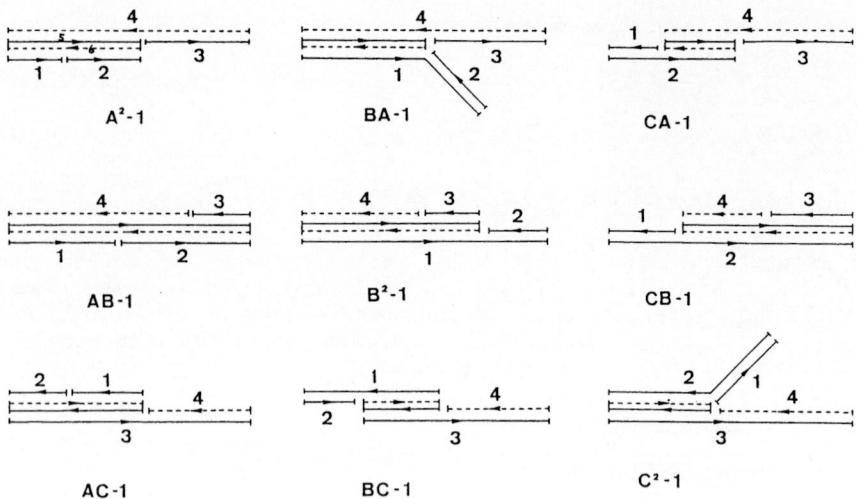

Fig.10.   nine configurations contributing to $(\Phi_1 * \Phi_2) * \Phi_3$.

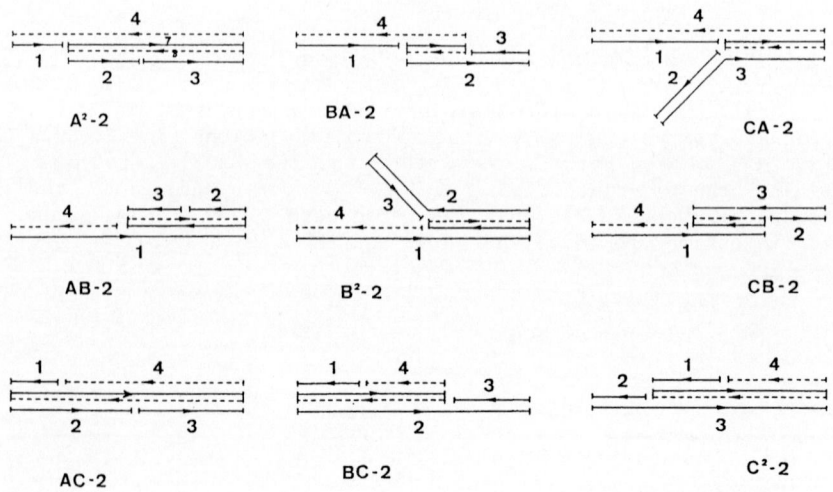

Fig.11.   nine configurations contributing to $\Phi_1 * (\Phi_2 * \Phi_3)$.

resultant effective 4-string vertex Δ is the δ-functional $V_0^{(4)}$ correspoinding to the configurations illustrated in Fig.11 in this case.

These δ-functional parts $V_0^{(4)}$ of $\Delta[Z_1,Z_2,Z_3,Z_4]$ for $(\Phi_1 * \Phi_2) * \Phi_3$ and $\Phi_1 * (\Phi_2 * \Phi_3)$ coincide with each other for the common arguments $Z_1 \sim Z_4$ in almost all the cases. For instance, consider the AA-type configurations, A^2-1 in Fig.10 and A^2-2 in Fig.11, which result from the A-type configurations both for the two 3-string vertices in $(\Phi_1 * \Phi_2) * \Phi_3$ or $\Phi_1 * (\Phi_2 * \Phi_3)$. Comparing the figures A^2-1 and A^2-2, we clearly see that they represent the same overlapping structure of the strings $1 \sim 4$ and thus the same δ-functional $V_0^{(4)}$. The two ghost factors GG in (6.5), each of which is associated with the interaction point of the 3-string vertex, are also clearly common between such common 4-string configurations.

However, nontrivial point is the c-number coefficient appearing after the dZ_5-integration in (6.5). It must be calculated more precisely by using the oscillator mode expression (5.4) for the 3-string vetex, and is found to be

$$\frac{1}{|\alpha_5|} \left[det(1 - \widetilde{N}^{\bar{5}\bar{5}} \widetilde{N}^{55}) \right]^{-\frac{1}{2}(d-2)} \mu(\alpha_1, \alpha_2, -\alpha_5) \mu(\alpha_5, \alpha_3, \alpha_4) \qquad (6.6)$$

with $\alpha_5 = \alpha_1 + \alpha_2 = -(\alpha_3 + \alpha_4)$ and

$$\widetilde{N}^{55}_{mn} = \sqrt{m} \ \bar{N}^{55}_{mn}(\alpha_5, \alpha_3, \alpha_4) \sqrt{n} \ ,$$
$$\widetilde{N}^{\bar{5}\bar{5}}_{mn} = (-)^m \sqrt{m} \ \bar{N}^{\bar{5}\bar{5}}_{mn}(\alpha_1, \alpha_2, -\alpha_5) \sqrt{n} \ (-)^n . \qquad (6.7)$$

Note that the determinant factor in (6.6) came from the contraction of the oscillator modes of the intermediate string 5 using the formula

$$\langle 0 | e^{\frac{1}{2} a^\dagger M a} \ e^{\frac{1}{2} a^\dagger N a} | 0 \rangle = \left[det(1 - MN) \right]^{-\frac{1}{2}} , \qquad (6.8)$$

and so the exponents $-\frac{1}{2}d$ and $-\frac{1}{2}(-2)$ represent the contributions from the bosonic oscillators $\alpha_n^{\mu(5)}$ and the ghost ones $c_n^{(5)}$ and $\bar{c}_n^{(5)}$, respectively (and also the OSp(d/2) symmetry of the 3-string vertex V). The c-number coefficient appearing for the term $\Phi_1 * (\Phi_2 * \Phi_3)$ has also the same form as (6.6) and (6.7),

$$\frac{1}{|\alpha_7|} \left[det(1 - \widetilde{N}^{\bar{7}\bar{7}} \widetilde{N}^{77}) \right]^{-\frac{1}{2}(d-2)} \mu(\alpha_2, \alpha_3, -\alpha_7) \mu(\alpha_7, \alpha_1, \alpha_4) \qquad (6.9)$$

by calling the intermediate string 7 in this case. As functions of $\alpha_1 \sim \alpha_4$, these two coefficients (6.6) and (6.9) look in fact very different, nevertheless they turn out to equal each other if (and only if) d = 26 !

This miracle is realized by a particular mechanism realizing duality. Indeed, in the light-cone gauge string field theory, the "s-t dual" amplitude for the decay process $4 \to 1+2+3$ is realized by the sum of two diagrams drawn in Fig.12. As is clear intuitively from the diagrams, the s (i.e., 1+2)-channel resonance is contained only in the diagram A^2-1 while the t(2+3)-channel resonance only in the diagram A^2-2. As a matter of fact, the former diagram A^2-1 contribute to the amplitude in a form $\int_0^\infty dT \widetilde{f}(T)$, which is converted into $\int_0^{x_0} dx f(x)$ ($0 < \exists x_0 < 1$) by a change of variable from the time interval parameter T to the so-called Veneziano variable x. [The relation between T and x is given through the Mandelstam mapping. See the EXERCISE below.] And the latter diagram A^2-2 yields $\int_0^\infty dT \widetilde{g}(T)$, converted into $\int_{x_0}^1 dx g(x)$. On the other hand the full s-t dual Veneziano amplitude has a form $\int_0^1 dx F(x)$ with an smooth function F(x) [$F(x) = x^{-\frac{t}{2}-1} (1-x)^{-\frac{t}{2}-1}$ for the ground state]. Therefore the fact that the light-cone gauge string field theory[26] realizes the duality, implies the functions f(x) and g(x) resultant from the two diagrams in fact give a single smooth function F(x); f(x) = g(x) = F(x). In particular, this implies the equality

$$f(x_0) = g(x_0) \qquad (6.10)$$

at the common boundary $x = x_0$ corresponding to the $T = 0$ diagrams of Fig.12. It is this eq.(6.10) that guarantees the equality of the two c-number coefficients (6.6) and (6.9), as inferred from the fact that the $T = 0$ diagrams of Fig.12 just correspond to the configurations A^2-1 of Fig.10 and A^2-2 of Fig.11.

To show this duality in the light-cone gauge string field theory, Cremmer and Gervais[28] actually proved the following identity explicitly:

$$\frac{1}{|\alpha_5|} \left[det \left(1 - \tilde{N}_T^{66} \tilde{N}^{55}\right)\right]^{-12} exp\left\{ \frac{T}{\alpha_5} - \sum_{r=1,2,6} \frac{\tau_0(\alpha_1,\alpha_2,\alpha_6)}{\alpha_r} - \sum_{r=5,3,4} \frac{\tau_0(\alpha_5,\alpha_3,\alpha_4)}{\alpha_r} \right\}$$

$$= \left| \frac{\prod_{i=1}^{4} dZ_i}{dV_{abc}\, dT} \right| \, e^{-\sum_{r=1}^{4} \bar{N}_{oo}^{(4)rr}} \, , \qquad (6.11)$$

$$dV_{abc} = \frac{dZ_a\, dZ_b\, dZ_c}{|Z_a - Z_b||Z_b - Z_c||Z_c - Z_a|} \quad , \quad \tau_0(\alpha_1,\alpha_2,\alpha_3) = \sum_{r=1}^{3} \alpha_r \ln|\alpha_r| \, , \quad (6.12a)$$

$$\left(\tilde{N}_T^{66}\right)_{mn} = e^{mT/\alpha_6} \tilde{N}_{mn}^{55} \, e^{nT/\alpha_6} \, , \qquad \alpha_6 = -\alpha_5 \, , \qquad (6.12b)$$

where $\bar{N}_{oo}^{(4)rr}$ is the $m = n = 0$ component of the Neumann function defined on the 4-string diagram A^2-1 of Fig.12. The LHS of this equation (6.11) is the momentum-independent factors appearing in calculating the amplitude $\tilde{f}(T)$ of the diagram A^2-1 in Fig.12, while the RHS, expressed by the Koba-Nielsen variables $Z_1 \sim Z_4$, give (the momentum-independent part of) $|dx/dT|\, f(x)$ after the projective gauge-fixing $Z_1 = \infty$, $Z_2 = 1$ $Z_3 = x$, $Z_4 = 0$. [See the EXERCISE below for the projective gauge-fixing.] Note that the LHS reduces exactly to our c-number coefficient (6.6) at $T \to 0$ if $d = 26$ and

$$\mu(\alpha_1,\alpha_2,\alpha_3) = exp\left(-\sum_{r=1}^{3} \frac{\tau_0(\alpha_1,\alpha_2,\alpha_3)}{\alpha_r}\right). \qquad (6.13)$$

The same form of formula as (6.11) of course holds also for the amplitude $\tilde{g}(T)$ of the diagram A^2-2 in Fig.12, and its LHS also reduces at $T = 0$ to the c-number coefficient (6.9). Since the RHS gives the same smooth function as (6.11), we can conclude the desired equality of (6.6) and (6.9), if (and only if) $d = 26$ and the measure function μ of the 3-string vertex is taken as (6.13). Thus we have proved the associativity $(\Phi_1 * \Phi_2) * \Phi_3 = \Phi_1 * (\Phi_2 * \Phi_3)$ for the AA-type configurations, A^2-1 in Fig.10 and A^2-2 in Fig.11.

For the other configurations in Fig.10 also, one can easily find the corresponding partner configurations in Fig.11 almost always and can

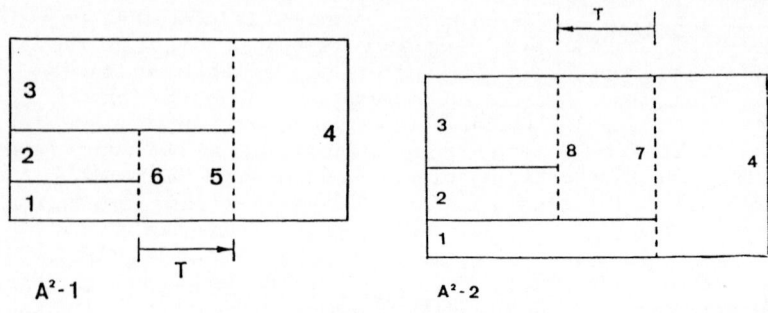

Fig.12. String diagrams contributing to the "s-t dual" decay amplitude.

prove the associativity for those configurations. [For instance, AB-1 in Fig.10 corresponds to BA-2 when $|\alpha_3| < |\alpha_2|$ and to CB-2 when $|\alpha_3| > |\alpha_1|$ in Fig.11.] However, the particular shape of configurations BA-1 and C^2-1 in Fig.10 and CA-2 and B^2-2 in Fig.11 have no counterparts in the other figure. These configurations occur when (α_1, α_2, α_3, α_4) have alternating signs (+,-,+,-) or (-,+,-,+), and were called "horn diagrams" in Ref.29. So the associativity ($\Phi_1 * \Phi_2$) $* \Phi_3 =$ $\Phi_1 * (\Phi_2 * \Phi_3)$ is violated for the cases of horn diagram configurations.

EXERCISE : i) It is known that the most general form of one-to-one conformal mapping of the upper half-plane onto itself is given by

$$z \rightarrow z' = \frac{Az + B}{Cz + D} \qquad (A, B, C, D \, ; \, real, \quad AD - BC > 0) \qquad (*)$$

and is called projective (or Möbius) transformation. Confirm that (*) is actually such a transformation.
ii) The ρ-plane of N-point (tree) string diagram is generally conformally mapped onto the upper half-plane of z by the Mandelstam mapping $\rho(z) = \sum_{r=1}^{N} \alpha_r \ln(z - Z_r)$, with real parameters $Z_1 \sim Z_N$ called Koba-Nielsen variables. Show that the string diagram does not uniquely specify the N parameters $Z_1 \sim Z_N$ but leaves a freedom to fix three of them at arbitrary values. [We call this projective "gauge-invariance".] iii) For N = 4 case, we can take $Z_1 = \infty$, $Z_2 = 1$ and $Z_4 = 0$, for instance, as a "gauge-fixing" and then the remaining variable is only Z_3 identified with the Veneziano variable x. Show that the 4-string diagram A^2-1 in Fig.12 with $\infty \geq T \geq 0$ corresponds through the Mandelstam mapping to $0 \leq x \leq^3 x_0$ and the diagram A^2-2 in Fig.12 with $0 \leq T \leq \infty$ to $x_0 \leq x \leq 1$. [α_1, α_2, $\alpha_3 > 0$, $\alpha_4 < 0$ for these diagrams.]

6.2. Distributive Law (6.2)

Let us turn to examine whether the distributive law (6.2) with respect to the (o o) product holds or not, or equivalently $\sum_{r=1}^{4} Q_B^{(r)}$ vanishes or not on the 4-string vertex $|V^{(4)}\rangle$. Recall that the 4-string vertex $V^{(4)}$ was given in the integral form (4.4) with a weight function $f(\sigma_0)$. The calculation of $(\sum_{r=1}^{4} Q_B^{(r)}) |V^{(4)}\rangle$ is performed quite similarly to the 3-string vertex case using the contour integral method. Then we find that the result becomes the following total derivative form [20]

$$(\sum_{r=1}^{4} Q_B^{(r)})|V^{(4)}\rangle = \frac{1}{2i} \int_{\sigma_-}^{\sigma_+} d\sigma_0 \, \frac{d}{d\sigma_0} \{ f(\sigma_0) C(z_0) C(z_0^*) | V_0^{(4)}(\sigma_0)\rangle \}$$

$$= \frac{1}{2i} \left[f(\sigma_0) C(z_0) C(z_0^*) | V_0^{(4)}(\sigma_0)\rangle \right]_{\sigma_0 = \sigma_-}^{\sigma_0 = \sigma_+} , \qquad (6.14)$$

if d = 26, $\alpha(0) = 1$ and the weight function $f(\sigma_0)$ is chosen as

$$f(\sigma_0) = \left| \frac{\prod_{i=1}^{4} dZ_i}{dV_{abc} \, d\sigma_0} \right| \exp\left(- \sum_{r=1}^{4} \bar{N}_{oo}^{(4)rr} \right). \qquad (6.15)$$

So the distributive law (6.2) is also violated slightly, i.e., only by the surface term at $\sigma_0 = \sigma_\pm$ shown in (6.14). Note, however, that the 4-string configurations at these end-points $\sigma_0 = \sigma_\pm$ become those as drawn in Fig.4. One immediately notices that they have exactly the same overlapping structures of 4 strings as the "horn diagram" configurations in the previous subsection. Therefore the δ-functional part $|V_0^{(4)}(\sigma_\pm)\rangle$ here coincides with the previous ones for the horn diagrams and the two ghost factors at the interaction point are also common between them. Further the c-nmuber coefficient $f(\sigma_0)$ given in (6.15) also coincides with the previous ones since the Cremmer-Gervais identity of the form

(6.11) holds also for the horn diagrams and the $\bar{N}_{oo}^{(4)rr}$ in the RHS of (6.11) clearly becomes the same as that in (6.15) for such common configurations. Thus we have proved the identity (4.7b) :

$$(\Phi_1 * \Phi_2) * \Phi_3 \, - \, \Phi_1 * (\Phi_2 * \Phi_3) \tag{6.16}$$

$$= (-)^{|1|+|2|+|3|} \left\{ Q_B(\Phi_1 \circ \Phi_2 \circ \Phi_3) - \left(Q_B\Phi_1 \circ \Phi_2 \circ \Phi_3 + (-)^{|1|}\Phi_1 \circ Q_B\Phi_2 \circ \Phi_3 + (-)^{|1|+|2|}\Phi_1 \circ \Phi_2 \circ Q_B\Phi_3 \right) \right\}.$$

The other associative laws (4.7c) and (4.7d) corresponding to higher order of g are also proved similarly. [See Sec.V E and F in Ref. 20.]

7. JACOBI IDENTITY (CLOSED-STRING)

Finally we come to the Jacobi identity (4.8c)

$$(\Phi_1 * \Phi_2) * \Phi_3 \, + (-)^{|1|(|2|+|3|)}(\Phi_2 * \Phi_3) * \Phi_1 \, + (-)^{|3|(|1|+|2|)}(\Phi_3 * \Phi_1) * \Phi_2 \, = 0. \tag{7.1}$$

for the closed-string case. We call these three terms, P, Q and R terms, respectively. Considering the case α_1, α_2, $\alpha_3 > 0$ ($\alpha_4 = -(\alpha_1 + \alpha_2 + \alpha_3) < 0$) for definiteness, the relevant configurations for these are represented by the diagrams P, Q and R in Fig.13 with the time interval T set equal to zero.

Using the expression (5.38) of the closed 3-string vertex and similar calculation as for the previous open-string case, we find that the P-term, for instance, is written in the form :

$$\left((\Phi_1 * \Phi_2) * \Phi_3 \right)[Z_4] = \alpha_4^{-1}\int [dZ_1 dZ_2 dZ_3] \, \Phi_1[Z_1]\Phi_2[Z_2]\Phi_3[Z_3] \, W \Delta_P[Z_1,Z_2,Z_3,\tilde{Z}_4], \tag{7.2}$$

$$\Delta_P[Z_1,Z_2,Z_3,Z_4] = \int_{-\pi}^{\pi} \frac{d\theta_P}{2\pi} \, \mathcal{P}_{1234} \, G(\sigma_{int}^{126}) G(\sigma_{int}^{534}) \, V_{0,\theta_P}^{(4)}[Z_1 \sim Z_4] \, D(\alpha_1 \sim \alpha_4 ; \theta_P), \tag{7.3}$$

$$W = \frac{1}{2} \sum_{r,s,t=1}^{3} \epsilon^{rst} \alpha_r^2 \, \pi_c^{0(s)} \pi_c^{0(t)} \quad , \quad \mathcal{P}_{1234} = \mathcal{P}^{(1)} \mathcal{P}^{(2)} \mathcal{P}^{(3)} \mathcal{P}^{(4)},$$

where the main body of Δ_P is again the 4-string overlapping δ-functional $V_{0,\theta_P}^{(4)}$ corresponding to the configuration of T = 0 diagram in.Fig.13. The important point in the closed-string case is that the θ_P-integration appears in (7.3) as an effect of the projection operator $\mathcal{P}^{(5)} = \int \frac{d\theta_P}{2\pi} \exp i\theta_P(L_+^{(5)} - L_-^{(5)})$ operated on the intermediate string 5 (or 6). The diagram P itself represents the configuration at $\theta_P = 0$ and the

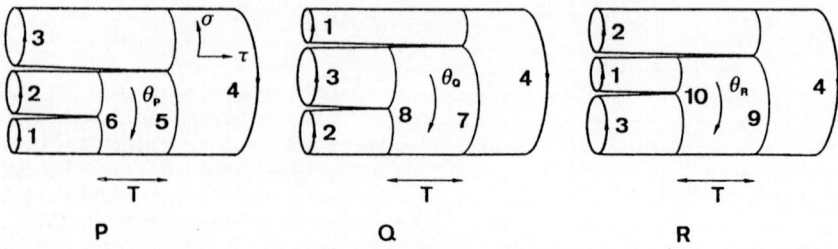

Fig.13.   The closed-string diagrams corresponding to the first (P), second (Q) and third (R) terms in the Jacobi identity (7.1).

configuration with non-zero θ_P is given by twisting the intermediate string 6 together with the strings 1 and 2 by an angle θ_P as indicated in Fig.13. $G(\sigma_{int}^{126})G(\sigma_{int}^{534})$ are the ghost factors at the two 3-string interaction points, and $D(\alpha_1 \sim \alpha_4 ; \theta_P)$ is a determinant factor

$$D(\alpha_1 \sim \alpha_4 ; \theta_P) = \frac{1}{|\alpha_5|} \left| det(1 - \tilde{N}_{\theta_P}^{66} \tilde{N}^{55}) \right|^{-(d-2)} \mu^2(\alpha_1,\alpha_2,\alpha_6) \, \mu^2(\alpha_5,\alpha_3,\alpha_4). \tag{7.4}$$

$$(\tilde{N}_{\theta_P}^{66})_{mn} = e^{im\theta_P} \tilde{N}_{mn}^{55} e^{in\theta_P}, \qquad \alpha_5 = \alpha_1 + \alpha_2 = -\alpha_6,$$

quite similar to the previous coefficient (6.6) of open-string case. Note that the power of the determinant factor is doubled compared with (6.6) since the right- and left-moving modes are now contributing to it equally. Similar expressions to (7.3) are obtained also for the Q and R terms by a suitable cyclic permutation of the indices 1,2 and 3.

We now show that the cancellations occur between $P \leftrightarrow Q$, $Q \leftrightarrow R$ and $R \leftrightarrow P$ depending on the regions of the twisting angles θ_P, θ_Q and θ_R, and the sum of the three terms vanish as a whole so that the Jacobi identity (7.1) holds.

Let us first exemplify this cancellation by considering the P-term with $\theta_P = 0$ and the Q-term with $\theta_Q = \pi$. Imagining the twisted configuration of the Q-diagram of Fig.13 by the angle $\theta_Q = \pi$, one immediately understands that the P-diagram at $\theta_P = 0$ and the Q-diagram at $\theta_Q = \pi$ represent the same configuration at $T = 0$ and hence the following equality of the 4-string δ-functionals:

$$i) \quad \mathcal{P}_{1234} \, V_{0,\theta_P=0}^{(4)} \Big|_P = \mathcal{P}_{1234} \, V_{0,\theta_Q=\pi}^{(4)} \Big|_Q \tag{7.5}$$

[It is important here that the discrepancy of the positions of the σ-coordinate origins of the external strings $1 \sim 4$ between these two configurations becomes harmless owing to the presence of $\mathcal{P}_{1234}$.] The two ghost factors are of course common between these two coincidental configurations ; that is, we see from the Fig.13 $G(\sigma_{int}^{126})|_P = G(\sigma_{int}^{714})|_Q$ and $G(\sigma_{int}^{534})|_P = G(\sigma_{int}^{238})|_Q$. However, we should note that the ghost factors in the Q term appear in the order $G(\sigma_{int}^{238})G(\sigma_{int}^{714})$ (as is understandable from the cyclic permutation $(1,2,3) \to (2,3,1)$ in eq.(7.3)), and hence we have a relative minus sign :

$$ii) \quad G(\sigma_{int}^{126}) \, G(\sigma_{int}^{534}) \Big|_P = - \, G(\sigma_{int}^{238}) \, G(\sigma_{int}^{714}) \Big|_Q . \tag{7.6}$$

Further, also for the determinant factor D, we have an equality

$$iii) \quad |d\theta_P| \, D(\alpha_1 \sim \alpha_4 ; \theta_P) = |d\theta_Q| \, D(\alpha_2,\alpha_3,\alpha_1,\alpha_4 ; \theta_Q), \tag{7.7}$$

if d = 26 and the measure function μ^2 is chosen as the square of the open-string's one (6.13), as a result of the Cremmer-Gervais identity generalized to the closed-string case.[21] These three equalitites i) $\sim$ iii) guarantee the cancellation of the P term with an interval $d\theta_P$ around $\theta_P = 0$ and the Q term with an interval $d\theta_Q$ around $\theta_Q = \pi$.

For the other configurations with more general twisting angles also, we can see the similar cancellations. A systematic way to see this is to use the Mandelstam mapping. Similarly to the open-string case, the ρ-plane (i.e., surface) of each closed-string diagram in Fig.13 is mapped onto the (entire) complex z plane by the Mandelstam mapping, taking the following form under the projective gauge fixing $z_1 = \infty$, $z_2 = 1$ and $z_4 = 0$:

$$\rho(z) = \alpha_2 \ln(1-z) + \alpha_3 \ln(z - \bar{z}_3) + \alpha_4 \ln z . \tag{7.8}$$

Only a difference from the open-string case is that the Kabo-Nielsen variables are now <u>complex</u>. Once the "gauge" of projective invariance is fixed, any closed-string diagram, characterized by two parameters T and θ in this 4-string case, corresponds one-to-one to a complex variable Z_3 through the Mandelstam mapping (7.8).

Analysing the mapping (7.8), we find the correspondence schematically drawn in Fig.14. The complex Z_3 plane splits into three regions P, Q and R which correspond to the closed-string diagrams P, Q and R in Fig.13, respectively. For instance, consider the string diagram P and fix the time interval T to a certain value $T_0 > 0$. Then, as the twisting angle θ_P is varied from 0 to 2π, the corresponding point Z_3 moves along a closed path once as indicated by the dotted line in Fig.14. Such closed paths approach to the boundary as T going to zero. So the solid lines along the boundaries of three regions (each infinitesimally inside each region) give such closed paths corresponding to the T = 0 closed-string diagrams P, Q and R in Fig.13. We have indicated the values of twisting angles $\theta_{P,Q,R}$ at some characteristic points. The triple-points, indicated by X and Y in Fig.14, correspond to the particular string configurations in which the two 3-string interaction points coincide as shown also in Fig.14.

Since a point on the boundary, for instance, between the P and Q regions, corresponds to a single 4-string configuration and it is realized as $T \to 0$ limits of both P-type and Q-type string diagrams, we can conclude the coincidence of the overlapping δ-functional $V^{(4)}_{0,\theta_P}|_P$ and $V^{(4)}_{0,\theta_Q}|_Q$ at corresponding angles θ_P and θ_Q. In this way we find the equalities on each of the three boundary segments separating P and Q, Q and R, R and P. The correspondence of the twisting angles realizing the same configurations is given as follows :

$$PQ : \quad \theta_P = \frac{\varphi}{\alpha_5} \quad \longleftrightarrow \quad \theta_Q = \pi - \frac{\varphi}{\alpha_7} \quad for \ |\forall\varphi| \leq \alpha_2\pi \ ,$$

$$QR : \quad \theta_Q = \frac{\varphi}{\alpha_7} \quad \longleftrightarrow \quad \theta_R = \pi - \frac{\varphi}{\alpha_9} \quad for \ |\forall\varphi| \leq \alpha_3\pi \ , \qquad (7.9)$$

$$RP : \quad \theta_R = \frac{\varphi}{\alpha_9} \quad \longleftrightarrow \quad \theta_P = \pi - \frac{\varphi}{\alpha_5} \quad for \ |\forall\varphi| \leq \alpha_1\pi \ .$$

Therefore the P term in one θ_P region is cancelled by the Q term in some θ_Q region and that in the other θ_P region is cancelled by the R

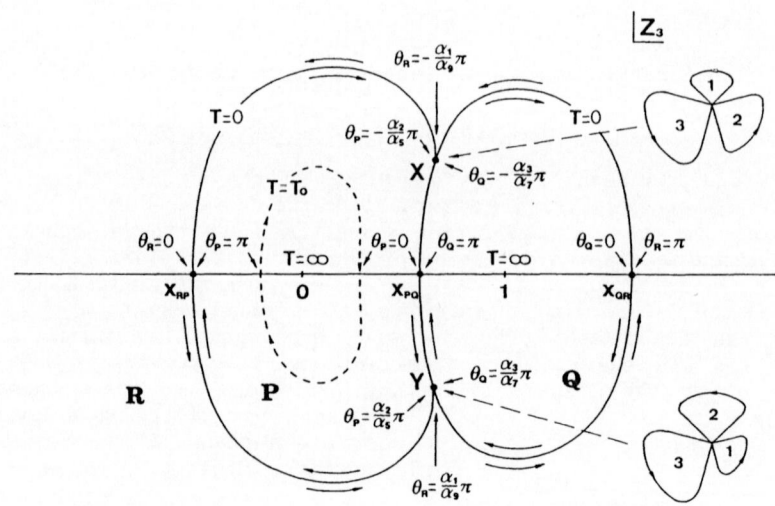

Fig.14. Three regions P, Q and R of the Koba-Nielsen variable Z_3, which correspond to the closed-string diagrams P, Q and R in Fig.13, respectively. [case α_1, α_2, $\alpha_3 > 0$]

term in some θ_R region. The Q term in the rest θ_Q region is cancelled by the R term in the rest θ_R region. Thus the three terms cacel completely as a whole.

Of course, to conclude this cancellation, we also need the equalities of the determinant factors between those corresponding two terms. This is also OK by the generalized Cremmer-Gervais identity if d = 26 and μ^2 is chosen as stated before.[21] [Incidentally, this is in fact a necessary condition for duality just like in the open-string case. The full dual amplitude of 4-closed-string is given by integrating with respect to the Z_3 variable over the whole complex plane. However, the contribution of each of the diagrams P, Q and R in Fig.13 covers only the regions P, Q and R in Fig.14, respectively. So only the sum of the three diagram contributions can give the full dual amplitude. For this to be the case, however, it implies that those three diagrams yield a single smooth function defined over the entire Z_3 plane and in particular coincident values at the boundaries in Fig.14. The last point is equivalent with the equalities of our determinant factors.]

Up to here the proof was for the case α_1, α_2, $\alpha_3 > 0$. The other cases are also proved quite similarly. [See Ref.21.]

EXERCISE : Confirm the correspondence shown in Fig.14 between the closed-string diagrams in Fig.13 and the Koba-Nielsen variable Z_3. Derive also the indicated values of the twisting angles $\theta_{P,Q,R}$ at the triple-points X and Y.

8. CONCLUSION

We have thus confimed the various identities presented in Sec.4 and hence the gauge invariance of our action. Further, though we were not able to touch in this talk, our gauge-fixed string field theory is known to satisfy the following :
i) At the tree level, the N-point on-shall amplitudes reproduce the correct dual amplitudes for arbitrary physical external states aside from the total α-conservation factor $2\pi\delta(\sum\alpha_r)$.
ii) In the zero-slope limit, the open-string system reduces correctly to the YM action ; namely [20]

$$\hat{S} = \int dx\, d\tilde{\alpha}\ tr\left[-\frac{1}{4}F_{\mu\nu}^2(x,\tilde{\alpha}) + i\delta_B\left(\bar{C}(x,\tilde{\alpha})\,G(x,\tilde{\alpha})\right)\right],$$

$$F_{\mu\nu}(x,\tilde{\alpha}) = \partial_\mu A_\nu(x,\tilde{\alpha}) - \partial_\nu A_\mu(x,\tilde{\alpha}) + i g_{YM}\left[A_\mu(x,\tilde{\alpha}), A_\nu(x,\tilde{\alpha})\right], \quad (8.1)$$

$$G(x,\tilde{\alpha}) = -\frac{1}{2}B(x,\tilde{\alpha}) + \partial^\mu A_\mu(x,\tilde{\alpha}) + (A^2,\ \bar{C}C\ terms),$$

where $\tilde{\alpha}$ is Fourier conjugate of α. Nontrivial α-dependence appears only in the last term of the gauge-fixing function $G(x,\tilde{\alpha})$.
iii) The action is found to have a symmetry which implies that the on-shell physical N-string amplitudes $\mathcal{T}_{phys}^{(N)}$ are essentially α-independent ; [30]

$$\mathcal{T}_{phys}^{(N)} = 2\pi\,\delta\left(\sum_{r=1}^{N}\alpha_r\right) \times T_N\,(\alpha\text{-indep.}) \quad (8.2)$$

formally at any loop order level and at least for N ≤ 26. Therefore our theory, closed- or open-string, is perfectly consistent theory at least as a classical string field theory. However the α-indepedent amplitudes T_N are seen [, for instance, by setting the external α_r's equal to zero,] to contain a (divergent) factor $[\int d\alpha_\ell]^L$ at L-loop level which is, we believe, harmlessly absorbed into the renormalization of loop expansion parameter. A real problem is that the amplitudes seem to contain an infinite multiple overcounting in connection with the modular invariance ; namely, the integration region with respect to the moduli parameter(s) is not restricted to the fundamental region but cover an infinite number of its copies also. Note, however, that this was shown for the amplitudes by setting the external α_r's equal to zero first.[30] There are some indications that this singular $\alpha_r = 0$ limit may not commute with the integrations of loop momenta, just like in the light-cone gauge string field theory case. So it is in fact still unclear yet whether this overcounting is actually occuring in HIKKO's theory.

A remarkable achievement of the string field theory up to now is its pregeometrical formulation;[5] the string feild action is given at the start by purely cubic action Φ^3 which is completely independent of any background geometry. The space-time with definite metric and the motion of string are simultaneously generated after the spontaneous condensation of string field. This interesting view certainly gave us a new insight. We should, however, analyse its structure further for the better understanding of its "geometrical" meaning ; we should find a variety of solutions of its classical field equation $\Phi * \Phi = 0$, find the full invariance group of the vertex functional or prove its general coordinate invariance more rigorously, and so on.

Various important problems are still left to be solved : First of all, the problem of α , or equivalently unitarity, should be clarified and solved. Extensions to superstrings and heterotic strings are still unclear[31] and not yet performed, although, for the open superstring case, Witten's field theory[32,33] seems to work well. Of course, much more challenging is to derive all the superstring field theory from the pregeometrical Φ^3 theory in d = 26 !

ACKNOWLEDGEMENT

I would like to thank H. Hata, K. Itoh, H. Kunitomo, K. Ogawa and K. Suehiro.

REFERENCES

1) A.A. Belavin, A.M. Polyakov and A.B. Zamolodchikov, Nucl. Phys. B241(1984) 333;
   V. Knizhnik and A.B. Zamolodchikov, Nucl. Phys. B247(1984) 83;
   D. Friedan, E. Martinec and S. Shenker, Nucl. Phys. B271(1986) 93.
2) D. Friedan and S. Shenker, Phys. Lett. 175B(1986) 287; Nucl. Phys. B281(1987) 509;
   E. Martincec, Nucl. Phys. B281(1987) 157;
   V. Knizhnik, Phys. Lett. 178(1986) 21;
   T. Eguchi and H. Ooguri, Phys. Lett. 187B(1987) 127; Nucl. Phys. B282(1987) 308;
   H. Sonoda, Nucl. Phys. B281(1987) 509;
   N. Ishibashi, Y. Matsuo and H. Ooguri, Mod.Phys.Lett.A2(1987) 119.
3) P.G.O. Freund, Phys. Lett. 151B(1985) 387;
   A. Casher, F. Englert, H. Nicolai and A. Taormina, Phys. Lett. 162(1985) 121.
4) W. Siegel, Phys. Lett. 149(1984) 157; 162; 151B(1985) 391; 396.
5) H. Hata, K. Itoh, T. Kugo, H. Kunitomo and K. Ogawa, Phys. Lett. 175(1986) 138;
   See also, G. T. Horowitz, J. Lykken, R. Rohm and A. Strominger, Phys. Rev. Lett. 57(1986) 283.
6) H. Hata, K. Itoh, T. Kugo, H. Kunitomo and K. Ogawa, Phys. Lett. 172B(1986) 186; 172B(1986) 195; 175B(1986) 138; Phys. Rev. D34(1986) 2360; D35(1987) 1318; D35(1987) 1356; Nucl. Phys. B283(1987) 433; Prog. Theor. Phys. 77(1987) 443.
7) A. Neveu and P.C. West, Phys. Lett. 168B(1986) 192; Nucl. Phys. B278(1986) 601.

8) E. Witten, Nucl. Phys. B268(1986) 253.

9) J.H. Schwarz in Lattice Gauge Theories, Supersymmetry, and Grand Unification; proceedings of the 6th Johns Hopkins Workshop on Current Problems in Particle Theory, Florence 1982 (Physics Department, Johns Hopkins University, Baltimore 1982); N. Marcus and A. Sagnotti, Phys. Lett. 119B(1982) 97.

10) M. Kato and K. Ogawa, Nucl. Phys. B212(1983) 443. See also K. Fujikawa Phys. Rev. D25(1982) 2584; S. Hwang, Phys. Rev. D28(1983) 2614.

11) C. Becchi, A. Rouet and R. Stora, Phys. Lett. 52B(1974) 344; Ann Phys. (N.Y.) 98(1976) 29.

12) T. Kugo and I. Ojima, Phys. Lett. 73B(1978) 459; Prog. Theor. Phys. Supplement 66(1979) 1.

13) R.C. Brower, Phys. Rev. D6(1972) 1655; J.H. Schwarz, Nucl. Phys. B46(1972) 235; P. Goddard and C.B. Thorn, Phys. Lett. 40B(1972) 235; R.C. Brower and K.A. Friedman, Phys. Rev. D7(1973) 535.

14) E. Witten, in Ref.8; A. Neveu, H. Nicolai and P.C. West, Phys. Lett. 167B(1986) 307.

15) T. Banks and M.E. Peskin, Proceedings of the Symposium on Anomaly, Geometry and Topology, Argonne Nat. Lab. March 1985; M. Kaku, CCNY preprint (1985).

16) W. Siegel and B. Zwiebach, Nucl. Phys. B263(1986) 105; T. Banks and M.E. Peskin, Nucl. Phys. B264(1986) 513; K. Itoh, T. Kugo, H. Kunitomo and H. Ooguri, Prog. Theor. Phys. 75(1986) 162.

17) T. Goto, Picture of Extended Elementary Particle (Iwanami, Tokyo 1978).

18) S.B. Giddings and E. Martinec, Nucl. Phys. B278(1986) 91; H. Hata, K. Itoh, T. Kugo, H. Kunitomo and K. Ogawa, Phys. Rev. D35(1987) 1318.

19) See however, e.g., A. Strominger, Lectures on Closed String Field Theory, delivered at ICTP School on Superstrings, Trieste, Italy, April 1987.

20) H. Hata, K. Itoh, T. Kugo, H. Kunitomo and K. Ogawa, Phys. Rev. D34(1986) 2360.

21) H. Hata, K. Itoh, T. Kugo, H. Kunitomo and K. Ogawa, Phys. Rev. D35(1987) 1318.

22) D.J. Gross and A. Jevicki, Nucl. Phys. B283(1987) 1; S. Samuel, Phys. Lett. 181B(1986), 255; K. Itoh, K. Ogawa and K. Suehiro, to appear in Nucl. Phys. B.

23) T. Banks, M.E. Peskin, C.R. Preitschopf, D. Friedan and E. Martinec, Nucl. Phys. B272(1986) 490.

24) C. Thorn, Nucl. Phys. B287(1987) 61.

25) T. Goto and S. Naka, Prog. Theor. Phys. 51(1974) 299.

26) S. Mandelstam, Nucl. Phys. B64(1973) 205. B69(1974) 77; B83(1974) 413; M. Kaku and K. Kikkawa, Phys. Rev. D10(1974) 1110; D10(1974) 1823. E. Cremmer and J.L. Gervais, Nucl. Phys. B90(1975)1653.

27) S. Mandelstam, first paper in Ref.26.

28) E. Cremmer and J.L. Gervais, in Ref.26.

29) H. Hata, K. Itoh, T. Kugo, H. Kunitomo and K. Ogawa, first paper in Ref.6.

30) H. Hata, K. Itoh, T. Kugo, H. Kunitomo and K. Ogawa, Phys. Rev. D35(1987) 1356.

31) H. Hata, K. Itoh, T. Kugo, H. Kunitomo and K. Ogawa, preprint KUNS 854, to appear in Prog. Theor. Phys.

32) E. Witten, Nucl. Phys. B276(1986) 291.

33) K. Suehiro, Kyoto preprint KUNS 857 HE9(TH) 87/04 S. Samuel, New York preprint CCNY-HEP-87/2; D.J. Gross and A. Jevicki, preprint BROWN-HET-603.

DISCUSSION I

– Liu:

The gauge algebra closes only on-shell. Can you foresee introducing auxilliary fields so that you can have gauge symmetry off-shell?

– Kugo:

I guess it could be possible to introduce auxilliary fields but is not necessary in order to show the gauge invariance of the action. Even in supergravity the inconvenience only appears when trying to find the ghost action. As it is, we have a consistent string theory.

– Kiritsis:

Do you know how to quantize your theory?

– Kugo

I have presented a gauge fixed action.

– Kiritsis:

How do you introduce the ghosts?

– Kugo:

$$\widehat{S}[\Phi] = \left[\Phi \cdot Q_B \Phi + \frac{2}{3} \, g\Phi^3 + \frac{2}{4} \, g^2 \Phi^4 \right]_{\Psi=0}$$

This is the gauge fixed, BRS–invariant action, obtained from the gauge invariant one. However, in this case there are no constraints on the internal ghost number. This field contains more components than the gauge invariant one, and the ghost components are already included. We cannot quantize canonically so we have used the Feynmann–path integral approach. We use the gaussian integration formula and then we can get the propagator of Φ as L^{-1}, and at the three level we can reproduce the dual amplitude for on–shell physical states.

– Kiritsis:

Can you compare your approach with Witten's theory? Can you recover the standard light cone theory?

– Kugo:

My work contains the quartic interaction for the open string, that enables us to reproduce the dual amplitudes. Witten's theory does not have a Φ^4 term but can also provide the dual amplitudes. In both theories the cubic interaction term have the same form but the content of the vertex functional is very different as has been shown.

The scattering amplitude for the unphysical modes is different, but the same for the physical modes, at the three level at least.

– Kiritsis:

Do you think these are just gauge-fixed versions of a more fundamental theory?

– Kugo:

May be. This is what Dr. Kaku has tried to emphasize, but nobody has been able to prove.

– Ganchev:

Continuing with the comparison between the two theories, can you explain the equivalence of the HIKKO algebra of closed strings and Witten's algebra of open strings?

– Kugo:

As you have seen our star product for closed strings obeys a distributive law and the Jacobi identity. In Witten's original theory the star product in the form of its graded commutator (Eq. 4.10) is what actually appears in his action. All of our closed–string identities are also valid in Witten's theory if we understand our star product here for the graded commutator.

$$[\Phi, \Psi] \;=\; \frac{1}{2}\left(\Phi \,*\, \Psi - (-)^{|\Phi||\Psi|}\Psi \,*\, \Phi\right)$$

The Jacobi identity is very important to show the gauge invariance of the action. I should note that in both theories the vertex functions are very different.

– Quackenbush:

Open strings contain closed strings as a natural consequence of their interactions. How is this manifested in the HIKKO model of open string field interactions, and how does it appear in Witten's scheme?

– Kugo:

Recently we published a paper in which we calculated 1–loop open string amplitudes. Here we took the limit to zero of the α–parameter of the open string, before performing the loop integrations. We found that the closed string pole gets dynamically generated.

However, I have begun to doubt the validity of these calculations. Taking the $\alpha \to 0$ limit of the external states may not be allowed before integration. It was found in a calculation that, in doing this, one obtaines the closed string amplitude in a form that has infinite multiple covering with respect to modular invariance. It covers the fundamental region and infinitely many copies. However, as has been shown by Giddings, Welpert and others, the amplitude in the light–cone gauge closed–string field theory covers the fundamental region only once. Nevertheless, if we take the $\alpha \to 0$ limit first in that theory, it also looks to cover infinitely many copies. This implies that the $\alpha \to 0$ limit is not commutative with the loop integration. Therefore, I believe that we must put the closed string in by hand in order to get gauge–invariance at the 1–loop level just like in light–cone gauge case.

In Witten's scheme, it is claimed that it also contains the closed strings dynamically generated. However I believe that it must be shown that unitary exists in the extended Fock space consisting of the closed and open string states.

– Lewellen:

So far SFT's have concentrated on reproducing scattering amplitudes in perturbation theory, which are more easily calculated with first quantization methods. Is your SFT formalism simple enough to permit the calculation of any new, non–perturbative results?

– Kugo:

Your question concerns what my intention is. I agree with you that it is better to do the perturbative expansion in the first quantized version. However, the aim of SFT is to reveal non–perturbative aspects of string theory. We hope we will be able to derive all known string theories from the pregeometrical SFT I showed you this morning. PSFT has no kinetic term, it has only an interaction term. If all those varieties of SFT are derivable from this unique PSFT then you can decide which one is preferable energetically or from the point of view of stability. Let me explain it in detail.

As we have seen closed SFT contains Einstein gravity. Its action is

$$S = \Phi \cdot Q_B \Phi + \frac{2}{3} g \Phi^3 \, , \tag{1}$$

whereas the gauge transformation takes the form

$$\delta\Phi \;=\; Q_B\Lambda \;+\; 2g\,\Phi \,*\, \Lambda \tag{2}$$

This is unsatisfactory, because it refers to flat Minkowski space through the BRS operator

$$Q_B \;=\; \int d\sigma\; i\pi_{\bar{c}}\left[\, -\, \frac{1}{2}(\eta^{\mu\nu}P_\mu P_\nu \;+\; \eta^{\mu\nu}X'_\mu X'_\nu) \;+\cdots\right] \tag{3}$$

which depends explicitly on the Minkowski metric $\eta^{\mu\nu}$.

– *Windey:*

May I interrupt you. The question was: "What can you imagine to produce that is not already done by standard 1st quantized methods, and should be intrinsically non–perturbative?"

– *Kugo:*

I agree that the sigma–model approach is better for doing perturbative calculations. On the other hand SFT is better for non–perturbative phenomena, e.g. for investigation of possible backgrounds.

– *Windey*

Do you think it will be simpler than going to infinite genus? In principle, you can use 1st quantized methods in a non–perturbative fashion if you formulate the theory on arbitrary genus surface including the infinite one.

– *Kugo:*

In principle it would be possible. However, imagine that you are given all the information on propagators and vertices for a flat background. Then, will you think it practically possible to find Schwarzschild solution, for instance with that information alone without knowing Einstein equation?

– *Windey*

Build the most general string theory, not necessarily starting with a flat background.

– *Ganchev:*

Can you continue on with the explanation?

– *Kugo*

We would like to reformulate SFT in such a way that the action is independent of any particular background metric. As we have seen the free part of the action depends on $\eta^{\mu\nu}$. On the other hand, the interaction term in (1)

$$\Phi^3 \;=\; \Phi \,\cdot\, (\Phi \,*\, \Phi) \tag{4}$$

199

and the field–dependent part of the gauge–transformation,

$$\delta\Phi = \Phi * \Lambda \tag{5}$$

are independent of the space–time metric since the $*$ product is constructed out of delta–functions alone. Also the constraint

$$P\Phi = \Phi \tag{6}$$

where P is the sigma–translation operator, is metric independent.

The action of the pre–theory is obtained from (1) by simply dropping the free part and rescaling the field ($g\Phi \rightarrow \Psi$):

$$S^{pre} = \frac{2}{3g^2} \Psi^3 \qquad (P\Psi = \Psi) \tag{7}$$

This is invariant under the local gauge transformation

$$\delta^{pre}\Psi = 2\Psi * \Lambda \tag{8}$$

This theory is called pre–geometrical, since it is not only independent of any space–time metric, but does not have any kinetic term either. So: no space–time notion and no motion!

The space–time with definite background metric appears only after string field condensation occurs:

$$< \Psi > = \Psi_0 \tag{9}$$

If we now split the field Ψ into classical (Ψ_0) and fluctuation (Φ) parts:

$$\Psi = \Psi_0 + g\Phi \tag{10}$$

and substitute (10) into (7), we find that a kinetic term has been generated for the fluctuation field:

$$\begin{aligned}
S^{pre} &= \frac{2}{3g^2} (\Psi_0 + g\Phi)^3 \\
&= 2\Phi \cdot (\Psi_0 * \Phi) + \frac{2g}{3} \Phi^3
\end{aligned} \tag{11}$$

Here we have used the classical equation of motion of the pre-theory

$$\Psi_0 * \Psi_0 = 0 \tag{12}$$

which the classical field Ψ_0 has to satisfy. If we now define a linear operator Q by

$$\Psi_0 * \Phi \equiv \frac{1}{2} Q\Phi \tag{13}$$

200

then (11) and (8) can be written as

$$S^{pre} = \Phi \cdot Q\Phi + \frac{2g}{3} \Phi^3 \tag{14}$$

$$\delta^{pre}\Phi = Q\Lambda + 2g\, \Phi * \Lambda \tag{15}$$

which are of the same form as (1) and (2) but now the operator Q plays the role of the BRS charge.

Q has the same properties as the BRS charge Q_B, i.e. it is nilpotent, satisfies the distributive law and the partial integration law.

At this point it is natural to ask whether it is possible to realize the original BRS operator Q_B in the form

$$\Psi_0 * \Phi = \frac{1}{2} Q_B \Phi \tag{16}$$

with some classical field Ψ_0. Here Q_B is realized as a differential operator. How such a differential operator can appear? The answer is related to the fact that Q_B does not change the string length parameter α, therefore

$$\Psi_0 \sim \delta(\alpha) \tag{17}$$

i.e. Ψ_0 has vanishing string length. So if we substitute Ψ_0 into the $*$ product formula and take the limit $\alpha \to 0$, Ψ_0 actually becomes a differential operator on Φ as illustrated by these pictures

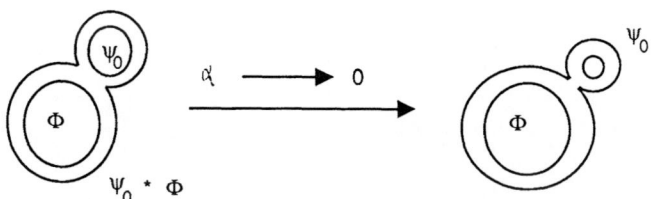

In this way we can reproduce the kinetic term for the flat case.

We can find other classical solutions Ref.[6] corresponding to non-trivial backgrounds. We can proceed as follows.

First we construct a field Γ which satisfies

$$\Gamma * \Phi = \left(N_{FP} + 1 - \alpha \frac{\partial}{\partial\alpha} \right) \Phi \tag{18}$$

$$N_{FP}\Gamma = -2\Gamma$$

Then, using (17) and the properties of Q_B we can show that

$$\Psi_0 = -\frac{1}{2} Q_B \Gamma \tag{19}$$

is a solution of the classical equation of motion (12). We have explicitly constructed many different Γ's satsfying (18). There are clearly many solutions to the classical equation of motion 12. All of them correspond to a consistent string theory with a non–trivial background metric.

– *Kiritsis:*

You have just shown that you get a Φ^3 term from the pre–SFT. What about the Φ^4 term?

– *Kugo:*

This derivation just applies for the closed string case. We do not have yet a pre–SFT for the open string case.

– *Liu:*

Both, conformal invariance in the Polyakov approach and $Q_B^2 = 0$ in the string field theory give D=26. Is there any connection between them?

– *Windey:*

That is just BRS invariance. Even in the 1st quantized case Q_B is conserved only in D=26.

– *Kugo:*

Q_B is conserved in any dimension, only the nilpotency of Q_B demands D=26.

– *Singh:*

For the 3 string interaction why do you only consider $\alpha_3 = \alpha_1 + \alpha_2$? Why not other cases like Witten's $\alpha_1 = \alpha_2 = \alpha_3$?

– *Kugo:*

I demand such a rule because it gives a gauge invariant action. Gauge invariance will be violated if you take another kind of vertex. There are no known vertices other than HIKKO's and Witten's meeting the gauge–invariance requirement.

DISCUSSION II

– *Liu:*

Standard lore in string theory is that strings only interact through 3–point vertex, why do you have elementary 4–point vertex in your SFT?

– *Kugo:*

In the open string case we really have a 4–point vertex which is necessary for gauge invariance. The same is true in the usual light cone gauge SFT. In this case it is clear that the 4–point vertex is necessary for proving Lorentz invariance (as well as duality). In our theory, as I have explained this morning, associativity is violated (in open SFT) by the 3–point vertex alone. Almost all configurations of figs. 10 and 11 cancel pairwise, but those diagrams which have branch points (we call them horn diagrams) violate associativity. To cure this violation of associativity we clearly need a 4–point vertex. The distributive law of the BRS operator for this 4–point vertex, almost holds, but is slightly violated by surface terms. The violation of the associativity by horn diagrams is compensated by this violation of the distributive law. In Witten's case his $*$ product satisfies associativity and he does not need to introduce any higher vertex.

– *Liu:*

Do you think that your theory is less stringy than Witten's?

– *Kugo:*

Yes, maybe. But the converse is true for the closed string case. Here instead of the associative law we need the Jacobi identity. If you extend Witten's theory to the closed string in a straightforward manner as Lykken and Raby did, the resultant $*$ product does not satisfy Jacobi identity (as was shown by Giddings, Martinec and Hikko). So you have to introduce a 4–point vertex, and the conjecture is that this algebra does not close. You will need a 5–point vertex, and so on.

– *Balog:*

My question is connected with what you explained us yesterday afternoon. I would like to see more explicitly how Ψ_0, the classical solution of the pregeometrical SFT is constructed?

– *Kugo:*

Ψ_0 was formally given by

$$\Psi_0 \;=\; \frac{1}{2}\, Q\Gamma$$

where you can use any BRS operator Q. The most practical way to construct Q corresponding to a general background is to use the non-linear sigma-model. The corresponding BRS operator satisfies nilpotency if Einstein equation holds as Professor Grisaru explained you. In addition we need the condition

$$\sum_{r=1}^{3} Q^{(r)} |V>= 0$$

which is the distributive law for the vertex. Although I don't know the profound meaning, in all cases we have considered whenever nilpotency holds, this condition holds too. This was proven recently by Kikkawa, Maeno and Sawada. For the explicit expression of the field Γ, I refer to our Phys. Lett. paper B175 (1986) 138.

– *Quackenbush:*

Can you explain us how to calculate scattering amplitude in your string field theory?

– *Kugo:*

The calculation method for the amplitudes in our covariant string field theory is essentially identical with the one appearinng in the light-cone gauge SFT.

Basically one goes to the path integral expression and evaluate it by the help of the Neumann function defined on the string diagram. See the details presented in the excellent lecture of Mandelstam at Santa Barbara. See also the calculation given in Sec. VIII in Ref. 20.

– *Lewellen:*

A key test of your closed SFT is whether you obtain a single covering of moduli space. How difficult is it in your formalism to calculate the one loop vacuum amplitude and see if you obtain the correct fundamental domain of the modular group?

– *Kugo:*

It is not difficult to calculate the 1-loop diagrams, but we have not calculated it yet. In our third paper Ref. 30 which appeared in Phys. Rev., we have shown that our closed SFT gives the wrong answer for the 1–loop amplitude, it gives an infinite multiple covering of moduli space. But it is true only if you put $\alpha = 0$ for the external strings before performing the loop integration.

– *Lewellen:*

This is why I am suggesting you to do the vacuum amplitude.

– *Kugo:*

We are considering for instance the two–point function. Or n–point function...

– *Lewellen:*

You can just do the zero–point function, the vacuum amplitude.

– *Kugo:*

It is not necessary for this case to reproduce single covering of the fundamental domain.

– *Lewellen:*

Why?

– *Kugo:*

In field theory the vacuuum energy is nothing else but zero–point energy. Since the string theory is equivalent to an infinite number of local fields, the vacuum energy is simply the sum of zero–point energies of all those fields. For that reason the vacuum energy will not give you a single covering of the fundamental region, but much more. But if you have external legs, you do cover the fundamental region but once. This is the reason why we are considering the two-point function, for instance.

– *Lewellen:*

But the vacuum amplitude has a physical meaning, it is the cosmological constant. For example, in the $O(16) \times O(16)$ model you have a cosmological constant which is big, but finite. And to get the right result, you need only one covering of moduli space.

– *Kugo:*

The vacuum graph is not the cosmological constant. The cosmological constant, in field theory effective action, appears in the form $\sqrt{-g}$ times cosmological constant. But to calculate the $\sqrt{-g}$ term, you have to calculate the 2–point function, 3–point function, etc. So anyhow, it is better to calculate the amplitude with external strings. We are still calculating that amplitude. We will know the answer to your question soon, but as yet we don't know.

– Schuler:

You showed us yesterday how to derive the kinetic term from a pure Φ^3 term. I wonder how this would look like in terms of the component fields?

– Kugo:

In terms of component local fields the Φ^3 interaction is highly non–local, it is only local in the string sense. This is the reason why this theory with the purely algebraic term can give you a kinetic term, by shrinking the classical string to zero length. As was shown in the previous discussion, the action of the classical solution is equivalent to a differential operator. So if you expand the Φ^3 term in terms of local fields defined on the centre of mass coordinate the interaction is very non–local.

– Liu:

Do you think it is possible to have gauge invariance of the SFT action with the 3–point vertex only, if you give up the requirements of associativity and commutativity of the $*$ product?

– Kugo:

No, I think it impossible.

CONSTRUCTION OF STRING AND SUPERSTRINGS

IN ARBITRARY SPACE-TIME DIMENSIONS

Costas Kounnas

Physique Théorique de l'Ecole Normale Supérieure

24 rue Lhomond, 75231 Paris cedex 05

FRANCE

ABSTRACT

String models with space-time chiral fermions and physically relevant interactions may easily be constructed in space-time dimensions lower than ten by using different versions of the underlying two-
dimensional superconformal theory. The fundamental string constraints
and their general solutions are presented for the case of closed strings
in which all the string internal quantum numbers are carried by free periodic and antiperiodic world sheet fermions. We also present a symmetry
breaking mechanism, for gauge symmetry and supersymmetry, which operates
on the string level. This mechanism defines new string solutions which
are characterized by a mass spectrum depending upon several breaking
parameters, therefore introducing new arbitrary continuous parameters
on strings.

1.   INTRODUCTION

(Super)-String models[1] have recently been constructed in dimensions D lower than their critical dimensions either by compactification[1,2-7] or by utilization of different versions of the underlying

two-dimensional (super)-conformal theory[8-17]. Many of the string models are defined directly in four-dimensional space-time[8-11] and yield an acceptable low energy phenomenology. They are based on a gauge symmetry group $G \supset SU(3) \times SU(2) \times U(1)$ and contain, in their massless spectrum, chiral families in complex representations of G. Moreover with the gravitational forces naturally present, the superunification and quantization of all known interactions is successfully achieved in the framework of any consistent and physically relevant string model. Thus string theories are, in this respect, the only serious candidates for the superunification of all known particle interactions.

It is at present widely recognized that a unified string theory[1] cannot be singled out uniquely on the basis of perturbative consistency[18]; many four-dimensional, in particular, models have been shown to pass the tests of invariance under local and global world-sheet super-reparametrizations[2-17]. The current belief is of course[13] that all these models correspond to classical solutions of only a few second-quantized theories, whose non-perturbative dynamics, when fully understood, will allow us to make a unique choice. In the meantime, however, it is important to continue the study of these different string vacua, in order to both a) obtain a phenomenologically viable and calculable model with realistic gauge group and chiral quark and lepton families, and b) provide a context in which to study general properties of string theories such as their low-energy symmetries, and the possibility they offer for resolving such long-standing problems like the generation of fermion masses, gauge hierarchies, etc.

A strong indication that all these new models correspond to classical solutions of only few second-quantized theories follows also from the fact that the asymptotic density of states of different string models possess some universal properties independent of the dimensions of spacetime in which the theory is formulated[19],

$$\rho(m) \xrightarrow[m \to \infty]{} c\, m^{-a} \exp bm$$

Indeed, b is a dimension and an essentially model-independent quantity. It only depends on the (super)-reparametrization properties of the world sheet and the value of the string mass scale $(\alpha')^{-1/2}$ ($\alpha' \equiv$ the Regge slope). This remarkable property follows simply from the modular invariance of the world sheet and shows that in short distances the string with the same reparametrization properties look the same[19]. The sub-leading contribution is dictated by the parameter α and it depends only on the dimensionality of the space ($\alpha = D$). The coefficient c is a high-ly model-dependent quantity. It depends, for instance, on the radius of compactification and/or the general method of construction of the model (orbifolds, fermionic construction, etc.). The universality of b sug-gests a classification scheme for all string models on the basis of the super-reparametrization properties of the world sheet. In particular b depends only on the number of world sheet supersymmetries. For closed strings with Regge slope α' we have:

(i) Type II strings $(1,1)$[20] with left and right superconformal in-variance (as indicated by the $(1,1)$ notation)

$$b(1,1) = 2\sqrt{2}\,\pi\sqrt{\alpha'} \qquad (1.1)$$

(ii) Heterotic type strings $(1,0)$[21] with left world sheet supersymmetry, left superconformal invariance and right conformal inva-riance

$$b(1,0) = (\sqrt{2}+2)\,\pi\sqrt{\alpha'} \qquad (1.2)$$

(iii) Bosonic type strings $(0,0)$[1] with left and right conformal in-variance. There is neither left nor right world-sheet supersymmetry[19]

$$b(0,0) = 4\pi\sqrt{\alpha'} \qquad (1.3)$$

From the physical point of view, the $(0,0)$ strings are not interesting because they do not contain space-time fermions in their spectrum. Furthermore, they seem to suffer from severe instabilities, e.g. tachy-onic states and/or unstable flat space-time metric at the quantum level.

A way of constructing string models $(1,1)$ or $(1,0)$, in less than ten dimensions, is to describe the quantum number densities on the string either by extra free world-sheet fermions[6,8-13], or by free world sheet bosons parametrizing some flat torus or orbi-fold[3-7,14,15,17]. The two formulations are thought to be equivalent[22], so that the choice is to some extent a matter of taste and convenience. Counting the number of distinct models, or finding the gauge group, is in some cases easier in the bosonic language, but for most other purpo-ses the fermionic language is advantageous. For instance, in the fermi-onic formulation, one may easily classify all possible world sheet supersymmetries[8], find the effective low-energy field theory of a string model[23,24], or prove that space-time supersymmetry implies the vanishing of the one-loop cosmological constant[10] and under some as-sumptions to all orders in loop expansion[25].

In these lectures, by introducing free fermions on the world sheet, we show how one may obtain consistent string models in dimensions lower than ten. In our approach (sections 2-6), in order to define a consistent D-dimensional string, we only employ the fundamental tools used to construct the known ten-dimensional close string models[1,20,21,26,27]. In other words, we systematically take account of all requirements and constraints following from[10]:

(1) World sheet reparametrization invariance. (Local diffeomorphism $\oplus$ conformal invariance).

(2) Local world sheet supersymmetry. (1,0) in heterotic case, (1,1) in type II case.

(3) Absence of global anomalies $\oplus$ multiloop modular invariance of the higher genus surfaces.

(4) The unitarity requirement of cluster decomposition which implies the factorization of the vacuum to vacuum amplitude.

The first and second constraints fix the number of bosonic and fermionic degrees of freedom while the third and fourth choose the properties of boson and fermion densities when they are parallel transported around some non trivial cycles existing in higher genus surfaces ($g \geqslant 1$).

In the remaining sections (7-8), we describe some symmetry breaking phenomena on strings, using a generalized (coordinate dependent) torus compactification[14,17]. This breaking mechanism (Higgs and SuperHiggs phenomenon) is nothing else but the generalization to strings of the Scherk-Schwarz[28] breaking mechanism in field theory.

2.   SUPER-REPARAMETRIZATION CONSTRAINTS

As we already mentioned in the introduction the constraints of super-reparametrization invariance uniquely fix the number of free world sheet bosons and fermions. In order to preserve the classical symmetries of the world sheet at the quantum level it is necessary to ensure the cancelation of conformal and gravitational anomaly[1,29]. This requirement leaves us with two matching anomaly conditions for the left- and right moving degrees of freedom separately. As mentioned above, the way we reduce the space-time dimensionality is by introducing free world-sheet fermions[30,8] instead of bosons. From the point of view of local anomalies, (central charges of left- and right- Virasoro algebras c_L, c_R), one left-(right-) moving boson gives equal contribution to the conformal and gravitational anomaly as two left-(right-) handed Weyl-Majorana fermions C_L(Boson) = 2 C_L(W-M Fermion).

In the absence of a world sheet supersymmetry, this fermionization[30] is straightforward and can be done without any obstruction.

However, when a world-sheet supersymmetry is assumed (super-reparametrization), we are in general faced with an obstruction, since some of the bosons are fermionized. It is necessary to examine when a non-linear supersymmetry can be defined among free fermions[8,31]. The answer to that problem is given in ref. 8], where it was shown that a system of free fermions (say left-moving) is supersymmetric, (left-supersymmetry), if and only if the free fermions belong to the adjoint representation of a semi-simple group (H_L). The proof stays as follows:

Let $\{\chi^A\}$ $A = 1,2,\ldots,n$ a system of left-moving Weyl-Majorana free fermions. The action of the system is then,

$$S = \frac{1}{2} \int dz\, d\bar{z}\ \chi^A \partial_{\bar{z}} \chi^A \tag{2.1}$$

and possesses a classical fermionic symmetry

$$\delta_\epsilon \chi^A = \epsilon\, \eta^{ABC} \chi^B \chi^C \quad \leadsto \quad \delta_\epsilon S = 0 \tag{2.2}$$

ϵ is a Grasmann variable. This symmetry is a supersymmetry if and only if the anticommutator of the currents which generate $\delta_\epsilon \chi^A$ give the energy-momentum density up to possible c-number anomaly:

$$\{ T_F(z),\, T_F(\omega) \}_+ = 2\, T_B(z)\, \delta(z-\omega) \quad \text{+c-number}$$

where

$$T_F = \frac{1}{3}\, \eta^{ABC} \chi^A \chi^B \chi^C, \quad T_B = \frac{1}{2}\, \chi^A \partial_z \chi^A \tag{2.3}$$

This condition is equivalent to requiring that the commutator of two supersymmetry transformations be a translation, as it should. Its validity can most easily be examined by means of the operator-product expansion

$$T_F(z)\, T_F(\omega) \underset{z \to \omega}{\sim} -\frac{2}{3}\, \frac{\eta^{ABC}\eta^{ABC}}{(z-\omega)^3} - 2\eta^{ACD}\eta^{BCD} \left\{ \frac{\chi^A(\omega)\chi^B(\omega)}{(z-\omega)^2} \right.$$

$$\left. -\frac{\chi^A(\omega)\partial_\omega \chi^B(\omega)}{(z-\omega)} \right\} + \eta^{ABE}\eta^{CDE}\, \frac{\chi^A(\omega)\chi^B(\omega)\chi^C(\omega)\chi^D(\omega)}{(z-\omega)}$$

$$\text{+ regular terms,} \tag{2.4}$$

where we have here used Wick's theorem, and the free fermion contraction

$$\chi^A(z)\, \chi^B(\omega) = \frac{\delta^{AB}}{(z-\omega)} \tag{2.5}$$

Since normal-ordered fermions anticommute it follows immediately that the right-hand side of eq. (2.4) agrees with eq. (2.3) if and only if

$$\eta^{ABE}\eta^{CDE} + \eta^{ACE}\eta^{DBE} + \eta^{ADE}\eta^{BCE} = 0$$

and

$$\eta^{ACD} \eta^{BCD} = \frac{1}{2} \delta^{AB} \tag{2.6}$$

The first condition is the Jacobi identity, implying that η^{ABC} are the structure constants of a Lie group H_L, while the second guarantees that H_L has no normal abelian subgroup. In the system $\{\chi^A\}, A = 1,2,\ldots,n$ the supersymmetry is non-linearly realized and its contribution to the left conformal anomaly is[8]

$$C_L(\{\chi^A\}) = \frac{1}{2} \dim H_L = \frac{1}{2} n \tag{2.7}$$

The dimension of the adjoint of H_L is fixed from the left anomaly cancellation once the number of bosonic degrees of freedom (dimension of spacetime D) is given. In fact the reparametrization ghosts b and c have conformal weights 2 and -1 respectively, and give $C_L = C_R = -26$[1,29]; the super-reparametrization ghosts β_L, γ_L of left supersymmetry with 3/2 and -1/2 conformal weights give $C_L = 11$, $C_R = 0$[1,29]. (In the presence of right moving supersymmetry the β_R, γ_R ghosts give $C_L = 0$, $C_R = 11$). These numbers are universal and uniquely fix the number of left and right moving free fermions once the space-time dimension is given.

Indeed if we assume (1,0) world sheet supersymmetry and space-time dimension D, (number of bosonic degrees of freedom $C_L(X^\mu) = D$, $C_R(X^\mu) = D$), the left and right moving fields must be chosen as follows[8,9,10]:

(i) <u>Right moving</u> (reparametrization)

string coordinates (bosons): $\partial_{\bar{z}} X^\mu$, $\mu = 1,2,\ldots D$ $\longrightarrow C_R = D$

extra fermions : $\bar{\psi}^A$, $A = 1,2,\ldots 2(26-D) \rightarrow C_R = 26-D$

reparametrization ghosts : b, c $\longrightarrow C_R = -26$

$$\rightsquigarrow C_{R,total} = 0. \tag{2.8}$$

(ii) <u>Left moving</u> (super-reparametrization)

string super-coordinates $\Big\}$ $\partial_z X^\mu$ 2-d superfields ψ^μ $\quad \mu = 1,2,\ldots, D \quad \longrightarrow C_L = \frac{3}{2} D$

extra fermions in the adjoint of $\Big\}$ χ^a $\quad a = 1,2,\ldots, 3(10-D) \longrightarrow C_L = \frac{3}{2}(10-D)$

super reparametrization ghosts $\Big\}$ b, c $\quad$ β_L, γ_L $\quad\quad \longrightarrow C_L = -26 + 11$

$$\rightsquigarrow C_{L,total} = 0. \tag{2.9}$$

In the case of left and right super-reparametrization, (1,1), the right moving spectrum is similar to the left moving one ; however, the H_R group is not necessarily the same as H_L.

Because the dimension of H_L (and H_R) is fixed to 3 (10-D) there are not many independent choices. In arbitrary dimension $D \leq 10$, one solution is[8]

$$H_L = \left[SO(3) \right]^{10-D}$$

(2.10)

This is the only solution for $D \geq 6$. For $D = 5$ there is another one $H_L = SU(4)$ and for $D = 4$ there are two other possibilities $H_L = SU(4) \times SO(3)$ and $H_L = SO(5) \times SU(3)$.

Even if the world-sheet field content is uniquely specified once $H_{L,R}$ is fixed, the string model is not yet defined globally. For topologically non-trivial world-sheets we must still specify the spin-structures of all world-sheet fermions[32,33,26,27,9-11], i.e. the properties of spinor fields under parallel transport around non-contractible loops of the surfaces, and we must verify that our choice is invariant under large diffeomorphisms not continuously connected to the identity, i.e. modular transformations. Different choices of spin structures correspond to different string theories.

3. GLOBALLY DEFINED STRINGS

In the previous section we have examined the consequences of the requirement of super-reparametrization in the presence of fermionic degrees of freedom. Here we derive some additional constraints which follow from the modular invariance and the factorization of higher genus surfaces $(g \geq 1)$[33].

We consider first a world-sheet with the topology of a torus. After a reparametrization and a Weyl transformation the torus can be represented by a flat parallelogram in the complex plane with sides 1 and corresponding to its two non-contractible loops. The one loop vacuum to vacuum string amplitude is written as

$$Z = \int \frac{d\tau d\bar{\tau}}{(Im\tau)^2} Z_B(\tau,\bar{\tau}) Z_S(\tau,\bar{\tau})$$

(3.1)

In (3.1) Z_B is the contribution of the X^μ coordinates and the reparametrization ghosts (left,right) b,c:

$$Z_B = \left| Im\tau \right|^{-\frac{D-2}{2}} \left| \eta(\tau) \right|^{-2(D-2)}$$

(3.2)

213

where $\eta(\tau) = q^{1/12} \prod_n (1 - q^{2n})$ is the Dedekind eta function with $q = \exp i\pi\tau$.

Also in (3.1) Z_s is the contribution of the world-sheet fermions and (super)reparametrization ghosts β, γ.

$$Z_S = \sum_{\text{spin Str.}} C\begin{bmatrix} a \\ b \end{bmatrix} Z_{long,3/2}\begin{bmatrix} a_v \\ b_v \end{bmatrix} \prod_{f=1}^{n_L + n_R} Z_F\begin{bmatrix} a_f \\ b_f \end{bmatrix}$$

(3.3)

where $Z_F\begin{bmatrix} a_f \\ b_f \end{bmatrix}$ is the contribution of the fermion f (the determinant of the chiral Dirac operator) which depends on its spin structure (a_f, b_f). Following standard conventions $a_f = 1$ or 0 denotes periodic or antiperiodic boundary conditions in the direction 1 ; or similar notation applies for b_f in the direction τ. More explicitly, using standard path integral techniques, we have

$$Z_F\begin{bmatrix} a \\ b \end{bmatrix} = \eta(\tau)^{-1/2} \theta\begin{bmatrix} a \\ b \end{bmatrix}^{1/2}(\tau)$$

(3.4)

where $\theta\begin{bmatrix} a \\ b \end{bmatrix}(\tau)$ are the well-known theta functions up to some phase convention

$$\theta\begin{bmatrix} a \\ b \end{bmatrix}(\tau) = \sum_{n \in Z} \exp\left[i\pi\tau(n + \tfrac{a}{2})^2 + 2i\pi(n + \tfrac{a}{2})\tfrac{b}{2} - in\tfrac{ab}{2} \right]$$

(3.5)

The phase convention, $-i\pi\tfrac{ab}{2}$, is chosen for later convenience ; with this choice $\theta\begin{bmatrix} a \\ b \end{bmatrix} = \theta\begin{bmatrix} a \\ b+2k \end{bmatrix}$, $k \in Z$, for any a. In the case we use periodic or antiperiodic boundary conditions (a,b = 1 or 0) we have

$$Z_F\begin{bmatrix} 0 \\ 0 \end{bmatrix} = tr\left[e^{i\tau 2\pi H_{NS}} \right] = \theta_3^{1/2}(\tau) \Big/ \eta^{1/2}(\tau)$$

$$Z_F\begin{bmatrix} 0 \\ 1 \end{bmatrix} = tr\left[(-)^F e^{i\tau 2\pi H_{NS}} \right] = \theta_4^{1/2}(\tau) \Big/ \eta^{1/2}(\tau)$$

$$Z_F\begin{bmatrix} 1 \\ 0 \end{bmatrix} = tr\left[e^{i\tau 2\pi H_R} \right] = \theta_2^{1/2}(\tau) \Big/ \eta^{1/2}(\tau)$$

$$Z_F\begin{bmatrix} 1 \\ 1 \end{bmatrix} = tr\left[(-)^F e^{i\tau 2\pi H_R} \right] = \theta_1^{1/2}(\tau) \Big/ \eta^{1/2}(\tau)$$

(3.6)

These formulae hold for left movers and should be complex conjugate for the right movers. H_{NS} and H_R are the Hamiltonians for antiperiodic a and periodic a respectively (a = 0, Neveu-Schwarz ; a = 1, Ramond).

$$H_{NS} = \sum_{z \in Z^{+} - \frac{1}{2}} z \left(\psi_z^{+} \psi_z + \psi_{-z} \psi_{-z}^{+} \right) - \frac{1}{24}$$

$$H_R = \sum_{n \in Z^{+}} n \left(\psi_n^{+} \psi_n + \psi_{-n} \psi_{-n}^{+} \right) + \frac{1}{12} \qquad (3.7)$$

$\psi_z = \psi_{-z}^{+}$ and $\psi_n = \psi_{-n}^{+}$ are fermionic oscillators with antiperiodic (0) and periodic (1) boundary conditions. The operator $(-)^{F}$ is equal to ± 1 according to whether the two-dimensional fermion number is even or odd.

Finally, $Z_{long,3/2} \begin{bmatrix} a_\psi \\ b_\psi \end{bmatrix}$ is the contribution of two of the components of ψ^μ and the world-sheet gravitino ghosts β, γ ; it is equal to 1 in the case of the torus and thus can be dropped. In higher genus surfaces $Z_{long,3/2}$ is not trivial and plays an important role in the multiloop factorization requirement[10]. A few remarks are in order: (1) In (3.3) the summation runs over all possible spin structure assignments that are consistent with two-dimensional supersymmetry, i.e. such that $T_F(Z)$ has well defined parallel transport properties. (2) The fermions ψ^μ, the supercurrent $T_F(Z)$ and the two-dimensional gravitino must all have the same spin structure which we denote by $\begin{bmatrix} a_\psi \\ b_\psi \end{bmatrix}$. (3) The measure $\frac{d\tau \, d\bar{\tau}}{(Im\, \tau)^2}$ as well as the bosonic contribution $Z_B(\tau, \bar{\tau})$ are invariant under the transformation

$$\tau \longrightarrow \tau + 1 \, , \qquad \tau \longrightarrow - \frac{1}{\tau} \qquad (3.8)$$

which generates the modular group of the torus ; the integration is over the uppercomplex plane moded out by the modular group.

Under the transformation (3.8) the spin structures transform into each other or into themselves up to phases. Explicitly:

i) $\theta \begin{bmatrix} a \\ b-a+1 \end{bmatrix}(\tau+1) = e^{i\pi \frac{a^2}{4}} \theta \begin{bmatrix} a \\ b \end{bmatrix}(\tau)$

ii) $\theta \begin{bmatrix} b \\ -a \end{bmatrix}(-\frac{1}{\tau}) = (-i\tau)^{-\frac{1}{2}} e^{i\pi \frac{ab}{2}} \theta \begin{bmatrix} a \\ b \end{bmatrix}(\tau)$

iii) $\eta(\tau+1) = e^{i\pi/12} \eta(\tau)$

iv) $\eta(-\frac{1}{\tau}) = (-i\tau)^{-\frac{1}{2}} \eta(\tau) \qquad (3.9)$

To ensure the one loop modular invariance we must therefore impose the following conditions on the coefficients $C \begin{bmatrix} a_i \\ b_i \end{bmatrix}$, $i = 1,2,\dots,n_L+n_R$

$$C \begin{bmatrix} a_i \\ b_i \end{bmatrix} = e^{\frac{i\pi}{8}(a \cdot a + 1 \cdot 1)} C \begin{bmatrix} a_i \\ b_i - a_i + 1 \end{bmatrix}$$

$$C \begin{bmatrix} a_i \\ b_i \end{bmatrix} = e^{i\pi \frac{a \cdot b}{4}} C \begin{bmatrix} b_i \\ -a_i \end{bmatrix} \qquad (3.10)$$

215

where a, b, $\mathbb{1}$ are vectors with $n_L + n_R$ components:

$n_L = n_R = D - 2 + 3 (10 - D) = 28 - 2D$ in type II case (1,1)

$n_L = n_R = 2 (26 - D) = 52 - 2D$ in non supersymmetric case (0,0) and

$n_L = 28 - 2D$, $n_R = 52 - 2D$ in heterotic case (1,0).

The dot product of two vectors is Lorentzian among left- and right-movers ; the global phase $e^{i n/8\, \mathbb{1} \cdot \mathbb{1}}$ in the first eq. (3.10) comes from the transformation properties of $\eta(\tau)$ function, it equals -1 for heterotic and $+1$ for the bosonic or type II strings. All other phases come from the transformation of the θ-function[34]. ($\mathbb{1}$ is a vector all of whose components are 1).

The modular invariance of the genus one (g = 1) surfaces is not sufficient to consistently define a string model[32,33]. Further constraints, which follow from the higher genus modular invariance and factorization, are necessary. At this point our analysis differs from that suggested by Witten and Seimberg[32] and which was used by Kawai, Lewellen and Tye[9]. For them, the additional constraints follow from the physical requirement of correct particle interpretation and correct connection between spin and statistics. For us, these requirements are not taken as assumptions but are proved[10] as consequences of the multi-loop modular invariance and factorization properties which the vacuum to vacuum amplitude must satisfy due to unitarity[33].

A complete spin-structure assignment to all fermions on a genus - g surface can be specified by 2g vectors

$$\begin{bmatrix} a^1, a^2, \ldots, a^g \\ b^1, b^2, \ldots, b^g \end{bmatrix} \quad \text{with,} \quad \begin{aligned} a^i &= (a^i_1, \ldots, a^i_{n_L + n_R}) \\ b^i &= (b^i_1, \ldots, b^i_{n_L + n_R}) \end{aligned}$$

(3.11)

where a^i and b^i are the boundary conditions corresponding to "a^i cycle" and "b^i cycle" of the canonical homology basis. Thus the g-loop partition function involves a coefficient factor $C\begin{bmatrix} a^1 \ldots a^g \\ b^1 \ldots b^g \end{bmatrix}$, and each fermion has 2^{2g} possible spin structures. By considering a limit in which the surface pinches off into a product of g tori one can argue that unitarity requires the factorization

$$C\begin{bmatrix} a^1 \ldots a^g \\ b^1 \ldots b^g \end{bmatrix} = C\begin{bmatrix} a^1 \\ b^1 \end{bmatrix} \cdots C\begin{bmatrix} a^g \\ b^g \end{bmatrix} \quad .$$

(3.12)

This is not all that multiloop amplitudes tell us. There are modular transformations that mix up the basis cycles of different tori[33]. By considering one such case at two loops one learns that[10]

$$C\begin{bmatrix} a \\ b \end{bmatrix} C\begin{bmatrix} a' \\ b' \end{bmatrix} = \delta_a \delta_{a'} e^{-i\pi \frac{a \cdot a'}{4}} C\begin{bmatrix} a \\ b+a' \end{bmatrix} C\begin{bmatrix} a' \\ b+a \end{bmatrix} \quad (3.13)$$

The phase of the above equation comes from the standard transformation of θ-functions with the exception of the sign $\delta_a \delta_{a'}$ which is due to the two-loop light-cone gauge correction factor[101]. (The contribution to the partition function of two space-time fermions ψ^1, ψ^2 and the two-dimensional gravitino ghosts β and γ). The light-cone correction factor is not known explicitly, however its modular transformation properties are fixed uniquely by the requirement that the ten-dimensional SO(32) heterotic string be modular invariant. The extra phase $\delta_a \delta_{a'}$ is of course absent in the case of the bosonic string. One finds that,

$$\delta_a = \exp i\pi \left(a(\psi_L^\mu) + a(\psi_R^\mu) \right)$$

with

$$a\left(\psi_{L,R}^\mu\right) = \begin{cases} 1 \text{ if } & \psi_{L,R}^\mu \text{ are periodic } (a_{\psi_{L,R}^\mu} = 1) \\ 0 \text{ if } & \psi_{L,R}^\mu \text{ are antiperiodic } (a_{\psi_{L,R}^\mu} = 0) \end{cases} \tag{3.14}$$

Eqs. (3.10, 3.13) express the consistency conditions of multiloop modular invariance and factorization in an algebraic form. We may now proceed further and find all possible solutions. Any solution is characterized by a different choice of C coefficients and will correspond, in general, to non-equivalent string models.

A comment is in order here, the one-loop modular invariance conditions (3.10) ensure the absence of all potential anomalies in the effective field theory of the string's massless modes. The two-loop condition (3.14) on the other hand ensures, as we will see, the spin-stastics connection and unitarity. This is a marvellous unification of principles: world-sheet (super-)reparametrization invariance encodes by itself some of the most profound consistency conditions of field theories.

4. GENERAL SOLUTIONS[10]

In this section we present the general solution of eqs. (3.10, 3.14). For simplicity we restrict ourselves to the case where all fermionic degrees of freedom are periodic or antiperiodic (a_i, b_i are equal to 0 or 1). It is then convenient to express the spin structure assignment as an ordered pair (α / β) where α and β are sets of fermions that are periodic in the direction 1 and τ of the torus, respectively; α and β are arbitrary subsets of the set F of all left- and right-moving fermions. From the definition of (α / β) it follows that

$\alpha \cap \beta$ fermions are with $\begin{bmatrix} 1 \\ 1 \end{bmatrix}$ spin structure

$\alpha - \alpha \cap \beta$ " $\begin{bmatrix} 1 \\ 0 \end{bmatrix}$ "

$F - \alpha \cup \beta$ " $\begin{bmatrix} 0 \\ 0 \end{bmatrix}$ "

$\beta - \alpha \cap \beta$ " $\begin{bmatrix} 0 \\ 1 \end{bmatrix}$ "

$$(4.1)$$

Thus it is straightforward to pass from the notation $\begin{bmatrix} a_i \\ b_i \end{bmatrix}$ to the (α / β) one. The contribution of (α / β) to the one-loop amplitude in terms of theta functions is:

$$(\alpha/\beta) \longrightarrow \Theta_1^{\alpha \cap \beta} \; \Theta_2^{\alpha - \alpha \cap \beta} \; \Theta_3^{F - \alpha \cup \beta} \; \Theta_4^{\beta - \alpha \cap \beta}$$

where

$$\Theta_i^X = \left(\frac{\theta_i}{\eta} \right)^{\frac{n_L(x)}{2}} \left(\frac{\bar{\theta_i}}{\bar{\eta}} \right)^{\frac{n_R(x)}{2}}$$

$$(4.2)$$

with θ_1, θ_2, θ_3 and θ_4 corresponding to $\theta \begin{bmatrix} 1 \\ 1 \end{bmatrix}, \theta \begin{bmatrix} 1 \\ 0 \end{bmatrix}, \theta \begin{bmatrix} 0 \\ 0 \end{bmatrix}, \theta \begin{bmatrix} 0 \\ 1 \end{bmatrix}$ respectively.

$n_L(x)$ = # of left moving fermions of the set X

$n_R(x)$ = # of right moving fermions of the set X $\quad\quad$ (4.3)

In terms of (α / β), the Z_S of eq.(3.3) takes a simple form

$$Z_S = \sum_{\alpha, \beta} C(\alpha/\beta) \cdot (\alpha/\beta) \quad\quad (4.4)$$

and our problem consists in determining the values of the coefficients $C(\alpha/\beta)$. In order to express the consistency conditions (3.10, 3.13) in terms of $C(\alpha/\beta)$, it is necessary to introduce a multiplication rule between two sets α, β. We define[10]:

$$\alpha \cdot \beta = \alpha \cup \beta - \alpha \cap \beta \quad\quad (4.5)$$

The unity of this rule is the empty set $\emptyset$ and α is equal to its inverse $(\alpha^2 = \emptyset)$. If F is the set of all fermions then $F \cdot \alpha$ is the complement of α. For later discussion it is also convenient to define the parity operator $(-)^X$ that counts the fermions in x modulo two

$$(-)^X f = \begin{cases} -f(-)^x & \text{if } f \in X \\ f(-)^x & \text{otherwise} \end{cases} \quad\quad (4.6)$$

It is also useful to associate to every set of fermions X the phases

$$\mathcal{E}_X = \exp \frac{i\pi}{8} \, n(x) \,, \quad n(x) = n_L(x) - n_R(x) \tag{4.7a}$$

$$\delta_X = \delta_{X_L} \, \delta_{X_R}$$

$$\delta_{X_L} = \begin{cases} -1 & \text{if } \psi_L^\mu \in X \\ 1 & \text{otherwise} \end{cases}$$

$$\delta_{X_R} = \begin{cases} -1 & \text{if } \psi_R^\mu \in X \\ 1 & \text{otherwise} \end{cases} \tag{4.7b}$$

The consistency conditions of eqs. (3.10, 3.13), translated in terms of $C(\alpha/\beta)$, take a simplified form:

$$C(\alpha/\beta) = \mathcal{E}_F \, \mathcal{E}_\alpha \, C(\alpha/F.\alpha.\beta) \tag{4.8a}$$

$$C(\alpha/\beta) = \mathcal{E}^2_{\alpha \cap \beta} \, C(\beta/\alpha) \tag{4.8b}$$

$$C(\alpha_1/\beta_1) \, C(\alpha_2/\beta_2) = \delta_{\alpha_1} \, \delta_{\alpha_2} \, \mathcal{E}^2_{\alpha_1 \cap \alpha_2} \, C(\alpha_1/\beta_1.\alpha_2) \, C(\alpha_2/\beta_2.\alpha_1) \tag{4.8c}$$

Together with eqs. (4.8a-c) we should ensure that the supercurrents $T_{F_L}(z)$ and $T_{F_R}(\bar{z})$ have the same periodicity properties as $\psi_L^\mu(z)$ and $\psi_R^\mu(\bar{z})$.

This requirement is summarized by the following conditions

$$(-)^X T_{F_L} = \delta_{X_L} \, T_{F_L} (-)^X \qquad (-)^X T_{F_R} = \delta_{X_R} \, T_{F_R} (-)^X \tag{4.8d}$$

In order to specify a solution to the equations (4.8a-d) and therefore obtain a particular string model, it is necessary to find a collection of sets $\{X_i\} \equiv \Xi$ satisfying eqs. (4.8) and also specify the values of $C(\alpha/\beta)$ with $\alpha, \beta \in \Xi$.

Following always ref. 10], Ξ is closed under the multiplication rule defined previously in eq. (4.5) and always contains $\emptyset$ and F. Thus $(\Xi, .)$ is a group which means that any element $\alpha \in \Xi$ can be expressed in terms of a basis element $\{b_0 \equiv F, b_1, \ldots, b_N\}$

$$\alpha = b_0^{m_0} \cdot b_1^{m_1} \ldots b_N^{m_N} \quad \text{with} \quad m_i = 0 \,, 1. \tag{4.9}$$

The consistency conditions (4.8a-c) imply the following restrictions among the basis elements $\{b_i\}, i = 0, 1, \ldots, N$ (for more details see ref. 10]):

$$n(b_i) = 0 \pmod 8$$
$$n(b_i \cap b_j) = 0 \pmod 4$$
$$n(b_i \cap b_j \cap b_k \cap b_\ell) = 0 \pmod 2 \tag{4.10}$$

The conditions (4.10) together with (4.8d) are necessary and sufficient to select all possible solutions (Ξ, C) of the string constraints. To completely specify a string model it is necessary to give the algebraic values of $C(\alpha/\beta)$ with $\alpha, \beta \in \Xi$. We choose for convention $C(\emptyset/\emptyset) = 1$. (This is always possible by a rescaling of the string coupling constant). Once eqs. (4.10) are valid, by using eqs. (4.8), we may prove the following important properties for $C(\alpha/\beta)$:

(i) $$C(\alpha/\beta) = \begin{cases} \pm 1 & \text{if } \alpha, \beta \in \Xi \\ 0 & \text{otherwise} \end{cases}$$

(ii) $$C(\alpha/\emptyset) = C(\emptyset/\alpha) = \delta_\alpha \quad , \alpha \in \Xi$$

Property (ii) together with eqs. (4.8) may be used as an alternative definition of the group Ξ .

(iii) All coefficientts $C(\alpha/\beta)$, $\alpha, \beta \in \Xi$ can be expressed in terms of the basis element coefficients $C(b_i/b_j)$. More explicitly, if $\alpha = b_{\alpha_1} \cdot b_{\alpha_2} \cdots b_{\alpha_A}$ and $\beta = b_{\beta_1} \cdot b_{\beta_2} \cdots b_{\beta_B}$ then

$$C(\alpha/\beta) = \Delta \prod_{i=1}^{A} \prod_{j=1}^{B} C(b_{\beta_j}/b_{\alpha_i}) \tag{4.11}$$

and Δ is a phase given by

$$\Delta = \left(\prod_{i=1}^{A} \delta_{b_{\alpha_i}}^{B-1} \right) \left(\prod_{j=1}^{B} \delta_{b_{\beta_j}}^{A-1} \varepsilon_{b_{\beta_j} \cap (b_{\alpha_1} \cdots b_{\alpha_A})}^{\ell} \right) \tag{4.12}$$

Thanks to the restrictions of eq. (4.10), the phase Δ is real ($\Delta = \pm 1$); this shows the necessity of those restrictions.

(iv) The $C(b_i/b_j)$ by themselves are not all independent but satisfy the conditions (from eqs. (4.8)):

$$C(b_i/b_j) = \varepsilon_{b_i \cap b_j}^{\ell} C(b_j/b_i)$$

$$C(b_i/b_i) = \varepsilon_F \varepsilon_{b_i} C(b_i/b_o) \tag{4.13}$$

From (iii) and (iv) it follows that for a given basis $\{b_i\}$, i = 0,1...N, there are $\frac{N(N+1)}{2} + 1$ independent signs we can choose at will. For instance the sign of $C(F/F)$ and $C(b_i/b_j)$ for i < j. Any arbitrary choice of the signs corresponds to a consistent string model. Some of the models are equivalent ; this is because the signs of $C(F/b_i)$, i = 0,1...N are just chirality conversions. This leaves us with $2^{N(N-1)/2}$ independent string models corresponding to a given basis $\{b_i\}$, i = 0,1...N.

We now give a simple interpretation of Ξ and C in terms of generalized GSO projections[35]. The Hilbert space of a string theory is a direct sum of different sectors of type $R^\alpha N^{F,\alpha}$, in which all fermions in the set α have Ramond (i.e., periodic) boundary conditions, while all remaining fermions have Neveu-Schwarz (i.e., antiperiodic) boundary conditions. We may write the contribution of the spin structure (α/β) as:

$$(\alpha/\beta) = (-)^{\beta} R^{\alpha} N^{F.\alpha} \tag{4.14}$$

meaning that one sums the contributions of all states in the sector $R^{\alpha} N^{F.\alpha}$, with a sign given by the value $(-)^{\beta}$ on this state ; this follows easily from the definition of (α/β) and eqs. (3.6, 4.2). Using the properties of $C(\alpha/\beta)$ (eqs. 4.8) we may rewrite Z_S of eq. (4.4) in the following suggestive form[10]

$$Z_S = \sum_{\alpha \in \Xi} \delta_\alpha \left[\prod_{i=0}^{N} \frac{1}{2} \left(1 + \delta_\alpha C(\alpha/b_i)(-)^{b_i}\right) \right] R^{\alpha} N^{F.\alpha} \tag{4.15}$$

The meaning of the subgroup Ξ and the coefficients $C(\alpha/\beta)$ should be now clear ; to each element $\beta \in \Xi$, there corresponds i) a sector $R^{\beta} N^{F.\beta}$ of the Hilbert space, and ii) a generalized GSO projection that leaves only those states in $R^{\alpha} N^{F.\alpha}$, whose β-parity is $(-)^{\beta} = C(\alpha/\beta)\delta_\alpha$. For a basis $\{b_0 \cdots b_N\}$ there are N+1 independent projections. (iii) the factor δ_α in front of $R^{\alpha} N^{F.\alpha}$ ensures the correct connection between spin and statistics.

It is remarkable that expression (4.15) summarizes the content of a general (D $\leqslant$ 10) string model. For instance, the Hilbert space of a consistent string theory is

$$\mathcal{H} = \bigoplus_{\alpha \in \Xi} \left(\prod_{\beta \in \Xi} \frac{1}{2} \left(1 + \delta_\alpha C(\alpha/\beta)(-)^{\beta}\right) \right) R^{\alpha} N^{F.\alpha} \tag{4.16}$$

The question that naturally arises is whether these theories are particular compactifications of the ten-dimensional superstrings, possibly in the presence of non-trivial background gauge fields[5]. Although this might well be the case, they still stand out by virtue of their simplicity. Indeed, as we saw, the complete spectrum of those models is known by virtue of eq. (4.15) ; their vertex operators, one-loop amplitudes, effective low-energy Lagrangian, etc., are all readily calculable since one uses little more than the old Neveu-Schwarz-Ramond formalism. If nothing else, they thus provide a playground in which the low energy properties of string theories can be tested. Besides, both of these D-dimensional and the ten-dimensional models presumably correspond to perturbatively stable vacua of a second-quantized string theory, and should in this sense be treated on an equal footing ; which of those vacua will be dynamically preferred, is a non-perturbative and highly non-trivial question for which we do not have at present any hint.

Using eq. (4.15) and the general properties of $C(\alpha/\beta)$ one may prove some general statements[10]. (i) the graviton, dilaton and antisymmetric tensor (pseudoscalar in D=4) are always present in the massless spectrum of any consistent string theory ; they cannot be projected out for any choice of $C(b_i/b_j)$. (ii) The presence of at least one spin 3/2 massless state (space-time gravitino) in the spectrum of a string model implies the absence of tachyonic states and the vanishing of the vacuum energy at the one-loop level (genus=1). (This is probably true for higher genus.)[25] In the next section I shall present some simple examples based on the heterotic construction (1,0).

5.   EXAMPLES OF FOUR-DIMENSIONAL SUPERSTRINGS

(A) The simplest N=4, D=4 superstring with (1,0) world sheet supersymmetry is defined by the basis $\{F,S\}$ with

$$F = \left\{ \psi_L^\mu, \chi_L^I, y_L^I, w_L^I ; \bar{\psi}_R^A \right\}, \quad \mu = 3,4 ; \ I = 1,2,\dots,6 ;$$
$$A = 1,2,\dots, 2(26-D) = 44$$
$$S = \left\{ \psi_L^\mu, \chi_L^I \right\}. \tag{5.1}$$

The group Ξ contains four elements $\Xi = \{\emptyset, F, S, F\cdot S\}$ and as a result the Hilbert space of the theory is given by:

$$\left(-R_-^S + N_-^S \right) \left(R_+^{F\cdot S} + N_+^{F\cdot S} \right) \tag{5.2}$$

The quantity $Z_B Z_S$ in eqs. (3.1,4.15) becomes

$$Z_B Z_S = |Im\tau|^{-1} \eta^{-24} \bar{\eta}^{-12} \left\{ -\tfrac{1}{2}\left(\theta_2^4 - \theta_1^4\right) + \tfrac{1}{2}\left(\theta_3^4 - \theta_4^4\right) \right\} \left\{ \tfrac{1}{2} \sum_{i=1}^{4} \theta_i^6 \, \bar{\theta}_i^{22} \right\} \tag{5.3}$$

In eq. (5.3) the part $-\tfrac{1}{2}\left(\theta_2^4 - \theta_1^4\right)$ gives the contribution of the space-time fermions and the part $\tfrac{1}{2}\left(\theta_3^4 - \theta_4^4\right)$ the one of space-time bosons. Due to the well known identity[34],

$$\theta_2^4 - \theta_1^4 = \theta_3^4 - \theta_4^4 \tag{5.4}$$

the number of space-time bosonic and fermionic degrees of freedom is equal at any mass level. The same identity ensures the vanishing of the cosmological constant at one-loop before the τ integration.

The massless states are given as a direct product of the form $|left\rangle \otimes |right\rangle$, namely

$$\left\{\, |\mu\rangle + |I\rangle + |4,\alpha\rangle \,\right\}_L \otimes \left\{\, |\nu\rangle + |adj\ SO(44)\rangle \,\right\}_R \qquad (5.5)$$

where

$$|\mu\rangle_L \otimes |\nu\rangle_R = \psi^\mu_{-1/2} |0\rangle_L \otimes a^\nu_{-1} |0\rangle_R$$

graviton,dilaton,antisymmetric tensor

$$|I\rangle_L \otimes |\nu\rangle_R = \chi^I_{-1/2} |0\rangle_L \otimes a^\nu_{-1} |0\rangle_R$$

six graviphotons of N=4

$$|4,\alpha\rangle_L \otimes |\nu\rangle_R = S^R_{-1/2} |0\rangle_L \otimes a^\nu_{-1} |0\rangle_R \qquad \left(S^R = (\psi^3_L, \psi^4_L, \chi^1_L, \ldots, \chi^6_L)^R_+ \right)$$

$\underline{4}$ gravitinos,$\underline{4}$ majorana fermions.($\underline{4}$ of SO(6))

$$|\mu\rangle_L \otimes |adj\ SO(44)\rangle_R = \psi^\mu_{-1/2} |0\rangle_L \otimes \bar{\psi}^A_{-1/2} \bar{\psi}^B_{-1/2} |0\rangle_R$$

gauge bosons in the adjoint of SO(44)

$$|I\rangle_L \otimes |adj\ SO(44)\rangle_R = \chi^I_{-1/2} |0\rangle_L \otimes \bar{\psi}^A_{-1/2} \bar{\psi}^B_{-1/2} |0\rangle_R$$

$\underline{6}$ scalar in the adjoint of SO(44)

$$|4,\alpha\rangle_L \otimes |adj\ SO(44)\rangle_R = S^R_{-1/2} |0\rangle_L \otimes \bar{\psi}^A_{-1/2} \bar{\psi}^B_{-1/2} |0\rangle_R$$

$\underline{4}$ gauginos in the adjoint of SO(44) $\qquad (5.6)$

We denote by $|0\rangle_L$ and $|0\rangle_R$ the left and right Neveu-Schwarz tachyonic vacuum ; $\psi^\mu_{-1/2}$, $\chi^I_{-1/2}$ are the left fermionic creation oscillators in N-S ; a^ν_{-1} is the (right moving) bosonic oscillator ; S^R denotes the spin-field which defines the corresponding Ramond left vacuum, acting on $|0\rangle_L$. The gauge group of this model is SO(44).

(B) As a second example we will consider a class of four-dimensional superstrings with N=2 space-time supersymmetries. They are defined by the basis $\left\{ F, S, b_1 \right\}$, with F and S as in (A) and with b_1 satisfying: $S \cap b_1 = \left\{ \psi^3_L, \psi^4_L, \chi^1_L, \chi^2_L \right\}$, $n(b_1) = 0 \pmod 8$. As a result b_1 introduces one extra projection (Z_2) which breaks the initial SO(6) symmetry, generated by $(\chi^1, \ldots, \chi^6)$, to SO(2)xSO(4) generated by $(\chi^1 \chi^2)$ and $(\chi^3 \chi^4 \chi^5 \chi^6)$. This reduces the number of supersymmetries from 4 to 2 and breaks the SO(44) gauge group to SO(n_1)xSO(n_2).($n_1 + n_2$ =44, n_1 = 2r + 8m ; r = 0,1,2., m = 0,1,2,...). The vanishing of the one loop cosmological constant is manifest here as well.

(C) A third example is some N=1, D=4 superstrings which are defined by the basis $\left\{ F, S, b_1, b_2 \right\}$ with b_2 satisfying: $S \cap b_2 = \left\{ \psi^3_L, \psi^4_L, \chi^3_L, \chi^4_L \right\}$. Now b_2 introduces one more fermion number projection (Z'_2) ; the initial SO(6) symmetry is broken into SO(2)³ which is generated by $(\chi^1 \chi^2)$, $(\chi^3 \chi^4)$ and $(\chi^5 \chi^6)$. This eliminates three out of the four gravitinos and defines a N=1 theory based on a gauge group $G = SO(n_o) \times SO(n_1) \times SO(n_2) \times SO(n_3)$. This family of N=1 superstrings contains a class of models with massless fermions in complex representations of $SO(n_o)$; in this class: $n_o = 2 + 4k$ (k = 1,2,...), $n_o + n_i = 0 \pmod 8$, $n_o + n_1 + n_2 + n_3 = 44$. Chiral massless fermions exist when at least one n_i is such that $n_o + n_i = 8$ or 16. For instance the choice $n_o = 10$ makes SO(10) the unified gauge group with massless fermions in the 16 representation of SO(10). For instance when G = SO(10) xSO(6)xSO(14)xSO(14) we have sixteen SO(10) families (16), while when G = SO(10)xSO(6)xSO(6)xSO(22) we have a chiral model with 32, 16 families.

Introducing the basis element $\{F, S, b_1, b_2, J_1, J_2 \ldots\}$, we may further reduce the number of families and break down the gauge group. In general, one may obtain a big number of semirealistic N=1 supersymmetric models based on a gauge group $H \supset SU(3) \times SU(2) \times U(1)$ with three or four families.

We would like to stress here an important observation that illustrates the main advantage of these theories, namely their calculability. Let (Ξ, C) be one arbitrary theory ; to every subgroup $\Xi' \subset \Xi$ there corresponds a naturally induced theory obtained by restricting the coefficients C to Ξ'. From eqs. (4.15 and 4.16) it follows easily that part of the Hilbert space of the theory (Ξ, C) is an exact truncation of the Hilbert space of (Ξ', C). To see how this remark can be useful, let $\xi = \{F, S, b_1, b_2, J_1 \ldots J_k\}$ be a basis for a N=1 supersymmetric theory; then the subbasis $\xi' = \{F, S, J_1, \ldots J_k\}$ corresponds to an induced N=4 theory. Now the tree-level Lagrangian of the massless modes of an N=4 theory is uniquely specified[36], up to the unknown coupling constants of the semi-simple components of the gauge group. Thus we can immediately obtain by truncation part of the low-energy Lagrangian of the original N=4 theory, including in particular the couplings of all vector multiplets ; this is not trivial because this Lagrangian is not polynomial. We can even determine the ratios of gauge-coupling constants, by noting that the vector bosons of $\otimes SO(n(R_i))$ come from the truncation of the SO(44) multiplets[23], which has a single arbitrary coupling. The interactions of the additional massless scalar multiplets, other than those left over by the truncation of the N=4 vector multiplets, cannot be similarly determined[24]; they are however considerably restricted by the fact that they can be obtained by truncating an N=2 supersymmetric theory. Indeed, in refs. 23,24], by using the connection of the N=1 superstrings with those of N=2 and N=4, the full low energy supergravity Lagrangian is derived at least for some N=1, D=4 superstring models. The low energy behaviour of those models possess interesting geometrical properties with some important physical implications. The reader is referred to the literature for a more complete discussion about their properties[23,24,37].

6. CLASSICAL DEGENERACY (DISCUSSION)

It is obvious that the number of consistent four-dimensional string theories is alarmingly big. In addition to the models described before, we can for instance envisage constructing fermionic models of similar texture but with generalized boundary conditions on the world sheet fermions[9,11]. Another source of four-dimensional theories is provided by asymmetric orbifolds[7]. Unfortunately there may be more entries to this list of string model construction mechanisms. Probably most of these theories are solutions of the same string field theory. Indeed, it is at present widely believed that there are just a few fundamental string theories, but with a lot of classically degenerate ground states. For instance all (space-time) supersymmetric solutions have zero vacuum energy and look degenerate[10,11]. At present, the dynamical reason which may select some of the degenerate solutions is unknown. Nevertheless, even our understanding of this classical string landscape and its symmetries, which might offer the key to the solution of long-standing problems of grand-unification, is still rather rudimentary.

Another problem is the breakdown of space-time supersymmetry. In general non-supersymmetric string solutions are in number at least as many as the supersymmetric ones. However, most of them are unstable as they generate non zero vacuum energy at the quantum level destabilizing the flat background. Are there any stable non-supersymmetric string models ? If one expresses a preference for spontaneous breakdown, then it is necessary to find a breaking mechanism which operates on the string level.

A first attempt to construct explicitly string models with spontaneous gauge and supersymmetry breaking was made recently using a Scherk-Schwarz type compactification from five dimensions[14]. In what concerns the breaking of supersymmetry, it is not clear yet if such models can be constructed by using world sheet fermions (see next section). In what concerns the breaking of gauge symmetries on the other hand, it defines a Higgs phenomenon directly at the string level[14]. What is important here, is that the resulting string model shows a mass spectrum which depends upon several <u>continuous</u> parameters and for particular values of these parameters, reproduces a more symmetric theory.

In the remainder of these lectures I will present you the way we introduce arbitrary parameters on strings by defining a Scherk-Schwarz type compactification on strings.

Before entering the details of the Scherk-Schwarz mechanism I want to comment here on the recent understanding about a more general Higgs phenomenon on strings. By using world sheet bosonic degrees of freedom it is possible to continuously deform the string vacuum by Lorentz boosting the charge lattice of string states[5]. Narain, Sarmadi and Witten[5] showed that this amounts to changing radii and background gauge and antisymmetric fields in the language of compactification. Very recently it has been shown that from the four-dimensional point of view, this is a standard Higgs phenomenon[15]; gauge symmetries are broken by vacuum expectation values of Higgs fields sliding along flat directions of scalar potential. The symmetry breaking scales are thus classically undetermined parameters[15], which may in particular be hierarchically smaller than the Planck mass. The "boosting" construction of ref. 15] is more general than the "Scherk-Schwarz" string compactification[14]. My choice of presenting the latter is just of pedagogical interest ; the one loop partition function in that case can be set, as we will see, in a very simple and closed form, strongly related to that of eq. (4.15).

7.   SYMMETRY BREAKDOWN (HIGGS PHENOMENON)

In this section we shall present a way of achieving spontaneous symmetry breaking directly at the string level, by constructing a conformal and modular invariant four-dimensional string analogous to the Scherk-Schwarz mechanism in field theory[28]. We, more specifically, construct four-dimensional strings whose mass spectrum depends upon several continuous parameters[14] and for particular values of these parameters, reproduces the symmetric phase of the theory.

We first remind the reader of some basic ideas about the Scherk-Schwarz mechanism in field theory[28]. There one starts from a Lagrangian in five dimensions, invariant under some gauge symmetries (global or local ones). The model is then compactified to four dimensions while giving non-trivial dependence on the fifth coordinate to all fields

$$\phi^i(x, x_5) = \left(e^{\,ieQx_5}\right)^i_j \sum_m e^{\frac{2i\Pi}{R} m x_5} \hat{\phi}^j_m(x) \qquad (7.1)$$

$\frac{2\Pi}{R} m$ is a quantized (m = 1, 2, ...) momentum, conjugate to x_5 coordinate, R is the radius of compactification and Q^i_j is a generator of the Lagrangian internal symmetry group in five dimensions and e is a charge free parameter. The extra x_5 dependence gives a modification to

226

the four-dimensional mass term arising from the kinetic terms. Taking massless scalar fields as an example we get

$$\partial_\alpha \Phi^i \partial^\alpha \Phi_i = \partial_\mu \hat{\Phi}^i \partial^\mu \hat{\Phi}_{i,0} - \hat{\Phi}^i_0 (e^2 Q^2)^j_i \hat{\Phi}_{j,0} + (\dots m \neq 0)$$

$$\alpha = 1, 2, \dots, 5 \quad , \quad \mu = 1, 2, 3, 4 .$$

(7.2)

The above equation shows how the mass terms can be generated by a non-trivial coordinate dependent torus compactification. The consistency of this mechanism is due to the fact that Q^j_i is a generator of a symmetry group in higher dimensions. The number of free parameters e_i introduced in this way can be equal to the rank of the gauge symmetry group G.

In order to find the analogue of this mechanism in string theory we need to construct five-dimensional string models. This can be easily done by using the fermionic construction we have explained in the previous sections. In this way we may obtain many inequivalent five-dimensional strings. We assume hereafter only periodic or antiperiodic boundary conditions for simplicity and we group all fermions in (complex) pairs for later convenience: $\psi^3_L + i \psi^4_L$; $\chi^I_L + i \chi^{I+1}_L$, $(\chi^6_L \equiv \psi^5_L)$, $\dots$; $y^I_L + i w^I_L$; $\bar{\psi}^A_R + i \bar{\psi}^{A+1}_R$, $(A = 1, 3, 5, \dots 41$, in the heterotic case, $(1,0))$. In this way we are reduced to work with complex (Weyl) world sheet fermions and the spin structure assignment is given by the vectors $(\vec{a}_L, \vec{a}_R)$, $(\vec{b}_L, \vec{b}_R)$ with $N_L + N_R$ components. $N_L = \dim \vec{a}_L = \dim \vec{b}_L$, $(N_R = \dim \vec{a}_R = \dim \vec{b}_R)$ is the number of complex left-(right-)moving fermions.

In five dimensions the one-loop partition function is given by

$$Z^{(5)}(\tau, \bar{\tau}) = \tau_I^{-3/2} |\eta(\tau)|^{-6} \sum_{\text{spin Str}} C \begin{bmatrix} \vec{a}_L & \vec{a}_R \\ \vec{b}_L & \vec{b}_R \end{bmatrix} Z_L \begin{bmatrix} \vec{a}_L \\ \vec{b}_L \end{bmatrix} (0, \vec{0} | \tau)$$

$$\cdot Z_R \begin{bmatrix} \vec{a}_R \\ \vec{b}_R \end{bmatrix} (0, \vec{0} | \bar{\tau})$$

(7.3)

with

$$Z_L \begin{bmatrix} \vec{a}_L \\ \vec{b}_L \end{bmatrix} (u_L, \vec{V}_L | \tau) = e^{-i\pi u_L} \left(\prod_{i=1}^{N_L} \frac{\theta \begin{bmatrix} a^i_L \\ b^i_L \end{bmatrix} (V^i_L | \tau)}{\eta(\tau)} \right)$$

and

$$Z_R \begin{bmatrix} \vec{a}_R \\ \vec{b}_R \end{bmatrix} (u_R, \vec{V}_R | \bar{\tau}) = e^{i\pi u_R} \left(\prod_{i=1}^{N_R} \frac{\bar{\theta} \begin{bmatrix} a^i_R \\ b^i_R \end{bmatrix} (V^i_R | \bar{\tau})}{\bar{\eta}(\bar{\tau})} \right)$$

(7.4)

where we define the θ-function with $\nu \neq 0$ as:

$$\theta \begin{bmatrix} a \\ b \end{bmatrix} (\nu | \tau) = \sum_{n \in \mathbb{Z}} e^{i\pi \tau (n + \frac{a}{2})^2 \tau + 2\pi i (n + \frac{a}{2})(\nu + \frac{b}{2}) - i\pi \frac{ab}{2}}$$

(7.5)

and $u_{L,R}, \vec{V}_{L,R}$ are, for the moment, arbitrary parameters. The coefficients C satisfy the relations we explained in the previous sections ensuring the modular invariance of eq. (7.3) as well as the correct factorization properties of its multi-loop extension.

We now compactify the fifth bosonic coordinate on a torus of radius R. We also introduce a non-trivial dependence upon the winding numbers m,n in the theta-functions. We shall see that in this way, we can define in strings the equivalent of a Scherk-Schwarz compactification. Let us denote by $Z_{mn}(\tau, \bar{\tau})$ the partition function of a compactified boson ($X^5(z, \bar{z})$), with widing numbers m,n[1]: ($\tau_I = Im \tau$)

$$Z_{mn} = R \tau^{-1/2} |\eta(\tau)|^{-2} \exp - \frac{\pi R^2}{\tau_I} |m - \tau n|^2 \qquad (7.6)$$

(In our normalization the radius of the X^5 torus is $\sqrt{2} \cdot R$).

Under a modular transformation

$$\tau \rightarrow \frac{a\tau + b}{c\tau + d} \quad , \quad \begin{bmatrix} a & b \\ c & d \end{bmatrix} \in SL(2, Z) \qquad (7.7)$$

the $Z_{mn}(\tau, \bar{\tau})$ transform as:

$$Z_{mn} \left[\frac{a\tau + b}{c\tau + d} \right] = Z_{m'n'}(\tau)$$

with

$$\begin{pmatrix} m' \\ n' \end{pmatrix} = \begin{bmatrix} d & -b \\ -c & a \end{bmatrix} \begin{pmatrix} m \\ n \end{pmatrix} = \begin{bmatrix} a & b \\ c & d \end{bmatrix}^{-1} \begin{pmatrix} m \\ n \end{pmatrix} \qquad (7.8)$$

This property shows the modular invariance of the torus compactification (summation over the winding numbers m and n)

$$Z^{torus}(\tau) = \sum_{m'n'} Z_{m'n'}(\tau) = \sum_{m,n} Z_{mn} \left[\frac{a\tau + b}{c\tau + d} \right] \qquad (7.9)$$

We introduce now the following "charge" dependent partition function defined as[14]:

$$Z(e_L^i, e_R^i | \tau) \equiv \tau_I^{-1} |\eta(\tau)|^{-4} \sum_{m,n} \sum_{spin\ St_2} Z_{mn}(\tau, \bar{\tau})$$

$$\cdot C \begin{bmatrix} \vec{a_L} & \vec{a_R} \\ \vec{b_L} & \vec{b_R} \end{bmatrix} Z_L \begin{bmatrix} \vec{a_L} \\ \vec{b_L} \end{bmatrix} (u_L, \vec{V_L} | \tau)$$

$$\cdot Z_R \begin{bmatrix} \vec{a_R} \\ \vec{b_R} \end{bmatrix} (u_R, \vec{V_R} | \bar{\tau}) \qquad (7.10)$$

where $u_{L,R}$, $\vec{V}_{L,R}$ are winding number depending charges defined as

228

$$\vec{V_L} = \vec{e_L}(m - \tau n) \, , \quad U_L = \vec{e_L} \cdot \vec{e_L} \, n(m - \tau n)$$

$$\vec{V_R} = \vec{e_R}(m - \bar{\tau} n) \, , \quad U_R = \vec{e_R} \cdot \vec{e_R} \, n(m - \bar{\tau} n) \tag{7.11}$$

Notice first that $Z(0,0|\tau)$ corresponds to the usual torus compactification from five to four dimensions[11]. This compactification gives rise to a four-dimensional string with two U(1) extra gauge group factors corresponding to $g_{\mu 5}$ and $B_{\mu 5}$ gauge bosons, $\mu = 1,2,\ldots 4$, ($g_{\mu\nu}$ $\mu,\nu = 1,2,\ldots 5$ is the five-dimensional metric tensor and $B_{\mu\nu}$ $\mu,\nu = 1,2\ldots 5$ is the five-dimensional antisymmetric tensor). On the other hand, when $(\vec{e_L}, \vec{e_R}) \neq 0$, $Z(e_L^i, e_R^i|\tau)$ is the partition function of a class of new string solutions with spontaneously broken gauge symmetries 14]. Indeed, due to the transformation properties of $U_{L,R}$, $\vec{V}_{L,R}$ under

$$\binom{m}{n} \to \begin{pmatrix} a & b \\ c & d \end{pmatrix} \binom{m}{n} \, , \quad \tau \to \frac{a\tau + b}{c\tau + d} \, , \quad \begin{pmatrix} a & b \\ c & d \end{pmatrix} \in SL(2,\mathbb{Z}) \tag{7.12}$$

namely:

$$\vec{V_L} \longrightarrow \frac{1}{c\tau + d} \vec{V_L} \, , \quad U_L \to U_L + \frac{c}{c\tau + d} \vec{V}^2$$

$$\vec{V_R} \longrightarrow \frac{1}{c\bar{\tau} + d} \vec{V_R} \, , \quad U_R \to U_R + \frac{c}{c\bar{\tau} + d} \vec{V}_R^2 \tag{7.13}$$

the generalized partition functions $Z_L[{:}](U_L, \vec{V_L}|\tau)$, $Z_R[{:}](U_R, \vec{V_R}|\bar{\tau})$ transform in the same way as $Z_L[{:}](0, \vec{0}|\tau)$, $Z_R[{:}](0, \vec{0}|\bar{\tau})$ under modular transformation. This property shows that the charge dependent partition function $Z(\vec{e_L}, \vec{e_R}|\tau)$ is a modular invariant function.

The next step is to demonstrate that $Z(\vec{e_L}, \vec{e_R}|\tau)$ is the partition function of a class of new string solutions with spontaneously broken symmetries. By using the trace representation of $\theta[{:}](v|\tau)$ we may express $Z(\vec{e_L}, \vec{e_R}|\tau)$ in a more convenient form. Indeed, following ref. 14] by using the trace representation of θ-functions and performing a Poisson resummation on the winding number m, $Z(\vec{e_L}, \vec{e_R}|\tau)$ may be written as a sum of products of the form[14]:

$$Z(\vec{e_L}, \vec{e_R}|\tau) = \sum_{Sectors} Tr \, g \, q^{L_0} \bar{q}^{\bar{L}_0} \, , \quad q = e^{2\pi i \tau} \tag{7.14}$$

where g is a fermion-number projection defined in the five-dimensional theory, and $L_0, \bar{L}_0$ are the Virasoro operators of the left and right moving sectors respectively. Still following ref. 14] we have

$$L_0 + \bar{L}_0 = L_0 + \bar{L}_0' + \frac{1}{2R^2} \left(m + \vec{e}_L \vec{Q}_L - \vec{e}_R \vec{Q}_R - \frac{1}{2} n (\vec{e}_L^2 - \vec{e}_R^2) \right)^2$$
$$+ \frac{1}{2} R^2 n^2 - (\vec{e}_L \vec{Q}_L + \vec{e}_R \vec{Q}_R) n + \frac{1}{2} (\vec{e}_L^2 + \vec{e}_R^2) n^2 \quad (7.15)$$

and

$$L_0 - \bar{L}_0 = L_0' - \bar{L}_0' + m n \qquad (7.16)$$

where L_0' and $\bar{L}_0'$ are the contribution of all terms not containing winding number dependence. Q_L^i, Q_R^i are U(1) charge operators defined by the world sheet complex fermions.

Two things are to be emphasized with respect to eqs. (7.14) and (7.15). At first, the mass level matching condition $L_c - \bar{L}_0 \in \mathbb{Z}$ is independent of the charges $\vec{e}_L$ and $\vec{e}_R$ and so are the fermion number projections. Therefore all the states, which are physical in the symmetric limit $e_L^i = e_R^i = 0$, remain physical for every value of e_L^i, e_R^i without any additional states being introduced. In this respect the theory gives precisely a spontaneous breaking of gauge symmetry: the degrees of freedom remain the same, only their masses change.

Secondly, the above construction introduces mass scales through $e_{L,R}^i$, which can be chosen arbitrarily small, while keeping the separation between the first and all other levels of $\mathcal{O}(\alpha'^{-1/2})$. This is a fundamental difference with respect to all previously constructed string models. There indeed, introducing a small scale, for example through a compactification radius R >> 1, always reduces the mass differences between levels from $\mathcal{O}(\alpha'^{-1/2})$ to $\mathcal{O}(\frac{\alpha'^{-1/2}}{R})$.

The interpretation of eq. (7.15) in terms of Scherk-Schwarz compactification is manifest to the soliton independent sector (n = 0). There indeed m represents the quantized momentum, conjugate of the X^5 coordinate, while $(\vec{e}_L \vec{Q}_L - \vec{e}_R \vec{Q}_R)$ is the mass shift arising from the X^5 dependent compactification, through the factor exp i $X^5 (\vec{e}_L \vec{Q}_L - \vec{e}_R \vec{Q}_R)$. The soliton-dependent mass square modification (n $\neq$ 0) is determined by modular invariance of the one-loop partition function. There is no analogue to the n $\neq$ 0 sectors in field theory: their very presence is a purely "stringy" phenomenon, and so are their mass spectra.

In order to study the symmetry breaking pattern of the four-dimensional string defined by the $Z(\vec{e}_L, \vec{e}_R | \tau)$ of eq. (7.10), it is sufficient to examine how the massless states change in the presence of non-zero e_L^i and e_R^i. In the tachyon free theories, the sector which becomes massless for e_L^i, $e_R^i = 0$ has m = n = 0 and, from eq. (7.15), its masses are:

$$M^2 = \frac{1}{2R^2} \left(\vec{e}_L \vec{Q}_L - \vec{e}_R \vec{Q}_R \right)^2 \qquad (7.17)$$

230

All gauge bosons with non-zero $(\vec{e}_L \vec{Q}_L - \vec{e}_R \vec{Q}_R)$ charge become massive ; at the same time some scalar fields with the same quantum numbers (Higgs bosons) provide the necessary longitudinal components of the previously massless gauge bosons. Indeed, for any gauge boson in a representation R, A^R_μ $\mu = 1,2,3,4$ there is a four-dimensional scalar $\Phi^R = A^R_5$ with the same quantum numbers. (The fifth component of a gauge boson of the uncompactified five-dimensional theory).

In ref. 14] it was suggested to use the string extension of the Scherk-Schwarz mechanism to break the space-time supersymmetries spontaneously. A possible choice for Q^1_L and Q^2_L in the heterotic construction would be the charge operators defined by $J^1_L = \chi'_L \chi^2_L$, $J^2_L = \chi^3_L \chi^4_L$ currents. Even though the $Z(\vec{e}_L, \vec{e}_R | \tau)$ is one-loop modular invariant function the resulting theory violates one of the fundamental requirements of string constraints. Namely, Q^1_L and Q^2_L do not respect the world sheet supersymmetry. (They do not commute with the left supercurrent $T_{F_L}(z)$). In fact, using the fermionic representation (χ^I_L, y^I_L, w^I_L) to describe the internal degrees of freedom it is <u>impossible</u> to define a local charge operator which commutes with the supercurrent. However in terms of compactified bosons $((y^I, w^I) \leftrightarrow \Phi^I)$ there are such operators, namely those which correspond to the currents:

$$\hat{J}^1_L = \chi'_L \chi^2_L + \Phi^1 \partial_z \Phi^2 - \Phi^2 \partial_z \Phi^1$$

$$\hat{J}^2_L = \chi^3_L \chi^4_L + \Phi^3 \partial_z \Phi^4 - \Phi^4 \partial_z \Phi^3 \qquad (7.18)$$

Although $\hat{J}^1_L$ and $\hat{J}^2_L$ have not well defined conformal weights, their charge operators are well defined. This explains the reason why $\hat{J}^1_L$ and $\hat{J}^2_L$ cannot be represented using local expression in the fermionic representation. (The fermions have always a well defined conformal weight).

The use of $\hat{Q}^1_L$ and $\hat{Q}^2_L$ operators to obtain a generalized partition function similar to $Z(\vec{e}_L, \vec{e}_R | \tau)$ is not so straightforward ; it is necessary to find the bosonic partition function in the presence of non trivial $\vec{V}(n,m)$ and $U(n,m)$ charge quantities. For the time being this generalization is not well known. Here, assuming the consistency of such a generalization, we may obtain the mass modification of the previously massless states. Indeed, in the light cone gauge, the bosonic part of the operator $\hat{Q}^i_L$ becomes irrelevant; in the limit $e^i_L = 0$, all massless states of a space-time supersymmetric model are obtained by the action of fermionic oscillators $\psi^\mu_{-1/2}$, $\chi^I_{-1/2}$ and "spin fields", on the Neveu-

Schwarz "left" vacuum. The bosonic part of $\hat{Q}_L^i$ gives relevant mass modification <u>only for the massive</u>, $\mathcal{O}(\alpha'^{-1/2})$, states. This fact allows us to study the symmetry breaking pattern of string models with spontaneously broken space-time supersymmetry. The sector becoming massless for $\vec{e}_L$, $\vec{e}_R = 0$ has m = n = 0 and from what we said before its masses are:

$$M^2 = \frac{1}{2R^2} \left(\vec{e}_L \vec{Q}_L - \vec{e}_R \vec{Q}_R \right)^2 \tag{7.19}$$

8. EXAMPLES OF SUPERSYMMETRY BREAKING

In order to illustrate the meaning of eq. (7.19) we will here examine two cases: the N=4 heterotic four-dimensional string with matter gauge group SO(42)xSO(2) and the N=8 four-dimensional type II superstring.

In the first case for $\vec{e}_L = \vec{e}_R = 0$ (exact supersymmetry), the massless states are given as a direct product of the form $|\text{left}\rangle \otimes |\text{right}\rangle$, namely:

$$\left\{ |\mu\rangle + |5\rangle + |I\rangle + |4,\alpha\rangle \right\}_L \otimes \left\{ |\nu\rangle + |5\rangle + |adj\, SO(42)\rangle \right\}_R$$

$$\mu, \nu = 3,4 \quad , \quad I = 1,2,\ldots,5 \tag{8.1}$$

where

$$|\mu\rangle_L = \psi^\mu_{-1/2} |0\rangle_L \;, \quad |5\rangle_L = \psi^5_{-1/2} |0\rangle_L \;, \quad |I\rangle = \chi^I_{-1/2} |0\rangle_L \;,$$

$$|4,\alpha\rangle_L = S^R_{-1/2} |0\rangle_L \;. \quad S^R \longrightarrow \left(\psi^3_L \psi^4_L \chi^1_L \chi^2_L \chi^3_L \chi^4_L \chi^5_L, \psi^5_L \right)^R_+$$

is the spin field which defines the Ramond left vacuum acting on the left Neveu-Schwarz tachyonic vacuum $|0\rangle_L$.

$$|\mu\rangle_R = a^\mu_{-1} |0\rangle_R \;, \quad |5\rangle_R = a^5_{-1} |0\rangle_R \;, \quad |adj\, SO(42)\rangle = \bar{\psi}^A_{1/2} \bar{\psi}^B_{-1/2} |0\rangle_R$$

with $\mu = 3,4$, A = 1,2,...42. $|0\rangle_R$ is the right Neveu-Schwarz tachyonic vacuum . This spectrum is composed of three supermultiplets:

(a) gravitational multiplet:

$|\mu\rangle_L \otimes |\nu\rangle_R$ graviton, dilaton, antisymmetric tensor (pseudoscalar)

$\left.\begin{array}{l} |I\rangle_L \otimes |\nu\rangle_R \\ |5\rangle_L \otimes |\nu\rangle_R \end{array}\right\}$ six graviphotons in the <u>6</u> of SO(6)

$|4,\alpha\rangle_L \otimes |\nu\rangle_R$ <u>4</u> gravitinos   and <u>4</u> majorana fermions of SO(6)

(b) vector multiplet (SO(2)):

$|\mu\rangle_L \otimes |5\rangle_R$ SO(2) gauge boson

$\left.\begin{array}{l}|I\rangle_L \otimes |5\rangle_R \\ |5\rangle_L \otimes |5\rangle_R\end{array}\right\}$ scalars in the $\underline{6}$ of SO(6)

$|\underline{4},\alpha\rangle_L \otimes |5\rangle_R$ $\underline{4}$ majorana fermions of SO(6)

(c) matter multiplet SO(42):

$|\mu\rangle_L \otimes |\,\text{adj SO(42)}\,\rangle_R$ SO(42) gauge bosons in the $\underline{861}$

$\left.\begin{array}{l}|I\rangle_L \otimes |\,\text{adj SO(42)}\,\rangle_R \\ |5\rangle_L \otimes |\,\text{adj SO(42)}\,\rangle_R\end{array}\right\}$ $\underline{861}$ scalars in the $\underline{6}$ of SO(6)

$|\underline{4},\alpha\rangle_L \otimes |\,\text{adj SO(42)}\,\rangle_R$ $\underline{861}$ majorana fermions in the $\underline{4}$ of SO(6)

$$(8.2)$$

By introducing for instance $e_L{}^1$, $e_L{}^2$ both in the Cartan subalgebra of the SO(4) acting on $\chi_L^1 \chi_L^2 \chi_L^3 \chi_L^4$, we get the supermultiplet masses summarized in Table I.

Note that the states (8.2b) provide the degrees of freedom necessary for the Higgs and super-Higgs effect of the gravitational sector. When dealing with N=4 matter, we can introduce other right charges which spontaneously break the SO(42) symmetry. Let us take for instance a charge e_R acting on the complex fermion $\psi^1 + i\psi^2$. This charge breaks the gauge group through an ordinary Higgs mechanism down to SO(2)×SO(40). The resulting spectrum is given in Table II.

The states $|5\rangle_L \otimes |5\rangle_R$ and $|5\rangle_L \otimes |\,\text{adj SO(40)}\,\rangle_R$ contain physical scalar fields, as well as the Higgs bosons necessary to provide the longitudinal components of the vectors corresponding to broken gauge-symmetries.

The second example is the N=8 left-right symmetric superstring theory. Here, the originally massless degrees of freedom giving rise to the spectrum of the four-dimensional N=8 supergravity are

$$\left\{|\mu\rangle + |5\rangle + |I\rangle + |\underline{4},\alpha\rangle\right\}_L \otimes \left\{|\nu\rangle + |5\rangle + |J\rangle + |\underline{4},\beta\rangle\right\}_R$$
$$\mu,\nu = 3,4\,, \qquad I,J = 1,2,3,4. \tag{8.3}$$

By introducing left charges $e_L{}^1$, $e_L{}^2$ and right charges $e_R{}^1$, $e_R{}^2$ of the two SO(4) acting on $\left(\chi^1\chi^2\chi^3\chi^4\right)_{L,R}$ we obtain the mass spectrum given in Table III.

Table I. Gravitational multiplet (a $\oplus$ b).

Spin	Mass	Degeneracy	# of physical Modes
2	0	1	2
3/2	$1/2\,|e_L^1 \pm e_L^2|$	2	16
1	$|e_L^1|$, $|e_L^2|$, 0, 0	2	16
1/2	$1/2\,|e_L^1 \pm e_L^2|$	2	4
0	0	2	2

Table II. Matter Multiplet.

Spin	Mass	Degeneracy	# of Physical Modes
1	0	781	1562
	$|e_R|$	80	240
1/2	$1/2\,|e_L^1 \pm e_L^2|$	1562	6248
	$1/2\,|e_L^1 \pm e_L^2 \pm 2e_R|$	80	640
0	0	1562	1562
	$|e_L^1|$, $|e_L^2|$	1562	3124
	$|e_R|$	80	80
	$|e_L^1 \pm e_R|$, $|e_L^2 \pm e_R|$	80	320

Table III. N = 8 Superstring.

Spin	Mass	Degeneracy	# of Physical Modes
2	0	1	2
3/2	$1/2\,|e_L^1 \pm e_L^2|$, $1/2\,|e_R^1 \pm e_R^2|$	2	32
1	$|e_L^1|$, $|e_L^2|$, $|e_R^1|$, $|e_R^2|$	2	24
	$1/2\,|e_L^1 \pm e_L^2 \pm e_R^1 \pm e_R^2|$	2	48
1/2	$1/2\,|e_L^1 \pm e_L^2|$, $1/2\,|e_R^1 \pm e_R^2|$	4	32
	$1/2\,|2e_L^1 \pm e_R^1 \pm e_R^2|$	2	16
	$1/2\,|2e_L^2 \pm e_R^1 \pm e_R^2|$	2	16
	$1/2\,|\pm e_L^1 \pm e_L^2 + 2e_R^1|$	2	16
	$1/2\,|\pm e_L^1 \pm e_L^2 + 2e_R^2|$	2	16
0	0	6	6
	$|e_L^1|$, $|e_L^2|$, $|e_R^1|$, $|e_R^2|$	2	8
	$|e_L^1 \pm e_R^1|$, $|e_L^2 \pm e_R^2|$	2	8
	$|e_L^1 \pm e_R^2|$, $|e_L^2 \pm e_R^1|$	2	8
	$1/2\,|e_L^1 \pm e_L^2 \pm e_R^1 \pm e_R^2|$	2	16

This spectrum is identical to that of ref. 28], as expected, and gives another confirmation that we are, indeed, dealing with the string analogue of the Scherk-Schwarz mechanism.

The previously proposed breaking mechanism can only be used to break symmetries in models with an even number of space-time supersymmetries. This is a major obstruction for the application of this mechanism to realistic models with chiral families which allow the presence of N=1 supersymmetry at most. However, this can be avoided by a suitable twist of the extra(fifth) dimension[17]. By using twisted boundary conditions for the fifth coordinate, (Z_2-orbifold compactification), we can introduce a suitable projection which reduces the number of supersymmetries down to one[17]. ·The resulting string model contains chiral fermion and gauge symmetry and supersymmetry are spontaneously broken at an arbitrary scale. Schematically we have:

$$\begin{pmatrix} N=1 \ , \ D=5 \\ \text{superstring} \end{pmatrix} \ \xrightarrow{\ \ T(x_5)/Z_2\ \ } \ \begin{pmatrix} N=1 \ , \ D=4 \\ \text{spontaneously} \\ \text{broken superstring} \end{pmatrix}$$

Specific examples of such models are constructed in ref. 17]. Here I just give the masses of the light states (massless in the limit $e_{L,R}{}^i = 0$)

$$M^2 \Big|_{m=n=0} = \frac{1}{4R^2} \left(1+h \right) \left(\vec{e_L}\vec{Q_L} - \vec{e_R}\vec{Q_R} \right)^2$$

where h = +1 in the twisted sector and h = -1 in the untwisted one. For more details we refer the interested reader to ref. 17].

9. CONCLUSION AND DISCUSSION

In these lectures I explained how we can reduce the critical dimensionality of the space-time by using free fermions on the world sheet. The constraint of super-reparametrization implies a non linear realization of the world sheet supersymmetry among free fermions, while the modular invariance and the factorization requirements fix the periodicity properties of the two-dimensional fermions, e.g. when the latter are parallel transported around non-trivial cycles of a higher genus world sheet surfaces. Different consistent choices define acceptable string solutions which are characterized by a different set of generalized GSO projections.

The number of consistent string solutions constructed in this way is very big, and, in addition to some different consistent constructions, their number becomes alarmingly big (like the symmetric or asymmetric orbifold constructions or other consistent compactifications).

One then may ask about the validity of a previous belief concerning the uniqueness of string theory. At present it is widely recognized that only few string field theories exist but with many non trivial classical solutions. This belief is enforced due to the fact that the density of states are, for high energies, identical for all possible string solutions with the same super-reparametrization properties. This means that in short distances there are only few different classes of strings, e.g. the type II class with (1,1) supersymmetry, the heterotic class with (1,0) supersymmetry and the non supersymmetric bosonic class (0,0).

It is then probable that, to a given class of string solutions, corresponds one string field theory. Even if this statement turns out to be true, we have to understand the non perturbative string dynamics which allow us to make a unique choice among the possible string solutions.

Since we know very little about non perturbative string phenomena, it is important in the meantime to continue the study of these different string vacua and, hopefully, obtain a phenomenologically viable model.

In the second part of these lectures we show the existence of string solutions whose mass spectrum depends upon continuous parameters. For certain values of these parameters one obtains a more symmetric solution. In fact we found the analogue in string theory of the Higgs mechanism in field theory ; gauge symmetries are broken by vacuum expectation values of Higgs fields sliding along flat directions of scalar potential.

Explicit construction of string solutions with spontaneous broken gauge symmetries are presented here by means of a generalized coordinate torus compactification from five to four dimensions. However, a more general Higgs phenomenon on strings exists[15], and can be easily examined in the lattice language, namely by using the possibility of continuously deforming the string vacuum by boosting the "Lorentzian" lattice of string states. Very recently we have demonstrated that this lattice deformation results into a standard Higgs phenomenon on strings[15].

In the last part of these lectures we have also proposed a Super-Higgs phenomenon on string level, e.g. spontaneous breakdown of some space-time supersymmetry at an arbitrary scale. Some technical problems remain in this direction. Namely, we do not know for the moment the complete one-loop partition function of the model. The knowledge of the latter, however, is necessary in order to examine, at the quantum level, the stability of the flat background in the broken phase of the theory. Hopefully this remaining problem as well as many other stringy problems will find their solutions in the near future.

I would like to thank UCB and LBL where part of this work was done.

REFERENCES

1.   Schwarz, J.H., Physics Reports 89, 223 (1982) ;
      Gross, D.J., Harvey, J.A., Martinec, E. and Rohm, R., Phys. Rev.
      Lett. 54, 502 (1985) ; Nucl. Phys. B256, 253 (1985) and B267, 75
      (1986).
      For recent review, see Green, M.B., Schwarz, J.H. and Witten, E.,
      Superstring Theory, Vols. I and II, (Cambridge University Press,
      1987).

2.   Candelas, P., Horowitz, G.T., Strominger, A. and Witten, E., Nucl.
      Phys. B258, 46 (1985).

3.   Dixon, L., Harvey, J.A., Vafa, C. and Witten, E., Nucl. Phys.
      B261, 678 (1985) ; B274, 285 (1986).

4.   Narain, K.S., Phys. Lett. 169B, 41 (1986).

5.   Narain, K.S., Sarmadi, M.H. and Witten, E., Nucl. Phys. B279, 369
      (1987) ;
      See also Ginsparg, P. and Vafa, C., Nucl. Phys. B289, 414 (1987).

6   Lerche, W., Lüst, D. and Schellekens, A.N., Nucl. Phys. B287, 477
      (1987).

7.   Narain, K.S., Sarmadi, M.H. and Vafa, C., Harvard Preprint HUTP-
      86/A089 (December 1986).

8.   Antoniadis, I., Bachas, C., Kounnas C. and Windey, P., Phys. Lett.
      171B, 51 (1986).

9.   Kawai, H., Lewellen, D.C. and Tye, S.H.H., Phys. Rev. Lett. 57,
      1832 (1986) and Nucl. Phys. B288, 1 (1987).

10.   Antoniadis, I., Bachas, C.P. and Kounnas, C., Nucl. Phys. B288, 87
      (1987).
      See also Kounnas, C., UCB-PTH-87/21, LBL-23506, Talk at the Inter-
      national Workshop on Superstring, Composite Structures and
      Cosmology, Univ.Maryland (March 1987).

11.   Antoniadis, I. and Bachas, C.P., CERN-TH.4767/87 or Ecole Poly-
      technique A.7920687.

12.   Kawai, H., Lewellen, D.C. and Tye, S.H.H., CLNS/87/780, Cornell
      Preprint.

13. Schwarz, J.H., CALT-68-1426 (1987), CALT preprint, Talk at the International Workshop on Superstrings, Composite Structures and Cosmology, Univ. Maryland (March 1987).

14. Ferrara, S., Kounnas, C. and Porrati, M., CERN-TH. 4800/87 (corrected version) ; Phys. Letter B197, 135 (1987).

15. Antoniadis, I., Bachas, C. and Kounnas, C., CERN-TH. 4863/87, to appear in Phys. Lett. B.

16. Tomboulis, E., UCLA/87/TEP/29.

17. Ferrara, S., Kounnas, C. and Porrati, M., UCB-87/41 or LBL-24096 or UCLA/87/TEP/33.

18. Witten, E., in Geometry, Anomalies and Topology, W.A. Bardeen and R.A. White eds., N.Y. World Scientific 1985.

19. Axenides, M., Ellis, S.D. and Kounnas, C., UCB-PTH-87/27 or LBL-2364.

20. Green, M.B. and Schwarz, J.H., Phys. Lett. 149B, 177 (1984) ; Phys. Lett. 151B, 21 (1985).

21. Gross, D.J., Harvey, J.A., Martinec, E. and Rohm, R., Phys. Rev. Lett. 54, 502 (1985) ; Nucl. Phys. B256, 253 (1985) ; Nucl. Phys. B267, 75 (1986).

22. This was established for toroïdal constructions by Bagger, J., Nemeschansky, D., Seiberg, N. and Yankielowicz, S., Harvard Preprint HUTP-86A088 (1986).

23. Ferrara, S., Girardello, L., Kounnas, C. and Porrati, M., Phys. Lett. 192B, 368 (1987) ;
Antoniadis, I., Ellis, J., Floratos, E., Nanopoulos, D.V. and Tomaras, T., Phys. Lett. 191B, 96 (1987).

24. Ferrara, S., Girardello, L., Kounnas, C. and Porrati, M., Phys. Lett. 194B, 358 (1987).

25. Arnaudon, D., Bachas, C., Rivasseau, V. and Vegreville, P., Ecole Polytechnique Preprint A771.0387 (March 1987).

26. Dixon, L.J. and Harvey, J.A., Nucl. Phys. B274, 93 (1986).

27. Alvarez-Gaumé, L., Ginsparg, P., Moore, G. and Vafa, C., Phys. Lett. 171B, 155 (1986).

28. Scherk, J. and Schwarz, J.H., Phys. Lett. 82B, 60 (1979) ; Nucl. Phys. B153, 61 (1979).
Cremmer, E., Scherk, J. and Schwarz, J.H., Phys. Lett. 84B, 83 (1979).

29. Polyakov, A.M., Phys. Lett. 103B, 207 (1981) ; Phys. Lett. 103B, 211 (1981) ;
Belavin, A.A., Polyakov, A.M. and Zamolodchikov, A.B., Nucl. Phys. B241, 333 (1984) ;

Friedan, D., Qiu, Z. and Shenker, S., Phys. Lett. 151B, 37 (1985);
Friedan, D., Proceeding of the Workshop on "Unified String
Theories", Santa Barbara, 1985.

30. Bardakci, K. and Halpern, M., Phys. Rev. D3, 2493 (1971).

31. Di Vecchia, P., Knizhnik, V.G., Petersen, J.L. and Rossi, P.,
Nucl. Phys. B253, 701 (1985) ;
Goddard, P. and Olive, D., Nucl. Phys. 257, 226 (1985).

32. Seiberg, N. and Witten, E., Nucl. Phys. B276, 272 (1986).

33. Alvarez-Gaumé, L., Moore, G. and Vafa, C., Commun. Math. Phys.
106, 1 (1986).

34. Mumford, D., Tata Lectures on Theta I and II, Birkhäuser (1983 and
1984) ;
Rauch, H.E. and Farkas, H.M., Theta Functions with Applications to
Riemann Surfaces, The Williams and Wilkins Co. (1974).

35. Gliozzi, F., Scherk, J. and Olive, D., Nucl. Phys. B122, 253
(1977).

36. Bergshoeff, E., Koh, I.G. and Sezgin, E., Phys. Lett. 155B, 71
(1985); De Roo, M. and Wagemans, P., Nucl. Phys. B262, 644 (1985).

37. Ferrara, S., Kounnas, C., Porrati, M. and Zwirner, F., Phys. Lett.
194B, 366 (1987) ;
Kounnas, C., Quiros, M. and Zwirner, F., UCB/PTH/87/39 or LBL-
24001.

DISCUSSION I

– *Liu:*

What are the contributions to the conformal anomaly from spin 1 or higher spin fields? Could you use them to construct new string theories?

– *Kounnas:*

Yes, for example you can have $N = 2$ supersymmetry on the world sheet with a $U(1)$ gauge boson. I do not believe that you can have fields of spin greater than 2. There are some models constructed by Toboulis which use $U(1)$ gauge fields coupled to fermions.

– *Liu:*

In familiar string theories, if $D \neq 26$, branch cuts appeare in one–loop scattering amplitudes, which are physically unacceptable. Do you encounter such branch cuts is your theories?

– *Kounnas:*

If all of the constraints I have described are not satisfied then you find unphysical cuts. If they are satisfied unitarity is guaranteed by the requirement of factorization.

– *Liu:*

In the heterotic string, the left movers are supersymmetric while the right movers are not. There is a mismatch between fermionic and bosonic degrees of freedom. What is happening?

– *Kounnas:*

In $N = 1$ supersymmetry one has both left and right gravitini, but here we have only a left handed one. The fact that one can consider such a purely left handed supersymmetry is particular in 2–d. This is not true for example in 4–d where we cannot choose the gravitino to be simultaneously Majorana and Weyl.

– *Diaz:*

Is there any consistency check on the assumptions that you presented to us this morning?

– Kounnas:

They are not assumptions, they are constraints you are forced to have. What you might ask is if you have more constraints than those. I do not know but I believe that you do not have extra constraints.

– Lewellen:

I would just like to make a comment. You might imply that your conditions of factorization and two–loop modular invariance are somewhat a more basic starting point than the condition of a physically sensible projection of states used in the original approach of Kawai, Tye and myself. We have recently proved this quite rigorously even for complex fermions with general boundary conditions. One can argue one condition over the other as a matter of taste, but as a practical matter, no new results are obtained.

– Kounnas:

For the case of complex fermions I will agree with you, but if real fermions are required, there is an additional constraint which arises for two loop modular invariance and not by just putting physical requirements on the one–loop amplitude.

– Lewellen:

This quartic constraint that you are refering to is certainly sufficient for two loop modular invariance but I think, quite possibly it is not necessary that is, even in this case the conditions arising in one–loop are probably sufficient.

What I am calling into question here is whether your expression embodying the requirement of two–loop modular invariance and factorization is in fact correct if your quartic condition is not satisfied. This involves subtle questions of how to impose the factorization condition and calculate modular transformations for products of real fermion determinants, many of which are identically zero.

– Balog:

Could you comment on the connection between fermionic and bosonic constructions of 4–d strings?

– Kounnas:

Locally world sheet fermions are equivalent to bosons. This may not be true globally, but in general the two approaches are more or less equivalent, at least in the case of complex fermions. The fermionic formulation is in general much simpler. For example reproducing some of the fermionic models in terms of bosons requires compactification on singular manifolds e.g. orbifolds.

– Lewellen:

I would like to add a comment. In one of our papers we explicitely related the free bosonic formulation to the complex fermionic one. Essentially the momentum lattice used to define the compactification of the bosonic fields is equivalent to the charge lattice of allowed states in the fermionic formulation. Real fermion models giving rise to gauge groups of rank less than 22 require some modding of the bosonic momentum lattice, i.e. same orbifold.

– Duff:

You implied that any four dimensional superstring obtainable in the bosonic formulation could also be obtained in the fermionic formulation and that one might as well use the fermion one because it is simpler. But as far as I am aware, those obtained by Calabi–Yau compactification have as yet no fermionic analogue. I mention this, not because I like Calabi–Yau but merely to emphasize that, in the absence of any complete classification one cannot advocate any one formulation as being superior. Do you agree?

– Kounnas:

Yes I do. In fact tomorrow I will also present an example which can be interpreted using interacting 2–d fermions and gauge bosons. What I am trying to stress is that even in the context of the fermionic formulation you can get things as chiral fermions which were believed to occur only in the Calabi–Yau compactifications.

– Windey:

I think the question that should be asked is as follows: is there a scheme that classifies all conformally invariant field theories in 2–d, and the answer so far is negative.

– Miele:

Today you spoke about the fermionic models. Could you explain the relation between anomaly cancellation and the choice of the particular gauge groups?

– Kounnas:

It is just a matter of counting. The contribution to the conformal anomaly is just one half the dimension of the semi–simple group. In 10–d you do not need anything extra. In 9–d you need a group of dimension 3 and the only solution is $SO(3)$. In 8–d you need a group of dimension 6 and the only solution is $SO(3) \times SO(3)$. In this way you can find the possible gauge groups in any dimension.

– Miele:

Is it possible for every SUSY theory to reformulate it in a non–linear way, so that bosonic and fermionic degrees of freedom do not match, or is this a peculiarity of 2–d?

– Kounnas:

No, there are also non–trivial realizations of SUSY in other dimensions, for instance you may have spontaneously broken SUSY, where you don't have an equal number of bosons and fermions. This happens because you have a non–linear term associated with the goldstino. But for a non-linear realization of the type that I have discussed I do not know an example in other than 2–d. Professor Ferrara may have a comment on this.

– Ferrara:

I think the answer is no. This happens only in 2–d because of bosonization.

– Kugo:

Why do you need world sheet SUSY? Don't you need 4–d SUSY?

– Kounnas:

I don't need 4–d SUSY. I do need local 2–d SUSY in order to eliminate the negative norm states in the theory.

DISCUSSION II

– *Liu:*

What do you mean by non–linear SUSY? Do you mean spontaneously broken SUSY?

– *Kounnas:*

No, two–dimensional supersymmetry is not broken. By non–linear supersymmetry I mean the following SUSY transformation

$$\delta\psi^{\alpha} \; = \; \varepsilon f^{abc}\psi^{b}\psi^{c}$$

which is obviously non–linear in the fermion fields. In 2–d because of bosonization you can write the term quadratic in the fermion fields as the derivative of a boson. In this language SUSY becomes linear.

– *Balog:*

How many distinct models can you construct using your prescription in 4–dimensions?

– *Kounnas:*

Perhaps Lewellen could comment on this issue.

– *Lewellen:*

There is a very large number but not near as many as one would initially suspect. Many seemingly distinct spin–structure assignements give rise in fact to the same string models. This redundancy can be understood by considering the equivalent bosonic formulation. The redundant models are related by a rotation of the bosonic momentum lattice. In 10–dimensions for example, there naively appear to be hundreds or thousands of models obtainable with the fermionic construction, but in fact there are only 9 distinct models. In 4–dimensions there naively seem to be millions of $N = 1$ supersymmetric rank 22 models but in fact there appears to be only a few thousands.

– *Kounnas:*

Let me continue on this question. If you include the spontaneous breakdown supersymmetry which I have discussed then we obviously have an infinite number of models since there are continuous parameters. These essentially correspond to the same models, however, since they are clearly dynamically connected. Similarly all of the 4 dimensional $N = 4$ models are vacua of the same theory as was shown by Narain, Sarmadi and Witten. Whether all of the 4 dimensional models are similarly connected is an open question which cannot be settled in the first quantized approach.

– *Duff:*

Why did you say that vacua with non–zero cosmological constant are unstable? You must tailor your definition of energy to the space-time in question. Minkowski space has zero Minkowski energy but anti–De Sitter space has zero anti–De Sitter energy and so as such is perfectly stable.

– *Kounnas:*

The cosmological constant is a field dependent object so you expect that it is going to relax to its minimal value. Your remark about energy is correct. In fact in $N = 8$ SUGRA it is easy to see that the energy is positive so what I said is true. In the $N = 4$ case I do not know the sign, so I cannot say anything more on it.

– *Quackenbush:*

In terms of low energy phenomenology what are the advantages and drawbacks of constructing superstring models in 4–dimensions instead of compactifying10–dimensional theories?

– *Kounnas:*

I think that the answer is clear after Professor Ferrara's lecture. In the 4–dimensional models we know how to obtain almost everything, the Yukawa couplings, Kahler potential, etc. In contrast these are quite difficult to obtain in orbifold or Calabi–Yau compactifications. Quite simply the principle advantage of the fermionic formulation of string theories in 4–dimensions is their calculability.

– *Schuler:*

What is known about incorporating interactions on the world sheet?

– *Kounnas:*

One kind of interaction that you may add are Thirring interactions (four–fermion interactions). Such cases have been analyzed extensively and it is known that they correspond to changing the radius of compactification of the equivalent bosonic coordinate. Another kind of interactions that you may add is gauge interactions. As I told you before Tomboulis gave an interpretation of the models we constructed with spontaneous broken SUSY in terms of fermions interacting with $U(1)$ gauge fields on the world sheet which have the effect of generating background charges. There maybe other kinds of interactions but we don't know how to use them so far.

– Kiritsis:

Do you know if your construction of string models with spontaneously broken SUSY, by compactifying a la Scherk–Schwarz one dimension, can be extended to compactifying more than one dimensions? In such a case you would expect a much bigger family of models.

– Kounnas:

Yes, it seems possible despite the fact that it has not been attempted so far. But I expect that in such a case you may obtain a low energy supergravity theory which is gauged. Unless the gauging is non-compact the theory is unstable (it does not have a bounded from below spectrum). But supergravities with non–compact gauging can be stable, despite the fact that there is no complete classification of the particular cases.

– Bantay:

What is the connection of your construction with Kac–Moody algebras?

– Kounnas:

In fact the fermions we use form representation of the Kac–Moody algebra. There is a paper of Windey in Communications of Mathematical Physics which makes this connection.

– Diaz:

When you are degenerating the Riemann surfaces I would expect you to have divergences. It that true?

– Kounnas:

In a consistent string theory without tachyons you don't get any divergences. In the vacuum to vacuum amplitude you get a contribution from each state of energy E, which are of the form $e^{-E.t}$ where t is the length of the tube that connects the two pieces of the degenerating surface. If you have a tachyon ($E < 0$) then you get a divergence as $t \to \infty$. But if $E \geq 0$, the contributions of the $E > 0$ states go to zero, whereas the contribution of the ground state ($E = 0$) becomes unity.

– Bantay:

Gliozzi has claimed in a paper that the $O(16) \times O(16)$ model in ten dimensions is not two-loop modular invariant. Does your construction include this model?

246

– *Kounnas:*

Yes, and it is certainly two-loop modular invariant. Gliozzi does not correctly include the world sheet gravitino contribution to the phase arising in two–loop modular transformations. One cannot just naively fix the light cone gauge for the two–loop amplitude.

SUPERGRAVITY ASPECTS OF SUPERSTRINGS

IN FOUR–DIMENSIONS

S. Ferrara

CERN, Geneva, Switzerland

and

University of California, Los Angeles, U.S.A.

Abstract

Supergravity theories which arise in the low–energy limit of four–dimensional superstrings are obtained and their general properties discussed.

In this lecture the general structure of the low–energy limit of four–dimensional superstrings will be reported.

As well known superstring theories reduce at low–energy to supergravity theories with a specific content of supermultiplets, depending on the particular supestring theory, the number of space–time supersymmetries and the choice of the gauge group.

In a recent development in the understanding of heterotic string compatctifications,[1,2] starting with the work of Narain, [3] it has been realized that stringy compactifications may exist which are rather more general than the previously considered ones, [4,7] based on gauge groups contained in G ε $E_8 \times$ E_8 or SO(32).[8] In particular, the four–dimensional gauge group can be a group of rank twenty–two which is larger than the gauge group obtained from studying the compactifications of the point–field limit supergravity Lagrangian in ten space–time dimensions.

For instance, in the N=4 case the gauge symmetry SO(32) $\times$ U(1)^6 $\times$ U(1)^6 may be enlarged to SO(44) $\times$ U(1)^6. For N=4 theories the generic Yang–Mills group G is a product of simply laced groups[9] of total rank 22. Other possible gauge groups, [5] beyond SO(44) are for example, $E_8 \times$ SO(28), $E_8 \times E_6 \times E_7 \times SU(2)$, $E_8 \times E_6 \times E_6 \times SU(3)$, $E_8 \times E_6 \times E_6 \times$ SO(4). All these N=4 theories have been shown[9] to be toroidal compactifications of the D=10 heterotic strings[1,2] where some of the background fields take a constant v.e.v. The low–energy limit of N=4 four–dimensional superstrings can be obtained by noticing the fact that the N=4 supergravity Lagrangian coupled to N=4 Yang–Mills matter multiplets is almost unique. The only freedom in the N=4 Lagrangian is the further gauging of the six graviphotons and some extra "phases" if the gauge group G is not simple.[10]

However, in the case of toroidal compactifications, the graviphotons correspond to trivial U(1) factors and for a simple group SO(44) no relative gauge coupling constants or phases are possible. In this case the N=4 low–energy effective theory is completely specified.

The main difference of this theory, to the one obtained by trivial torus compactification and studied in Section II, is the absence[11] of the neutral vector multiplets whose scalar T fields correspond to the breathing modes of the internal manifold.

In the case of compactifications à le Narain[3] the radius of compactification is fixed and no scale symmetry related to the breathing mode is present. The N=4 multiplets whose scalar components were the g_{ij} and b_{ij} fields have become "charged" and they now belong to the Cartan subalgebra of the simple gauge group SO(44).

Let us discuss in some detail the symmetries of the N=4 supergravity theory with gauge group SO(44).

The scalar fields of these theories parameterize the coset space[12]

$$\frac{SO(6, N)}{SO(6) \times SO(N)} \times \frac{SU(1,1)}{U(1)} \tag{1}$$

where $N =$ dim G and $SU(1,1)/U(1)$ is the coset space of the pure N=4 supergravity sector. The non–compact symmetry is realized non–linearly on scalar fields and as a duality transformation on the 6+N vector fields. This symmetry is broken by the G gauge coupling terms. The only case where the non–compact symmetry is unbroken is for the gauge group $G = U(1)^{22}$. In this case the non–compact symmetry becomes SO(6,22) which is precisely the group associated with the Lorentzian lattices considered by Narain.[3]

Recently the Narain construction has been extended to theories with residual N=2 and N=1 supersymmetries.[13,14,15] In these constructions the even–lattices considered before are generalized to odd Lorentzian "charged lattices". A simpler way of constructing four–dimensional string models is to introduce 2D–free fermions as internal degrees of freedom, carrying a nonlinear realization of world–sheet supersymmetry.[16] Modular invariance and factorization strongly constrains the permitted choices on the fermionic spin structures.

If we consider a N=4, 4D model, N=2 models can be constructed with a Z_2 projection [13,14] with the following characteristics:
(a) There is a sector of the N=2 model which is obtained by a Z_2 truncation of the corresponding N=4 model. We call this sector the "untwisted sector" (UTS) in analogy with Z_2 orbifolds.
(b) There is an additional sector, necessary for modular invariance, which cannot be obtained by Z_2 truncation of the N=4 model. This sector is by itself Z_2 symmetric and we call it "twisted sector" (TS). N=2 models can be obtained with an extra Z_2' projection.

In some of the N=2 and N=1 models, the TS sector contains only massive states, and therefore does not contributed to the effective low–energy theory. In those cases, the entire N=2 and N=1 effective theories are given once the N=4 those cases, the entire N=2 and N=1 effective theories are given once the N=4 theory is specified and the action of Z_2 and $Z_2 \times Z_2'$ defined. However, for these theories no net number of chiral families survive in four dimensions. To get chiral families, one must include a massless twisted sector. In this case the N=2 and the N=1 theories cannot be obtained by a truncation of the N=4 theory. Nevertheless the N=1 theory can be obtained, in some cases, by a single Z_2 truncation from the N=2 theory[17] once the interactions of twisted and untwisted N=2 states are known.

Let us examine the quantum numbers and properties of the massless twisted and untwisted states in the case of the "minimal model" based on the simple gauge group SO(44) for the N=4 theory. Similar considerations can be applied to all other models.

The massless states of the N=4 theory are the N=4 supergravity multiplet and N=dim G vector multiplets. The N=4 theory has a global SU(4) – SO(6) symmetru in which the SO(6) indices can be regarded as "compactified" internal indices.

The N=4 supergravity multiplet has fields[10]

$$e_{a\mu'}, \Psi_{\mu i'}, A_{\mu ij'}, \chi^i, s, (i,j = 1,\dots,4) \tag{2}$$

and the gauge multiplet has fields[10]

$$V_\mu^A, \lambda_i^A, \phi_{ij}^A, \quad A = 1,\dots,\dim G . \tag{3}$$

The graviphotons and the matter scalar fields are in the six–dimensional representation of O(6), whole the fermions are in the 4–dimensional representation of SU(4).

To get a Z_2 and $Z_2 \times Z_2'$ truncation of the N=4 theory, we have to show how the Z_2 projections act on the gauge group SO(44), as well as on the SU(4) indices. Let us pick up a Z_2 in SU(4) as follows;

$$Z_2: \ \alpha \ = \ \begin{pmatrix} 1 & & & \\ & 1 & & \\ & & -1 & \\ & & & -1 \end{pmatrix}, \ \alpha^2 = 1 . \tag{4}$$

The Z_2 truncation of the N=4 theory, which retains only Z_2 singlets, will give a N=2 theory since there are only two gravitinos which are Z_2 singlets. The action of Z_2 on the gauge group G is obtained as follows: we decompose SO(44) into $SO(N_A) \times SO(N_B)$ with $N_A + N_B = 44$. Modular invariance also requires: $N_A = 2r + 8m$ $(N_A, N_B \neq 0)$, r ε (0,1,2), m ε (0,1,2,3,4,5). Then

$$n_V = (n_A, 1) + (1, n_B) . \tag{5}$$

Under the Z_2, α element $n_a \to n_a, n_B \to -n_B$, where n_V, n_A, n_B are the vector representations of SO(44), SO(n_A), SO(n_B) respectively. Eqs. (4) and (5) define the action of Z_2 on the Adj G and on the 6–dimensional representation of SO(4).

$$\alpha : Adj \ G \to Adj \ G_A \ + \ Adj \ G_B \ + \ \alpha(n_A, n_B)$$
$$\alpha : I_6 \to i_2, \alpha i_4 \tag{6}$$

where $i_2 = (0,1)(2,3)$ and $i_4 = (0,2)(0,3)(1,2)(1,3)$.

The scalar fields which are Z_2 inert are therefore[18]

$$(Ad\ G_A)_{i_2}\ ,\ (Ad\ G_B)_{i_2}\ ,\ (n_A, n_B)_{i_4}\ . \tag{7}$$

The first two are the "complex" scalar partners of the vector fields of $SO(n_A) \times SO(n_B)$ in N=2 vectyor multiplet, while the third ones are supermultiplets in the (vector, vector) representation. The S field is the scalar partner of one combination of the $A_{\mu(0,1)}, A_{\mu(2,3)}$ vectors while the other remaining combination is the N=2 graviphoton.

In the N=2 theory the "superpotential term" is cubic and of the form (YYV) where Y is the hypermultiplet and V is the vector multiplet.[17] The only part which remains to be determined in order to specify the theory is the Kahler manifold of the dim $G_a + \dim G_B + 1$ vector multiplets and the quaternionic manifold of the $n_a n_B$ hypermultiplets.

These manifolds can be obtained by a truncation of the N=4 manifold (see Eq. (1)) under the Z_2 action.

We here only report the result. The Kahler manifold is[18,19]

$$\frac{SU(1,1)}{U(1)} \times \frac{SO(2,\ \dim G_A\ +\ \dim G_B)}{SO(2)\ \times\ SO(\dim G_A\ +\ \dim G_B)} \tag{8}$$

while the quaternonic manifold is

$$\frac{SO(4, n_A n_B)}{SO(4)\ \times\ SO(n_A n_B)}\ . \tag{9}$$

We note that the Kahler manifold given by (8) is indeed compatible with N=2 local supersymmetry,[17] while the quaternionic manifold in (9) is one of the spaces considered by Wolf.[17] The only freedom of the N=2 theory is the relative strength of the two gauge couplings of the two factor groups $SO(n_A)$ and $SO(n_B)$.

To further truncate the theory to a N=1 supersymmetric Lagrangian we have to perform another Z_2 projection. For this purpose we introduce a second Z_2 embedded in SU(4) as follows:[14]

$$Z_2': \beta \begin{pmatrix} 1 & & & \\ & -1 & & \\ & & 1 & \\ & & & -1 \end{pmatrix}\ . \tag{10}$$

Now the $Z_2 \times Z_2'$ group has elements $\alpha, \beta, \alpha\beta,\ 1$. The $Z_2 \times Z_2'$ projection breaks N=4 supersymmetry to N=1 since only one gravitino is surviving the second Z_2' projection.

To see how the $Z_2 \times Z_2'$ group acts on the gauge group we decompose SO(44) into the four factors[16]

$$SO(44) \rightarrow \prod_{i=0}^{3} SO(n_i)\ ,\ \sum_i n_i\ =\ 44 \tag{11}$$

$$(n_0 = 2 + 4k, \ n_0 + n_1 = 0 \bmod 8) \text{ so that}$$

$$n_V = n_0 + n_1 + n_2 + n_3 \qquad (12)$$

under $Z_2 \times Z_2'$ we have

$$
\begin{aligned}
n_0 &\to n_0 \\
n_1 &\to \alpha n_1 \\
n_2 &\to \beta n_2 \\
n_3 &\to \alpha\beta n_3
\end{aligned}
\qquad (13)
$$

For $\alpha = 1$ we would get back to the N=2 theory. Now under the action $Z_2 \times Z_2'$ the adjoint and six–dimensional representations od SO(44) and S)(6) transform as follows;

$$Adj \ G \to \sum_i Adj \ G_i + \alpha(n_2, n_3) + \alpha(n_0, n_1) + \beta(n_1, n_3) + \beta(n_0, n_2)$$

$$+ \ \alpha\beta(n_0, n_3) + \alpha\beta(n_1, n_2) \qquad (14)$$

$$I_6 \to \alpha i_1 + \beta i_2 + \alpha\beta i_3 \qquad (15)$$

where $i_2 = (0,2)(1,3), i_3 = (0,3)(1,2)$. Therefore, the $Z_2 \times Z_2'$ scalar singlets are the fields with label

$$[(n_1, n_3) + (n_0, n_1)]i_1 \qquad (16)$$

from the N=2 vector multiplets and the fields with label[18]

$$[(n_0, n_3) + (n_1, n_2)]_{i_3} \ , \ [(n_0, n_2) + (n_1, n_3)]_{i_2} \qquad (17)$$

from the hypermultiplets. The three sets of chiral multiplets given by Eqs. (16) and (17) exhaust the massless states of the untwisted sector of the N=1 superstring models based on gauge group

$$\prod_i SO(n_i) \ .$$

To determine the effective Lagrangian, one has to calculate the Kahler potential, the superpotential and the gauge kinetic function which define the N=1, 4D standard supergravity theory.[20]

The Kahler manifold of the chiral multiplets can be obtained by a Z_2 projection of the N=2 manifolds or $Z_2 \times Z_2'$ projection of the N=4 manifold. The N=2 Kahler manifold of the N=2 vector multiplet "disintegrates" as follows:[18,19]

$$\frac{SO(2, \dim G_A + \dim G_B)}{SO(2) \times SO(\dim G_A + \dim G_B)} \to \frac{SO(2, n_2 n_3 + n_0 n_1)}{SO(2) \times SO(n_2 n_3 + n_0 n_1)} \qquad (18)$$

The quarternionic manifold of the N=2 hypermultiplets decomposes according to

$$\frac{SO(4, n_A n_B)}{SO(4) \times SO(n_A n_B)} \to \frac{SO(2, n_0 n_3 + n_1 n_2)}{SO(2) \times SO(n_0 n_3 + n_1 n_2)}$$

$$\times \frac{SO(2, n_0 n_2 + n_1 n_3)}{SO(2) \times SO(n_0 n_2 + n_1 n_3)} \ . \qquad (19)$$

253

The Kahler potential takes the form

$$J = \sum_{A=0}^{2} J_a(y_i) \tag{20}$$

where

$$J_a = -\log\left(1 - |y_a^\alpha|^2 + \frac{1}{4}|y_a^{\alpha^2}|^2\right) \tag{21}$$

is a "canonical" form[21] of the $SO(2,n)/SO(2) \times SO(n)$ Kahler potential.

The superpotential is trilinear and couples the three different manifolds as:[18,19]

$$g(y_0, y_1, y_2) = T_{\alpha\beta\gamma} y_o^\alpha y_1^\beta y_2^\gamma \tag{22}$$

where α, β, γ are gauge indices and $T_{\alpha\beta\gamma}$ are suitable $SO(n_i)$ tensor coefficients.

The gauge kinetic function is simply given by

$$f_{AB} = \delta_{AB}\, S \tag{23}$$

and it is the same as in Calabi–Yau type compactifications.

These models do not have ordinary fermion families. A massless twisted sector for N=2 superstring models can only exist[8] if n_A or $n_B = 8$ or 16, while for N=1 superstrings it can occur if $n_0 + n_i = 8$ or 16 for at least one i. When a massless twisted sector occurs in the theory, the low–energy supergravity Lagrangian cannot be obtained by a N=4 truncation. However, in the case of N=2 models, the only arbitrariness lies in the structure of the quaternionic manifold of the massless hypermultiplets. For N=1 theories the situation is even more complicated because there are more functions to be determined. Fortunately, the N=1 theory can be obtained by a Z_2 truncation of the N=2 theory in the case in which the N=1 massless "twisted" chiral multiplets are obtained by a Z_2 projection of the N=2 twisted chiral hypermultiplets. To give just an example, this situation occurs in the models obtained by the following $Z_2 \times Z_2'$ projections:

$$
\begin{array}{ccccc}
SO(44) & \xrightarrow{} & SO(16) \times SO(28) & \xrightarrow{} & SO(10) \times SO(6) \times SO(14)^2 \\
& Z_2 & & Z_2 \times Z_2' &
\end{array} \tag{24}
$$

This N=1 model contains sixteen chiral families in the spinorial representation of $SO(10)$ with gauge quantum numbers $(16, 4 + \bar{4}, 1, 1)$. The knowledge of the twisted state interactions can be obtained by integrating out the massive states which are coupled to the massless ones.[22]

This integration can determine the structure of the quaternionic manifold of the N=2 theory and by a further Z_2 projection the structure of the Kahler manifold of the N=1 chiral theory alluded before. We cannot give here the details. The integration over the massive string states has the effect of "deforming" the hypermultiplet quaternionic manifold in a non–symmetric non–homogeneous space which is an SU(2)–quotient[23] of the projective quaternionic space

$$HP^n = \frac{SP(4+n)}{SP(1) \times SP(3+n)} \tag{25}$$

where n is the total number of (twisted and untwisted) hypermultiplets. If we set the twisted hypermultiplet to vanish then this SU(2) quotient gives back the Wolf–manifold given by Eq. (4.9). The N=1 theory, obtained by Z_2 truncation of this SU(2) quotient has a Kahler manifold of the form

$$\frac{SU(1,1)}{U(1)} \times \frac{SO(2,N_0)}{SO(2) \times SO(N_o)} \times M \tag{26}$$

where M is a non–symmetric, non–homogenuous Kahler manifold with Kahler potential given by[22]

$$J_M = -2\log \left(e^{-(J_1+J_2)/2} - |z|^2 \right) \tag{27}$$

where J_1, J_2 are given by Eq. (4.21) and y_1, y_2 are the scalar fields coming from the truncation of the N=2 twisted hypermultiplets. "z" denotes the scalar fields of the twisted chiral multiplets. The superpotential for the twisted sector is triliner and of the for (ZZY_0) as it comes from the Z_2 projection of the N=2 "Yukawa couplings". If $z = 0$, J_M reduces to a pair of $SO(2,n) \times SO(n)$ manifolds, while if we set $Y_1 = Y_2 = 0$, J_M reduces to a non–compact CP_n manifold. The theory with one chiral family sector whose properties have been outlined has a semipositive definite potential. Supersymmetry breaking terms can be introduced[24] with the property that they lead to vanishing vacuum energy. Most interestingly the supertrace mass–square formula gives a vanishing result for this particular class of 4D–models with Higgs–fields living on $SO(2,n)/SO(2) \times SO(n)$ manifolds; this is to be contrasted with standard no–scale models[25] based on (non–compact) CP^n manifolds which suffer a quadratic divergence in the effective point–field theory.

References

1. D.J. Gross, S.A. Harvey, E. Martinec, R. Rohm, Phys. Rev. Lett. **54**, 502 (1985).

2. D.J. Gross, J.A. Harvey, E. Martinec, R. Rohm, Nucl. Phys. **B260**, 569 (1985); Nucl. Phys. **B267**, 75 (1986).

3. K.S. Narain, Phys. Lett. **169B**, 418 (1986).

4. P. Candelas, G. Horowitz, A. Strominger and E. Witten, Nucl. Phys. **B258**, 46 (1985).

5. L. Dixon, J.A. Harvey, C. Vafa, E. Witten, Nucl. Phys. **B261**, 1678 (1985); Nucl. Phys. **B474**, 285 (1986).

6. E. Witten, Nucl. Phys. **B258**, 75 (1985).

7. L.E. Ibañez, H.P. Nilles, F. Quevedo, CERN–TH **4661/87** and **4673/87** (1987).

8. M.B. Green and J.H. Schwarz, Phys. Lett. **149B**, 117 (1984).

9. K.S. Narain, M.H. Sarmadi, E. Witten, Nucl. Phys. **B279**, 369 (1987).

10. E. Cremmer, S. Ferrara, J. Scherk, Phys. Lett. **74B**, 61 (1978);
    E. Bergshoeff, I.G. Koh, E. Sezgin, Phys. Lett. **155B**, 71 (1985);
    M. de Roo, P. Wagemans, Nucl. Phys. **B262**, 644 (1985).

11. M. Muller, E. Witten, Phys. Lett. **182B**, 28, (1986).

12. J.P. Derendiger, S. Ferrara, *"Supersymmetry and Supergravity '84"*; (B. de Wit, P. Fayet, P. van Nieuwenhuizen, eds.) (World Scientific, Singapore), p. 159.

13. H. Kawai, D.L. Lewellen, S.-H. H. Tye, Phys. Rev. **D34**, 3794 (1986); Phys. Rev. Lett. **57**, 1852 (1986); Cornell Preprints CLNS **87/351**, **87/760** (1987).

14. I. Antoniadis, C. Bachas, C. Kounnas, Ecole Polytechnique Preprint, **A761**.

15. W. Lerche, D. Lüst, A.N. Schellekens, CERN–TH **4590/86**.

16. I. Antoniadis, C. Bachas, C. Kounnas, P. Windey, Phys. Lett. **171B**, 51 (1986).

17. B. de Wit, P.G. Lawers, A. van Proeyen, Nucl. Phys. **B255**, 569 (1985); E. Cremmer et al., Nucl. Phys. **B250**, 385 (1985); E. Cremmer, A. van Proeyen, Class Quantum Grav. **2**, 445 (1985); J.A. Wolf, J. Math. Phys. **14**, 1033; J. Bagger, E. Witten, Nucl. Phys. **B222**, 1 (1983).

18. S. Ferrara, L. Girardello, C, Kounnas, M. Porrati, Phys. Lett. **B193**, 368 (1987).

19. I. Antoniadis, J. Ellis, E.F. Floratos, Q.V. Nanopoulos, T. Tomaras, Phys. Lett. **B191**, 96 (1987).

20. E. Cremmer, S. Ferrara, L. Girardello, A. van Proeyen, Nucl. Phys. **B212**, 413 (1983).

21. E. Calabi, E. Visentini, Ann. Math. Vol. **71**, no. 3 (1960).

22. S. Ferrara, L. Girardello, C. Kounnas, M. Porrati, Phys. Lett. **B194**, 358 (1987).

23. K. Galicki, Com. Math. Phys. **108**, 117 (1987); K. Galicki, Stony Brook Preprint, ITP–SB **86**, 96 (1986).

24. S. Ferrara, C. Kounnas, M. Porrati, F. Zwirner, Phys. Lett. **B194**, 366 (1987).

25. E. Cremmer, S. Ferrara, C. Kounnas and D.V. Nanopoulos, Nucl. Phys. **241B**, 406 (1984); Nucl. Phys. **247B**, 373 (1984).

CHAIRMAN: S. Ferrara

Scientific Secretaries: A. Morelli and J. Quackenbush

DISCUSSION

– *Liu:*

The argument that we can have only $N < 8$ supersymmetry, or alternatively, space–time dimension < 11 is based on the assumption that we cannot have spin > 2. Yet in string theory, you can have arbitrarily high spin. How can you reconcile the difference?

– *Ferrara:*

The bound $D = 11$ is based on the fact that we cannot construct consistent interactions for massless particles with spin higher than 2. In string theory, you have an infinite number of massive states of arbitrarily high spin, but the massless states are limited to those of spin ≤ 2. If I were to increase the dimension, I would increase the spin of the massless states. It is a miracle that string theory gives the same massless states for which you can construct constintent interacting field theories. This bound is equivalent to the bound $N \leq 8$ in 4–dimensions, this is because the number of spinorial charges in $D = 4$, $N = 8$ is the same as in $D = 11$, $N = 1$ theories.

$D = 10$ is the maximum dimension for which you have coupling of massless spin 2 to massless matter fields (the Yang–Mills field). So strings saturate the maximum dimension for such couplings.

– *Liu:*

Is the cosmological constant zero in the $N = 1$ model you presented today?

– *Ferrara:*

The effective low energy Lagrangian has a scalar potential which is known. You can show that all the extrema have unbroken supersymmetry and vanishing cosmological constant. In string theory, you have vanishing cosmological constant even when you include the massive states. In the point field limit, you get a modified potential that if you compute the minimum of the potential, you find you have supersymmetry preserving extrema and vanishing cosmological constant.

– *Duff:*

A comment on maximum spin two for massless particles: very recently Fradkin and Vaseliev claimed to have consistent interactions for gauge fields of spin greater than two. The only price you pay is to have fields of arbitrarily high spin. However, one suspects that all but the spin < 2 gauge invariances will be spontaneously broken. This means that, as far as the vacuum is concerned, we still have massless spins bounded by two, and hence the usual restriction of $N < 8$ in four dimensions.

– Miele:

Could you explain why a point like quantum gravity theory becomes renormalizable if you introduce ghost fields?

– Ferrara:

If you count the powers of momentum appearing in the propagators and vertices, the addition of ghosts increases the negative power of propagators. In the R^2 term, the addition of massive spin 2 ghosts changes the P^{-2} behaviour of the propagator to P^{-4}, making the theory power counting renormalizable. Actually, one can show that the R^2 term which you have to include is a term which contains the full Reimann tensor. Using the Gauss–Bonnet theorem, one can show that the only R^2 terms which you must consider are the squares of the Reimann tensor and the scalar curvature. If you add the square of the scalar curvature alone, the theory does not become power–counting renormalizable. If it did, that would solve the problems of quantum gravity because one can show that the R^2 term, with the proper sign, does not introduce any ghost. It is the square of the Reimann tensor which introduces the spin-2 ghosts. It is the very presence of the spin-2 ghosts which makes the theory power counting renormalizable. It is a theory which has essentially 2 gravitons - a massless graviton and a massive spin–2 ghost.

– Toppan:

Does the effective $D = 10$ supergravity theory derived from superstrings present the same type of problems; namely, does it violate unitarity at the Planck scale? Recent work by the Padua group claims that it is possible to generalize the Green–Schwarz mechanism in order to have an effective supersymmetric Lagrangian, but this Lagrangian seems to have ghosts with Planck masses.

– Ferrara:

What I considered this morning were just lowest order terms. When you construct the effective action, you integrate over the massive modes. In principal, you have a power series expansion in α', and so you have infinitely many curvature terms. I only considered the lowest order terms; those with two derivatives in bosons and one in fermions. If you want to include the Green—Schwarz mechanism you have to add an R^2 term to the Lagrangian. There are several arguments. First, if you consider the theory as only an effective theory, the R^2 term is not unique as it can be changed by field redefinition. The second argument is that even if you introduce a ghost through the R^2 term, it actually only appears at an energy which is beyond the validity of the approximation. When you include the R^2, in principle, you are only looking at the massless states. You are not entitled to look at the massive states as they may be an artifact of the approximation; something like the Landau ghost in quantum electrodynamics. It could disapear in the presence of higher order terms. It is probably not revelant that you have a ghost which has a mass on the order of M_p in the effective theory, as you are just examining low order terms which can be modified by field redefinitions.

– *Kiritsis:*

Is there way in which you can write down the effective action for the massless twisted sector just from supergravity considerations or do you have to calculate it from string scattering amplitude?

– *Ferrara:*

When you have an arbitrary twisted sector, there is no general way of getting an effective action. However, there is one case which we consider, in which you can get the effective action of the massless states, that is the case in which the $N = 1$ twisted sector is a Z_2 truncation of the $N = 2$ twisted sector. In the $N = 2$ theory, there is no freedom concerning the Yukawa couplings. The only arbitrariness lies in the quaternionic manifold of the massless hypermultiplet. That manifold can be obtained by integrating out the massive modes of the string. So, in the case of the twisted sector, the $N = 2$ theory is not a truncation of the $N = 4$ theory, but the $N = 1$ theory can be obtained as a truncation of the $N = 1$ theory.

- *Morelli:*

At what scale below the Planck scale, do you expect radiative effects due to string loops to appear in the effective super gravity theory?

– *Ferrara:*

You expect them to appear as you approach the Planck scale, but there may be some nonperturbative effects which may show up and which may even be essential to determine the physics at low energies. String theory just provides a natural ultraviolet cutoff which regularizes gravity at short distances. This is a perturbative argument. However, you may argue that summation of string loops might produce string effects at much larger length scales.

– *Liu:*

Is your $N = 1$ model anomaly–free?

– *Ferrara:*

These models are anomaly–free by construction; this is another miracle of string theories. The modular invariance of string theories ensures the cancellation of local anomalies at the level of the massless states of the effective field theory. That was shown in the Green–Schwarz mechanism where the choice of the groups $SO(32)$ and $E_8 \times E_8$ can be understood at the level of field theory by the cancellation of gravitational and Yang–Mills anomalies or at level of string theories by looking at the modular invariance. It is also true for lower dimensional theories that the one–loop modular invariance assures that the chiral spectrum of the massless states is anomaly free, provided you include all the massless states. In the models I presented in my talk, the twisted sector is non chiral and therefore anomaly free. The only chiral part is the twisted sector, which cannot be obtained from field theory as there are massless states which come from the string and as I have shown, this is also anomaly free. There are other compactifications in which the twisted and untwisted sectors are both anomalous, but the combination is anomaly free. In any case, provided you include all the massless states, these effective theories are guaranteed to be anomaly free.

– *Kiritsis (comment):*

Warner and Scheleters showed that absence of global anomalies in $D = 2$ guarantee anomaly freedom in space–time.

– *Szcherba:*

Could you give the precise definition of the Kahler manifold, and could you make a short comment on the reasons why this geometrical structure is important for nowdays supergravity theory?

– *Ferrara:*

A Kahler manifold is a complex manifold where the metric can be written locally as the second derivative of a scalar function, or

$$g_i^j = \frac{\partial^2}{\partial Z^i \partial Z_j^*} K(Z, Z^*)$$

this scalar function, $K(Z, Z^*)$, is known as the Kahler potential. If you consider standard theories with two derivatives in bosons and one derivative in fermions, the couplings of $N = 1$ chiral multiplets (Z^i, χ_L^i), containing scalar fields Z^i and a fermion field χ_L^i, are such that they are governed by the Kahler geometry. This is an old result which, in the case of global supersymmetry, is due to Zumino, and which has a trivial generalization to supergravity. The non–linear interaction of these scalars is dictated by the geometry of the Kahler manifold in the sense that if you write the supersymmetric Lagrangian with a nonpolynomial interaction of the scalars, the bosonic part of the Lagrangian has the form

$$g_i^j \partial_\mu Z^i \partial_\mu Z_j^*$$

where g_i^j has to form I previously mentioned here, the Kahler potential is an arbitrary function. For any choice you can construct an $N = 1$ theory.

– Schuler:

I would like to hear you comment on the statement of Professor Glashow that $SU(4)^6$ is the best gauge group in four 4 dimensions.

– Ferrara:

No comment.

REVIEW LECTURES

LIGHT-QUARK SPECTROSCOPY FROM CHARMONIUM DECAY

Clemens A. Heusch

Institute for Particle Physics
University of California
Santa Cruz, CA 95064

INTRODUCTION

Hadron phenomenology inspired by quantum chromodynamics (QCD) has made great progress in explaining, in a semi-quantitative way, the spectroscopy and decay rates of mesons containing heavy (b, c) quarks. Light (u, d, s) quark spectroscopy was vital for the early successes of the SU_6 quark model; these early successes were, however, never permitted to grow into a quantitatively descriptive, much less a predictive, theory of light quarks and antiquarks bound together by gluons, in a rigorous QCD framework. In the present lecture, we restrict ourselves to meson spectroscopy in the low-mass region $\lesssim 2.2$ GeV/c^2, and to the attempts to understand their mass and symmetry structure. We point up some particularly vexing open questions and problems. We then review the information that has recently become available from heavy quarkonium (mainly charmonium) decays into light-quark-based mesons. It turns out that these decays, observable largely in the center-of-mass frame, with large counting rates and low multiplicities, are able to permit valuable insights into the quark content and symmetry structure of this regime of u, d, s-based mesons. The lecture is organized as follows:

2. Open questions in the lowest-mass $q\bar{q}$ nonets.

3. The use of charmonium decays to define projection operators of quark content and symmetry structure.

4. Information available from hadronic and radiative $c\bar{c}$ decays: a case-by-case review.

5. Do gluonia show up in radiative decays?

6. Exotic candidates: do $c\bar{c}$ decays have unique information to contribute?

7. A score sheet.

Let me mention a *caveat*: we will deal almost exclusively with the vector ground state of charmonium, J/ψ.

OPEN QUESTIONS IN THE LOWER-MASS $q\bar{q}$ NONETS

We restrict our interest to mesons consisting of u, d, s quarks and antiquarks. A harmonic oscillator-type interquark potential with spin-$\frac{1}{2}$ quarks will then lead, for relative orbital angular momenta $\ell = 0, 1, 2$, to a sequence of bound states that

The Superworld II
Edited by A. Zichichi
Plenum Press, New York, 1990

can be easily accommodated in terms of SU_3 multiplets of given spin and parity.[1]
A QCD-inspired version of the appropriate non-relativistic wave equation

$$(H_0 + V)\psi = E\psi \tag{2.1}$$

will add to the harmonic oscillator kinetic term

$$H_0 = \sum_{1,2} m_i + \frac{p^2}{2m_i} \tag{2.2}$$

four terms that make up an effective potential V.[2] There is, first, a confinement
term

$$H^{(1)} = -\left(c + br - \frac{\alpha_s}{r}\right)\left(-\frac{\lambda_i \lambda_j}{4}\right) \; ; \tag{2.3}$$

α_s is the strong coupling constant; λ_i, λ_j are the quark helicities. Next, a hyperfine
splitting term

$$H^{(2)} = \frac{\alpha_s}{m_i m_j}\left\{\frac{8\pi}{3}\vec{s}_i \cdot \vec{s}_j \; \delta^3(\vec{r}_{ij}) + \frac{1}{r_{ij}^3}\left[\frac{3\vec{s}_i \cdot \vec{r}_{ij}\vec{s}_j \cdot \vec{r}_{ij}}{r_{ij}^2} - \vec{s}_i \cdot \vec{s}_j\right]\frac{\lambda_i \lambda_j}{4}\right\} \; , \tag{2.4}$$

where $\vec{s}_i$ and $\vec{s}_j$ are the quark spins, $\vec{r}_{ij}$ their separation. Third, there is a spin-orbit
term

$$H^{(3)} = \frac{\alpha_s}{r_{ij}^3}\left(\frac{1}{m_i} + \frac{1}{m_j}\right)\left(\frac{\vec{s}_i}{m_i} + \frac{\vec{s}_j}{m_j}\right)\cdot \vec{\ell}_{ij}\frac{\lambda_i \lambda_j}{4}$$
$$- \frac{1}{2r_{ij}}\frac{\partial}{\partial r_{ij}}H^{conf}\left(\frac{\vec{s}_i}{m_i} - \frac{\vec{s}_j}{m_j}\right)\cdot \vec{\ell}_{ij} \; . \tag{2.5}$$

Lastly, there will be a phenomenological quark mixing term to take into account.
The calculation of the process

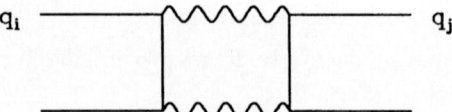

proceeds *via* standard field theory methods once a set of parameters and assumptions from, say, a bag model are adopted.[3]

Leaving this last term out for the time being, it is a relatively straightforward
task to construct the lowest-mass multiplets explicitly: the well-known pseudoscalar
and vector nonets emerge for $\ell = 0$. For each angular momentum > 0, there are
four nonets, denoted by $^{2s+1}\ell_j$:

$$^3\ell_{\ell-1}, \; ^3\ell_\ell, \; ^3\ell_{\ell+1}, \text{ and } ^1\ell_\ell \; .$$

For masses below 2 GeV, a reasonable set of parameters, when inserted in
eqs. (2.2)–(2.5) such as to fit a few signal states, leads to a scheme as shown in

Fig. 1.[*] Early semiquantitative identification of known meson states with the solutions of the simple oscillator potential had led to a belief that a refinement of both measurement and theory would soon lead to full confirmation by a scheme such as the one outlined here.

Even then, there were two flies in the ointment: for one thing, the space wave function $R_{n\ell}$ for the system has radial nodes ($n = 1, 2, \ldots$), leading to the expectation of a repetition of the scheme shown in Fig. 1 at higher energies hard to specify. For another, the basic tenets of QCD, including the self-coupling of the gluon, lead to the expectation of additional states: while frequently sharing the space-time properties of the $q\bar{q}$ states in Fig. 1, they contain valence gluons in the absence or presence of (one or several) $q\bar{q}$ pairs. There is no clear idea of how to calculate such "gluonia" or quark–glue hybrid eigenstates' masses. So, they may well show up such as to confuse the picture, or even to mix with legitimate $q\bar{q}$ configurations.

Such states have never been unequivocally identified (see below), but their presence may complicate the parametrization of wave functions once thought straightforward.

In the following, we will make use of the fact that the charmonium decays

$$J/\psi \to \text{hadrons},$$
$$\to \gamma + \text{hadrons},$$
$$\eta_c \to \text{hadrons}$$

start from well-defined kinematics and symmetry states, and therefore permit inferences about the properties of their decay products, to look into some of the open questions perceived in low-mass meson spectroscopy. Moreover, due to the fact that the $c\bar{c}$ system's decay is a highly local interaction, with

$$\Gamma(c\bar{c} \to X) \sim |\psi(0)|^2 \ ,$$

we do not expect it to favor high orbital or radial quantum numbers; we will try to study the QCD underpinnings of the relevant spectroscopy by examining the $\ell = 0$ and $\ell = 1$ multiplets of Fig. 1, their masses, wave functions, and decay characteristics (if any).

Figure 2 lists the spin/parity/charge conjugation (J^{PC}) combinations accessible to a $q\bar{q}$ system with $\ell = 0, 1$; it compares the allowed configurations for $q\bar{q}, gg, q\bar{q}g$, and $q\bar{q}q\bar{q}$ for $J = 0, 1, 2$, for further reference.

$\ell = 0, \ J = 0$: Pseudoscalars, $J^{PC} = 0^{-+}$

This nonet has puzzled phenomenologists ever since the first SU_3 mass relationships were written down. The two physical isosinglets, η and η', are apparently not at all as "ideally mixed" as is the case, to a good approximation, for the 1^{--} vector and 2^{++} tensor nonets.[5]

One reason for this may be connected with the inability of the $s\bar{s}$ component of η to decay strongly *via* the usual diagram

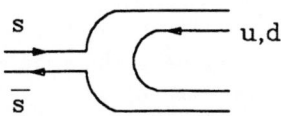

---

[*] For current nomenclature of mesonic states, see Table 1.

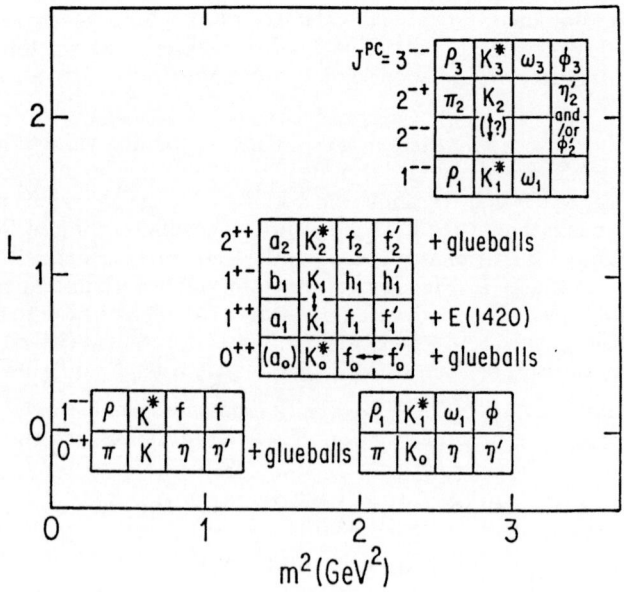

Fig. 1. An optimistic picture of "full occupancy" for $\ell = 0,1$ $q\bar{q}$ mesons was given by J.L. Rosner (Ref. 4) at the last Hadron Spectroscopy Conference. Much of the information contained in this table is tentative, as shown in the text. The $L = 1$, $J^{PC} = 1^{+-}$ states are not discussed in the text, since J/ψ decays contain very little information on them.

Table 1. Nomenclature of L=0,1 Mesons

n,L	J^{PC}	$I = 0$ Mesons	$I = 1$ Meson	$I = \frac{1}{2}$ Meson
1,0	0^{-+}	$\eta(549)[\eta]$ $\eta'(958)[\eta']$	$\pi(140)[\pi]$	$K(494)[K]$
	1^{--}	$\omega(783)[\omega]$ $\phi(1020)[\phi]$	$\rho(770)[\rho]$	$K^*(892)[K^*]$
1,1	0^{++}	$f_0(975)[S^*]$ $f_0(1300)[\epsilon]$	$a_0(980)[S]$	$K_0^*(1350)[K]$
	1^{++}	$f_1(1285)[D]$ $f_1(1420)[E]$	$a_1(1270)[A_1]$	$K_1(1280)[Q_K]$
	2^{++}	$f_2(1270)[f]$ $f_2(1525)[f^*]$	$a_2(1320)[A_2]$	$K_2^*(1425)[K^*]$
2,0	0^{-+}	$\eta(1275)$ $\eta'(1420)$	$\pi(1300)$	$K(1460)$

Note: In the text, we leave out the mass designation whenever there is no danger of confusion. The traditional names are given in square brackets. The last line gives the tentative assignment of the lowest-mass radial recurrence. The 1^{+-} octet, not discussed in the text, has not been included here either.

J^{PC}	$q\bar{q}$	gg	$q\bar{q}g,$	$q\bar{q}q\bar{q}$
0^{++}				
0^{+-}			X	
0^{-+}				
0^{--}	X	X	X	
1^{++}				
1^{+-}		X		
1^{-+}	X	X		
1^{--}				
2^{++}				
2^{+-}	X	X		
2^{-+}	X	X		
2^{--}		X	X	

Fig. 2. Space-time properties of low-mass meson systems based on various quark and gluon constituents. Meson nonet quantum numbers *in*accessible to various quark and gluon configurations are marked with an X. Note that 4-quark states are available for *all* J^{PC} combinations.

but will rather have to go by way of the disconnected

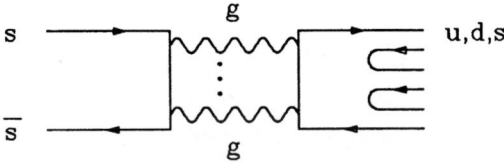

configurations, or decay electromagnetically. In fact, the large $\gamma\gamma$ branching ratio and the G-parity violation in the 3π decay show that it chooses the latter possibility. The higher-mass η', on the other hand, is close to $K\overline{K}$ threshold.

Next, there are too many low-mass pseudoscalars observed for one nonet. Does that mean there is evidence for a radially excited nonet? After all, this is the lowest-mass radial excitation we might expect. If not, do they constitute evidence for gluonic degrees of freedom, or for a 4-quark state?

Further, recall that the low-mass pseudoscalar mesons are at the evidentiary basis of the so-called U-1 problem,[6] due to the absence of net quantum numbers in the U(1) group: the QCD Lagrangean has extra terms

$$\partial_\mu j^\mu \sim \varepsilon_{\lambda\mu\nu\sigma} F_{\lambda\mu} F^{\nu\sigma} + \dots ;$$

they are proportional to the axial current divergence: this is non-perturbative, "soft" physics, for which we have no calculational prescription. If we had a theory fully symmetric under $SU_2 \otimes SU_{3_{color}} \otimes SU_{3_{flavor}}$, then (if the u, d, s quarks shared mass eigenvalues) the π, K, η would be massless Goldstone bosons. This is radically different from, say, the vector mesons or the nucleon octets, etc., the masses of which are largely determined by the QCD forces that bind the quarks.

Lacking theoretical guidance, we are in dire need of the best experimental evidence on symmetry structure, quark content, and the potential emergence of new degrees of freedom.

$\ell = 0$, $J = 1$: Vectors, $J^{PC} = 1^{--}$

This is probably the best understood of the low-mass meson nonets. The singlet-octet decomposition of the isoscalar members of the nonet is

$$
\begin{aligned}
|\mathbf{1}\rangle &= \frac{1}{\sqrt{3}} \left(|u\bar{u}\rangle + |d\bar{d}\rangle + |s\bar{s}\rangle \right) \\
&= \sqrt{\tfrac{2}{3}} \, |\omega\rangle + \sqrt{\tfrac{1}{3}} \, |\phi\rangle \\
|\mathbf{8}\rangle &= \frac{1}{\sqrt{6}} \left(|u\bar{u}\rangle + |d\bar{d}\rangle - 2 |s\bar{s}\rangle \right) \\
&= \sqrt{\tfrac{1}{3}} \, |\omega\rangle - \sqrt{\tfrac{2}{3}} |\phi\rangle \ ,
\end{aligned}
$$

This means the vector singlet will hadronize into ω twice as frequently as into ϕ; the octet will do the converse. Isoscalar vector mesons are therefore excellent quark content indicators, and we will use them as such.

Further, the vector dominance relations describe the coupling ratios of the neutral vector mesons to the photon. J/ψ decays into two vector mesons are restricted by

$$
J/\psi \nrightarrow V^0 V^0 \quad \text{(for isoscalar } V^0)
$$

by C invariance. In the SU_3 limit, decay into *all* pairs of vector-mesons is forbidden (for strong decays)

$$
\begin{aligned}
J/\psi &\nrightarrow \rho^+ \rho^- \ , \\
J/\psi &\nrightarrow K^* \overline{K}^* \ .
\end{aligned}
$$

These signals therefore denote electromagnetic decays. Lastly, the allowed decays of the pseudoscalar charmonium

$$
\eta_c \to V V
$$

have a $\rho : \omega^0 : \phi^0 : K^*$ ratio predicted by SU_3 symmetry. For the spin-parity analysis of decaying states, it is further important that the two-body decays $\phi \to K\overline{K}$ and $\rho \to \pi\pi$ lead to a straightforward J^P determination by final-state angular correlations. Lastly, perturbative QCD makes specific predictions[7] about the helicity structure of the process

$$e^+ e^- \to \gamma^* (J/\psi) \to VV \ .$$

$\ell = 1, \ J = 0$: Scalars, $J^{PC} = 0^{++}$

This is the least understood of all the low-mass nonets. There is considerable confusion about the mass spectrum, and indeed about the identity of all the members of the multiplet: only the non-strange octet members, the $(I = 0)$ $f_0(975)$ $[S^*]$ and the $(I = 1)$ $a_0(980)$ $[\delta]$ are well established; but their status as bona fide $q\bar{q}$ states has been disputed. The a_0 and f_0 appear to be narrow, and almost mass-degenerate, the $K^*(1350)[\kappa]$ and $f_0(1300)[\varepsilon](1SU_3)$ are wide, their masses ill-determined. Moreover, the mass difference between the isodoublet and the isosinglet in the octet appears to be ~ 400 MeV, vs. about 100 in other octets. If, as has been suggested, the $a_0(980)$ and $f_0(975)$ are in effect 4-quark $(s\bar{s}q\bar{q})$ states, the mass degeneracy is explainable, but not the latter mass difference.

There are more apparent inconsistencies: The width ratio $\Gamma(K_0^*)/\Gamma(a_0) \gtrsim 6$ contradicts the flavor symmetry prediction of 2. The small width $\Gamma(f_0 \to \pi\pi)$ appears to favor ideal mixing with $f_0 \approx |s\bar{s}\rangle$, but the Gell-Mann-Okubo mass rule suggests f_0 to be a $\mathbf{1}(SU_3)$ and $a_0 \in \mathbf{8}(SU_3)$. This would imply a large $q\bar{q}$ $(q = u, d)$ component in the f_0 wave function.

Also, there have been speculations that $f_0(975)$ is nothing but a $K\overline{K}$ threshold effect.[8] The most vexing, and potentially most exciting, question we have of the scalar meson spectrum concerns the search for the scalar gluonium: $|gg\rangle$ $(J^P = 0^{++})$ can be made in an $\ell = 0$ state, unlike the scalar $q\bar{q}$, so that $m |gg\rangle^{0^{++}} < m(q\bar{q})^{0^{++}}$ is well possible for the scalar ground states. Can we locate the lowest-mass gluonium in this fashion?

Further, note a theoretical complication of unknown phenomenological consequence: instantons—topological quasiparticles with the quantum numbers of the vacuum—can tunnel in an uncontrollable fashion between the vacuum and the scalar sector.[9] This process may vitiate all mass calculations, mainly for the $\ell = 0 |gg\rangle$ sector, and increase widths beyond recognizability. Clearly, our knowledge of the scalars needs all the help it can get!

$\ell = 1, \ J = 1$: Axial Vectors $J^{PC} = 1^{++}$

This relatively well documented nonet has nonetheless its open questions: the isosinglet $f_o(1285)[D]$ and $f_1(1420)[E]$ have been reported to be next to ideally mixed. But there is considerable confusion about the prevalent decay channels $(\eta\pi\pi, \ K\overline{K}\pi, \ \gamma\rho)$ in this mass range: Is there also a radially excited η in the D mass region? Is there another radial excitation, or are there ι-related structures,[*] in the E region?

This question may turn out to be of considerable importance: two massless, transverse gluons do *not* form a 1^{++} state (nor a 1^{--} vector!). This fact may help to disentangle any potential confusion.

For the isovector, there is the thorny question of the mass and width of $a_1(1270)$ $[A_1]$: the mass structure reported from $\pi\pi$ phase shift analyses does not agree with that determined from semileptonic τ decay.[10] Any further evidence would be helpful here.

---

[*] where $\iota(1460)$ is the pseudoscalar gluonium candidate discussed below.

This channel has received a lot of attention in recent years due to the prediction of perturbative QCD, that much of the hadronization of transverse gg systems will happen here. Its respectable members—the isovector $a_2(1321)[A_2]$, the almost ideally mixed isoscalars $f_2(1270)[f]$ and $f_2(1525)[f']$—have been joined by more non-strange states: the gluonium candidates $f_2(1720)[\Theta]$, a reported set of closely spaced, broad states in the mass range 2100–2350 observed in hadronic production of $\phi\phi$ systems, $[g_T]$, and presumably the unexplained $X(2230)[\xi]$.

There is a rash of new evidence bearing on the quark $vs.$ gluon content of the various f_2 states, on their separation from scalar structures in the same mass range, and on the possible confirmation of the $X(2230)$ discovery in hadronic interactions.

It is clearly imperative that the confused situation in these lowest multiplets be cleared up. Questions to be addressed throughout the spectrum of low-mass mesons are these:

(1) Why does nature appear to favor pure $u\bar{u} + d\bar{d}$ or $s\bar{s}$ states ("ideal mixing") for isospin singlet states? Indeed, does she?

(2) How much weight do we have to allot to quark-mixing/disconnected diagrams of the type

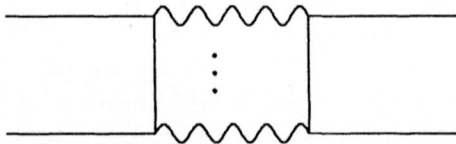

as ostensibly evidenced in the decays $\phi \to \pi\pi\pi$, $f_0[S^*] \to \pi\pi$?

(3) What is the evidence that gluon hadronization occurs such as to indicate, preferentially, transverse (massless) gluons?

(4) Can we pin down the fraction of 2- or 3-gluon-mediated decays that occur via doubly disconnected ($i.e.$, doubly OZI-suppressed) diagrams?

(5) What is the evidence that some observed isoscalar mesons may, in fact, be gluonia?

In addition, a look back at Fig. 2 shows several "exotic" states not accessible to either $q\bar{q}$ or gg configurations. Is there evidence in $(c\bar{c})$ decay for recent claims that a $J^{PC} = 1^{-+}$ hybrid has been observed in the $\eta\pi^o$ final state, a (possibly related) $J^{PC} = 1^{--}$ state in the $\phi\pi^o$ channel? A look at the data will give some, but not all answers.

The Use of Charmonium Decays to Define "Projection Operators" of Quark Content and Symmetry Structure

Much of the experimental difficulty associated with the clean determination of meson states stems from the preparation of the sample: strong-interaction production experiments frequently present formidable challenges: large multiplicities, tight forward bundling of secondaries, difficulties in particle identification and precise kinematical reconstruction at high energies make the extraction of mass, width, and quantum numbers of strongly decaying meson systems often problematical; $c\bar{c}$

production and annihilation, on the other hand, can proceed *via* a huge cross section in the process

$$e^+e^- \rightarrow \gamma^* \rightarrow J/\psi \rightarrow \text{ hadrons}$$
$$\rightarrow \text{ photon } + \text{ hadrons}$$
$$\rightarrow \text{ photon } + \eta_c \,,$$
$$\hookrightarrow \text{ hadrons}$$

from well-defined initial states. Final-state recognition is eased by experimentation in the rest system; multiplicities are low, and often contain easily identifiable hadrons. The scheme is well suited for an experimental definition of projection operators. Recall that, mathematically, these Hermitean operators project any vector in Hilbert space onto the coordinate axis defined by some given unit vector

$$\hat{P}(\vec{r}) = \int u_n(r) u_n^+(r') dr'$$

with $\hat{P}^2 = \hat{P}$, eigenvalues 0 or 1. We can define a phenomenological analogue: starting from a precisely defined initial state J/ψ (or η_c), we note its quantum numbers

$$J^{PC} I^G = 1^{--} 0^- \text{ (or } 0^{-+} 0^+) \,.$$

It is a unitary singlet **1** $(SU_3)_{\text{color, flavor}}$ state. Its helicity is ± 1 (or 0). The principal decay diagrams of J/ψ then help to define the operators:

a) three-gluon decay

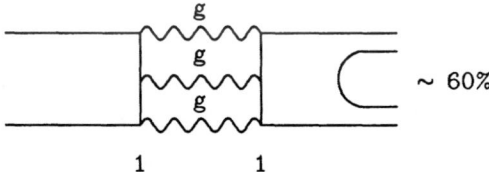

b) electromagnetic decay

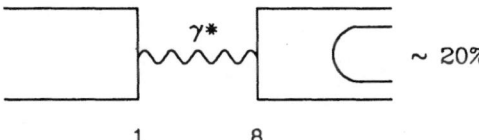

c) radiative decay

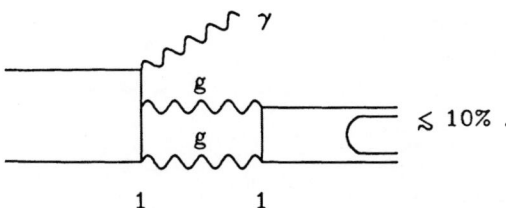

273

The latter includes the specific

d) decay *via* the $J^{PC} = 0^{-+}\eta_c$ state

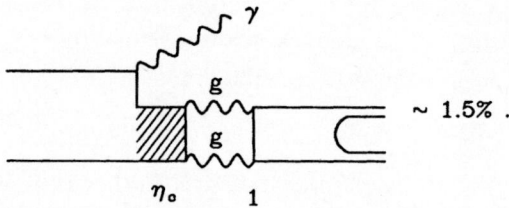

~ 1.5% .

Add to this, in next higher order, the doubly disconnected gluonic decay diagrams,

e) doubly disconnected

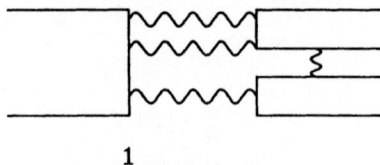

f) doubly disconnected, quark rearrangement

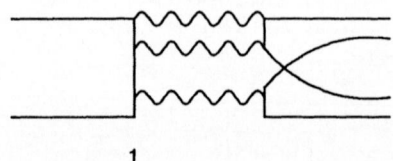

potentially leading to final states consisting of two separate OZI forbidden meson systems (so that we guess, roughly, $\Gamma_{e,f} \lesssim \frac{1}{10}\Gamma_a$); with this, we probably have accounted for all important contributions.

Using the symmetry group labels in the diagrams and the quark lines, we can use diagrams a)–f) for a definition of projection operators in J/ψ decay. In the η_c case, we take, in order of allowed-ness,

g) 2g decay

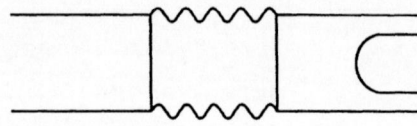

h) doubly disconnected diagram

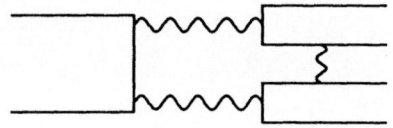

i) quark rearrangement

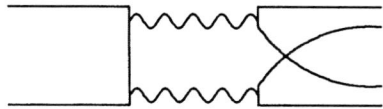

j) electromagnetic decay

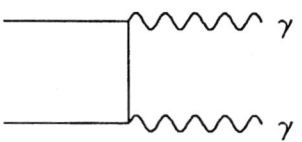

g) and i) are experimentally indistinguishable, except for symmetry breaking effects that affect relative rates for different quark flavors. There is no significant hadronization of the electromagnetic decay, diagram j).

When investigating the questions (1) through (5) above, we use the following "projection operators": for a check on the quark content of system X, use graphs *a),b)*; choose an ideally mixed vector or tensor to accompany X. To leading order, the occurrence of X jointly with ω or $f_2(1270)$, projects out the u, d quark content, with $\phi, f_2(1525)$, the strange quark component. For a determination of unitary symmetry labels (as required for a gluonium), choose graphs *a),c)*: both start from a singlet on the right-hand side; find singlet or octet status of one final state meson by determining the other. To project onto gluonium candidates, choose graph c), and proceed to the criteria in the section on gluonia, below. To single out symmetry-breaking (not just mass-breaking) effects, choose graph *b)*. To project out given spin-parity states, we will have to resort to more mundane methods of detailed final-state analysis. A brief mnemonic of the straightforward choices is shown in Table 2.

Note that the η_c decay graphs have not been incorporated in these projection operators, largely because sufficiently large clean data samples are not available.

INFORMATION AVAILABLE FROM HADRONIC AND RADIATIVE J/ψ AND η_c DE-CAYS: A CASE-BY-CASE REVIEW

Recalling the set of questions at the end of Section 2, and the methods summarized in Table 2, let us choose specific configurations to attack the weak points in our $\ell = 0, 1$ meson spectroscopy.

J/ψ hadronic decay into two mesonic systems is easily observable in these channels shown in Table 3.

The leading radiative decay graph *c)* will let the **1** (SU$_3$) system X decay into the 2-body modes

$$X \to PP$$
$$\to PV$$
$$\to PT$$
$$\to VV$$

If $J^{PC}(X) = 0^{-+}$, we expect it to decay into the specific 3-body final states $\eta\pi\pi$, $K\overline{K}\pi$, but not $\pi\pi\pi$.

Table 2. Definition of "Projection Operators"

"Operator"	Diagram(s)	Example	Remarks
$P_{1(SU_{3,f})}$		$J/\psi \to (\omega,\phi)\theta$ $J/\psi \to \gamma f_2$	If $\Gamma_\omega \approx 2\Gamma_\phi$
$P_{8(SU_{3,f})}$		$J/\psi \to \rho\pi$	For isoscalars, if $2\Gamma_\omega \approx \Gamma_\phi$
$P_{e.m.}$		$J/\psi \to \pi\eta$	
$P_{q\bar{q}}$		$J/\psi \to \omega f_2(1280)$	
$P_{s\bar{s}}$		$J/\psi \to \phi f_2(1525)$	
P_{gg}		$J/\psi \to \gamma\iota(1460)$	
P_{I+1}		$J/\psi \to \rho a_0(980)$	
$P_{I=0}$		$J/\psi \to \omega f_0(975)$	
P_{J^P}		$J/\psi \to \phi KK$ $J/\psi \to \gamma KK$	Use classical methods of spin-parity or particle-wave analysis

Note: To illustrate the methods, our approach is shown for leading diagrams only.

Table 3. Two-body Hadronic Decays of J/ψ

Decay	May Proceed via Graphs	SU$_3$ Allowed?
$J/\psi \to PV$	a,b,c	yes
$\to SV$	a,b,c	yes
$\to AV$	a,b,c	yes
$\to VT$	a,b,c	yes
$\to PP$	b	no
$\to VV$	b	no

Note: P stands for pseudoscalar, V for vector, T for 2^{++} tensor, A for axial vector, S for scalar.

We now apply the tools defined above, to check on the principal data sets available from general-purpose detectors on ground-state charmonium decay—about 6M events from MARK III at SPEAR, 8M events from DM2 at DSI. The detectors have somewhat different quality criteria;[11] still, with one notable exception, discussed below, their conclusions largely are in good agreement. Referring to the questions raised in section 2, what can we make of these data?

Pseudoscalars

In the pseudoscalar nonet, the isoscalars (η, η') have caused excitement due to the very different couplings they show to the two-gluon channel P_{gg}. In fact, the relative abundance $J/\psi \to \gamma(\eta'/\eta/\pi^0) \propto /0.42 : 0.086 : 0.004$ led to an early assumption that the η' wave function contain a gg component.

Let us choose the operators P_{qq} in the $J/\psi \to VP$ channel, where we assume ideal ω/ϕ mixing. Then of the 9 VP combinations containing non-strange mesons, $(\rho/\omega/\phi, \pi/\eta/\eta')$, 4 contain only isoscalars (and are therefore accessible to strong decays). It is straightforward to write down[12,13] the amplitudes for each of the nine PV channels in terms of the strong amplitude, modified by a nonet symmetry-breaking amount for each strange quark in the final state, and an electromagnetic amplitude for the isospin-breaking channels (such as $\rho\eta$, $\omega\pi^0$,...). If, in addition, we amplify the singlet-octet mixing scheme to contain a potential gluonium component, a global fit to the observed data concluded (for $X_{\eta,\eta'} = u, d$ quark component of η, η'; $Y_{\eta,\eta'} =$ strange quark component, $Z_{\eta,\eta'} =$ "something else", maybe gg in the wave function) from the early MARK III results [14] for the wave functions

$$|\eta\rangle = X_\eta |q\bar{q}\rangle + Y_\eta |s\bar{s}\rangle + Z_\eta |gg\rangle$$
$$|\eta'\rangle = X_{\eta'} |q\bar{q}\rangle + Y_{\eta'} |s\bar{s}\rangle + Z_{\eta'} |gg\rangle$$

that $X_\eta^2 + Y_\eta^2 \simeq 1$, and that $X_{\eta'}^2 + Y_{\eta'}^2 = 0.65 \pm 0.18$; this left room for a sizeable $|gg\rangle$ component in the η', in accord with its prominence in $J/\psi \to \gamma\eta'$. However, there is independent information available from $\gamma\gamma \to X$, which scans quark charges in an obvious way, and does not couple to $|gg\rangle$. Recent data from that channel[15] and an analysis of the full set of MARK III data[16] now make the need for a gluonic component disappear in both η and η'; with $Z = 0$, the wave functions revert to the simple two-component mixing scheme with

$$X_\eta = 0.71 \pm 0.04, \quad X_{\eta'} = 0.58 \pm 0.04,$$
$$Y_\eta = -0.71 \pm 0.07, \quad Y_{\eta'} = 0.99 \pm 0.08,$$

favoring a mixing angle of some $\Theta_P \simeq -19°$.[4] While this result is satisfactory in itself, and simplifies the phenomenology of the 0^{-+} nonet, it fails to give any indication for the reason why Θ_P is so far from ideal.

Moreover, it does away with a number of phenomenological approaches to the inclusion of the pseudoscalar state(s) observed at somewhat higher mass[17] into the η–η' mixing scheme. Does this leave a naturally explainable place for the additionally observed pseudoscalars?

Recall that we expect radially excited η and η' states, and, possibly, a gluonium in the range 1.2–1.5 GeV/c^2. Pseudoscalars in this mass range will decay into $K\overline{K}\pi, \eta\pi\pi$. Resonance structure has been variously reported in both of these channels, with repeated assertionsthat a $K\overline{K}\pi$ pseudoscalar [18,19] was seen at the mass of the $f_1(1420)[E](J^{PC} = 1^{++})$ around 1.42 GeV/c^2. Rosner[4] suggests that $\eta(1280)$ and $\eta(1420)$ be seen as the $n = 2$ η and η' states. That leaves the closely higher-mass $\iota(1460)$ as a potential gluonium candidate. What is the evidence from $c\bar{c}$ decay?

Let us resort to our "projection operators" to see whether J/ψ decays can shed light on this scenario. First, take $P_{q\bar{q}}$ and $P_{s\bar{s}}$ in the form $\Gamma(J/\psi \to \omega(K\overline{K}\pi), \omega(\eta\pi\pi))$ and $\Gamma(J/\psi \to \phi(K\overline{K}\pi), \phi(\eta\pi\pi))$, respectively, and check for enhancements. The corresponding distributions are shown in Figs. 3 and 4. We see that there is an excellent signal at ~ 1400 MeV/c^2 in both $\omega\eta\pi^+\pi^-$ and $\omega K\overline{K}\pi$, hinting, at first glance, at a (u, d) quark-based structure (an s-based structure will not easily decay into $\eta\pi\pi$)—and indeed the signal is barely visible in $\phi\eta\pi\pi$, not at all in $\phi K\overline{K}\pi$. (But note, there is no doubt about the existence of a 1^{++} state decaying into $K\overline{K}\pi$ and $\eta\pi\pi$, in the same mass region—as observed in an elegant experiment by the TPC/2γ Collaboration;[20] see below.) At ~ 1280 MeV/c^2, there is structure visible in $\omega/\phi(\eta\pi\pi)$, probably *some* in $\phi(K\overline{K}\pi)$—but all indications are that these structures, with their clear quark correlations, would rather fit in with the known axial vectors D and E (see below).

In the absence of a complete spin-parity analysis of the enhancements seen in the $K\overline{K}\pi$ and $\eta\pi\pi$ systems, let us next project the J/ψ decay data onto the gg sector, which is known to favor 0^{-+} over 1^{++}. Figs. 3(a) and 4(a) show the radiative decays

$$J/\psi \to \gamma(K\overline{K}\pi),$$
$$\to \gamma(\eta\pi\pi),$$

for comparison with the hadronic decays. In the $\gamma(K\overline{K}\pi)$ channel, a broad and important enhancement comprises the E(1420) *and* $\iota(1460)$ regions: it is too broad at the top to be fitted with only one Breit-Wigner resonant term, and provides the *only* incontrovertible evidence for the ι state in its upper-mass range, centered at ~ 1460 MeV/c^2.[21] Its spin-parity assignment is certainly $J_{PC} = 0^{-+}$—a statement we are not ready to make about the lower-mass part of the same enhancement— although we believe it to be so: no prominent axial vector state has been identified so far in the radiative channel, which favors the hadronization of massless, transverse

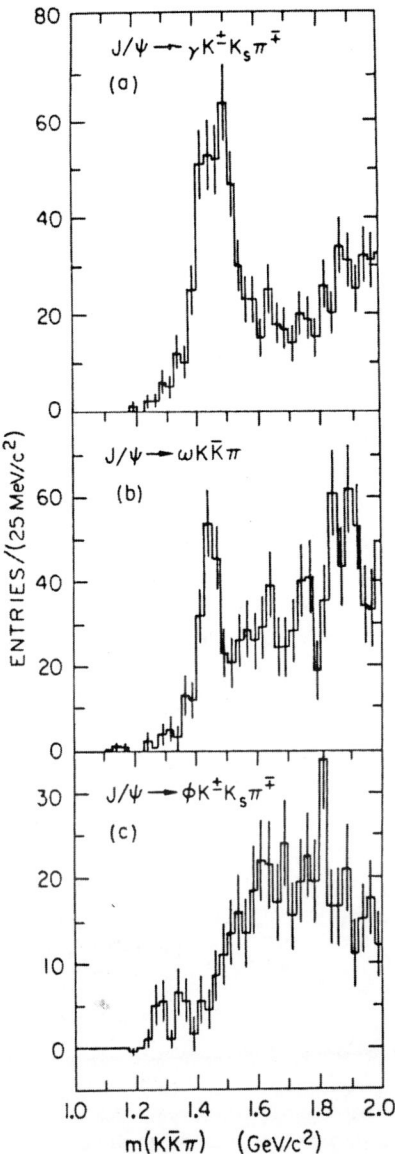

Fig. 3. Invariant-mass distributions for $K\overline{K}\pi$ systems emitted, in J/ψ decays, together with (a) a photon, (b) an ω meson, (c) a ϕ meson (MARK III).

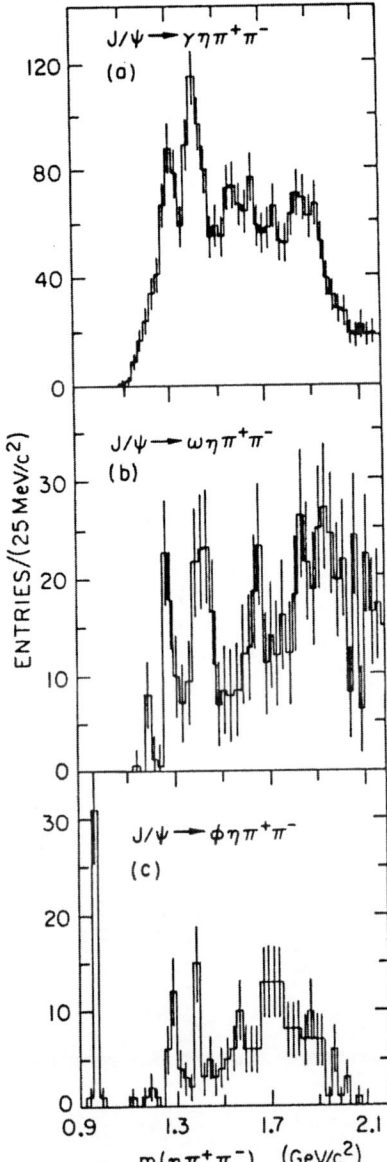

Fig. 4. Invariant-mass distributions for $\eta\pi^+\pi^-$ systems observed in J/ψ decays, together with the vectors γ, ω, ϕ.

gluons. For $\gamma(\eta\pi\pi)$, a prominent enhancement is seen at ~ 1390 MeV/c^2. If we assume this to be a $u\bar{u}, d\bar{d}$-based state related to the $s\bar{s}$-based state seen in the $\gamma(K\overline{K}\pi)$ channel, the two may make up the SU$_3$ singlet demanded by graph $d)$ for the 2-gluon annihilation: their slight mass difference is then due to the availability of close-by hadronic resonances. If, in addition, ongoing analysis efforts find that these enhancements indeed share $J^{PC} = 0^{-+}$, then it is tempting to identify them as the $\eta'^{(2)}$, quite distinct from the hadronically produced E(1420). They are probably

Table 4. Amplitudes for $\eta_c \rightarrow VV$ decays, for ideal ω/ϕ mixing; g_1 couples η_c to $1 \otimes 1$, g_8 to $8 \otimes 8$

Mode	Amplitude	Relative Rate for $g_1 = g_8$
$\eta_c \rightarrow \omega\omega$	$g_8 \sin^2 \Theta + g_1 \cos^2 \Theta$	1
$\eta_c \rightarrow \phi\phi$	$g_8 \cos^2 \Theta + g_1 \sin^2 \Theta$	1
$\eta_c \rightarrow \omega\phi$	$\sin \Theta \cos \Theta (g_8 - g_1)$	0
$\eta_c \rightarrow K^* \overline{K}^*$	$\sqrt{2} g_8$	4
$\eta_c \rightarrow \rho\rho$	$-\sqrt{3} g_8$	3

identical with the $J^P = 0^{-+}$ state decaying into $(\gamma\rho)$, also originally observed in radiative J/ψ decay at a mass somewhat below $\iota(1460)$;[22] a sensitive investigation into its possible decay into $\gamma\phi$ might further firm up our ideas of this quark-based state.[23]

Moving to somewhat higher masses, there is the curious appearance of very prominent pseudoscalar enhancements in the radiative decay channels

$$J/\psi \rightarrow \gamma V \bar{V} \quad (V = \rho, \omega, \phi, K^*) \ .$$

Appearing just above threshold for all six channels (see Fig. 5), the peaks decaying into $\rho\rho$ and $\omega\omega$ are in a mass range within reach of $\iota(1460)$. This fact motivated a coupled-channel analysis of all the MARK III data available,[24] implying a very hefty overall rate for $J/\psi \rightarrow \gamma\iota$ (including $\iota \rightarrow \rho^+\rho^-, \rho^0\rho', \omega\omega$), of $B(J/\psi \rightarrow \gamma\iota)$ of $\sim 0.5\%$. New results show that sizeable effects also show up in the $K^* \overline{K}^*$ and $\phi\phi$ cases, at very different masses (see Figs. 5(d),(e),(f)). This makes a quark re-scattering interpretation a more likely candidate for explaining a common dynamical feature.[25]

Vector Mesons

The lack of spectroscopic problems in the vector meson nonet permits a look at two precision measurements that bear on symmetry structure and spatial wave function for these systems: The DM2 Collaboration has recently studied all four decays[26] of η_c into VV modes:

$$B(\eta_c \rightarrow \rho^0\rho^0) = (8.7 \pm 2.6 \pm 1.3)10^{-3}$$
$$(\eta_c \rightarrow \phi\phi) = (3.1 \pm 0.7 \pm 0.4)10^{-3}$$
$$(\eta_c \rightarrow K^{*0}\overline{K}^{*0}) = (4.3 \pm 1.9 \pm 2.0)10^{-3}$$
$$(\eta_c \rightarrow \omega\omega) \leq (7 \times 10{-3} \ (90\% C.L.)$$

The decay of the SU_3 singlet state η_c predicts, in the limit of exact SU_3 symmetry, the relative decay rates for the ideally mixed vector nonet as seen in Table 4. The last column shows the expected relative rates for the case of exact nonet symmetry, $g_1 = g_8$, after removal of the momentum dependence $\sim p_V^3$. The results, even without considering mass breaking effects, are quoted to be in fair agreement with nonet symmetry. Once we weight the observed channels by the charge states available, they yield an acceptable $g_1 = 0.75 \pm 0.21$, $g_8 = 0.47 \pm 0.05$.

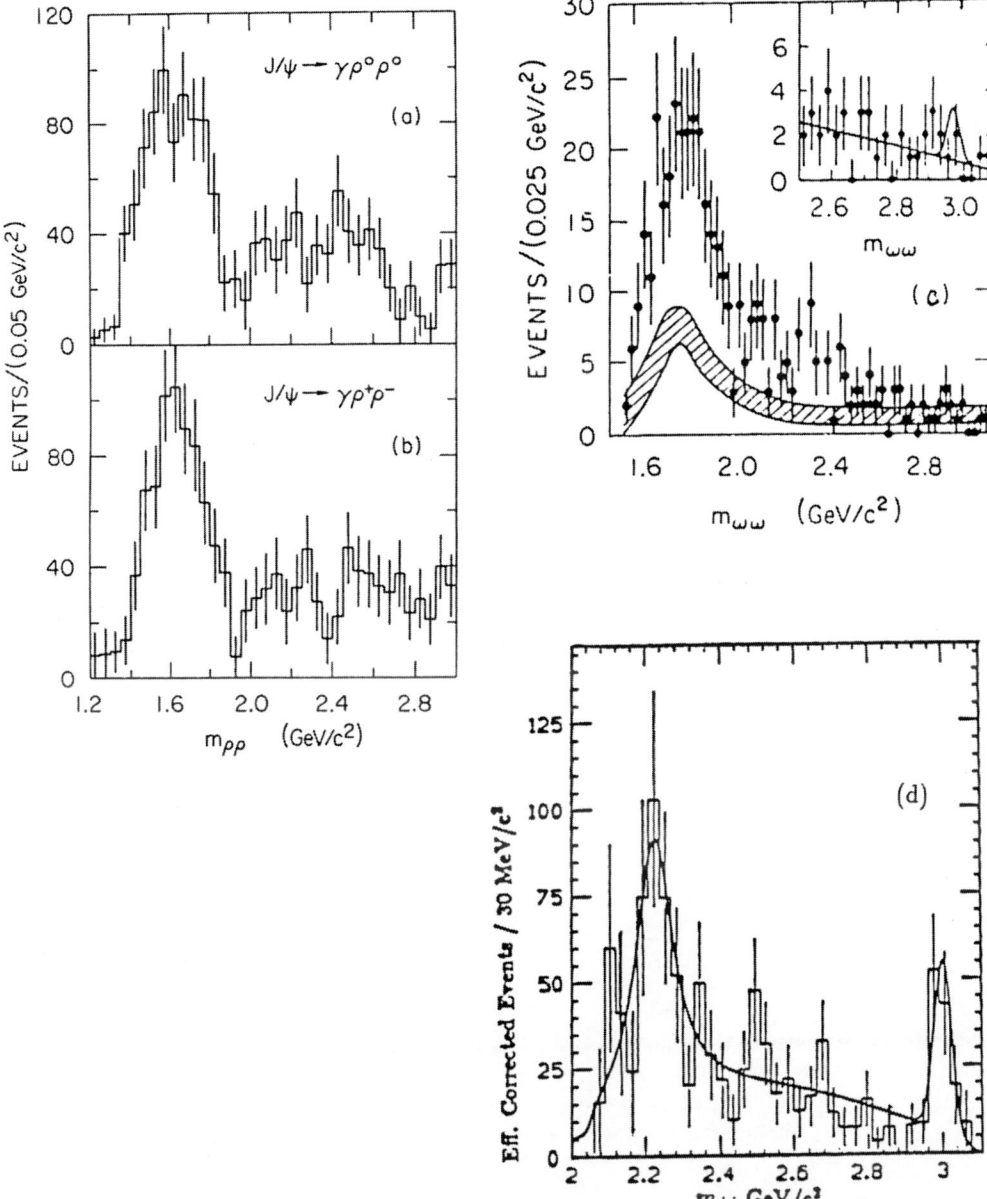

Fig. 5. Vector–vector enhancements above the threshold in the radiative decay process $J/\psi \to \gamma VV$, for (a) $\rho^0\rho^0$, (b) $\rho^+\rho^-$, (c) $\omega\omega$, (d) $\phi\phi$ (all from MARK III data; similar results are in preparation for $K^{*+}K^{*-}$ and $K^{0*}\overline{K}^{0*}$). All these enhancements appear to have $J^{PC} = 0^{-+}$. The DM2 collaboration has similar results (cf. Ref. 28).

This is a remarkable piece of consistency evidence, although we suspect that diagram h comes into play in the process

$$\eta_c \to \omega\phi \; ;$$

the MARK III result,[27] $B(J/\psi \to \gamma\eta_c,\ \eta_c \to \phi\omega) < 0.1 \times 10^{-4}$ is based on clear evidence for the $\gamma\phi\omega$ final state in the η_c mass region, with the implication that

281

$g_1 \neq g_8$. (Note, also, that preliminary MARK III results disagree with the $\rho\rho$ and $\phi\phi$ rates; a final experimental settlement remains outstanding.)

In another recent development, MARK III measurements[29] of the electromagnetic decays $J/\psi \rightarrow VV$ permit the measurement of charge form factors. The decay of the C eigenstate J/ψ into pairs of chargeless vector mesons ($\rho^0\rho^0, \phi\phi, \omega\omega$) is forbidden, but the $\rho^+\rho^-, K^{*0}\overline{K}^{*0}$, and $K^{*+}K^{*-}$ final states are permitted via graph b. In the absence of electric dipole and quadrupole moments (a natural approximation for spin-aligned quark pairs with $\ell = 0$), the results for the branching ratios can be used to evaluate charge form factors for the vector mesons in question

$$B(J/\psi \rightarrow V\overline{V}) = \frac{\beta^3}{4}|F_{00}(s)|^2 B(J/\psi \rightarrow \mu\mu) \, .$$

Table 5 shows the form factors evaluated at the s values corresponding to the vector meson poles.

Mode	Branching Ratio	$|F_{00}(m_V^2)|^2 \times 10^2$	$\frac{|F_{00}(m_V)^2|^2}{|F_{00}(m_P)^2|^2}$
$J/\psi \rightarrow \rho^+\rho^-$	$9.4 \pm 1.0 \pm 2.5$	$7.8 \pm 0.8 \pm 1.2$	6.55
$J/\psi \rightarrow K^{*+}K^{*-}$	$6.7 \pm 1.2 \pm 2.5$	$6.7 \pm 1.2 \pm 2.5$	3.21
$J/\psi \rightarrow K^{*0}\overline{K}^{*0}$	$2.9 \pm 0.4 \pm 0.6$	$2.9 \pm 0.4 \pm 0.6$	3.26

The last column shows a comparison between spin-aligned vector and spin-cancelling pseudoscalar $q\bar{q}$ mesons. The measured difference in charge distributions attests to the remarkable information content of the clean process studied. It makes sense that the "size ratio" of ρ to π, dominated by the color magnetic term, is larger than in the systems containing the heavier s quark, K^* to K. The numerical equality for the two kaon charge states, if not fortuitous, is remarkable.

Scalar Mesons

Scalar mesons in the 0.9 to 2 GeV/c^2 mass range are expected to decay prominently into two pseudoscalars. Evidence from strong interaction experiments gives the lowest-mass states, the isoscalar $f_0(975)$ and the isovector $a_0(980)$, of opposite G parity, thus letting the former decay into $\pi\pi, K\overline{K}$, the latter into $\eta\pi, K\overline{K}$. Both have small apparent widths, of 34 and 55 MeV/c^2, respectively.

We will use our $P_{s\bar{s}}$ projection operators, looking for strange quark content in singly disconnected processes via the reactions

$$J/\psi \rightarrow \phi\pi^0\pi^0 \, , \quad \phi\pi^+\pi^- \, ,$$
$$J/\psi \rightarrow \phi\pi^0\eta \, ,$$
$$J/\psi \rightarrow \phi\eta\eta \, .$$

The first and third of these may indicate the presence of isoscalars via graphs a),b), the second that of the isovector $a_0(980)$ via (the isospin-forbidden, therefore electromagnetic) graph b) only. The MARK III data were subjected to a coupled-channel fit[30] searching for $f_0(975)$ in the $\phi\pi^0\pi^0$, $\phi\pi^+\pi^-$, and ϕK^+K^- channels; they show a sizeable signal, in the $\phi\pi^0\pi^0$ and $\phi\pi^+\pi^-$ channels (cf. Fig. 6), but show remarkably wider structure for $f_0(975)$ than the width given in the Particle Data Book;

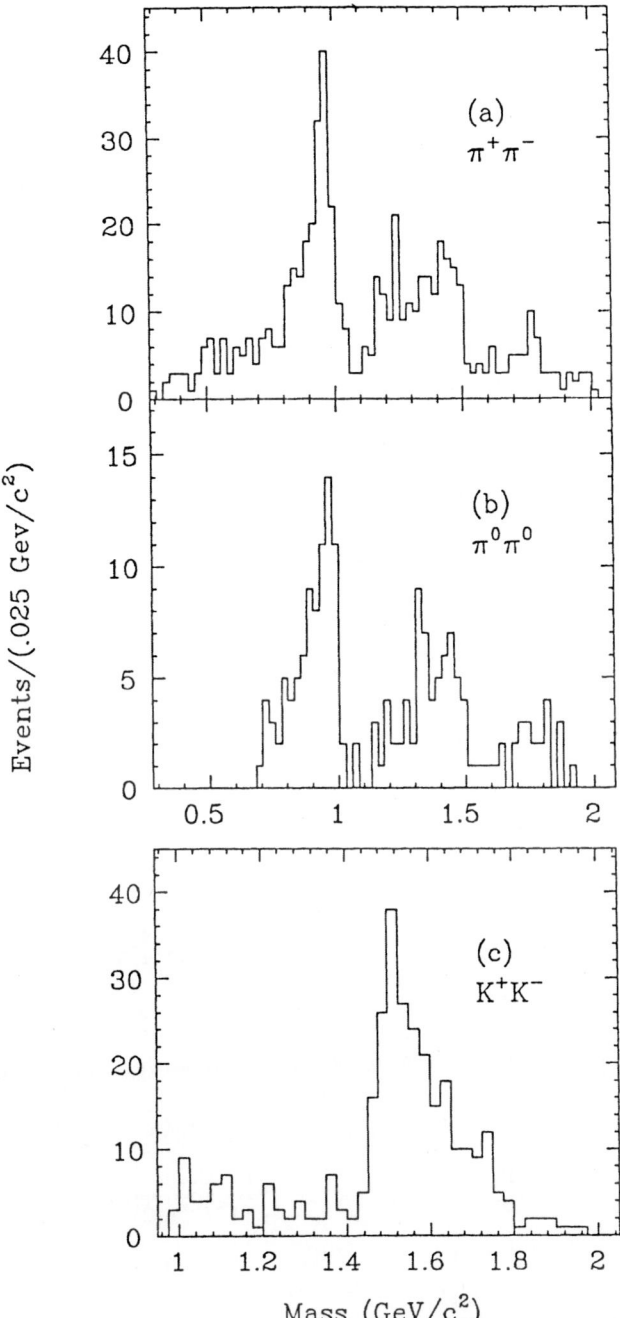

Fig. 6. Mass distribution for $\pi\pi$ or $K\overline{K}$ systems emitted together with a ϕ meson in exclusive J/ψ decays. The $f_0(975)$ state is dominant in the $\pi\pi$ systems, but is below threshold for $K\overline{K}$. The low-mass $\pi^0\pi^0$ spectrum was truncated, *via* a $p(\pi^0)$ cut, to avoid backgrounds from $J/\psi \rightarrow \phi\eta(\eta \rightarrow \pi^0\pi^0\pi^0)$ decays. Note, however, that the Crystal Ball collaboration set a stringent upper limit on radiative J/ψ decay toward a $\pi^0\pi^0$ state with $0.5 \leq m \leq 1.0$ GeV/c^2 and $\Gamma \leq 100$ MeV/c^2, of 1.3×10^{-5}. A coupled-channel fit to these three channels finds a large width to $f_0(975)$.

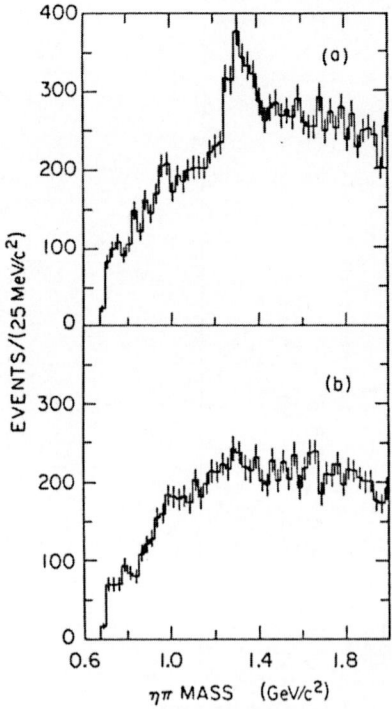

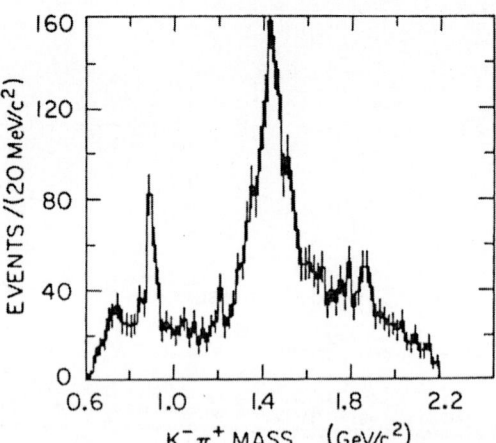

Fig. 7. Invariant-mass distributions emitted in J/ψ decays, into the channel $J/\psi \to \pi\pi(\eta\pi)$. (a) two of the π's, couple to form a ρ mass, (b) no two π's are compatible with a ρ mass (MARK III).

Fig. 8. $(K\pi)$ systems emitted in conjunction with K*(894) in hadronic J/ψ decays. There is no indication of the reported (broad) $I = \frac{1}{2}$ scalar $K(1300)$, while $K^*(894)$ and $K_2^*(1425)$ are well observed (MARK III).

$\Gamma_{\pi\pi} > 200$ MeV/c^2. It is the effect of the kinematic threshold of $K\overline{K}$ production that causes the appearance of the narrow cusp structure. Inclusion of the channel $\phi K^+ K^-$ (Fig. 6c) causes the underlying normal hadronic width to emerge.

There is no discernible signal in the (electromagnetic) $\phi\pi^0\eta$ channel.[31] Despite the consequent suppression, we infer that, while the $f_0(975)$ is largely $s\bar{s}$-quark based, $a_0(980)$ shows little $s\bar{s}$ affinity. We regard the observation of $f_0(975)$ in this situation, with a normal hadronic width, as a strong argument against its interpretation as a structure more complex and extended than an $s\bar{s}$ pair.

For a further check on the isovector $a_0(980)$, we use the operator $P_{q\bar{q},I=1}$ by means of the reaction

$$J/\psi \to \rho\eta\pi .$$

Only a small signal is visible in Fig. 7 for a_0 production, with

$$B(J/\psi \to \rho a_0)B(a_0 \to \eta\pi) < 4.4 \times 10^{-4} .$$

Similarly, very modest structure can be seen at the $f_0(975)$ mass when we apply the

284

operator $P_{q\bar{q},I=0}$ (see Fig. 10b below) *via* the reaction

$$J/\psi \to \omega\pi\pi \ ,$$

reinforcing our view that $f_0(975)$ is mostly s-quark based. The relative clarity of the f_0 *vs.* a_0 appearance in these channels may well indicate a different dynamical origin of the two states. This would be well compatible with recent data from $\gamma\gamma \to$ scalar formation.[32] Still, the major common feature of both isosinglet scalars is the relative weakness of their appearance when compared with appropriate pseudoscalars.

To check on other presumed members of the scalar nonet, look at the $K\pi$ mass spectrum recoiling against the vector $K^*(892)$ (Fig. 8): no indication of structure is seen between the well-known vector and tensor K^* states; note, however, that $K_0^*(1350)$ may be too broad to lead to a recognizable signal here.

Applying again the operator $P_{s\bar{s}}$ to the higher-mass region, Figs. 6(a),(b) show similar activity in the 1300 to 1400 MeV/c^2 region, but a spin-parity analysis has yet to be done—so we have no indication whether they can be related to the **1** singlet $f_0(1300)$. The $\eta\eta$ channel shows structure (Fig. 9(c)) that may well constitute evidence for a new scalar state, $f_0(1590)$ [or $G(1590)$]; it will be discussed in the next-to-last section below. The reported branching fraction is $\sim 3.4 \times 10^{-4}$.

For completeness, it should be mentioned that the LASS Collaboration[33] suggests the existence of a scalar state for which it observes evidence in the reaction $K^-p \to (K_S K_S)$, as the SU_3 singlet instead of $f_0(975)$.

Axial Vector Mesons: $J^{PC} = 1^{++}$

To address the question of ideal mixing of the $J^{PC} = 1^{++}$ isosinglets, apply the $P_{s\bar{s}}$ and $P_{q\bar{q}}$ operators onto data involving $V(K\overline{K}\pi)$ and $V(\eta\pi\pi)$ final states. Figures 3 and 4 had shown conclusive evidence for structure in the $(\phi,\omega)\left(\begin{smallmatrix} K\overline{K}\pi \\ \eta\pi\pi \end{smallmatrix}\right)$ channels at $K\overline{K}\pi$ or $\eta\pi\pi$ masses around those of $f_1(1285)$ and $f_1(1420)$. However, there is no $f_1(1285)$ signal in $\omega(K\overline{K}\pi)$ (speaking in favor of its $s\bar{s}$ content), and no $f_1(1420)$ signal in $\phi(K\overline{K}\pi)$; this would argue in favor of ideally mixed axial vector isoscalars—with the odd feature that the heavier state appears to be based on non-strange quarks. It is compatible also with the recent results on $\gamma\gamma^* \to K\overline{K}\pi$, which confirm a $J = 1$ character for the 1420 MeV/c^2 enhancement, but see a signal that couples to $\gamma\gamma$ more strongly than an $s\bar{s}$ structure would permit.[20]

The trouble with this interpretation is the evidence from the $V(\eta\pi^+\pi^-)$ channels, where some activity in the two f_1 regions recoils against both ω and ϕ—with $\omega(\eta\pi\pi)$ furnishing the most prominent signal (recall that $\eta\pi^+\pi^-$ is not likely to stem from $s\bar{s}$ hadronization) at a mass some 30 MEV/c^2 below the $f_1(1420)$ region. The only way out of this can of worms appears to be a tentative scenario that can be verified exclusively by a full phase shift analysis[35] of at least the numerically significant signals in all $V\left(\begin{smallmatrix} K\overline{K}\pi \\ \eta\pi\pi \end{smallmatrix}\right)$ channels: in addition to an approximately ideally mixed 1^{++} nonet, there are two radially excited η, η' states in this region. The existence of these pseudoscalars would also help clear up the puzzling evidence in the radiative channel: it might eliminate (in $\gamma(\eta\pi\pi)$) the only candidates for quantum numbers in the radiative process that cannot be formed out of transverse gluons; it would provide an explanation for the outsize shape of $\iota(1460)$; and it might reconcile the mass differences between the peaks around 1.4 GeV/c^2 in $\eta\pi\pi$ and $K\overline{K}\pi$. It could further accommodate the reported pseudoscalar signal in $J/\psi \to \gamma(\gamma\rho)$.

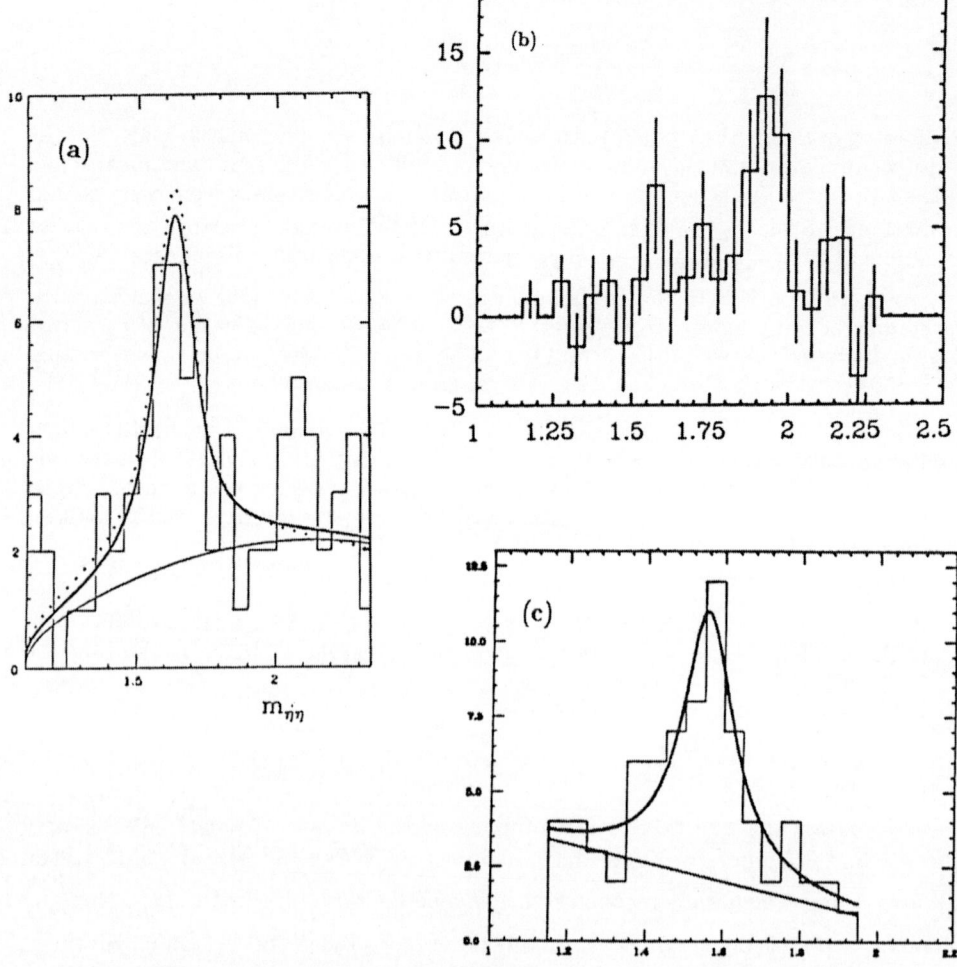

Fig. 9. Invariant-mass distribution for $\eta\eta$ systems emitted in J/ψ decays jointly with (a) a real photon, (b) an ω meson, (c) a ϕ meson. The statistics are not good enough to make a spin-parity determination for the enhancement at $m(\eta\eta) \simeq 1590$ MeV/c^2.[34] Note that the fit to (a) does not yield the precise parameters of $G(1590)$, but is centered some 35 MeV/c^2 below.

Note also that this scenario would not be in conflict with any of the known evidence from hadron-induced production experiments, and would indeed reconcile some of the conflicting evidence presented there.

For the isovector $a_1(1270)$ with its dominant $\rho\pi$ decay, there is unfortunately no hope to disentangle it from the tensor $a_2(1320)$ signal: the latter is seen to dominate the $\eta\pi$ distribution in Fig. 7, so that its fivefold more abundant decay into $\rho\pi$, which is kinematically less well defined, will obscure any 1^{++} signal due to $a_1(1270)$ decay.

Tensor Mesons $J^{PC} = 2^{++}$

The tensor nonet shows up most prominently in the radiative channel $J/\psi \to \gamma T$, as expected from perturbative QCD calculations. However, the leading 3-gluon decay of J/ψ also permits detailed studies of quark content and symmetry structure

286

via the set of reactions

$$J/\psi \to VT \ .$$

Evidence for almost ideal singlet-octet mixing can be gleaned from an application of the operators $P_{q\bar{q}}$, $P_{s\bar{s}}$ to the final state candidates

$$J/\psi \to \begin{pmatrix} \omega \\ \phi \end{pmatrix} \begin{pmatrix} \pi\pi \\ K\overline{K} \end{pmatrix} \ .$$

This is done in Figs. 10 and 11, respectively. Let us check on ideal mixing between the isoscalars $f_2(1270)$ and $f_2(1525)$: The Particle Data Book mentions $B(f_2(1270) \to \pi\pi) = (84.3 \pm 1.2)\%$, $B(f_2(1270) \to K\overline{K}) = (2.9 \pm 0.2)\%$—certainly indicative of a fairly pure u, d quark base; for $f_2(1525)$ it only says the $K\overline{K}$ decay is dominant. Our (u, d) and s quark projection operators show, in Figs. 10 and 11, the quality of this nonet's separation of u, d- from s-quark-based isoscalars: the MARK III data[36] permit the inferences

$$\frac{B(J/\psi \to \omega f_2(1270))}{B(J/\psi \to \phi f_2(1270))} = 14 \pm 7$$

$$\frac{B(J/\psi \to \phi f_2(1525))}{B(J/\psi \to \omega f_2(1525))} > 4.8 \quad (90\% \text{CL.}) \ .$$

Although not quantitatively stringent, this is strongly supported by the small rate for $J/\psi \to \gamma f_2(1525)$, limited by the poorly measured $\omega f'$ decay $(f_2(1525) \to \pi\pi)$.[37] Within the few percent admixture due to quark mixing *via* gluonic intermediate states ($H^{(4)}$ of the meson Hamiltonian), we will feel justified in using $f_2(1270)$ and $f_2(1525)$ as ideally mixed tensor mesons for a pragmatic definition of the $P_{q\bar{q}}$, $P_{s\bar{s}}$ operators. The projections of the K^+K^- mass spectra onto the $(q\bar{q})$ (Fig. 11a) and $(s\bar{s})$ (Fig. 11b) bases show a pattern that gives clear evidence for another structure, $f_2(1720)$, to be associated mainly with the $q\bar{q}$ basis; indeed, there is a distinct two-peak structure in the radiative channel (Fig. 11a). Yet, the analogous projection of this $\pi^+\pi^-$ invariant-mass spectrum onto the light-quark bases produces little if any signal (Figs. 9(a),(b)). The MARK III Collaboration reports a clear $J^{PC} = 2^{++}$ assignment for $f_2(1720)$, and identified this state with the signal originally observed by the Crystal Ball Collaboration[38] in the mode $J/\psi \to \gamma\eta\eta \to 5\gamma$ and called $\Theta(1660)$.

The latest evidence from MARK III and DM2 is more compatible with the notion that the $\eta\eta$ enhancement initially called Θ (and reportedly preferring a $J^{PC} = 2^{++}$ assignment over 0^{++} by 270 : 1) is *not* the same state as $f_2(1720)$ observed decaying into K^+K^-, $K_S K_S$, and $\pi^+\pi^-$, with a clear $J^{PC} = 2^{++}$: Fig. 9(a) shows the new $\eta\eta$ mass spectra,[39] together with the evidence concerning spin and parity: although the evidence is not entirely compelling, $J^{PC} = 0^{++}$ is an entirely acceptable assignment. This, however, makes the state suspiciously similar to the signal observed in the hadronic channel $\pi^- p \to \eta\eta n$ by the GAMS Collaboration.[40] Fig. 9(c) shows a preliminary set of MARK III data on the hadronic decay $J/\psi \to \phi\eta\eta$[34] with a fit using the GAMS parameters of this state (G1590). Although only a new set of MARK III data is expected to settle the question, we believe it is fair to say that $f_2(1720)$ has not been identified in the final state $\eta\eta$; and that the Crystal Ball observation *may* be the first glimpse of $G(1590)$.[41]

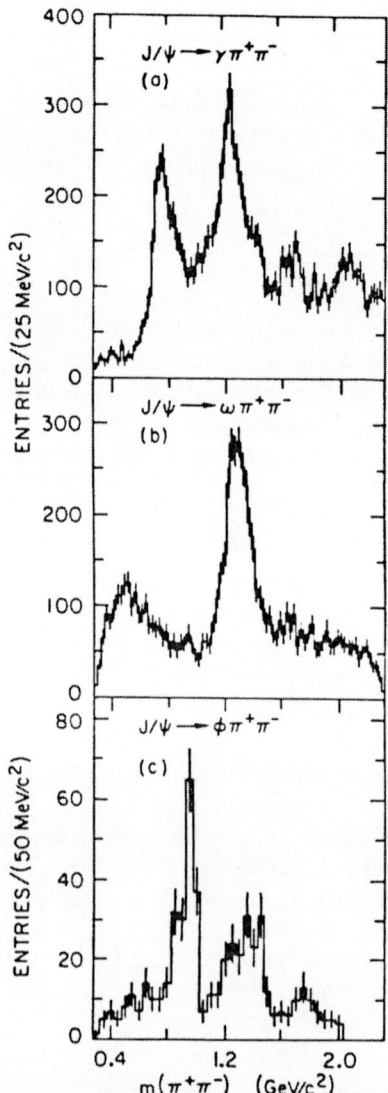

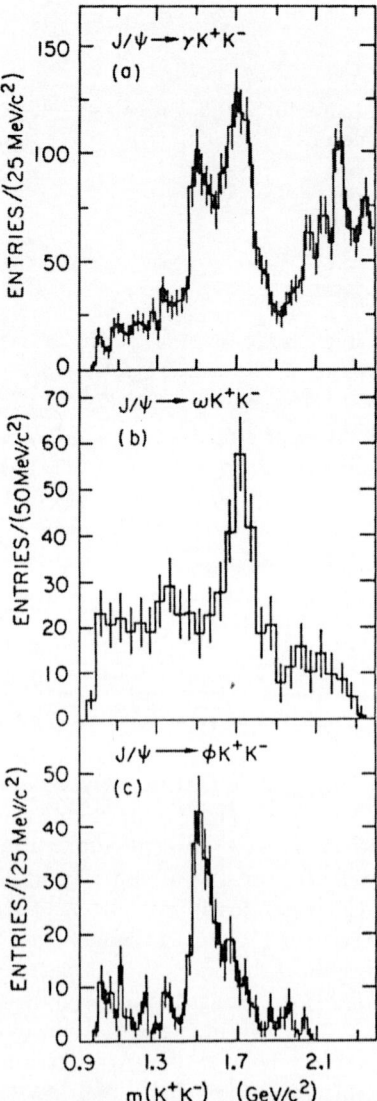

Fig. 10. $\pi^+\pi^-$ invariant mass distributions from J/ψ decays into (a) $\gamma\pi^+\pi^-$, (b) $\omega\pi^+\pi^-$, (c) $\phi\pi^+\pi^-$. Note the qualitatively different features of these three graphs: (a) is dominated by ρ feedthrough (one γ is lost in the decay $J/\psi \to \pi^+\pi^-\pi^0$, $\pi^0 \to 2\gamma$), $f_1(1275)$, $\Theta(1720)$, and an unidentified state around 2.1 GeV/c^2; (b) indicates a $q\bar{q}$ basis for $f_1(1275)$; (c) affirms the $s\bar{s}$ basis for $f_0(975)$, while the remaining structure remains uninterpreted (MARK III).

Fig. 11. K^+K^- systems emitted together with the vectors (a) γ, (b) ω, (c) ϕ in exclusive J/ψ decays. Note how P_{qq} (b) and P_{gg} (c) pick out different systems observed in the radiative (a) decay. Also note the $X(2230)$ observation by P_{gg} only (a). Similar spectra exist for $K_S K_S$ systems from MARK III and DM2.

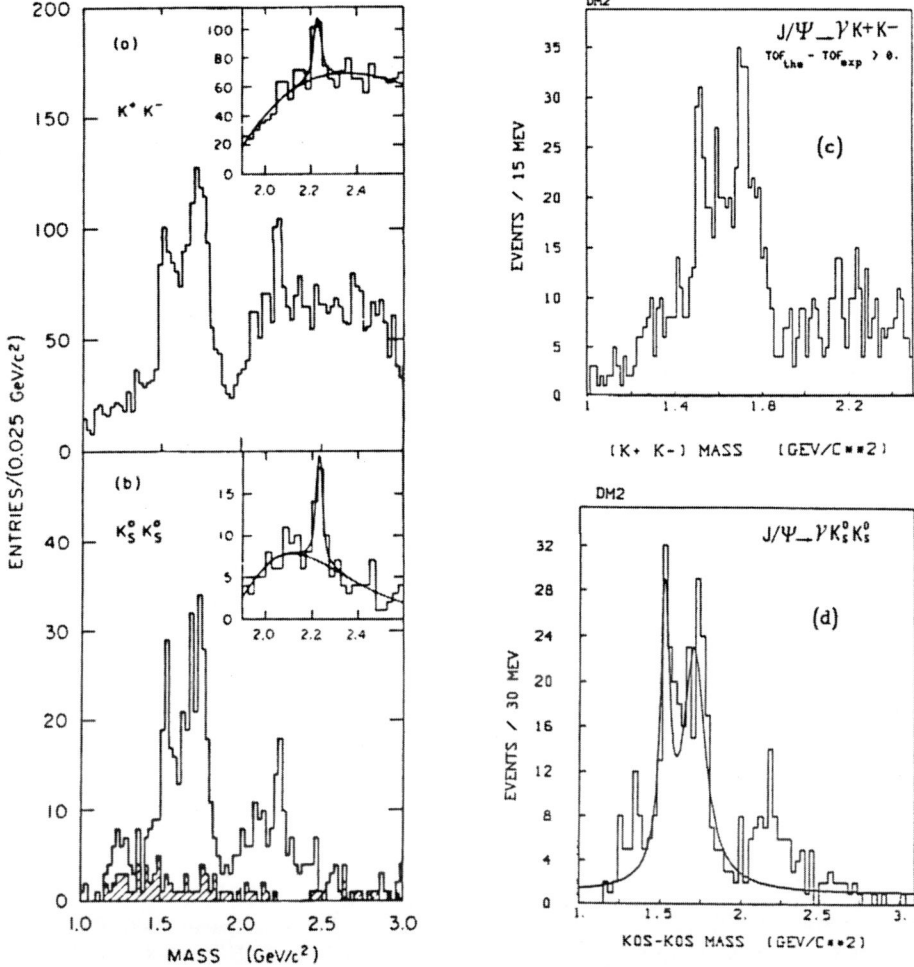

Fig. 12. (a),(b): MARK III observations of $[\xi] \sim X(2230)$ in $J/\psi \to K^+K^-, \gamma K_S K_S$; (c),(d): DM2 distribution of K^+K^-, $K_S K_S$ mass spectra in the same reaction. Notice difference in scales.

At higher mass, Fig. 11a shows the state ξ (now $X(2230)$), observed first by the MARK III Group[42] but not seen in the DM2 detector (as illustrated in Fig. 12).[28] Its quantum numbers appear to be 2^{++} (or possibly 4^{++}, less likely). It is seen in the K^+K^- and $K_S K_S$ final states in J/ψ radiative decay, but we do not find it in the hadronic decay modes $J/\psi \to \phi K\overline{K}$ or $\to \omega K\overline{K}$. Its most intriguing feature, a narrow width $\Gamma_\xi < 30$ Mev/c^2, remains unchallenged.

Recent evidence has given hints of narrow structure in this mass range from several hadronic reactions: the LASS Collaboration reports[33] on a $K_S K_S$ enhancement with $J^{PC} = 2^{++}$ or 4^{++} in the reaction $K^- p \to K_S K_S n(K^+K^- n)$—hinting at an $s\bar{s}$-based state. The GAMS experiment observes[43] a narrow enhancement at ξ mass, decaying into $\eta'\eta$.[44] Could they be manifestations of $X(2230)$? Recall that the reaction $J/\psi \to \gamma X$ defines the projection operator $P_{1(SU_3)}$ (Table 2), but a $|\eta\eta'\rangle$ state is predominantly $8(SU_3)$, as pointed out by Lipkin.[45] Consequently, a look at the $\eta\eta'$ mass spectrum in the MARK III data on $J/\psi \to \gamma\eta\eta'$ has been verified as structureless,[36] leading to a quoted upper limit of

$$B(J/\psi \to \gamma\xi, \ \xi \to \eta\eta') \leq 10^{-5}.$$

If, on the other hand, we were to identify the LASS signal with $X(2230)$, the resulting $s\bar{s}$ interpretation of the state would need a large L value ($L = 3$) to explain the narrow width: but we stated above that the highly local radiative decay process disfavors high L values. In other words, $X(2230)$ remains a question mark. A gluonium interpretation is still not ruled out, but the lack of confirmation in other decay channels is worrisome, and awaits further data.

Lastly, we recall the observation by a BNL-CCNY team[46] of strong enhancements, above threshold, in the $\phi\phi$ system of the interaction

$$\pi^- p \to \phi\phi n \to 4Kn \ .$$

The group can best fit the data by means of three broad resonant $\phi\phi$ states it calls $g_T^{(1,2,3)}$, claiming gluonium candidacy due to the disconnected production graph

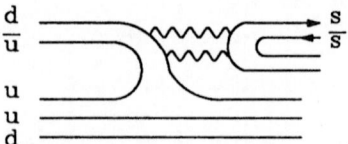

Given the $J^{PC}(g_T) = 2^{++}$ assignment for all three candidate states, the radiative decay channel is a natural place to look for confirmation of this existence. Both MARK III[47] and DM2[48] have reported on extensive analyses of the process $J/\psi \to \gamma\phi\phi$; we have shown one resulting mass spectrum in Fig. 5d: the predominant J^{PC} value of the threshold enhancement is 0^{-+}. Further, we would expect a $s\bar{s}$-based resonance to show up, quantum numbers permitting, in the $\eta\eta$, K^+K^-, $K_S K_S$, and/or $K^* \overline{K}^*$ final states. In fact, Figs. 9(a),11(a) and unpublished MARK III data on the remaining channels do not support the emergence of broad tensor states at g_T mass values in radiative J/ψ decay.

DO GLUONIA SHOW UP IN RADIATIVE DECAYS?[49]

Having made the rounds of the $\ell = 0,1$ quark-based nonets in the mass range accessible to charmonium decay, we use the knowledge gained to ask the most critical questions our review might shed some new light on: Does our projection operator P_{gg} find evidence for the existence of *gluon*-based hadrons? Using the term gluonium for a state built up out of *two* valence gluons (and leaving "glueball" as a generic term for n-gluon condensates), we recall that diagrams c and e above are expected to favor gluonium formation.

A priori, any meson state emerging more prominently from J/ψ radiative decay than in hadronic interactions must be seen as a gluonium candidate, *if*

- it is a singlet in SU$_3$ (color, flavor);

- it has $J^{PC} = 0^{++}, 0^{-+}, 2^{++}$ (the preferred assignments in the mass range < 2.2 GeV/c^2, for states built out of two transverse gluons);

290

- its decay modes are compatible with *a priori* flavor-insensitive couplings of the gluons;

- it finds no natural home within the appropriate $q\bar{q}$-based nonet.

- it is not prominently observed in $\gamma\gamma$ collisions (see below).

In the absence of firm theoretical guidance concerning masses and widths, we remain aware of heuristic arguments that place the scalar gluonium ground state somewhere between 0.5 and 1.5 GeV/c^2 with indeterminate width (cf. the section on scalar quarkonia), the pseudoscalar ground state between 1.2 and 1.6 GeV/c^2, the 2^{++} tensor ground state between 1.5 and 2.2 GeV/c^2—both of the latter with widths that range from 10 MeV/c^2 to the normal hadronic range, ~ 150 MeV/c^2.

Figure 2 had shown that no exotic signature inaccessible to $q\bar{q}$ configurations can be expected at these masses. Furthermore, the absence of proven $J^{PC} = 1^{++}$ (or 1^{-+}) states in radiative J/ψ decay prejudices us to concentrate on states that can be formed from two massless (*i.e.*, transverse) gluons.

From these selective criteria, what candidate states remain viable?

Scalars first: Recall that $(gg,\ \ell = 0)$ permits formation of a scalar object, whereas $(q\bar{q},\ \ell = 1)$ is needed for the lowest-mass available quark-based scalar. Is there a possible signal at masses that are below m_ρ (*a fortiori*, below $m(S_0(975))$)? Take the P_{gg} operator and look at $J/\psi \to \gamma\pi\pi$, $\phi\pi\pi$. The dipion distributions in Figs. 10(a),(c) and in Figs. 6(a),(b) do not support the existence of an identifiable state at low mass values. An unexplained enhancement at masses ~ 1.2 GeV/c^2 does show up in the $\pi^+\pi^-$ and $\pi^0\pi^0$ spectra, but we do not yet have a detailed analysis of this feature.

Let me remind you, however, that the possibility cannot be ruled out that the original Crystal Ball[38] $\eta\eta$ enhancement and $G(1590)$[40] from the GAMS group are in fact the same object. Although unseen in other channels, Bjorken's argument[50] in favor of $(gg) \to \eta\eta$ decays may make it a scalar gluonium candidate.

Pseudoscalars next: there are definitely too many pseudoscalars to have us see them accommodated in the $q\bar{q}$ nonets of lowest mass. While several of the enhancements in the $K\overline{K}\pi$ and $\eta\pi\pi$ mass spectra (Figs. 3 and 4) are likely candidates for a first radial excitation of η and η', our preferred scenario (described above) is the identification of $\iota(1460)$—seen only in $J/\psi \to \gamma(K\overline{K}\pi)$, but in three charge modes there—as a pure and unadulterated gluonium. To reinforce (or discredit) this notion, the MARK III group is presently performing a detailed spin-parity analysis of the relevant radiative as well as hadronic channels.

We wish to dissociate this $\iota(1460)$ from the states that are seen to decay to $(\gamma\rho)$— at somewhat lower masses—and to $\rho\rho$, $\omega\omega$, somewhat more massive. While the latter do originate from two-gluon hadronization, they appear to be VV threshold enhancements arising in this channel due to the specific helicity structure of this local interaction.

2^{++} Tensors finally: this channel, preferred for higher-mass two-gluon composites, contains three gluonium candidates: $\Theta(1720)$, $\xi(2230)$, and the $g_T(2300 - 2500)$ states. Θ—or, better, $f_2(1720)$, reviewed in the tensor section above, is produced in radiative as well as hadronic J/ψ decays; its strong appearance in the latter channel (preferentially in $J/\psi \to \omega K\overline{K}$, Fig. 11b) indicates a quark base. Moreover, while the relative signals in the $\omega K\overline{K}$ and $\phi K\overline{K}$ channels are compatible with $f_2 \in 1(SU_3)$, the observed decay channels are not: the $f_2(1720)$ peaks in Figs. 11(b),(c) translate into

$$\frac{\Gamma(f_2 \to \pi^+\pi^-)}{\Gamma(f_2 \to K\overline{K})} = \frac{1}{6} \neq \frac{3}{4}\left(\frac{P\pi}{P_K}\right)^5 > 1 \,,$$

as predicted by naive QCD. To be sure, no apt home can be found for $f_2(1720)$ in the basic $q\bar{q}$ multiplets, so its status remains ambiguous and awaits a larger data set as well as a final resolution of the $f_2 \to \eta\eta$ vs. $G \to \eta\eta$ questions.

Equally ambiguous is the candidacy of ξ, or $X(2230)$, observed only in the K^+K^- and $K_S K_S$ final states. We fail to find signals in the $\pi\pi$, $\eta\eta$, or $K^*\overline{K}^*$, $\phi\phi$ plots: the latter two channels definitely prefer $J^{PC} = 0^{-+}$ for enhancements at this mass. The arguments that tend not to let us dismiss its candidacy are two: the small width (< 30 MeV/c^2), and the fact that P_{gg} definitely discriminates against higher-spin states ($L = 3$) which have been invoked to explain this width.[51]

In the same energy range, Etkin, et al.[46,52] have discussed the case for gluonium interpretation of the three g_T states which best fit their $\phi\phi$ mass spectrum, as discussed in the previous section. While a look at Fig. 5d shows the lack of corroborating evidence for the emergence of these states in radiative J/ψ decay— and therefore the failure of our P_{gg} operator to make them stand out—we prefer to interpret the strong turn-on of the $\phi\phi$ signal in this disconnected topology along different lines: while $J/\psi \to \gamma\phi\phi$, just as in the $\rho\rho$, $\omega\omega$, and $K^*\overline{K}^*$ final-states, sees its highly local two-gluon hadronization prefer the $J^{PC} = 0^{-+}$-configuration, the more peripheral gg production in the process $\pi^- p \to \phi\phi n$ could plausibly hadronize preferentially in the $J^{PC} = 2^{++}$ channel. This implication of the different helicity structure of the two processes is presently under investigation.

An additional interesting way of comparing quark affinity to gluon affinity of a given meson is provided by its possible observation in $\gamma\gamma$ collisions via the graph

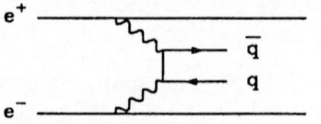

Gluonium production in this reaction is clearly suppressed, since it requires a quark loop:

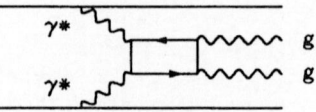

The comparison can be used as a gluonium quality criterion[53,54]

$$R_\gamma(X) = \frac{\Gamma(J/\psi \to \gamma X)}{\Gamma(\gamma\gamma \to X)} \,.$$

Large values for $R_\gamma(X)$ make X a good gluonium candidate. Small values rule it out. Similarly, the comparison of the action of the P_{gg} and the $P_{q\bar{q}}$ operators defines

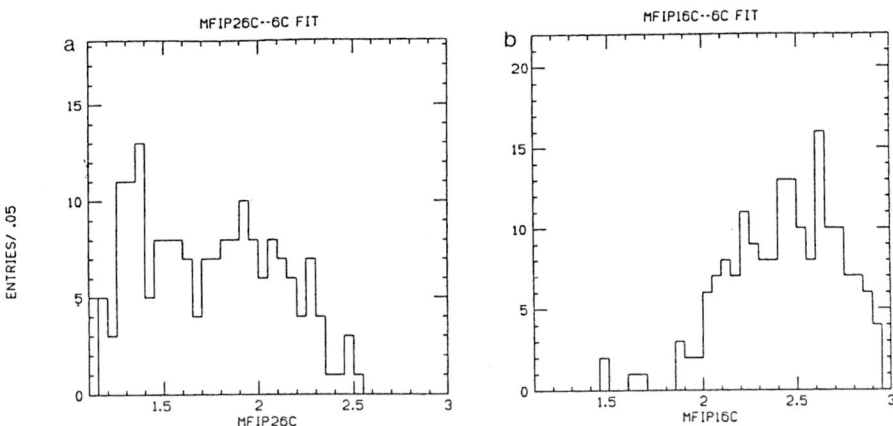

Fig. 13. Invariant-mass distributions for $(\phi\pi^0)$ systems recoiling against a π^0 in the hadronic decay $J/\psi \to \phi\pi^0\pi^0$. (a) The higher-momentum π^0 was chosen as the recoiling particle. (b) The lower-momentum π^0 is the recoil system.

a figure of merit:

$$R_V(X) = \frac{\Gamma(J/\psi \to \gamma X)}{\Gamma(J/\psi \to V^0 X)}.$$

Again, the larger $R_V(X)$, the more credible a gluonium candidacy of the state X. A quantitative comparison, including, for "calibration" purposes, $\eta'\eta$, and π^0, makes $\iota(1460)$ clearly stand out as a most promising gluonium candidate.

EXOTIC CANDIDATES: INFORMATION FROM J/ψ DECAY?

We showed in Fig. 2 that certain J^{PC} configurations are not allowed for $q\bar{q}$-based (or even gg-based) mesons. 0^{+-} would need a qqg or $q\bar{q}q\bar{q}$ interpretation; 1^{-+} is accessible to gg or the above. Recent hadron physics evidence[55] exists suggesting the observation of a $J^{PC} = 1^{-+}$ state decaying into $\eta\pi$. If we accept current theoretical speculation,[56] it might be a manifestation of the $10-10^*SU_3$ representation of a four-quark state. This argument receives some support from another recent sighting of an odd bird,[57] a $\phi\pi^0$ enhancement that, although of $J^{PC} = 1^{--}$, cannot be classed a $q\bar{q}$ state due to the absence of a $\omega\pi$ signal in its production reaction

$$\pi^- p \to (\phi\pi^0)n .$$

This state might be the isovector non-strange state of opposite charge conjugation predicted by the $q\bar{q}q\bar{q}$ algebra.

Notwithstanding our prejudice that 4-quark states are not likely to be copiously available from as local an interaction as charmonium decay, we investigated our data for the $\phi\pi$ state (the $\eta\pi^0$ state again falls victim to our inability to see all-neutral final states). The full sample of 6.3 M MARK III J/ψ events was searched via the

$P_{I=1}$ operator, using the hadronic decay reaction

$$J/\psi \to \pi^0(\phi\pi^0)$$

from $K\overline{K}(4\gamma)$ final states. The distributions for $(\phi\pi^0)_{\text{high}}$ and $(\phi\pi^0)_{\text{low}}$ masses are shown in Fig. 13. While a quantitative limit is yet to be extracted, it is clear that there is no signal for the reported $I = 1$ vector state at a level $\lesssim 10^5$, or a factor of some 10^3 below the leading hadronic VP decay $J/\psi \to \rho\pi$ of 1.27%.

A SCORE SHEET

I will leave you with a simplified score sheet which demonstrates both the usefulness and the limitations of charmonium decay studies as a means of shedding light on the basic questions posed by low-mass meson spectroscopy—the simplest locus for the study of light-quark bound-state dynamics.

First, general comments: Recent charmonium decay data have added significantly to our understanding by clarifying the following features (which must still be seen as open to improvement by a more complete set of data):

- Isoscalar mesons like to appear, when there is no added complication that infringes on the system as mass eigenstates, such that the quark basis is either all non-strange or all strange.

- Admixture of valence gluons in the wave function appears to be uncalled for by present evidence. Specifically, gg configuration mixing with $q\bar{q}$ states has *not* been established—contrary to earlier indications.

- There is no established low-mass, quark-based meson that has singlet status in SU_3 (flavor). Experimental evidence does not exclude $\Theta(1720)$ to be a candidate for this situation.

- Radiative J/ψ decay evidence for hadronization of two-gluon systems yields no conclusive evidence for the action of longitudinal gluons: the observed light-meson spectrum is limited to channels accessible to two-transverse (massless)-gluon systems.[58]

- All established low-mass ($m \lesssim 1.7$ GeV/c^2) mesons can be accommodated in $\ell = 0,1$ octets, with the exception of $\iota(1460)$ and $\Theta(1730)$—*if* we admit the appearance of the first pseudoscalar radially excited nonet at masses just above 1.2 GeV/c^2.

- Doubly disconnected decay processes definitely exist, but their contributions appear to be no stronger than ~10% of singly disconnected topologies.

Making use of charmonium decay data, the evidence on $\ell = 0,1$ multiplets has permitted the following upgrading of our information and understanding process.

$J^{PC} = 0^{-+}$ The isoscalars have no significant admixture beyond $\eta^{(1)}$ and $\eta^{(8)}$ or η and η'. Their mixing angle is $\Theta_P \approx 20°$, much larger than previously reported. Isoscalars in the 1.2–1.4 GeV/c^2 range may qualify as ($n = 2$) radial excitations. Even heavier isoscalars decaying into VV are candidates for threshold effects of gluon fragmentation.

$J^{PC} = 1^{--}$	The isoscalars are close to ideally mixed. SU_3 predictions about the couplings of all vector pairs to the SU_3 singlet state η_c are borne out as relative ratios.
$J^{PC} = 0^{++}$	The isoscalar state $f_0(975)$ is largely $s\bar{s}$-based; its width is at least as large as Γ_ρ. The (mostly) u,d-based isoscalar (at higher mass?) is either not seen or to be identified with $G(1590)$; the isodoublets are not seen here. There is no significant new information on the isotriplet $a_0(970)$.
$J^{PC} = 1^{++}$	The isoscalars appear to be mixed such that the mostly $s\bar{s}$-based state $f_1(1275)$ is lighter than the mostly u,d-based $f_1(1420)$. They are not to be confused with $J^{PC} = 0^{-+}$ states decaying into the same final particle combinations, at similar masses. Mass questions about the isovector could not be resolved.
$J^{PC} = 2^{++}$	This nonet again has close-to-ideally mixed isoscalars. Reported new tensor states ($\Theta(1720)$, $X(2230)$, $g_T^{1,2,3}$) must have a different dynamical origin. Isodoublets and isotriplets are well established.

Gluonia

Our best scenario for the pseudoscalar spectrum makes $\iota(1460)$ a candidate for a pure gluonium. Its only observed decay into $K\overline{K}\pi$ is due to mass breaking in the *a priori* flavor-neutral hadronization process.

In the 2^{++} tensor spectrum, Θ remains a candidate simply because it finds no other home. It has disturbing features (quark-affinity) in this context. $X(2230)[\xi]$ is ill understood, but remains a candidate. The g_T states probably have a different dynamical origin.

Scalar gluonia remain elusive. $G(1590)$ may be identical with the state that caused the original $\Theta \to \eta\eta$ observation, and *could* be gg based. The low-energy $\pi\pi$ spectrum looks unpromising.

Exotics:

Radiative decay data do not provide corroborating evidence for the emergence of $q\bar{q}g$ or $q\bar{q}q\bar{q}$ states.

All of the resulting features are approximate only; where data are still scarce, a most-plausible scenario was chosen. Fuller sets of data and more refined analysis may well modify this, but are hardly likely to change the overall picture that is emerging.

ACKNOWLEDGMENTS

It is a pleasure to thank Prof. Nino Zichichi for yet another excellent Subnuclear School in Erice, and Mrs. Maria Zaini, as well as Drs. Alberto Gabrieli and Pinola Savalli, for making everything function inside the wonderful scientific and human atmosphere he created in this enchanting town at the crossroads of three civilizations.

REFERENCES

1. For a description of the basic non-relativistic quark model, see R.H. Dalitz, in *High Energy Physics*, Les Houches 1965 Lectures, C. de Witt and M. Jacob, eds., Gordon & Breach, New York (1966) pp. 251 ff. The QCD-inspired version was first worked out by A. de Rujula, H. Georgi, S.L. Glashow, *Phys. Rev.* **D12**, 147 (1975).

2. See the excellent review article by N.A. Törnqvist, *Acta Phys. Pol.*, **B16**, 513 (1985); also: N. Isgur, *in The New Aspects of Subnuclear Physics*, A Zichichi, ed., Plenum Press, New York, (1980).

3. Such calculations were explicitly done for low-mass pseudoscalar mesons: J.F. Donoghue, and H. Gomm, *Phys. Lett.* **121b**, 49 (1983).

4. See also the summary talk of Hadron '87 by J.L. Rosner in the Proceedings volume, KEK 87-7(1987).

5. Ideal mixing corresponds in the standard mixing scheme (here illustrated for the case of the pseudoscalars),

$$\eta = \cos\theta\ \eta^{(8)} + \sin\theta\ \eta^{(1)}\ ,$$
$$\eta' = -\sin\theta\ \eta^{(8)} + \cos\theta\ \eta^{(1)}\ ,$$

to an angle $\theta \approx 35°$. With the usual $\eta^{(8)}, \eta^{(1)}$ quark assignments, it would lead to

$$\eta = \frac{1}{\sqrt{2}}(u\bar{u} + d\bar{d})\ ,$$
$$\eta' = s\bar{s}\ .$$

6. G. t'Hooft, *Phys. Rev. Lett.* **37**, 8 (1976); *Phys. Rep.* **142**, 357 (1986); R. Crewther, *Phys. Lett.* **70B**, 349 (1977); *Riv. Nuovo Cim.* **2**, 63 (1979).

7. S.J. Brodsky and G.P. LePage, *Phys. Rev.* **D24**, 2848 (1981).

8. N.N. Achasov *et al.*, *Sov. J. Nucl. Phys.*, **32**, 566 (1980).

9. V. Novikov *et al.*, *Nucl. Phys.* **B165**, 67 (1980); for a lucid explanation of the instanton concept, read chapter 7 of S. Coleman, *Aspects of Symmetry* (Selected Erice Lectures), Cambridge Univ. Press (1985).

10. B. Barish and R. Stroynowski, *Phys. Rep.* **157**, 1 (1988).

11. The MARK III detector is described in D. Bernstein *et al.*, *Nucl. Instrum. & Methods* **226**, 301 (1984); the DM2 detector in J.E. Augustin *et al.*, *Physica Scripta* **23**, 623 (1981).

12. J.L. Rosner, *Phys. Rev.* **D27**, 1101 (1983).

13. H.E. Haber and J. Perrier, *Phys. Rev.* **D32**, 2961 (1985).

14. R. Baltrusaitis *et al.*, (MARK III Collaboration), *Phys. Rev.* **D32**, 2883 (1985).

15. H. Aihara *et al.*, *Phys. Rev. Lett.* **57**, 51 (1986).

16. J. Adler *et al.*, (MARK III Collaboration), Contribution to the EPS Conference on High Energy Physics, Uppsala (1987), to be published; Z. Ajaltouni *et al.*, (DM2 Collaboration), *Contributions to the 1987 Lepton-Photon Symposium*, Hamburg.

17. Numerous such schemes have been suggested; see, *e.g.,* F. Caruso *et al.,* *Z. Phys.* **C30**, 493 (1986). S.C. Chao *et al., Phys. Lett.* **B172**, 253 (1986).

18. S.V. Chung *et al., Phys. Rev. Lett.* **55**, 779 (1985).

19. A. Ando *et al., Phys. Rev. Lett.* **57**, 1296 (1986).

20. H. Aihara *et al., Phys. Rev. Lett.* **57**, 2500 (1982); see also G. Gidal *et al.,* (MARK II Collaboration) *Phys. Rev. Lett.* **59**, 2016 (1987).

21. This signal is observed in the charge modes $K_s K^{\pm} \pi^{\mp}$, $K^+ K^- \pi^0$, $K_s K_s \pi^0$ by the MARK III Collaboration. (J. Richman, CalTech thesis (1983), unpublished), and by the DM2 Collaboration (J. Augustin, *et al.,* LAL-85/27 1985)).

22. This state (see J. Richman, previous ref.) was originally identified with $\iota(1460)$; its important radiative width ($\Gamma(X) \to \gamma\rho^0 = 1.9 \pm 0.7$ MeV) was taken as an argument in opposition to the gluonium interpretation of $\iota(1460)$. See J. Donoghue in *Particles and Fields 1981*, C.A. Heusch and W.T. Kirk, eds. AIP, New York (1982).

23. F. Close, *in Quarks and Hadronic Matter*, Yukon Advanced Studies Institute (1984), originally proposed this test of the ι wavefunction, in the context of vector-dominance relations between photon and vector mesons.

24. N. Wermes, *Proc. 5th Conference on Physics in Collision*, Autun, France, World Scientific (1986).

25. Note that S-wave $q\bar{q}$ scattering lengths would lead naturally to an appearance of 0^{-+} characteristics. A quantitative evaluation is presently in progress.

26. The recent DM2 results (D. Bisello *et al., Contributions to the 1987 Lepton-Photon Symposium*, Hamburg, and L. Stanco, Orsay preprint LAL-87-40) present the most consistent data sample. Note that the decay $\eta_c \to \phi\phi$ permitted the MARK III Collaboration to confirm the identity of the state by way of a straightforward spin-parity analysis (R. Baltrusaitis *et al., Phys. Rev. Lett.* **52**, 2126 (1984).

27. L. Köpke (MARK III Collaboration), *Proceedings of the XXIIIrd International Conference on High Energy Physics*, S. Loken, ed., World Scientific, Singapore (1986).

28. J.E. Augustin *et al.,* LAL 85/27 (1985).

29. D.M. Coffman *et al.,* (MARK III Collaboration), SLAC-PUB-4460 (to be published).

30. W. Lockman (MARK III Collaboration), *Proceedings, 1986 San Miniato Workshop* (to be published).

31. W. Lockman (MARK III Collaboration), *Proceedings, 1986 Lake Louise Conference on Intersections of Nuclear and Particle Physics*.

32. H. Kolanski and P. Zerwas, DESY Preprint 87-175 (1987).

33. D. Aston *et al.,* DPNU 87/15; SLAC-PUB-4279 (1987); to be published in *Nucl. Phys. B*.

34. L. Köpke (MARK III Collaboration), SCIPP/MARK III Memo (1986). Unpublished.

35. See the contributions of C. Heusch and A. Seiden to the MARK III Pow-Wow; SLAC-Report 323 (1988).

36. J. Adler *et al.,* (MARK III), to be published.
    T. Bolton, Ph.D. thesis, M.I.T. (1988); unpublished.

37. R. Baltrusaitis *et al.*, (MARK III), *Phys. Rev.* **D35**, 2077 (1987).

38. C. Edwards *et al.*, (Crystal Ball Collaboration), *Phys. Rev. Lett.* **48**, 458 (1982).

39. J. Adler *et al.*, (MARK III), Contribution to the *Proceedings of the EPS Conference*, Uppsala (1987), G. Dubois, editor. Note that the MARK III data are severely limited due to the absence of a neutral trigger; this is presently being installed.

40. D. Alde *et al.*, *Nucl. Phys.* **B269**, 485 (1986).

41. Should this scenario turn out to be correct, the confusing nomenclature of these states would obviously be redefined.

42. R. Baltrusaitis *et al.*, (MARK III), *Phys. Rev. Lett.* **56**, 107 (1986).

43. D. Alde *et al.*, *Phys. Lett.* **177B**, 120 (1986).

44. The interest of the final state $\eta\eta'$ in the context of gluonium searches has been discussed by S.S. Gershtein *et al.*, *Z. Phys.* **C24**, 305 (1984).

45. H.F. Lipkin, *Phys. Lett.* **109B**, 326 (1982).

46. A. Etkin *et al.*, *Phys. Lett.* **165B**, 217 (1985).

47. J. Adler *et al.*, (MARK III Collaboration), as quoted by G. Dubois *in Proceedings of the EPS Conference*, Uppsala (1987),

48. L. Stanco *et al.*, (DM2 Collaboration), LAL 87-42 (1987).

49. For recent gluonium reviews, see, e.g., F.E. Close, R/XL-87-072 to be published in Rep. Progr. Phys.; F. Couchot, LAL 87-40 (1987); C.A. Heusch, *Proceedings, Multiparticle Symposium*, Seewinkel, World Scientific, Singapore (1986); M.S. Chanowitz, *in Hadron '87*, KEK, Tsukuba (1987).

50. J.D. Bjorken, *Proceedings, 1979 SLAC Summer School*, A. Mosher ed., SLAC Report 224 (1980). See also M. Chanowitz, Ref. 53, and S. Gershtein *et al.*, Ref. 44.

51. S. Godfrey, *Phys. Lett.* **141B**, 439 (1984).

52. S.J. Lindenbaum, *Comments Nucl. Part. Physics* **13**, 285 (1984).

53. M. Chanowitz (*Proc. VIth International Workshop on Photon-Photon Collisions*, World Scientific, Singapore (1984)) defined this relative gluon affinity as "stickiness" $S = \left(\frac{ms}{k \cdot (\psi \to \gamma X)}\right)^3 \frac{\Gamma(Y \to \gamma X)}{\Gamma(X \to \gamma\gamma)}$. This measure can serve as a comparison of states with equal quantum number: for $J^{PC} = 0^{-+}$, $S(\iota) : S(\eta') : S(\eta) = (> 65) : 4 : 1$; for $J^{PC} = 2^{++}$, $S(\Theta) : S(f') : S(f) = (> 20) : 3 : 1$.

54. The values for R_γ and R_V given by C.A. Heusch in Ref. 49 have to be updated using recent MARK II data on $\gamma\gamma \to \iota(1460)$ limits: G. Gidal, *et al.*, *Phys. Rev. Lett.* **59**, 2016 (1987).

55. F. Binon (GAMS Collaboration), *Proceedings of the Hadron '87 Conference*, KEK, Tsukuba (1987); a similar signal may have been seen by W.D. Apel *et al.*, *Nucl. Phys.* **B193**, 269 (1983).

56. S.I. Bityukov *et al.*, *Phys. Lett.* **B188**, 383 (1987).

57. F.S. Close, H.J. Lipkin, RAL 87/046 (1987), and M. Boutemeur, *in Hadrons, Quarks, and Gluons*, J. Tran Thanh Van, ed., Editions Frontières, Paris (1987).

58. This feature was postulated long ago: T. Barnes, *Z. Phys.* **C10**, 275 (1981). Note, however, that preliminary data of both the DM2 and MARK III Collaborations show indications of a non-pseudoscalar enhancement at $m(4\pi) \approx 1285$ MeV/c^2, in the radiative decay process $J/\psi \to \gamma\pi^+\pi^-\pi^+\pi^-$. If this state is to be identified with $f_1(1285)$, the argument has to be modified from "absence" to "suppression" of states not accessible to two transverse gluons.

DISCUSSION

- *Ito:*

Do you have any new data to show on the X(2.2) and could you expand on the remark you made this morning about the lack of satisfactory explanation for this object?

- *Heusch:*

The $\xi(2.230)$ was discovered in the reaction $J/\psi \rightarrow K^+K^-$ and $J/\psi \rightarrow \gamma K_s K_s$ by the MARK III experiment at SPEAR in 1983. The observation was subsequently confirmed by MARK III in a different run at the J/ψ in 1985, which essentially doubled the statistics. The DM2 experiment at DCI does not observe the $\xi(2.230)$ in either the K^+K^- or the $K_s K_s$ modes. The DM2 experimental data set consists of $8.6 \times 10^6 J/\psi's$, while MARK III has $5.8 \times 10^6 J/\psi$ events. A preliminary spin-parity analysis by MARK III, presented at the 1986 Berkeley Conference, indicates that $J \geq 2$. The relation between possible structure absorbed at 2.23 GeV by the GAMS experiment at Serpukhov and CERN in the reaction $\pi^- p \rightarrow \eta\eta' n$ and by the LASS collaboration at SLAC in the reaction $K^- p \rightarrow K_s K_s \Delta$ and the $\xi(2.230)$ remains unclear. The $\xi(2.230)$ has been unsuccessfully searched for in $p\bar{p} \rightarrow \xi \rightarrow K^+K^-$ by experiments at Brookhaven and CERN.

Structure seen in the 2.2 GeV mass region by the DM2 and MARK III experiments in the reaction $J/\psi \rightarrow \gamma\phi\phi$ does not appear to be related to the $\xi(2.233)$, as both groups report that the $\phi\phi$ structure has negative parity ($J^{PC} = 0^{-+}$), while the parity of the $\xi(2.230)$ is constrained to be positive by the $K_s K_s$ decay. It should be noted that the 0^{-+} assignment for the $\phi\phi$ state at 2.2 GeV in radiative J/ψ decay indicates that this state is unlikely to be associated with the "g_T" states seen by Prof. Lindenbaum's Brookhaven experiment in the reaction $\pi^- p \rightarrow \phi\phi n$. The structure near threshold in $J/\psi \rightarrow \gamma\phi\phi$ is similar to effects seen in the related reactions $J/\psi \rightarrow \gamma\rho\rho$ and $J/\psi \rightarrow \gamma\omega\omega$ by the MARK III and DM2 experiments.

- *Ito:*

What are the future plans of MARK III?

- *Heusch:*

The MARK III collaboration would like to take data for several more years. The SPEAR injection line is being upgraded to allow compatibility with the SLC.

Work has been done to upgrade the main drift chamber, and a new vertex chamber has just been installed to replace the old inner drift chamber. An effort is also being made to bring up a neutral trigger. However, the immediate future plans of MARK III depend on the progress of the SLC. A three-month run is scheduled for late winter. The running energy has yet to be determined by the collaboration; possibilities include a run at the $J/\psi(3097)$, the $\psi(3685)$, the $\psi(3770)$ for D physics, the 4.0-4.2 GeV region for D_s ("F") physics, and the 5–6 GeV region for charmed baryon studies.

– *Singh:*

Would you kindly explain the exact nature of the U(1) problem?

– *Heusch:*

The U(1) problem is difficult to satisfactorily explain, so I will try to give you the flavour of the problem. It has to do with the fact that the U(1) group has no inherent quantum numbers, and so all hell can break loose in the presence of specific gluonic interactions one has to introduce. Many people have attempted to solve the problem without success; it has been the nemesis of theorists who try to understand pseudoscalars. It is related to the presence of a term naturally occurring in the QCD Lagrangian which is proportional to the divergence of the axial current. If you would like a more sastisfactory theoretical explanation, I suggest you read Sidney Coleman's lectures in past volumes of this Subnuclear School's Proceedings*, or that you discuss it with Professor Glashow. For both scalar and pseudoscalar particles you are liable to run into trouble due to the interpretability of the pseudoscalar mesons as Goldstone bosons in the limit of exact $SU(3)$ symmetry for massless quarks, and to the scalar/pseudoscalar connection implied by chiral symmetry.

---

*(e.g., S. Coleman in: "The Whys of Subnuclear Physics, A. Zichichi, ed. Plenum Press, New York, 1979).

QUANTUM COSMOLOGY AND SUPERSTRINGS

Dimitri V. Nanopoulos

Physics Department
University of Wisconsin–Madison
Madison, WI 53706

ABSTRACT

Superstrings may turn out to be the Theory of Everything. If so, they should provide, among other things, a satisfactory (quantum) cosmology. Recent studies of the quantum cosmology of superstrings indicate that we are getting a consistent picture *only if* (i) superstrings "live" in $D = 10$ space–time dimensions *and* (ii) six dimensions are compactified and thus leaving $D = 4$ space–time dimensions for us to live! In this review we provide a detailed expose of these recent developments on the (quantum) cosmological front of superstrings.

INTRODUCTION

Although closed string theory is the only serious candidate for a Theory of Everything including gravity[1], we do not yet know which string model Nature has chosen. Different closed string models[2-8] may be characterized as unitary realizations of the left– and right–moving Virasoro algebras with central charges $c = 26$ or $c = 15$ if there is a world sheet supersymmetry, which are constrained by multiloop modular invariance. It is a matter of opinion whether all these realizations of the Virasoro algebra should be regarded as different ground states of the same string theory, or as intrinsically different theories. This rather universal approach to string theories make it apparent that there is no intrinsic difference between external space–time and internal degrees of freedom. There is no necessity to interpret the internal degrees of freedom as any sort of compactified space–time. Indeed, recent developments[4-8] have shown that we can formulate consistent superstring theories in any number of space–time dimensions ($4 \leq D \leq 10$). It is not inconceivable that the choice between the bosonic formulation of superstrings (start with $D = 10$ space–time dimensions and compactify down to $D = 4$ on a suitable manifold or orbifold) and the fermionic formulation of superstrings (work in $D = 4$ space–time dimensions with extra free–fermions on the world–sheet) is going to be a matter of convenience (like the choice between different coordinate systems!). Nevertheless, the "magic" of $D = 10$ (nowadays interpreted as the maximal possible space–time dimensions that a consistent superstring theory may be formulated) and our "old friend"

The Superworld II
Edited by A. Zichichi
Plenum Press, New York, 1990

$D = 4$ is still with us. Is there any way that we can understand how these specific space–time dimensions are picked-up? Recent attempts to answer this fundamental question will be the main theme of this review.

In general relativity, powerful theorems[9] indicate that our Universe must have started from an initially singular state, where the laws of physics break down. In particular, it is unclear how any boundary condition necessary for a description of a dynamical system could have been imposed at the initial singularity. Hence the origin of our Universe appears to be a mysterious, almost random occurence. However, quantum corrections and/or short-distance modifications of the classical general relativity could change this picture and produce an unambiguous history for the universe. Currently popular string theories[1] are examples of the latter possibility, although it appears that at the classical point field theory level singularities may still not be avoided[10].

Quantization of gravity is well known to be riddled with problems, but at least formally one can write down quantum averages as path integrals[11]. In such an approach, one can write a quantum averaged wave functional for the whole Universe, and the question is then only to impose a reasonable boundary condition. This has been done in an attractive proposal by Hartle and Hawking[12], who argue that in the Euclidean formulation of the path integral the boundary condition of the Universe is that initially it has no boundary; that is, the Universe can be created out of nothing[13]. Because there is no boundary, there is no initial singularity, which is the great virtue of this proposal. However, one then needs to explain how our Universe got to be Lorentzian, a problem that has proven to be rather difficult[11].

It is very likely that the very concepts of space and time become foggy at very early times where general relativity must be replaced by a more fundamental theory such as some string theory[1]. String theories will lead to a point-like field theory with an effective action that contains an infinite series of terms. As such they however represent one step beyond the simplest general relativity, and one may then ask what is the fate of quantum cosmology in such effective theories.

In the present review we will consider the quantum cosmology of string-based effective theories in D dimensions, following closely Ref. 14. In Section 2 we recapitulate the basic ideas behind the Hartle-Hawking approach and discuss how they can be applied in effective field theories. In Section 3 we first show that in string theories, the transition from Euclidean to Lorentzian regime can not take place in four dimensions but that in the minisuperspace models the topology of space must be $S^m \times S^n$. We then solve the relevant Wheeler-DeWitt equations[15] in D dimensions in the presence of the dilaton, which always appears in string theories, and show how the Euclidean solution, i.e., the gravitational instanton, evolves dynamically to the Lorentzian regime. The scale factors of the two product spaces separate with one exponentially expanding and the other shrinking, and we discuss the consistency condition to be satisfied in order that the effective action remains reliable.

In Section 4 we show how the Ricci flatness of the compact space leads to its stability and discuss the relation between the scale factor of the compact space and the dimensions of S^m and S^n. In Section 5 we analyze the thermodynamic stability of our model which further restricts the space-time dimensions which are viable. We show, using the results of the previous sections, that the only consistent solution of the Wheeler-DeWitt equation exists in D=10. Moreover, the dimension of the compact space is shown to be six while the large space time has the dimension four. In Section 6 we analyze the result of the string one-loop correction and show that our perturbation analysis is valid. We also calculate the dynamical evolution of the dilaton ϕ and obtain the time variation of the gauge couplings. We show that the asymptotic value of the gauge coupling is one (in Planck units). Finally, Section 7 contains some discussion and our conclusions.

QUANTUM COSMOLOGY IN D DIMENSIONS

Hartle-Hawking Proposal for Boundary Conditions

In quantum mechanics, all information about the Universe is contained in the wave function $|\Psi>$ of the Universe, and the time evolution of the physical observables in a D-dimensional Universe may be determined by specifying the matter distribution and the space-time geometry on a (D-1)-dimensional space-like surface Σ. The amplitude of the Universe to be in such a state is given by the wave functional[11,12]

$$\Psi[h_{IJ}, \phi] = < h_{IJ}, \phi | \Psi > \tag{2.1}$$

where h_{IJ} $(I, J = 1 \ldots, D - 1)$ is the induced (D-1) metric on Σ, and ϕ is the distribution of matter and gauge fields.

In field theory the transition amplitudes can be calculated by integrating over all field configurations, weighted by their action, that connect the initial and final states. Likewise, as long as the Universe can be described in terms of local quantum fields, the evolution of the whole Universe may be evaluated by calculating the transition amplitude between two (D-1) surfaces Σ and Σ' (which we shall take to be surfaces of constant cosmological time):

$$< \Sigma' | \Sigma > = < h'_{IJ}, \phi' | h_{IJ}, \phi > = \int D[g_{MN}] D[\phi] e^{iS}. \tag{2.2}$$

Here the integral is over all D-metrics g_{MN} $(M, N = 1, \ldots, D)$ and matter fields ϕ which match (h_{IJ}, ϕ) and (h'_{IJ}, ϕ') at Σ and Σ', respectively. It is obvious that the formal expression (2.2) should be taken to represent only an approximation, because at very short distances gravity should be replaced by a more fundamental theory not plagued with problems on non-renormalizability, such as some string theory. It could be that in such a fundamental theory one could lose the very notion of space-time metric. However, at scales larger than the Planck scale we are likely to encounter some point-like local field theory (not necessarily Einstein gravity) as the effective theory of everything, and it is in this sense that the expression (2.2) should be understood.

Given the wave functional (2.1) of the Universe, for a complete specification of the quantum state of the Universe, one needs to impose a boundary condition. The aesthetically appealing proposition of Hartle and Hawking[12] is that one should integrate in the path integral over compact non-singular Euclidean D-dimensional metrics with compact boundaries, so that as the cosmological time $t \to 0$ the boundary $\Sigma(t)$ vanishes. This settles the problem of specifying the boundary conditions, since at the initial time there is no boundary. This is in sharp contrast to choosing a Lorentzian metric where at the initial time Σ

remains spatially infinite and has an infinite curvature. The wave functional of the Universe is then given by

$$\Psi[h_{IJ},\, \phi] = \int\limits_C D[g_{MN}]D[\phi]e^{-S_{eff}} \tag{2.3}$$

where C is the class of compact D-metrics and matter fields which match h_{IJ} and ϕ on Σ, and $S_{eff}[g_{MN},\, \phi]$ is the effective Euclidean action.

The Hartle-Hawking proposal for the boundary conditions makes sense also in the case that at very small distances the field theoretical description is no more adequate. It states that the emergence of space-time is not dependent on pre-existing conditions; this may or may not be true in the Theory of Everything, but it certainly is a feasible possibility, and we will adopt this proposal throughout this paper. However, let us first clarify what we mean by "initial time".

Time is a parameter[16] which correlates the state of a system under consideration with that of a "generic clock". For this definition of time to be operationally valid, the system and the clock should not be strongly coupled through any kind of interactions and the clock should be a "classical object" (in the same sense that ultimately any measuring apparatus must be a classical object so that there is no infinite regression of "collapse of wave functions"). When the scale of the Universe is small and the quantum gravitational effects are important such that the "whole Universe" is a quantum object, then both these criteria fail. An alternative definition of cosmological time is provided by the empirical fact that the Universe is expanding after the initial "big bang". Hence, "cosmological time" may be defined as a monotonically increasing function of the volume factor $\sqrt{D-1}h \left(\equiv (det\, h_{IJ})^{1/2} \right)$ of the Universe, i.e., $t \equiv f(\sqrt{D-1}h)$. This definition of time agrees with the classical concept of time (as parameter which gives a unique ordering of events) when the topology of the space is Lorentzian. In classical general relativity we allow the metric of space-time to change with time but not the topology. In quantum mechanics however, as we shall see in our particular model, the topology of the space-time is also allowed to change. In our model the Universe starts with a Euclidean metric and there is a quantum tunneling to the Lorentzian metric. Hence the classical notion of time and the definition of cosmological time ($\sqrt{D-1}h$) agree as long as the Universe is in the Lorentzian regime only.

The Universe as we know it is not Euclidean but Lorentzian, and the wave functional (2.3) should be such that it makes a dynamical transition (in contrast to mere analytic continuation) from the Euclidean to the Lorentzian regime. The birth of the Universe would then proceed by tunneling from nothing[13] via a gravitational instanton, resulting in a Big Bang but avoiding the initial singularity.

It is highly remarkable that recent developments in particle physics and cosmology support the idea that the Universe at its birth may as well have the quantum numbers of the vacuum, thus facilitating its tunneling from nothing[13]. Namely, a period of inflation can generate all the entropy observed in the present Universe, while B-violating interactions present in Unified Theories may create the apparent baryon asymmetry in the Universe. Furthermore, it is well known that the Universe seems to be devoid of net electric charge and that it does not have a net angular momentum. Clearly there is no evidence against a tunneling from nothing to our Universe. As we shall see in Section 3.2, the existence of higher dimensions may be of crucial importance for such a program to be realized.

Minisuperspace Models

The computation of the wave functional directly from the path integral (2.3) is an involved task and must generally be done numerically (for examples, see ref. 17). However, the wave functional obeys a zero energy Schrödinger equation (Wheeler-DeWitt equation), which can be solved in the semi-classical approximation. One can write down the D-metric in the neighborhood of Σ as

$$ds^2 = -(N^2 - N_I N^I)dt^2 + 2N_I dx^I dt + h_{IJ} dx^I dx^J \tag{2.5}$$

where N and N_I $(I = 1, \ldots, D - 1)$ are functions of t. Under diffeomorphisms of Σ, the lapse function N moves the point x^M normal to Σ, while the shift functions N^I move x^M along Σ. The general coordinate reparametrization invariance equals invariance under diffeomorphisms of Σ, so that one has to impose the constraint equations[18]

$$\frac{\delta S_{eff}}{\delta N} \equiv H = 0$$

$$\frac{\delta S_{eff}}{\delta N^I} \equiv H^I = 0 \tag{2.6}$$

In quantum mechanics these constraints must be imposed as operator equations:

$$< h_{IJ}, \, \phi | \frac{\delta S_{eff}}{\delta N} | \Psi > \equiv < h_{IJ}, \, \phi | H | \Psi > = 0$$

$$< h_{IJ}, \, \phi | \frac{\delta S_{eff}}{\delta N^I} | \Psi > \equiv < h_{IJ}, \, \phi | H^I | \Psi > = 0. \tag{2.7}$$

If we choose to work in the $| h_{IJ}, \, \phi >$ representation, we replace the canonical momenta

$$\Pi_{IJ} \equiv \frac{\delta \mathcal{L}_{eff}}{\delta \dot{h}_{IJ}}, \quad \Pi_\phi \equiv \frac{\delta \mathcal{L}_{eff}}{\delta \dot{\phi}} \tag{2.8}$$

by the operators $-i\delta/\delta h_{IJ}$ and $-i\delta/\delta\phi$, respectively.

In making the quantum correspondence we are faced with the question of operator ordering which also determines the choice of the measure which is used to define the inner product $< \Psi_1, \, \hat{A}\Psi_2 >$ of the Hilbert space spanned by the wave function. Since we are interested in the semiclassical trajectories we find the WKB solution of the form $\Psi \sim e^{iS}$ and the expectation value of our dynamical variables is given by the gradient of the phase S. Therefore the choice of operator ordering is irrelevant to us and we shall choose the simplest measure which makes the Hamiltonian hermitian.

Equation (2.7) contains the dynamics of the system in the following sense: if we shift the surface $\Sigma(t)$ along N by a small amount δt then the induced (D-1) metric h_{IJ} on $\Sigma'(t + \delta t)$ also changes accordingly such that the operator equation (2.7) remains valid. Therefore all information about the dynamical evolution of space-time is contained in the way the (D-1) space-like surface Σ is embedded in the D-dimensional manifold and the induced metric h_{IJ} on Σ.

Equation (2.7) in the $|h_{IJ}, \phi>$ representation is called the Wheeler-DeWitt equation: $H\Psi[h_{IJ}, \phi] = 0$. In Section 3 we shall derive this equation in a minisuperspace model of string cosmology and solve the resulting differential equation for $\Psi[h_{IJ}, \phi]$. The boundary condition that we shall impose is derived from the Hartle-Hawking boundary condition which states that the (D-1) dimensional "surface area" $\sqrt{D-1}h$ in the Euclidean action $S = \int i dt d^{D-1}x \sqrt{D-1}h \mathcal{L}$ vanishes at the initial "time". Therefore the wave-function (2.3) converges to a constant (which may be normalized to one) as $\sqrt{D-1}h \to 0$. At times below the Planck time, even if the notion of volume element could be ambiguous, we shall apply the Hartle-Hawking boundary condition in the sense of an extrapolation at the Planck time: $\Psi(\sqrt{h} \to 0) \to 1$.

Note that although N and N^I are dynamical quantities in that they depend on cosmological time, their time derivatives do not appear in the action. Therefore N and N^I are not the true dynamical variables of the theory but are Lagrange multipliers of the constraint equations (2.6). Usually Lagrange multipliers are determined by the theory and contain dynamical information about the system. In classical general relativity however the topology of the metric is not allowed to change. There the value of the lapse function N determines a gauge degree of freedom and can be absorbed in rescaling the time parameter. In our analysis we find that there is a change in the topology of the Universe (from Euclidean to Lorentzian) where N changes from imaginary to real (whose magnitude can be chosen to be 1 by the choice of gauge).

Although a priori the space need not be limited to having either an Euclidean or a Lorentzian metric, it can be shown that, based on consistency arguments for removing ghosts from the world sheet string amplitudes[19], the space-time metric can have at most one negative sign. This means that the only two viable possibilities for the topology of the space are Euclidean or Lorentzian. In accordance with the Hartle-Hawking boundary conditions, we initially choose $N = i$, so that the metric is Euclidean, and the phase of the wave functional is real. The transition from the Euclidean to the Lorentzian regime, where the phase is imaginary, is possible only when the effective action vanishes. Hence, although time has no meaning in the Euclidean space, one can speak of dynamical evolution of the Universe even here in terms of a sequence of points in the configuration space which corresponds to the highest transition amplitudes. This, of course, is the standard interpretation of an instanton solution.

Schematically, we can present this by writing the transition amplitude as $< 0|S|L > = < 0|E >< E|S|L >$, where 0, E and L stand respectively for nothing, Euclidean Universe and Lorentzian Universe, and here S is the S-matrix. The amplitude $< 0|E > \neq 0$ by virtue of the Hartle-Hawking boundary condition, and later we shall show how also $< E|S|L > \neq 0$ in a class of string theories.

QUANTUM COSMOLOGY IN STRING THEORIES

The Effective Action

As is well known, various anomaly free superstring theories[1] are currently popular candidates for a full quantum theory of gravity and matter. Certain phenomenological issues dictate that[1] these string theories should be based on closed strings, and they were first formulated in ten dimensions. Recently string models have also been formulated directly in four dimensions[4-8]. The bosonic part of the action for a closed string is a two-dimensional sigma model and is given by[20]

$$S = \frac{1}{4\pi\alpha'} \int d^2z \{ \sqrt{\gamma} [\gamma^{ab} g_{MN}(X) \partial_a X^M \partial_b X^N] +$$

$$\epsilon^{ab} B_{MN}(X) \partial_a X^M \partial_b X^N - \frac{\alpha'}{2} \sqrt{\gamma}^{(2)} R\phi(X) \} \tag{3.1}$$

where α'^{-1} is the string tension, $X^M (M = 1, \ldots, D)$ is the space-time coordinates of the string, γ^{ab} is the world sheet metric and $^{(2)}R$ is the associated curvature. The equations of motion for the massless background fields g_{MN}, B_{MN} and the dilaton ϕ are obtained by imposing conformal invariance on the world sheet. This is equivalent to the vanishing of the beta functions, which to lowest order in α' read

$$\beta^g_{MN} = R_{MN} - \frac{1}{4} H^2_{MNR} - \nabla_M \nabla_N \phi + O(\alpha') = 0$$

$$\beta^B_{MN} = \frac{1}{2} \nabla^L H_{LMN} + \frac{1}{2} \nabla^L \phi H_{LMN} + O(\alpha') = 0$$

$$\beta^\phi = -^{(D)}R + \frac{1}{2} H^2_{MNR} + 2\nabla^2 \phi + (\nabla\phi)^2 + O(\alpha') = 0 \tag{3.2}$$

where $H_{MNP} = 3\nabla_{[M} B_{NP]}$. The effective field theoretical action $S_{eff}[g_{MN}, B_{MN}, \phi]$ can then be constructed by requiring that the equations of motion $\delta S_{eff}/\delta\phi = 0$, etc. coincide with the vanishing of the beta functions (3.2). To order α' one then finds

$$S_{eff} = -\int d^D x \sqrt{|g|} \left\{ \frac{^{(D)}R}{2\kappa^2} + \frac{4}{(D-2)\kappa^2} \left(\frac{\nabla_M \phi}{\phi} \right)^2 + \right.$$

$$\left. + \frac{\kappa^2}{12} \phi^{-\frac{8}{D-2}} H^2_{MNR} + \frac{\kappa^2}{2} \phi^{-\frac{4}{D-2}} [F^2_{MN} - R^2_{MNPQ} + 4R^2_{MN} + R^2] \right\}$$

$$+ \text{surface terms} \tag{3.3}$$

where in (3.3) the metric was redefined to get rid of the dilaton-curvature coupling and $\kappa^{-\frac{1}{2}} = M_P$ is the D-dimensional Planck scale. In superstring theories the fermionic terms are related to the bosonic terms in (3.3) by supersymmetry. Cancellation of the anomalies also necessitates[21] the modification of H_{MNP} by the Chern-Simons terms, but these will not play any role in the following. It is interesting to notice that the supersymmetrization[22] of the Lorentz-Chern-Simons terms, in the case of the heterotic string, needed for the cancellation of anomalies, leads to the R^2 terms of the ghost free Gauss-Bonnet form, which are badly needed for allowing compactification from 10 to 4 dimensions[23].

The surface terms need to be added to cancel the total divergence terms which would otherwise arise due to the compact nature of the spaces we shall consider.

The effective tree level action (3.3) is an expansion in powers of α', and the lowest order Einstein term $^{(D)}R$ will be modified by the appearance of the higher order invariants, as is already indicated by (3.3). The equations of motion (3.2) have been constructed[24] up to $O(\alpha'^3)$, and the resulting effective action would be quite complicated. In addition, the tree level effective action can also be modified by string loop corrections[25]. The lowest order terms include the curvature $^{(D)}R$, the dilaton ϕ and the field strength H_{MNR} of B_{MN}. The first order terms are the Yang-Mills and the curvature squared terms which are $O(\alpha'/L^2)$ where L is a typical scale factor associated with smallest spatial dimensions. Hence the lowest order terms dominate in the expansion of the effective action as long $\alpha'/L^2 \ll 1$. If we choose to work in units of the D-dimensional Planck scale M_P, which is related[26] to α' through $M_P^2 = \phi^{-4/(D-2)}\alpha'^{-1}$, the domain of validity of the small α'-expansion is given by

$$\frac{\phi^{-\frac{4}{D-2}}}{L^2} \ll 1 \tag{3.4}$$

In the domain (3.4) we can study the quantum cosmology of D-dimensional strings based on the lowest order terms in the tree level effective action (3.3). Although the higher order terms may play important roles in e.g. cancelling the conformal instability found in the full path integral (2.3), here they should be of no consequence.

It is not obvious what role supersymmetry should play in the evolution of the ground state of the Universe, which ostensibly is not supersymmetric. On the other hand, supersymmetry is probably crucial for the string loop corrections to the effective action. We shall discuss the implications of the string loop corrections in Section 6. The supersymmetric extension of the Wheeler-DeWitt equation has been considered in ref. 27.

The Need for Extra Dimensions

The path integral (2.3) is likely to be dominated by the most symmetric metrics. A heuristic argument why this should be so is that the $(D-1)$ surface area $\sqrt{h}$ in the action $S_{eff} \sim \int N d^D x \sqrt{h} \mathcal{L}$ is smallest when the D-space is maximally symmetric. Hence the path integral would be stationary along the trajectory containing the maximally symmetric metrics. We shall therefore restrict ourselves to the minisuperspace of homogenous and isotropic metrics, given by

$$ds^2 = -N^2 dt^2 + \sum_n a_n^2(t) d\sigma_{G_n}^2 \tag{3.5}$$

where $d\sigma_{G_n}^2$ is the line element of G-dimensional maximally symmetric space, and $\sum_n G_n = D - 1$. In other words, we choose the topology of our minisuperspace to be $R \times S^{G_1} \times \cdots \times S^{G_n}$ so that $N^I = 0$ in Eq. (2.5). The scale factors $a_n(t)$ should obey the cutoff condition (3.4). We re-emphasize here that there exist consistency arguments that indicate that there can be at most one time-like variable in string theories[19]. Hence one lapse function N in (3.5) is sufficient.

As required by homogeneity and isotropy at the microscopic scale we shall assume that B_{MN} takes the classical background field value $H_{MNR} = 0$. As we shall see, this is also justified because the prefactor $\phi^{-\frac{8}{D-2}}$ in (3.3) will decrease during the dynamical evolution. Such a constraint on H_{MNR} is also essential in the Calabi-Yau compactification[28] of the heterotic string. Moreover, we will consider only the bosonic sector since we have assumed that the Universe is homogenous and isotropic at the microscopic scale and therefore the fermionic fields should have zero expectation value compatible with their classical equations of motion.

Let us first consider the simplest possibility that the topology is $R \times S^{D-1}$ and there is only one scale factor $a(t)$. The Wheeler-DeWitt equation is obtained by applying the constraint equation (2.7) to the effective action (3.3) and choosing the $|a, \phi >$ basis for the operators such that the canonical momenta Π_k $(k = a, \phi)$ are given by $< a, \phi|\Pi_k|\Psi >= -i\frac{\partial}{\partial k}\Psi[a, \phi]$. Fixing the operator ordering by requiring, as discussed before, the Hamiltonian to be Hermitian, and assuming homogeneity and isotropy so that $(\nabla_M \phi)^2 = -N^{-2}\dot{\phi}^2$, the Wheeler-DeWitt equation (2.7)–(2.8) will read:

$$< a, \phi|\frac{\delta S_{eff}}{\delta N}|\Psi >= H\Psi$$

$$= \left\{ -\left(a\frac{\partial}{\partial a} \right)^2 + \alpha \left(\phi\frac{\partial}{\partial \phi} \right)^2 + \bar{R} \right\} \Psi(a, \phi) = 0 \qquad (3.6)$$

where $\alpha = \frac{1}{4}(D-2)^2(D-1)$ and $\bar{R} = 4(D-1)(D-2)a^{2(D-1)\,(D-1)}R$. Factorizing $\Psi(a, \phi) = \psi(a)\Phi(\phi)$, the equation (3.6) reduces to:

$$\left\{ \left(a\frac{\partial}{\partial a} \right)^2 + \rho - \bar{R} \right\} \psi(a) = 0 \qquad (3.7a)$$

$$\alpha \left(\phi\frac{\partial}{\partial \phi} \right)^2 \Phi(\phi) = -\rho\Phi(\phi) \qquad (3.7b)$$

where ρ is the eigenvalue of the dilaton energy.

The WKB solution of (3.7a) is

$$\psi(a) = \exp \int dx\sqrt{\bar{R} - \rho} \qquad (3.8)$$

where $x = \ln a$. Now applying the Hartle-Hawking boundary condition $\Psi(a \to 0) \to 1$, we find that since $\bar{R} \sim a^{2D-4} \to 0$ as $a \to 0$, the boundary condition demands that $\rho \leq 0$ (since the wave function must remain real). The WKB solution of (3.7) is

$$\Psi(a, \phi) = \exp \left\{ \int dx\sqrt{\bar{R} + |\rho|} + \frac{|\rho|}{\sqrt{\alpha}} \ln \phi \right\} \qquad (3.9)$$

One can see that since both the terms under the square root sign are always positive, the phase of the wave function can never become imaginary and the wave function will never go into the Lorentzian regime. This means that there is no tunneling solution from nothing to our Lorentzian Universe. We note that our interpretation of the Euclidean metric is different from those of Hartle and Hawking. They appear to view the Euclideanization of the metric as a mathematical device akin to Wick rotation which makes the path integral convergent. Physical results about the Universe are obtained by analytic continuation of the Euclidean wave function into the Lorentzian regime. In our analysis we shall see that the Euclidean regime connects nothing to the Lorentzian Universe and we shall interpret the Euclidean solution as a gravitational instanton.

On the basis of the foregoing analysis we conclude that for a space with $R \times S^{D-1}$ topology there is no tunneling solution from nothing to our Lorentzian Universe, and therefore we must discard this case. It also means that we must abandon four dimensions in favor of higher dimensional theories, or add some new degrees of freedom to the four dimensional effective theory. Such extra (fermionic) degrees of freedom are present in the four dimensional formulation of string theories[4-8]. However, some of these D=4 string theories are bound to be equivalent to the higher dimensional compactified string theories.

In string quantum cosmology arbitrary additions (positive cosmological constant, etc.) are not possible since the form of the effective action is uniquely dictated by the vanishing of the beta functions (3.2). Moreover, since the effective action is a series in the slope α', the form of the Wheeler-DeWitt equation (3.6) is valid only as long as the leading terms in the action are larger than the higher order terms. Our results would have to be modified if the overall sign of the energy momentum tensor T_{oo} changes due to the inclusion of the higher order terms. However, we shall see in Section 7 that the perturbation expansion of the action (3.3) is strictly valid and hence the higher order terms in (3.3) cannot change the sign of T_{oo}. We shall also see that the string one-loop correction terms have δT_{oo} of the same sign as T_{oo}, and therefore they do not change the results of this section. Hence we abandon four dimensions (or, rather $R \times S^3$) and move on to higher dimensions, where the D-dimensional space can have a maximally symmetric product topology.

$R \times S^n \times S^m$

The next most symmetric solution has the topology $R \times S^n \times S^m$, where $n + m = D - 1$ ($n, m \neq 0$). With this choice (see Eq. (3.5)), the effective action (3.3b) reduces to

$$
S_{eff} = - \int N dt dx^{D-1} a^n b^m \left\{ {}^{(D-1)}R - \frac{1}{N^2}(n^2 - n)\left(\frac{\dot{a}}{a}\right)^2 \right.
$$
$$
\left. - \frac{1}{N^2}(m^2 - m)\left(\frac{\dot{b}}{b}\right)^2 - \frac{2mn}{N^2}\left(\frac{\dot{a}}{a}\right)\left(\frac{\dot{b}}{b}\right) + \frac{4}{N^2(D-2)}\left(\frac{\dot{\phi}}{\phi}\right)^2 \right\} \text{(3.10)}
$$

where a and b are respectively the scale factors of the n and m dimensional subspaces. To write down the Wheeler-DeWitt equation, we find the relations between the time derivatives of a, b and ϕ to their canonical momenta given by (2.8):

$$
\left(\frac{\dot{a}}{a}\right) = -\frac{N}{2(D-2)a^n b^m}\left[\frac{m-1}{n}a\Pi_a - b\Pi_b\right]
$$
$$
\left(\frac{\dot{b}}{b}\right) = -\frac{N}{2(D-2)a^n b^m}\left[\frac{n-1}{m}b\Pi_b - a\Pi_a\right]
$$
$$
\left(\frac{\dot{\phi}}{\phi}\right) = \frac{-N(D-2)}{8a^n b^m}\phi\Pi_\phi
$$

$$\text{(3.11)}$$

The Wheeler-DeWitt equation is then

$$\left[(m^2 - m)\left(a\frac{\partial}{\partial a} \right)^2 + (n^2 - n)\left(b\frac{\partial}{\partial b} \right)^2 - 2mn\left(a\frac{\partial}{\partial a} \right)\left(b\frac{\partial}{\partial b} \right) \right.$$

$$\left. + 4(D-2)mn(a^n b^m)^{2(D-1)}R + \frac{(D-2)^2}{4}mn\left(\phi\frac{\partial}{\partial \phi} \right)^2 \right]\Psi = 0 \qquad (3.12)$$

We seek a solution to (3.12) by separating the variables. Let $\Psi[a,\, b,\, \phi] = U(x,y)\Phi(\phi)$ with $x = \ln\left(\frac{a}{b} \right)$, $y = \ln\left(a^{\frac{n}{m}}b \right)$. Then (3.12) reads

$$\left[\partial_x^2 - \alpha\partial_y^2 + V(x,y) - \lambda^2 \right] U(x,y) = 0$$

$$\left(\phi\frac{\partial}{\partial \phi} \right)^2 \Phi(\phi) = -\gamma^2\lambda^2\Phi(\phi) \qquad (3.13)$$

where λ^2 is proportional to the eigenvalue of the dilaton energy momentum tensor, and

$$\alpha = \frac{n}{m(D-2)}, \quad \gamma^2 = \frac{4(D-1)}{(D-2)mn} \qquad (3.14)$$

and

$$V(x,y) = \beta^{(D-1)}R, \quad \beta = \frac{4mn}{D-1}(a^n b^m)^2 \qquad (3.15)$$

At very early times a and b are small, and the scalar curvature is $^{(D-1)}R \simeq a^{-2} + b^{-2}$. Therefore at small t the curvature term $V(x,y)$ can be ignored compared to the dilaton energy term λ^2, and the solution to (3.13) is

$$\Phi(\phi) = C_+ e^{i\gamma\lambda\ln\phi} + C_- e^{-i\gamma\lambda\ln\phi} \qquad (3.16)$$

where C_+ and C_- are constants, and

$$U(x,y) = A_o\exp\left[\lambda_1 x + \left(\frac{\lambda_1^2 - \lambda^2}{\alpha} \right)^{\frac{1}{2}} y \right] + B_o\exp\left[-\lambda_1 x - \left(\frac{\lambda_1^2 - \lambda^2}{\alpha} \right)^{\frac{1}{2}} y \right] \qquad (3.17)$$

where λ_1 is a constant of integration.

We now impose the Hartle-Hawking boundary condition on the solution (3.16)–(3.17). As we have already remarked, the effective action that we have used has a cutoff whose scale depends upon the value of ϕ through the equation (3.4). Although we can make our theory valid within scales much smaller than the Planck scale by choosing a large enough value for ϕ, we shall take the initial value of $\phi = 1$ and consequently our theory has a cutoff at the Planck scale which is the only fundamental scale in string theories. Although our field theory effective action has a cutoff at the Planck scale we shall extrapolate our solution below the Planck regime to select the trajectory which is consistent with the Hartle-Hawking boundary condition.

When the volume factor $\sqrt{h} \sim a^n b^m \to 0$, let the scale factor a go to zero as b^q where $q > 0$. Now imposing the condition $\Psi(\sqrt{h} \to 0) \to 1$ on (3.17) we obtain the constraint

$$q = \frac{\lambda_1 - \Delta}{\lambda_1 - \frac{n}{m}\Delta} > 0 \tag{3.18}$$

where $\Delta \equiv \left[\frac{\lambda_1^2 - \lambda^2}{\alpha}\right]^{\frac{1}{2}}$ and can be taken to be positive without loss of generality. Since the phase of the wave function (3.17) has to be real (the wave function is in the Euclidean regime) we have $\lambda_1 \geq \lambda > 0$. This condition together with (3.18) implies that $0 \leq q \leq 1$. Now the classical trajectory is the one where the phase $\sim \int \sqrt{h} N d^D x \mathcal{L}$ is smallest. Thus as long as $a, b < 1$ (in units of Planck length) the volume factor $\sqrt{h} \sim b^{nq+m}$ is a minimum when $q = 1$. Therefore the classical trajectories are the ones for which $a = b$ below the Planck scale. Now applying the condition $\Psi[a = b \to 0] \to 1$ to the wave function (3.17) we get $\lambda = \lambda_1$. Hence our solution at very early times ($\sqrt{h}$ small) reads simply

$$\Psi = \exp(\pm \lambda x \pm i\lambda\gamma \ln \phi) \tag{3.19}$$

where the we have set the initial value is ϕ to be $\phi_0 = 1$. We see that at early times the wave function (3.19) has a real phase, and therefore the lapse function $N = i$ (and the metric in Euclidean). The evolution of the classical quantities $<a>$, $<b>$ and $<\phi>$ can be determined from the phase $S = -i(\lambda x + i\lambda\gamma \ln \phi)$, and their trajectories are given by integrating $\nabla_k S$ ($k = a, b, \phi$) from the initial point ($a = b = 0$, $\phi = 1$). Setting $\nabla_k S = <\Pi_k>$ we obtain from (3.10) the evolution equations for $a = <a>$, $b = <b>$ and $\phi = <\phi>$:

$$\frac{\dot{a}}{a} = \frac{\lambda}{2n} a^{-n} b^{-m}$$

$$\frac{\dot{b}}{b} = -\frac{\lambda}{2m} a^{-n} b^{-m}$$

$$\frac{\dot{\phi}}{\phi} = i \left[\frac{(D-1)(D-2)}{mn}\right]^{\frac{1}{2}} \lambda a^{-n} b^{-m} \tag{3.20}$$

For large times $t > M_P^{-1}$ these have the solutions

$$a = \exp\left(\frac{\lambda(t-1)}{2n}\right)$$

$$b = \exp\left(-\frac{\lambda(t-1)}{2m}\right)$$

$$\phi = \exp\left(i\frac{1}{4}\left[\frac{(D-1)(D-2)}{mn}\right]^{\frac{1}{2}} \lambda(t-1)\right). \tag{3.21}$$

Substituting the equation of motion for $\phi(t)$ into the wave function (3.19) one can see that the wave function has a real phase consistent with the fact that the metric is Euclidean.

We have set $a = b$ at the Planck time. Hence the n-dimensional subspace grows while the m-dimensional subspace shrinks, and the total volume remains constant $V = a^n b^m = (M_P^{-1})^{D-1}$. Physically such a situation is attractive, as it entails that all that "happens" in the Euclidean regime is the deformation of the originally fully symmetric solution. We should also stress that the separation of the scale factors is exponential, so that there is some hope of identifying later the large space with our Universe and the small space with an initially Planck size internal space.

The Transition from Euclidean to Lorentzian Regime

The solution (3.19) is valid only at very small scales when $\lambda^2 \gg (a^n b^m) \times {}^{2(D-1)} R$. To find the behavior of the wave functional at larger scales we seek for a WKB solution that matches (3.19) and its first derivative for small a and b. The reason we seek for the WKB solution instead of solving the equation exactly is that we want the semiclassical trajectories of $< a >$, $< b >$ and $< \phi >$ which can be obtained from the gradient of the phase of the WKB solution. This is given by

$$U(x, y) = C(x, y) \exp \int \sqrt{\lambda^2 - \bar{R}} dx \qquad (3.22)$$

where $C(x, y) \simeq (\lambda^2 - \bar{R})^{-1/4}$ is a slowly varying prefactor expressing the fact that in the semiclassical limit that we are considering, the probability is inversely proportional to the velocity. The matching of the derivatives of (3.22) and of the approximate solution (3.19) tells us that the trajectory of the peak of the wave function is restricted to the plane $y = y_o$ (i.e., $V = const \simeq 1$) for all scales, and

$$\Psi[a, b, \phi] = \exp \left\{ \int dx \sqrt{\lambda^2 - \bar{R}(x, y_o)} - i\lambda\gamma \ln \phi \right\} \qquad (3.23)$$

where $\bar{R}(x, y_o) \propto (a^{-2} + b^{-2})$. Hence the initial exponential separation of the two scale factors a and b slows down as $\bar{R} \to \lambda^2$ and stops on the hypersurface $t = t_o$ where $\bar{R} = \lambda^2$. Strictly speaking the WKB solution (3.23) will be invalid on this surface because of the singularity of the prefactor in (3.22). The point where $\bar{R} = \lambda^2$ is the turning point familiar in WKB analysis where one can match the Euclidean and the Lorentzian solutions.

To do this, let us write the classical values at a and b on the surface t_o as

$$a = \exp(z/n), \quad b = \exp(-z/m) \qquad (3.24)$$

where

$$z = \int\limits_1^{t_o} dt \sqrt{\lambda^2 - \bar{R}(t)} \qquad (3.25)$$

is the expansion factor whose numerical value depends on the values at λ^2 and t_o. We shall not commit ourselves to any particular numerical value, but assume that the dilaton energy term λ^2 and the scale at the hypersurface t_o (in Planck units) are large enough to separate the scale factors a and b by many orders of magnitude at the surface t_o.

Let us now match to (3.23) the WKB solution of the Wheeler-DeWitt equation in the region $\bar{R} > \lambda^2$. Such solution is

$$\Psi_L[a,\, b,\, \phi] = \exp i \left\{ \int dx \sqrt{\bar{R} - \lambda^2} - \gamma \lambda \ln \phi \right\} \qquad (3.26)$$

where the subscript L denotes the fact that the wave function now has a complex phase and the metric is therefore Lorentzian (i.e., $N = 1$). This change in the topology of the metric is a quantum transition which is not allowed in classical general relativity, where the metric evolves in time but does not change its topology (signature). For an earlier discussion of this issue, see ref. 29.

Hence at time $t > t_o$ the topology of the Universe is $M^{n+1} \times K^m$, where M^{n+1} is a $n+1$ dimensional maximally symmetric Lorentzian space with a scale factor $a \simeq \exp(z/n)$, and K^m is an internal space with a small scale factor $e^{-z/m}$. However, the subsequent evolution of M^{n+1} based on $K^m \simeq S^m$ is not realistic as the volume of the compact space actually starts to decrease. From particle physics considerations we also know that it would be desirable to have an internal space which is compact with a constant radius. Otherwise, a varying internal radius would imply varying fundamental constants which, for example, may have catastrophic consequences on the production of primordial helium during nucleosynthesis[30]. So, we need to have a constant internal radius. We will now turn to such compactified solutions of K^m.

MATCHING OF COMPACTIFIED SOLUTIONS

In the D=10 heterotic string theory[2] a possible compacitified solution to the equation of motion[28] is $M^4 \times K^6$, where M^4 is four dimensional Minkowski space and K^6 is a Calabi-Yau space, which is Ricci-flat. Such compactifications were obtained by requiring one unbroken supersymmetry in four dimensions, and an integrability condition[28] is satisfied if $H_{MNP} = 0$. It could be that in the present context it would be possible to demonstrate the occurance of compactification by taking into account also the fermionic sector and considering the resulting Wheeler-DeWitt equations, but it may also turn out that compactification is something that happens only when high orders in α' are taken into account. We may also speculate that compactification is somehow triggered by the shrinking of b. However, here we will simply assume that the m-dimensional internal space becomes Ricci flat about the time when the topology of the metric changes from Euclidean to Lorentzian. Ricci flatness is crucial for the stability of the internal space, as we will see below.

Let us now assume that the internal space is completely Ricci flat at the time t_* and the wave functional is given by

$$\Psi_L = \exp i \left\{ \int dx \sqrt{\bar{R} - \lambda_*^2} dx - \gamma \lambda_* \ln \phi \right\} \qquad (4.1)$$

where now

$$\bar{R} = \frac{4mn}{D-1}(a^n b^m)^{2(D-1)} R \simeq \frac{4mn}{D-1} a^{2n-2} b^{2m} \qquad (4.2)$$

as the scalar curvature of $M^{n+1} \times K^m$ with K^m Ricci flat is $\sim a^{-2}$. In (4.1) and (4.2) we have accounted for possible shifts in energy during compactification by letting $\lambda \to \lambda_*$. It is easy to verify that now

$$\frac{\dot{b}}{b} = -\frac{N}{2(D-2)a^n b^m} \left\{ \frac{(n-1)}{m}(2m) - (2n-2) \right\} \bar{R} \frac{\partial S_{eff}}{\partial \bar{R}} = 0 \qquad (4.3)$$

so that for $t > t_*$ the radius of K^m is constant. One finds also that $a \sim t$ and that

$$\frac{\dot{\phi}}{\phi} = \frac{\lambda \gamma}{4} \left[\frac{(D-1)(D-2)}{mn} \right]^{\frac{1}{2}} a^{-n} b^{-m} \qquad (4.4)$$

which has the solution

$$\phi = \phi_o \exp(\chi_* t^{-2}) \qquad (4.5)$$

where χ_* is a constant.

Let us now consider what can happen during compactification. We assume that before compactification the scale factors have reached the asymptotic values $a(t_o)$ and $b(t_o)$ given by Eq. (3.24). The process of compactification is likely to be fast, as the relevant time scale should be given by $b(t_o)^{-1}$, and cannot act coherently at distances $\gg b(t_o)$. Therefore we will also assume that during compactification the scale $a(t_o)$ ($\gg b(t_o)$) does not change. Instead, b may and in fact will change, and we will denote the constant value of b after compactification by b_*.

We assume that the gravitational energy stored in the curvature term $\sim a^n b^{m(D-1)} R$ stays constant when the internal space becomes Ricci flat, i.e., as $^{(D-1)}R = (a^{-2} + b^{-2}) \to^{(D-1)} R = a^{-2}$. Therefore equating the gravitational energy at times t_o and t_* (before and after the Ricci flattening process) we get:

$$b_*(t > t_*) = \exp \left[\frac{z}{m} \left(2 \frac{(m+n)}{mn} - 1 \right) \right]. \qquad (4.6)$$

The scale factor changes discontinuously (by "quantum jump") during the tunneling from the Euclidean to Lorentzian regime. We shall use the relation (4.6) later in order to determine the values of the dimensions D, m and n. Our assertion that neither gravitational ($\sim a^n b^{m(D-1)} R$) nor the dilaton energy ($\sim \lambda^2 a^{-n} b^{-m}$) is either dissipated or transmitted is consistent with the fact that the phase of the wave function $\sim \sqrt{\bar{R} - \lambda^2}$ should match at the turning point:

$$(a_o^n b_o^m)^2 (a_o^{-2} + b_o^{-2}) - \lambda^2 = (a_*^n b_*^m)^2 (a_*^{-2}) - \lambda_*^2 = 0. \qquad (4.7)$$

We note here that due to the exponential expansion of the scale factor a, the large space in the Lorentzian regime has curvature $\sim a^{-2} \sim 0$ which may be the solution to the flatness problem[31] of the Big Bang cosmology.

CONSTRAINTS ON SPACETIME DIMENSIONS

Thermodynamic Stability

We have assumed that the $D - 1$ space has microscopic homogeneity and isotropy. The dilaton field is therefore constant over the $(D - 1)$ space and hence there is no temperature or entropy associated with the dilaton (or the other matter fields). It is well known[32] however that spaces with 'horizon' have a 'Hawking temperature' associated with the gravitational degrees of freedom.

At time $t < t_o$ (in the Euclidean regime) the space we have considered has a product topology $R \times S^n \times S^m$ where S^n is an n-dimensional de Sitter space with scale factor $a \sim \exp\left(\frac{z}{n}t\right)$ and S^m is an exponentially contracting space with a scale factor $b \sim \exp\left(-\frac{z}{m}t\right)$. The Hawking temperature associated with the subspace S^m (the process is identical for S^n) may be calculated from[33] the transition amplitude

$$< b_2 t_2 | b_1 t_1 > = \int d[b] \exp(iI) \tag{5.1}$$

where the action is over all possible $b(t)$ which take the value b_1 at t_1 and b_2 at t_2. But $< b_2, t_2 | b_1, t_1 > = < b_2 | \exp iH(t_2 - t_1) | b_1 >$ where H is the Hamiltonian. If we set $t_2 - t_1 = -i\beta$ and $b_1 = b_2$ and sum over b_1 we obtain

$$Tr \exp(-\beta H) = \int d[b] \exp(iI) \tag{5.2}$$

where now the path integral is taken over all $b(t)$ which are periodic with period β in imaginary time. The left hand side is just the partition function of the system with gravitational degree of freedom $b(t)$, with a temperature β^{-1}. Now the path integral will be dominated by the paths $b(t)$ which satisfy the equation of motion $b(t) \sim \exp\left(-\frac{z}{m}t\right)$. The temperature β^{-1} for which the Greens function is periodic in imaginary time is $T_b = \left(\frac{z}{2\pi m}\right)$. Using the same procedure one can determine the temperature of the subspace S^n to be $T_a = \left(\frac{z}{2\pi n}\right)$. Now since the scale factors $a(t)$ and $b(t)$ change exponentially with time, there is no chance of establishing thermal equilibrium between S^m and S^n which will therefore have different temperatures T_a and T_b. The gravitational entropy density of the whole space will be $S_{tot} = S_g^a + S_g^b$, where $S_g^a(S_g^b)$ is the gravitational entropy density of $S^n(S^m)$. During the process of expansion of S^n and contraction of S^m the total volume, energy and the matter energy density T_{oo} remain constant $\left(\frac{d}{dt}(T_{oo}V) = 0\right)$. Since the Hamiltonian $H = Hg + H\varphi = 0$, the energy density of the gravitational modes[34] $\rho_g = -T_{oo} = $ constant (and since $T_{oo} > 0$, $\rho_g < 0$). Therefore during the process of expansion/contraction there is a transfer of energy from S^m to S^n so that the total energy and the energy density stays the same. In this transfer of energy between spaces of different temperatures there is a generation of entropy of mixing so that $dS_{tot} > 0$. Therefore using $T_a dS_a = d(\rho_g V_a) + p_g dV_a$ and a similar relation for dS_b we have

$$dS_{tot} = \frac{1}{V_a T_a}(p_g + \rho_g)dV_a + \frac{1}{V_b T_b}(\rho_g + p_g)dV_b > 0. \tag{5.3}$$

Using the equation of state for massless fields in n spatial dimensions $p_g = \frac{1}{n}\rho_g$, and setting $V_a \propto a^n$ and $V_b \propto b^m$ we get

$$dS_{tot} = 2\pi T_{oo}(m - n) > 0. \tag{5.4}$$

In the usual inflationary models the constant energy density is related to the pressure through the relation $\rho = -p$ and therefore $Tds = pdV + d(\rho V) = 0$ and the expansion is adiabatic (the energy increases by the work done against negative pressure). Since in our model we do not have this equation of state the process in our case is not adiabatic and the entropy change dS_{tot} is strictly greater than zero.

Consistency Conditions on Dimensions

Let us now return to (4.6). We see that b_* differs many orders of magnitude from the D-dimensional Planck scale unless

$$n = \frac{2m}{m - 2} \tag{5.5}$$

In particular, b_* will also differ from the four dimensional Planck scale, which goes as b_*^{D-5}, unless (5.5). If b_* is much different from the Planck scale, the four dimensional gauge coupling will turn out to be unrealistic. Hence we impose (5.5) to get a physically meaningful theory.

It is then easy to see that the only integer solutions to (5.5) are given when (n, m) equals $(3,6)$, $(4,4)$ or $(6,3)$. Now from equation (5.4) we see that we have the added constraint that $m > n$. With this constraint the only solution of (5.5) is

$$D = 10, \quad n = 3, \quad m = 6. \tag{5.6}$$

It is quite remarkable that out of the many possibilities we find that the dimension of our space-time is indeed determined to be four. We also have the independent confirmation that the heterotic string theory is consistent only when $D = 10$.

STRING LOOP CORRECTIONS

In string theories at the tree level the vacuum energy is zero (SUSY is unbroken) for any value of the dilaton ϕ. Thus ϕ labels the degenerate vacua. Since there is no potential term for ϕ at the tree level all equations of motion obtained so far are invariant under the transformation $\phi(t) \to \phi(t) + \phi_o$. At the one loop level there have been general arguments as well as explicit calculations[25] which show that there will be a potential for ϕ of the form $V(\phi) \sim C\phi^\alpha$ (with $\alpha = \frac{4}{D-2}$). With this potential we shall calculate the dynamical evolution as well as the asymptotic value of ϕ which is of interest for the following reasons. We have already remarked that the tree level effective action (3.3) is an expansion in $\left(\frac{\phi^{-4/D-2}}{b^2} \right)$ and in order for this perturbation expansion (on which our entire analysis is based) to be valid ϕ must satisfy the criterion $\frac{\phi^{-4/D-2}}{b^2} \ll 1$. Also ϕ is the coupling strength for the loop amplitudes therefore it is desirable that ϕ remain small.

Lastly, the coefficient $\phi^{-4/D-2}$ of the F^2 term determines the gauge coupling strength as $\frac{1}{g^2} \sim \phi^{-4/D-2}$. We know on phenomenological grounds that $g \sim 1$. Therefore the asymptotic value of ϕ should be consistent with this. With the inclusion of the potential term, the dilaton sector of the Lagrangian (3.3) now reads:

$$\mathcal{L} = N\sqrt{D-1}g \left(\frac{+1}{N^2} \left(\frac{\dot{\phi}}{\phi} \right)^2 - C\phi^{\alpha}. \right) \tag{6.1}$$

We can convert the kinetic term into the cannonical form by the substitution $\sigma = \ln \phi$. The equation of motion for σ obtained from (6.1) then reads:

$$\frac{1}{N^2}\ddot{\sigma} + \frac{1}{N^2}\frac{1}{a^n b^m}\frac{d}{dt}(a^n b^m)\dot{\sigma}^2 + \alpha C e^{\alpha\sigma} = 0 \tag{6.2}$$

when the Universe is in the Euclidean regime $(t < t_o)$ the lapse function $N = i$ and the volume $a^n b^m = $ constant. Therefore the equation of motion (6.2) reduces to

$$\ddot{\sigma} = \alpha C e^{\alpha\sigma}. \tag{6.3}$$

It is clear that in the Euclidean regime the effective potential is upside down $(V_{Euclidean} = -Ce^{\alpha\sigma})$ and $\sigma(t)$ increases rapidly by rolling down the potential hill. At small t, where we have $\phi_o = 1$, we can solve the linearized form of (6.3) to obtain

$$\sigma(t) \sim \frac{1}{\alpha^2 C}\exp(\alpha\sqrt{C}t), \tag{6.4}$$

from which we get for $(t < t_o)$

$$\phi(t) \sim \exp\left\{ \frac{1}{\alpha^2 C}\exp(\alpha\sqrt{C}t) \right\}. \tag{6.5}$$

At the time $t = t_o$ the Universe tunnels into the Lorentzian regime. For $t > t_o$ we can set the lapse function $N = 1$ and we have already obtained $b = 1$. The scale factor a will in general have a power law solution $a \sim t^m$ where m is determined by the equation of state of the matter generated. Therefore the equation of motion (6.2) now reduces to

$$\ddot{\sigma} + \frac{m}{t^m}\dot{\sigma}^2 + C\alpha e^{\alpha\sigma} = 0. \tag{6.6}$$

Since we want the asymptotic solution we can drop the middle term $\sim t^{-m}$. It is clear that now the potential term has the right sign and as $t \to \infty$, σ will roll down the potential $V(\sigma) \sim e^{\alpha\sigma}$ towards the value $\sigma \to 0$. Again expanding the potential term in (6.6) in a power series and retaining only the linear term we obtain the time evolution of $\sigma(t)$ to be

$$\sigma(t) = \frac{1}{\alpha^2 C}\exp(i\alpha\sqrt{C}t). \tag{6.7}$$

We see that $\sigma(t)$ oscillates at the bottom of the potential well. The frequency of oscillation $V \sim \alpha\sqrt{C}$ and would be small for a small cosmological constant C. This oscillatory solution would be damped by the anharmonic terms in the potential $Ce^{\alpha\sigma}$ as well as by the gravitational decay of σ, which we have neglected. Therefore at large t we expect $\sigma(t)$ to settle down at the minima of the potential $\sigma = 0$. Thus the expectation value $< \sigma(t \to 0) >= 0$ and $< \phi >= 1$. Consequently, at large t the value of the gauge coupling constant converges to 1.

Now the cosmological constant C need not at this stage be fine-tuned to the value of the observed cosmological constant since the potential term $C\phi^\alpha$ represents only one of the terms in the string one-loop correction and if one believes that SUSY is not broken at this one-loop level, then this term should be canceled by a corresponding term in the fermionic sector and in the end one should have a vanishing cosmological constant. Therefore at this stage one can assume C to be a number $O(1)$, which has to be provided by the string theory. From equation (6.5) one can see that the cutoff condition $\phi^{-\frac{4}{D-2}}/b^2$ is easily satisfied (since $b \sim \exp\left(-\frac{z}{m}\right)t$ in the Euclidean regime). We note that even though the one-loop term $C\phi^\alpha$ is quite large during the Euclidean era, it has the same sign as the T_{oo} term (which is denoted bt ρ in Sec. 3.2 and λ^2 in Sec. 3.3), and therefore our previous analysis based on the tree level action (3.3) remains valid and is in fact strengthened since we need a large value of $\lambda^2 (\equiv T_{oo} + \delta T_{oo})$ in order to get condition (5.5) which fixes the dimension m and n, and in order to obtain a large enough exponential expansion of $a \sim \exp\left(\frac{\lambda}{2n}t\right)$ which solves the flatness problem.

DISCUSSION AND CONCLUSIONS

For $t \gg t_o$ the dilaton field oscillations die out and $\phi \sim 1$. Hence in the present scenario the values of both the breathing mode (b) and the dilaton are fixed; in low energy superstring models[35] these are related to the fields S and T, whose values are therefore fixed by quantum cosmology. However, this fixing can only be approximately true, because there is no matter in our model Universe, and this is why the scale factor $a \propto t$. Note that the curvature term in the Einstein equation will not be important because of the exponential expansion of $a(t)$ in the Euclidean region which has a potential for solving the flatness problem of the Standard Big Bang Cosmology, as was mentioned in Section 4.

The early behavior of the scale factor ($\alpha \propto t$) will however be modified. This unrealistic situation will be remedied at large t when the background matter fields, which we have set to zero, thermalize. If we define temperature by $aT \sim tT \sim 1$, a typical time scale for thermalization would be[36] $T \sim 1/t \sim 10^{16} \, GeV$.

The rapid change of scale factors with time gives rise to particle creation[37], and in fact these particles are thermalized already at their creation. Hence eventually quantum effects will be overwhelmed by classical behavior, and our Universe starts to evolve according to the classical four dimensional Hamiltonian $H = 6(\dot{a}/a)^2 + 6k/a^2 - T_{oo} = 0$, where T_{oo} will be the familiar energy momentum tensor of radiation-matter.

The quantum aspect of our cosmological picture gives rise to three major results. The first was that the dimension of space-time should be larger than four, as was discussed in Section 3.2. . Such a result, although almost trivial, is worth emphasizing because this is the one ingredient that pushed us to ten dimensions. This is our second main result. The consistency of quantum cosmology as outlined in Sections 3 and 4 with compactification requires that D=10. It is intriguing that quantum cosmology picked out just D=10, where we know that quantum theories of superstrings exist. Moreover, it is equally remarkable that compactification in arbitrary dimensional Ricci flat internal space, required by stability of that space, along with thermodynamic stability criteria, leads us precisely to six dimensional internal space and to a large four dimensional space-time.

The third main point, which is directly related to the first, is that the initially Euclidean Universe can change its topology and become Lorentzian. Such Euclidean Universes can be considered as gravitational instantons, which are tunneling solutions from nothing to a finite, non-singular Lorentzian Universe. Hence before the Big Bang there was an instanton. In this manner our quantum cosmology provides a dynamical explanation for the signature and dimensionality of the observable space-time. Of course, the question "What happened before the Big Bang?" is obsolete because the words "before" and "after" do not carry the same operational meaning in the Euclidean regime as they do in the Lorentzian regime.

It is worth emphasizing that our treatment of superstring quantum cosmology was appropriately done by using the bosonic formulation[2-3] of superstrings. If we want to find out what is so special about $D = 10$ or $D = 4$, we have to leave D arbitrary and let dynamics choose the "lucky" numbers. Clearly the bosonic formulation of strings fits best. On the other hand the equivalence (?) between bosonic and fermionic formulations, alluded in the introduction, indicate that corresponding results may hold true also in the fermionic formulation. Indeed, some recent results[38,39] concerning the universality of the mass spectrum in closed string models, obtained in the fermionic formulation of strings look very encouraging. As is well-known[1] the asymptotic expansion for the level density in a string model starts with a term.

$$n(m) \sim C m^{-\alpha} e^{bm} \qquad (7.1)$$

for large masses m, where α is a number, $b = 0(\sqrt{\alpha'})$ and $C = 0(\sqrt{\alpha'})^{\alpha}$. Very recent calculations[38,39] show that

$$\alpha = D = (\# \text{ space} - \text{time dimensions}) , \qquad (7.2)$$

independent of the type of superstring! and

$$b = \begin{cases} \frac{\sqrt{C-2}}{\sqrt{6}} \pi \sqrt{\alpha'} & \text{for} \quad N = 0 \text{ world sheet supersymmetry} \\ \frac{\sqrt{C-3}}{\sqrt{6}} \pi \sqrt{\alpha'} & \text{for } N = 1 \text{ world sheet supersymmetry} \end{cases} \qquad (7.3)$$

independent of space–time dimensions!

In other words,

$$b = \begin{cases} (2 + \sqrt{2})\,\pi\sqrt{\alpha'} & \text{for the heterotic string} \\ 2\sqrt{2}\pi\sqrt{\alpha'} & \text{for type II superstring} \end{cases} \tag{7.4}$$

independent of the number the space–time dimensions in which the superstring theory is formulated. This apparent *double* universality (α, b) of the superstring asymptotic mass spectrum may have some profound consequences. For example, the D–independence of b–(7.4) express the fact that all possible heterotic (type II superstring) models could still be variants of the same (another) theory, etc. ...

It goes without saying that there is a lot to be understood both in superstrings and in quantum cosmology. Still, it is encouraging that such a naive and rough and amagalmation presented here already may shed light in such a profound question as: Why do we live in 4–dimensions? For sure we have not heard the last word from the superstring quantum cosmology front and it may as well be that we even misheard the first ones.

Acknowledgements

I would like to thank K. Enqvist and S. Mohanty for an enjoyable collaboration that lead to the work in Ref. 14, on which this review is based. Once more, Professor Zichichi and his crew have given us an excellent opportunity to discuss physics in a relaxed, friendly and civilized atmosphere. Thank you all, for your successful efforts to keep us happy.

This work was supported in part by DOE grant DE-AC02-76ER00881 and in part by the University of Wisconsin Research Committee with funds granted by the Wisconsin Alumni Research Foundation.

REFERENCES

1. See, e.g.: M.B. Green, J.H. Schwarz and E. Witten, Superstring Theory (Cambridge University Press, 1987) and references therein.

2. D.J. Gross, J.A. Harvey, E. Martinec and R. Rohm, Phys. Rev. Lett. 54 (1985) 502; Nucl. Phys. B256 (19850 253; Nucl. Phys. B257 (1986) 75.

3. L. Dixon, J.A. Harvey, C. Vafa and E. Witten, Nucl. Phys. B261 (1985) 651; Nucl. Phys. B274 (1986) 286.

4. I. Antoniadis, C. Bachas, C. Kounnas and P. Windey, Phys. Lett. 171B (1986) 51.

5. K.S. Narain, Phys. Lett. 169B (1986) 41; W. Lerche, D. Lüst and A.N. Schellekens, Nucl. Phys. B287 (1987) 477.

6. H. Kawai, D.C. Lewellen and S.H.H. Tye, Phys. Rev. Lett. 57 (1986) 1832; Nucl. Phys. B288 (1987)1.

7. I. Antoniadis, C. Bachas and C. Kounnas, Nucl. Phys. B289 (1987) 87.

8. I. Antoniadis and C. Bachas, CERN Preprint TH.4767 (1987).

9. See e.g., G.F.R. Ellis and S.W. Hawking, "The Large Scale Structure of the Universe", Cambridge University Press (1973).

10. I. Antoniadis, G.F.R. Ellis, J. Ellis, C. Kounnas and D.V.Nanopoulos, Phys. Lett **191B** (1987) 393.

11. S.W. Hawking in: Relativity, Groups and Topology II, Les Houches 1983, Session XL, ed. by B.S. DeWitt and R. Stora, North Holland (1984).

12. J.B. Hartle and S.W. Hawking, Phys. Rev. D 28 (1983) 2960.

13. E. Tryon, Nature 246 (1983) 396; A. Vilenkin, Phys. Lett. 117B (1982) 25, Phys. Rev. D27 (1983) 2848, ibid. D30 (1984) 509; ibid D33 (1986) 3560.

14. K. Enqvist, S. Mohanty and D.V. Nanopoulos, Phys. Lett. **192B** (1987) 327 and University of Wisconsin–Madison preprint MAD/TH/87-9.

15. B.S. DeWitt, Phys. Rev. 160 (1967) 1113; J.A. Wheeler, in *"Battelle Rencontres"*, ed. by C.DeWitt and J.A. Wheeler (Benjamin, NY 1968).

16. See for example in "Gravitation", C.W. Misner, K. Thorne and J.A. Wheeler, Freeman (1973).

17. For a review see e.g., J.B. Hartle, UCSB preprint TH-71 (1985).

18. For a review, see K. Kuchar in "Quantum Gravity 2, A Second Oxford Symposium", ed. by C.J. Isham, R. Penrose and D.W. Sciama, Clarendon Press (1981).

19. For a recent concise discussion see S. Weinberg, Univ. of Texas preprint UTTG-22-86 (1986).

20. C.G. Callan, E.J. Martinec, M.J. Perry and D. Friedan, Nucl. Phys. B 262 (1985) 593.

21. M.B. Green and J.H. Schwarz, Phys. Lett. 149B (1984) 117.

22. S. Cecotti, S. Ferrara, L. Girardello and M. Porrati, Phys. Lett. 164B (1985) 46.

23. B. Zwiebach, Phys. Lett. 156B (1985) 315.

24. For a recent review see: M.D. Freeman, C.N. Pope, M.F. Sohnius and K.S. Stelle, CERN preprint TH-4639 (1987) and references therein.

25. C. Lovelace, Nucl. Phys. B273 (1986) 413; W. Fischler and L. Susskind, Phys. Lett. 173B (1986) 262; C.G.Callan, C. Lovelace, C.R. Nappi and S.A.Yost, Princeton preprint (1986).

26. M.Dine and N. Seiberg, Phys. Rev. Lett. 55 (1985) 366; V. Kaplunovsky, Phys. Rev. Lett. 55 (1985) 1036.

27. P.D.D'Eath, Phys. Rev. D29 (1984) 2199; R.C. Furlong and H. Pagels, Rockefeller Univ. preprint RU 86/B1/185 (1986).

28. P. Candelas, G.T. Horowitz, A.Strominger and E. Witten, Nucl. Phys. 258 (1985) 46; E. Witten, Nucl. Phys. 258 (1985) 75.

29. D. Brill, in "Magic Without Magic", ed. by J.R. Klauder, W.H. Freeman (1972).

30. E.W. Kolb, J. Perry and T.P. Walker, Phys. Rev. $\underline{D33}$ (1986) 869.

31. A. Guth, Phys. Rev. $\underline{D23}$ (1981) 347.

32. G.W. Gibbons and S.W. Hawking, Phys. Rev. $\underline{D15}$ (1977) 2738.

33. G.W. Gibbons and S.W. Hawking, Phys. Rev. $\underline{D15}$ (1977) 2752.

34. G. Horowitz and D. Weil, Phys. Rev. $\underline{D33}$ (1986) 567.

35. E. Witten, Phys. Lett. 155B (1985) 151; M. Dine, R. Rohm, N. Seiberg and E. Witten, Phys. Lett. 156B (1985) 55; E. Cohen, J. Ellis, K. Enqvist and D.V.Nanopoulos, Phys. Lett. 161B (1985) 85.

36. J. Ellis and G. Steigman, Phys. Lett. 189B (1980) 186.

37. L. Parker and S.A. Fulling, Phys. Rev. D9 (1974) 314; Ann. Phys. 87 (1974) 176.

38. I. Antoniadis, J. Ellis and D.V. Nanopoulos, Madison preprint MAD/TH/87–23, CERN preprint CERN–TH.4824/87.

39. M. Axenides, S.D. Ellis and C. Kounnas, Berkeley preprint UCB–PTH–87/27.

DISCUSSION

– *Schuler:*

Could you comment on the recent claims that $k = 0$ or $\Omega = 1$ based on counting the number of galaxies?

– *Nanopoulos:*

There are two different observations. One of those is a better method to determine the number density of galaxies by taking bigger and bigger samples (this is the Loh-Spielar counting), and from this they get $\Omega = 0.9 \pm 0.2$.

Secondly, it seems that not only our galaxy but a lot of galaxies around us are moving towars some attractor with very high peculiar velocities (600-700 km/sec). In order to explain that, you need Ω to be close to one. Of course these measurements have to be confirmed by other observations too.

– *Molten:*

Even though a number like the Hubble constant, which is a basic astrophysical quantity, has a factor of two uncertainty, how much should we believe the constraints on particle physics coming from cosmology?

– *Nanopoulos:*

If you noticed in the formulae I have written in my trasparencies, I was careful enough to write explicitly the dependence on the Hubble parameter so that it can vary from 50 to 100 Km $sec^{-1}Mpc^{-1}$. Whenever you are talking about an astrophysical constraint you have to take this possible uncertainty into account.

– *Martinelli:*

If the top quark is about 100 GeV, is there any constraint from energetic "solar" neutrinos?

– *Nanopoulos:*

Let us look at the diagram for photino annihilation.

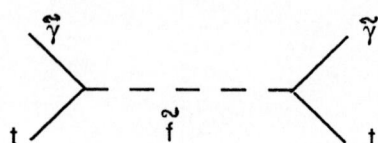

Of course, other things like electron etc. could be produced instead of the top quark. We are interested in the top quark because it can then have a chain of decays producing neutrinos. In order then for the top quark not to be produced you must have $m_{\tilde{\gamma}} < m_t$. In these theories $m_\gamma \simeq \frac{1}{6} m_{1/2}$ (the SUSY breaking parameter). If you have $m_t \simeq 40$ GeV then $m_{1/2} \simeq 240$ GeV and looking at the figure you get a constraint. If instead $m_t = 100$ GeV then $m_{1/2} \leq 600$ GeV. The fluxes of the "solar" energetic neutrinos may be unobservably small, making constraints more difficult to apply.

– *Zichichi:*

Could you please say something on the extremely low value of the Cosmlological Constant. In spite of the present "particle physics" knowledge ($\Lambda_c \simeq (10^{12}\ GeV)^4$) you need $\Lambda \simeq 0$?

– *Nanopoulos:*

In models with a gauge symmetry you need to break it (except for EM and QCD) and still keep the theory renormalizable. The way to do this is through spontaneous symmetry breaking. That is instead to have a potential like this:

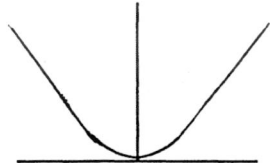

you have one like this

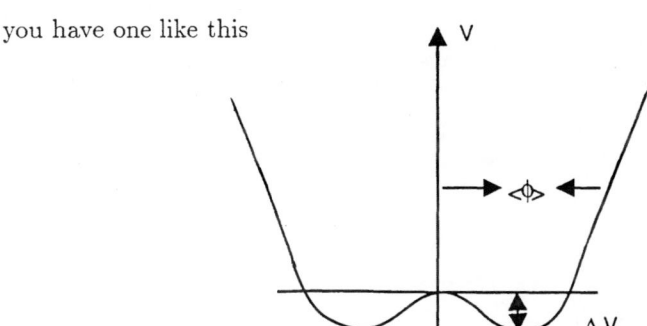

where ΔV is of the order of $< \phi >^4$. For instance in usual $E - W$ interaction $\Lambda_{COS} \simeq \Delta V \sim M_W^4$ whereas in GUTS $\Delta V \sim (10^{15}\ GeV)^4$. This is a big problem. In SUSY theories due to cancellation between bosons and fermions $\Lambda = 0$ if SUSY is exact. Of course if you break SUSY you are going to shift Λ from zero. So the problem remains. If we go to N=1 SUGRA then the potential is of the form

$$V \;=\; e^G[G_i((G_j^i)^{-1}G^j - 3)]$$

where G is a function, $G(Z, Z^*)$ and $G_i \equiv \frac{\partial G}{\partial Z_i}$. In this sense the SUGRA potential is highly constrained. In order to have V = 0 apart from fine tuning you can choose

$$G \;=\; 3\ell n\,(Z + Z^*).$$

so that $V \equiv 0$ and you get the so-called "no scale" model where the potential is everywhere flat. The interesting thing about such models is that they give you $M_W \sim e^{-1/\alpha}M_{p\ell}$ so that you can understand why $M_W << M_p$. Also this kind of theory comes out a low energy theory from some string models. People have found a symmetry responsible for the vanishing of the cosmological constraint in that case. This is a non-compact symmetry, SU(1,1). Another idea is to use some kind of Peccei–Quinn symmetry which allows the cosmological constant to relax dynamically to zero. But it is fair to say that a complete solution to the problem is still lacking.

– Windey (comment):

There are other ways to get the cosmological constant to be zero in the context of superstring models which rely on modula invariance. In such cases, without the need of supersymmetry but just a symmetry of moduli space called "Atkin-Lehner" symmetry one can arrange Λ to be zero. This has been pointed out by G. Moore.

– Nanopoulos:

Of course this is at the string level when you have the next symmetry breaking at low energy you may generate a cosmological constant.

– Windey:

Can you elaborate on the "dark matter detector"?

– Nanopoulos:

The basic idea is the following: you have some superconducting material which when it absorbs, say a photino, it is heated locally, generating a destruction of superconductivity. By detecting that, you can measure the energy released in the sample, and using the cross-section of the interaction between the photino and the detector you can put some bound on the photino mass. This you can do for every dark matter particle.

Of course, there are other kinds of such detector.

328

– Quackenbush:

Would you elaborate on some of the processes that would generate a baryon asymmetry?

– Nanopoulos:

As I was saying in the morning

$$\frac{n_B}{n_\gamma} \simeq 10^{-10} \implies \frac{n_B - n_{\overline{B}}}{n_B + n_{\overline{B}}} \sim 10^{-10}$$

which means that you have an extra quark for every 10 billion $q\bar{q}$ pairs. What particle theorists are thinking about, is to start with a symmetric universe, $N_q = N_{\bar{q}|_{t=0}}$.

In order to create an asymmetry, you need, first of all, a baryon number violating process. Then you need also a particle-antiparticle asymmetry and also a time reversal asymmetry (otherwise if you are creating a baryon here, then by time reversal you are creating an antibaryon somewhere else). Another important thing is to be out of equilibrium because due to CPT invariance $m_q = m_{\bar{q}}$, so that if you are at equilibrium their Boltzman factors are the same). The three conditions above were first written down by Sakharov in 1967. These conditions are satisfied in standard GUTS. For example, consider the following diagrams

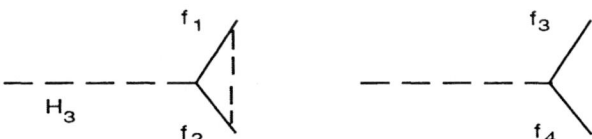

which are the decay of a superheavy Higgs. You notice that I need to have two channels with different baryon number so that $B_{f_1-f_2} \neq B_{f_3-f_4}$; and the interference of the two graphs produces the appropriate C and CP violations. Then it is straightforward to calculate

$$\frac{n_B}{n_\gamma} \sim \frac{n_x}{n_\gamma} \cdot \varepsilon_{CP} \cdot \left(\frac{m_b}{m_W}\right)^2 \sim (10^{-7} - 10^{-9})$$

$$\downarrow \qquad \downarrow \qquad \downarrow$$

$$(10^{-2})\,(10^{-1} - 10^{-3})\,(10^{-4})$$

Of course, this is not the only scenario for baryon number assymetry generation.

– Miele:

Could you repeat the considerations about the upper bound of the number of neutrinos, because you gave us two different bounds?

– Nanopoulos:

As I mentioned in the morning you can work out a formula for the Helium abundance which depends on three experimental parameters. Now one group (YOSST) has measured certain values of these parameters and it gets $N_\nu < 4$ while the other group (EENS) finds $N_\nu < 5.5$ using values for these parameters which for a particle physicist are very close to the previous ones. So the matter can be settled only by a better experimental determination of these parameters.

– Liu:

Do we expect astrophysical events which will give evidence for SUSY and/or Higgs?

– Nanopoulos:

As I mentioned before, dark matter can be consisting of photinos or other SUSY particles which may be detected by the dark matter detector. Of course, if you detect dark matter it does not mean that you will find SUSY particles because it may be heavy neutrino or axions or other things. But the possibility is there. As far as the E-W Higgs is concerned I don't know any dramatic astrophysical consequence apart from the Weinberg-Linde bound in the standard (non-SUSY) electroweak model, $m_H \geq 10$ GeV, (otherwise you don't have a stable vacuum).

– Klemm:

The photino solves the "dark matter problem" on large scale. Mr. Nanopoulos calculated the present density to be $\rho_{\tilde\gamma} = (\frac{1}{2} - 2) \times \rho_{crit}$.

What is the effect of the photino at small scales, what is the effect of the photino to the distribution of angular velocity of visible objects around the centre of mass of galaxies?

– Nanopoulos:

To state the problem let us look at the velocity distribution around the centre of mass of a galaxy. If it were a sphere of radius R it would look like

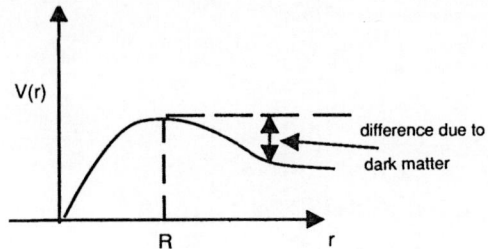

where close to the centre, it starts linearly. Because you have also some dark matter instead of going down, the curve remains almost flat. There is a big difference between the photino which is "cold dark matter" (when it decouples its mass is much higher than the decoupling temperature) and "hot dark matter" like the neutrinos (where the mass is much less than the decoupling temperature). The difference above is important, for example, "hot dark matter" particles, because they are relativistic, when they decouple, they spread around producing a smooth distribution. Thus here we expect large structure to form first and the collapse of small scale structure. On the contrary, photinos are very slow after decoupling, so they don't smooth out at small scale creating in such a way a hierarchical structure.

Of course this question has to be settled by appropriate experiments.

– *Diaz:*

What is the best explanation for galaxy formation?

– *Nanopoulos:*

I am going to plot $\frac{\delta\rho}{\rho}$ (the energy density perturbation) versus time (or R (t)):

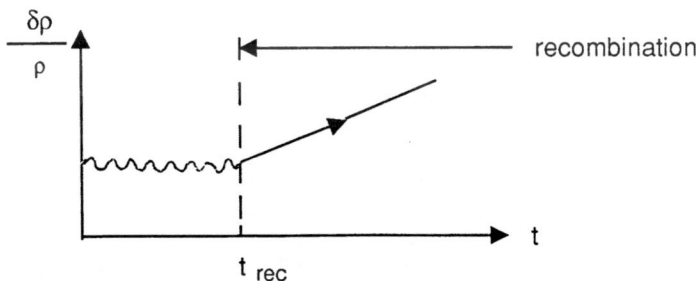

I put the recombination time on the diagram to separate the matter dominated era ($t > t_{rec}$) from the radiation dominated era ($t < t_{rec}$). In the radiation dominated era, if you have, say $\frac{\partial\rho}{\rho} \simeq 10^{-4}$, this is going to fluctuate without increasing because when matter is starting piling up somewhere its interaction with radiation smooth it out. When $t \geq t_{rec}$ we can easily calculate $\frac{\delta\rho}{\rho} \sim t^{1/2}$ and after a while it will reach $\frac{\delta\rho}{\rho} \simeq 1$ in which case you have the beginnig of galaxy formation. The recombination time which determines when the galaxies are formed, depends on the type of matter you consider. Photinos for example, decouple earlier than the regular baryonic matter which decouple around 10^5 years, and thus avoiding, or postponing the stringent limits on $\frac{d\rho}{\rho}$ coming from upper bounds or $\Delta T/T$.

– Windey:

How do you estimate the scale of galaxies from the previous scenario?

– Nanopoulos:

You have to use the Jeans limit. There are two competing factors. One is the gravitational force that tries to clump matter together and then there is the expansion of the universe, that tries to smooth out inhomogeneity. You can take both factors into account; given the perturbation spectrum, you can estimate the minimum mass (the Jeans mass) for a galaxy, $m \sim \lambda^3$ where λ is the characteristic scale of a galaxy. Of course, the above is estimated in perturbation theory, so you have to be in the linear region in the graph when $\frac{\delta \rho}{\rho}$ is still quite small. When the non-linear effects come into play you cannot do much.

THE NEW PHYSICS IN EUROPE

PHYSICS AT LEP, AND THE L3 EXPERIMENT

Harvey Newman

High Energy Physics
California Institute of Technology
Pasadena, California 91125

1. INTRODUCTION

Several "generations" of electron-positron, proton-proton and proton-antiproton colliding beam accelerators have been developed over the last twenty years. Each decade has brought a factor of 10 to 100 in the attainable energy. With the increased energy have come a host of major discoveries bearing on the nature of the fundamental constituents and their interactions. Examples are the discovery of the large hadron cross section at the CEA [1], the tau lepton [2] and charmonium intermediate states [3] at DORIS and SPEAR, gluons at PETRA [4], and the W and Z at CERN [5].

So it may be in the next accelerator generations. This is illustrated in Table 1 which shows the major colliders existing now, or expected to operate within the next twenty years. LEP, the LHC, SSC and the ELOISATRON will allow us to explore the 100-200 GeV energy scale (10^{-17} cm distance scale) by the early 1990's, the 1 TeV scale before the year 2000, and beyond 1 TeV in the next millennium.

The LEP physics program is unusual in the great diversity of experimental tests of known theories, and the potential for new physics discoveries which it will bring. This situation has been set up by the remarkable successes of the Glashow, Weinberg, Salam (GWS) "standard" picture of the unified electroweak interaction [6], the missing Higgs, and the strikingly absent top quark. Experiments at LEP have the opportunity to prove one of the following two scenarios:

- The GWS picture, including the gauge cancellations in W-pair production, is internally consistent to high precision. This may indicate that the GWS theory is exact, in the same sense that quantum electrodynamics (QED) provides an exact description of electromagnetic interactions.

- Precision determinations of the Z^0 and W masses, and other electroweak effects such as the muon pair forward-backward charge asymmetry at the Z^0 and at other center of mass energies, deviate from the standard model predictions. This may be associated with the absence of the top quark, even beyond the 200 GeV range at LEP II, and it could indicate the existence of new physics phenomena, up to the 1 TeV scale.

During my talk, I will emphasize that experimenters at LEP, and at L3 in particular, may be able to make the critical determination of which scenario is correct. The design principles of L3 — very high resolution for leptons and photons, combined with good resolution for jets — may provide an important advantage in the determination. Apart from the obvious requirement of high (design) luminosity at LEP, the ability to discern non-standard electroweak effects may require great care in the experimental measurements of masses and asymmetries, in order to reduce the systematic errors to a minimum.

Physicists at LEP may find the neutral Higgs, and/or the top quark, to complete the standard electroweak picture. They may also find new quarks or leptons, additional gauge bosons, "supersymmetric" particles, or charged Higgs which indicate the existence of extensions of the minimal standard electroweak (QED) and strong (QCD) interaction theories as we know them today. Most importantly, they may discover new particle families which are until now unpredicted, and which do not fit into any of the known theoretical schemes.

2. LEP

LEP is shown schematically in Fig. 1, superimposed on an aerial view of the region stretching from the Geneva airport to the foot of the Jura mountains. The main parameters of LEP are shown in Table 2 [7]. The vacuum pipe of LEP is scheduled to be closed, and injection into the completed machine to start, in July 1989.

Some of LEP's main features — its components and its progress — are summarized below.

2.1 Civil Engineering

The 26.6 km long LEP tunnel is excavated at depths of 50 to 150 m below the surface. Each of the four LEP experiments (ALEPH, OPAL, L3 and DELPHI) is situated in a 21 m diameter experimental cavern. Three of the caverns are 70 m long transverse to the beam, while L3's hall is 53m along the beam.

All but 3.5 km of the tunnel passes through relatively soft sandstone ("molasse"). This part of the machine was excavated in 1984-1986 by the use of three full face boring machines, which were able to proceed at an average rate of approximately 25 m per day, with peaks of 60 m per day. The last 3.5 km has been much more difficult, since it passes through fractured limestone under the Jura, below the sandstone layer. The excavation of the last 300 m of the tunnel has proceeded at a typical rate of 10 m per week, with some interruptions. Rock and adhesive have to be injected to seal fissures in the rock where water would enter, prior to blasting. The early decision to move LEP, make it smaller (26.6 instead of 30 km), and incline it by 1.4% to minimize the section under the Jura (Fig. 2), has been an essential step towards its completion in 1989. The excavation of the LEP tunnel was expected to be completed in February, 1988.

2.2 Magnet System

By the fall of 1987, nearly all of the 3370 dipoles, 776 quadrupoles, 504 sextupoles, and more than 500 correctors were complete and tested. The prototype for the 8 superconducting quadrupoles, for the low beta insertions at the intersection regions, was tested in the spring of 1987. Delivery of the 8 production superconducting quads is proceeding, on a schedule that will end in the fall of 1988.

One of the novel features of LEP's magnet system is the use of dipoles composed of thin stamped steel lamina embedded in concrete [8]. This design is a cost efficient solution to the problem of providing the low magnetic fields ($\sim$ 100 Gauss at injection) imposed by the requirement to keep synchrotron radiation losses in LEP within manageable limits.

2.3 Radio Frequency System

The initial complement of 128 RF copper cavities, and sixteen 1 Megawatt klystrons to feed them, were all delivered and tested by late 1987. A string of 16 cavities, assembled and fed exactly as they will be in LEP, has been successfully tested for more than a year.

One of the most striking features in the LEP design is the use of a spherical, higher-Q "storage" cavity coupled to each cylindrical 5-cell accelerating cavity. Each spherical-cylindrical cavity pair is connected by a coupler which produces "beats" between the two resonators, at the bunch crossing frequency. As a result the presence of the full electromagnetic field intensity in the accelerating cavities is synchronized with the beam passage. The energy lost to the cavity walls is thereby reduced by 40%.

One of the principal features designed into LEP, is the ability to go up in energy by the step-by-step addition of additional RF accelerating cavities. This is outlined in Table 3, as approved by the CERN Council for LEP Phase I and II. As shown in the table, all energy upgrades will be achieved by the use of superconducting cavities. In particular, four cell Niobium cavities have been tested reliably at 7 MV/m.

A scenario for going beyond 186 GeV center of mass energy, which requires new civil engineering to accommodate RF stations surrounding more than two intersection regions, is shown in Table 4.* The magnet lattice of LEP will not be the limitation up to 250 GeV in the center of mass.

2.4 Injection System

The new LEP preinjector (LPI), including 200 MeV and 600 MeV LINAC's and the 600 MeV electron-positron accumulator (EPA) has been tested successfully in 1986. Both the CERN PS and SPS have been adapted to accelerate electrons and positrons, as shown in Fig. 3. The CERN PS has accelerated electrons and positrons to 3.5 GeV. The SPS accelerated positrons to 12 GeV in the fall of 1987 — the highest possible energy with the present SPS RF system. New RF components, to take the beam energy to the final LEP injection energy of 20 GeV, are now being installed.

2.5 LEP Crash Program

In September 1987, installation of the main magnets and other components for beam handling and control began in the first of LEP's eight 2.8 km long sectors. The goal is to have beam injected and transported to a dump just short of the Point 2 experimental cavern (where L3 is located), by July 1988.

2.6 Outlook for Luminosity

All indications are that LEP will meet its scheduled date for first circulating beams in the summer of 1989.

The LEP design is considered quite conservative [9]. There are no barriers foreseen in achieving the design luminosity of $\sim 2 \times 10^{31}$ cm^{-2} sec^{-1} at 47 GeV beam energy, based on 4 bunches (2 per beam) of 0.75 mA each. The luminosity is expected to rise to approximately 10^{32} cm^{-2} sec^{-1} at the W^+W^- threshold in LEP Phase II, given sufficient RF power.

The bunch currents are limited by a short range bunch-cavity interaction. Use of superconducting cavities would improve the situation, with bunch current limits of ~ 1.25 mA, giving a luminosity three times higher. A feedback system, already tested

---

* Note that this table is a personal, technical view and is not under active consideration by CERN management.

TABLE 1

HIGH ENERGY COLLIDERS

e^+e^-	E_{cm}	Operation Dates
PETRA	12-46.8 GeV	1978-Nov. 1986
PEP	29 GeV	1980-1987
TRISTAN	to 60 GeV	1987 started
SLC	90-100 GeV	1988 start
LEP	40-200 GeV	1989 start

$p\bar{p}$ & pp		
CERN SPS ($p\bar{p}$)	540-900 GeV	1982-now
FNAL ($p\bar{p}$)	to 1.8 TeV	end 1986 start
SSC (pp)	40 TeV	1997 ?
LHC (pp)	18 TeV	1996 ?
ELOISATRON (pp)	200 TeV	

ep		
HERA	≥ 315 GeV	1990

TABLE 2

LEP -- Main Machine Parame
(Phase I)

- Reference circumference (including sagitta in dipoles) — 26658.883376 m
- Slope of the median plane of the machine — 1.42%
- Lattice type — FODO
- Phase advance/period — 60 or 90 degrees
- Number of bunches per beam — 4
- Number of interaction points — 4 + 4
- Equipped experimental areas (P2,P4,P6,P8) — 4
- Ratio horizontal/vertical β-values at IP — 25
- RF frequency — 352.209042 MHz
- Revolution time — 88.92446 μs
- Harmonic number — 31320
- Nominal klystron output power (total) — 16 MW
- Injection energy — 20 GeV
- Beam energy with nominal luminosity (copper cavities) — 55 GeV
- Maximum energy (zero luminosity, Cu cavities) — ~60 GeV
- Momentum compaction factor (60°/period lattice) — 3.866×10^{-4}
- Nominal betatron wave numbers (tunes) (id.)
    - horizontal — 70.35
    - vertical — 78.20
- Space between quadrupoles at interaction points — ±3.5 m
- Horizontal dispersion at interaction points — 0.0 m

TABLE 2 (Continued)

Data for 60°/period lattice at 55 GeV:

-	Luminosity in physics interaction region	$1.7 \times 10^{31} \text{ cm}^{-2} \text{ s}^{-1}$
-	Beam-beam tune shift (max. possible value)	0.032
-	Circulating current per beam (assumption)	3 mA
-	Particles per bunch	4.16×10^{11}
-	Natural bunch length ($J_3 = 2$)	15.6 mm
-	Relative rms energy spread with design damping	0.98×10^{-3}
-	Synchrotron energy loss	263 MeV/turn
-	Quantum lifetime (assumption)	1440 min
-	Required circumferential RF voltage	364 MV
-	Synchrotron power (two beams)	1.6 MW

TABLE 3

LEP I $\Rightarrow$ LEP II

MACHINE LIMIT: 250 GeV

Phases I and II as Approved in '87

(No New Civil Engineering)

- 128 Copper Cavities, 1.5 MV/m, 16 MW
  at Regions 2 and 6: To $E_{cm} = 110$ GeV

- Add 32 Superconducting Cavities, 7 MV/m,
  2 MW More: To $E_{cm} = 130$ GeV

- Add 32 More S/C Cavities; Fill Space
  in Regions 2 and 6: To $E_{cm} = 154$ GeV

- Replace All 192 Cavities By S/C Ones,
  at 7 MV/m, 16 MW: To $E_{cm} = 186$ GeV

at PEP, will also help increase the maximum bunch currents and hence the luminosity. But, in the interests of conservatism, the improvements expected from the feedback system are not included in the quoted luminosity figures.

3. EXPERIMENTS AT LEP

The four LEP experiments' relative locations, and the configuration of the experimental halls and shafts are shown in Fig. 4. The principal features of each experiment are summarized in Table 5, and the features of the Mark II and SLD at the SLC are shown for comparison.

3.1 ALEPH, OPAL and DELPHI

As shown in the table three LEP experiments — ALEPH, OPAL and DELPHI — emphasize precise particle tracking, and particle identification [10]. Identification of hadrons is to be achieved in ALEPH and OPAL through the use of many dE/dx samples in their relatively large radius central tracking chambers. DELPHI seeks to maximize its particle identification capabilities through the use of Ring Imaging Cherenkov (RICH) counters in the central and forward regions, in addition to dE/dx in its central time projection chamber (TPC). DELPHI also obtains very high spatial resolution for particles (and secondaries) in its High Density Projection Chamber.

3.2 The L3 Experiment

It is evident from Table 5 that the design of L3 is quite complementary to the others. In L3's design, every effort has been made to achieve the best possible resolution for electrons, muons and photons, combined with excellent resolution for hadron jets.

L3's design choices are predicated on the fact that over the last 25 years, some of the most important discoveries in experimental physics have been associated with precision measurements of photons and leptons. Examples are the discovery of the J, the Υ, the W and Z^0 particles. The main purpose of L3 is to use the high resolution of the detector for photons and electrons to search for new phenomena in the energy region up to the W^+W^- threshold, which will be made accessible by LEP. A second important purpose is to perform the most precise tests of the nature of electroweak interactions [6] over a large energy range.

3.3 The L3 Detector [11,12]

A perspective view of the L3 detector is shown in Fig. 5. Civil engineering of the hall was completed at the beginning of 1987, and the installation of the infrastructure to receive the experiment was then begun. By the end of 1987, the major cranes in the hall and at the surface were installed, and the infrastructure was completed. A full scale preassembly of major subdetector modules at the surface will begin early in 1988, to avoid problems or delays in the actual installation, between November 1988 and April 1989.

Particles emerging from the interaction region will pass through a very precise "Time Expansion Chamber" (TEC) that will reconstruct their initial directions and energies. Following the TEC, an array of single crystals (each weighing 1 kilogram) of Bismuth Germanate (BGO), will provide precise energy measurements of electrons and photons. Outside the BGO crystal array, a 600 ton uranium and brass "hadron calorimeter" will absorb strongly interacting particles and measure their energy and direction.

High energy muons which penetrate and escape from the hadron calorimeter will emerge into a large volume occupied by precise drift chambers, where they will be momentum analyzed.

Fig. 1. Aerial view of CERN and the surrounding region, with the LEP and SPS rings drawn in.

341

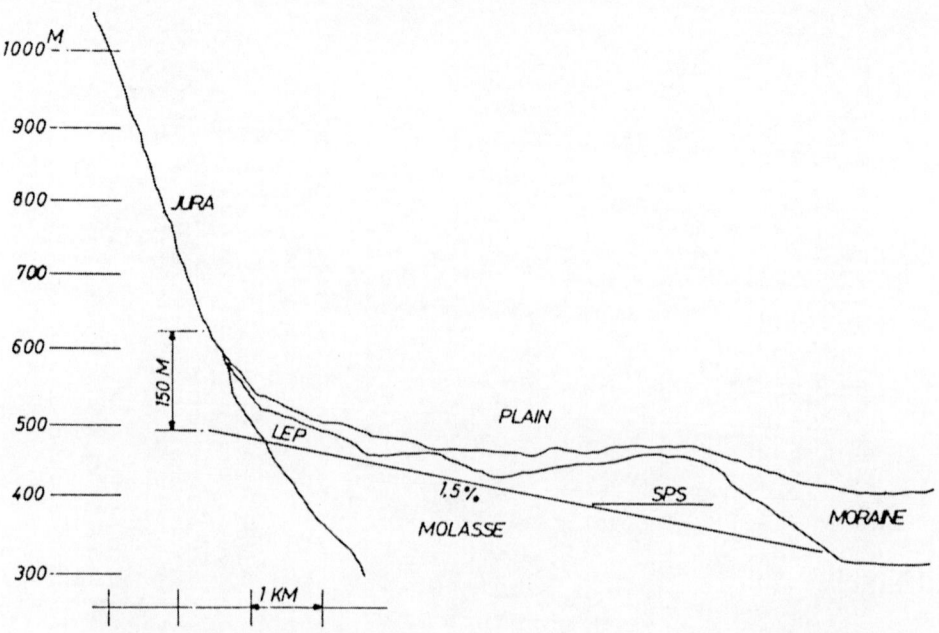

Fig. 2. Simplified geological section of the region at the foot of the Jura, showing the SPS, and LEP inclined by 1.5%. The vertical scale is magnified by a relative factor of 10. The maximum depth of the LEP ring below the surface, 150 meters, is indicated.

The performance and status of the principal detector components are summarized below. In addition to these systems, there is a precise luminosity monitor composed of proportional chambers and BGO crystals, which will measure small angle Bhabha scattering events. A ring of scintillation counters inside the hadron calorimeter barrel will be used as part of the fast trigger, and to help veto cosmic rays offline.

3.3.1 Magnet

The major elements of the L3 detector are housed in the world's largest solenoidal magnet (Fig. 6), which generates a 5 kG uniform magnetic field throughout a volume of 1300 m^3. The magnet coil is composed of 1000 tons of aluminum, and the return yoke is composed of 6600 tons of iron. The welding of all 28 of the 16 m, 40 ton aluminum coil packages was completed on schedule in July 1987. By the end of 1987, the lower 3 octants of the return yoke, composed of 14 m-long iron bars, were complete.

3.3.2 Time Expansion Chamber (TEC)

The precise inner track detector, developed by ETH (Zurich) uses 2-atmosphere pressure, flash ADC readout, and a very cool (low diffusion) gas. The chamber runs with a low electric field and hence a low drift velocity to achieve a single track resolution of 50 μm averaged over the drift cell, with excellent two track resolution.

Tests with a prototype chamber in the MARK J experiment were completed in December 1986. The L3 chamber has an inner ring of 12 segments with 16 sense wires each, and an outer ring of 24 segments with 62 sense wires each. Fourteen of the sense wires in the outer segments are used for charge division, to measure the coordinate along the beam direction.

Based on 50 μm coordinate resolution and 64 measurements along a track, the chamber will be able to determine the sign of the charge of a 50 GeV/c particle in the central region with 3σ significance.

3.3.3 Bismuth Germanate (BGO) Electromagnetic Calorimeter

The electromagnetic calorimeter is composed of BGO crystals each approximately 2 × 2 cm^2 (front face) by 3 × 3 cm^2 (back face) by 24 cm long. The central region is covered by two half-barrels (see Fig. 7) composed of 3840 crystals each. This is supplemented by two end caps of approximately 2,000 crystals each. Each crystal is read out with two 1.5 cm^2 silicon photodiodes that operate undisturbed in L3's 5 kilogauss magnetic field, and with very low noise (0.8 MeV equivalent) charge sensitive preamplifiers.

Beam tests at Cornell (100 and 180 MeV), and CERN (2-50 GeV) over the last three years have confirmed that the crystals will provide the necessary high energy resolution if precise calibration is maintained. The test results are shown in Fig. 8. The resolution at high energies in the 22 radiation length crystals approaches 0.5%. The low energy resolution of 5% at 100 MeV is limited by electronic noise in the photodiodes and preamps.

The last BGO crystals for the first half barrel were delivered to CERN in April 1987 from the Shanghai Institute of Ceramics. Production of crystals, of uniformly high quality in terms of transparency, light output, and dimensional tolerances of 0.1 mm, rose to 350 crystals per month at that time. Production at Shanghai is now proceeding at a rate of 400 crystals per month, with delivery of the last crystals for the second half barrel scheduled for February 1988. Production of the end cap crystals will then begin.

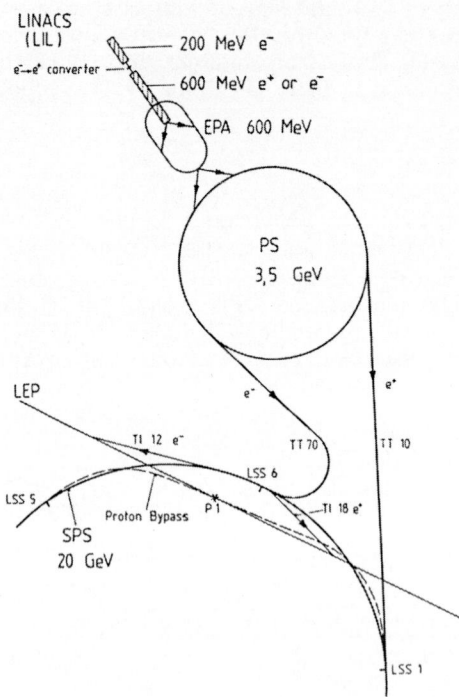

Fig. 3. The LEP injection scheme, which involves both the PS and SPS.

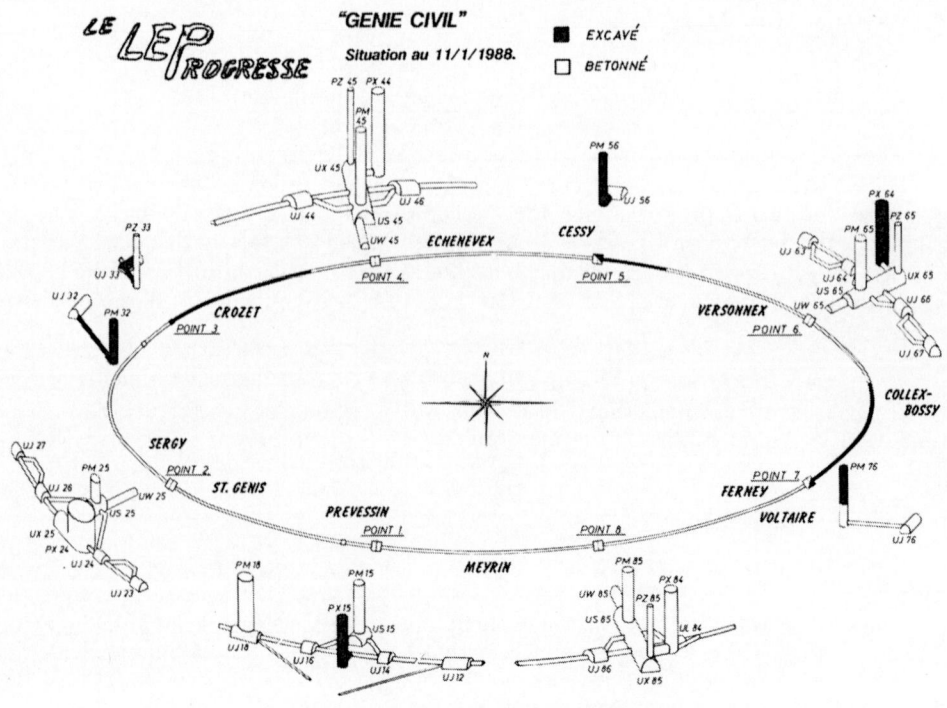

Fig. 4. Layout and progress of the LEP tunnel, experimental areas and shafts. The shaft configuration at each experimental site is shown enlarged.

Construction of the first half barrel, including crystal acceptance tests, photodiode installation, precalibration with cosmic rays, insertion into the thin walled (0.2 mm) carbon fiber support structure, and mounting and debugging of electronics was completed in July 1987. The half barrel was then moved to an SPS test beam, and mounted on a rotating table (see Fig. 9), where each crystal was calibrated at 2, 10 and 50 GeV.

3.3.4 BGO Calibration with a Radio Frequency Quadrupole (RFQ) Accelerator

Achieving the high resolution shown in Fig. 8 in practice, during physics runs at LEP, will hinge on accurate and frequent gain calibrations in situ. L3's sensitivity to a number of signatures for new physics processes, such as Higgs production, depends on the resolution of the BGO array, and hence on the quality of the calibration. After four years' study of potential calibration techniques, we have found that a low energy Radiofrequency Quadrupole (RFQ) proton accelerator may be used to bombard a target of lithium or boron. This produces a high intensity flux of 17 MeV photons by radiative capture, which acts as a calibration source. The photon flux from the RFQ is sufficient for calibrations more than once per day, or on a continuous basis by pulsing the RFQ at a low duty cycle during running at LEP. The complete procedure between LEP data-taking runs will take approximately 1 hour, for a high statistics precise calibration.

The RFQ technique is an essential part of the overall BGO calibration procedure. Because of the large number of crystals, LEP's high beam energy, and the fact that L3 is underground, e^+e^- interactions and cosmic rays occur at too low a rate for daily calibrations. Radioactive sources have too low an energy to be measured cleanly in the presence of electronic noise in the photodiodes. Light sources may help monitor the electronics and light transparency of the crystals, but do not provide a test of the scintillation properties.

The RFQ technique will be combined with less frequent cosmic ray runs, analysis of e^+e^- production by the two photon process, measurements of high energy Bhabha events, and continual monitoring with a system of Xenon flash tubes and light fibers to maintain the calibration of each crystal.

The plan to install the RFQ in the L3 experiment is shown in Fig. 10. The RFQ will be situated below the LEP machine, and the neutralized hydrogen atom beam will enter the detector at an angle of 21.6 degrees with respect to the beam line. After focusing and neutralization, the beam will have a small emittance, so that a narrow vacuum tube will suffice. The small beam tube will pass through the forward region of the L3 detector (including the BGO end caps which will be installed after LEP startup). The RFQ vacuum tube — which is entirely separate from the 10^{-9} Torr LEP vacuum system — will terminate at a target which is situated next to the end wall of the TEC vertex detector.

Because the RFQ target position is off center, so that photons enter some crystals at angles up to 55° with respect to the normal to the front surface, we originally planned to move the RFQ from one side of the detector to the other during LEP shutdowns. Simulations of the showering and absorption of the photons, however, has shown that even the worst case photons give a good energy spectrum, and will provide a good calibration.

3.3.5 Hadron Calorimeter

The hadron calorimeter consists of 144 barrel modules arranged in 9 rings, and end caps, as illustrated in Fig. 11. Each barrel module consists of one stainless steel plate and 58 uranium plates (each 5 mm thick, covered with a thin Cu-Ni alloy sheet) interleaved with proportional chamber sampling layers. Each module is approximately

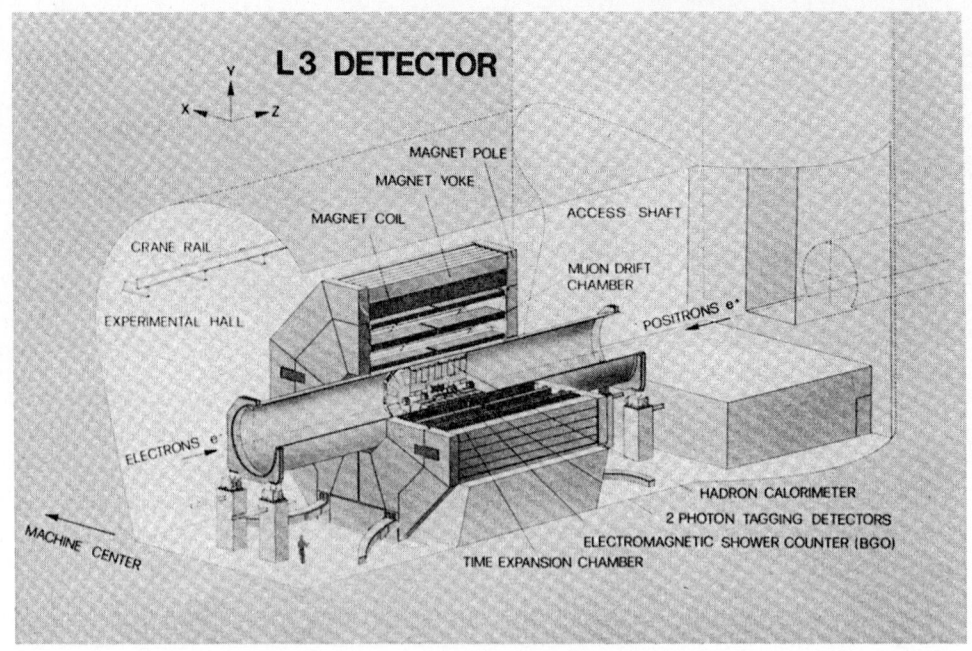

Fig. 5. The L3 detector at Point 2. A cutaway view is shown, indicating each of the major subdetectors.

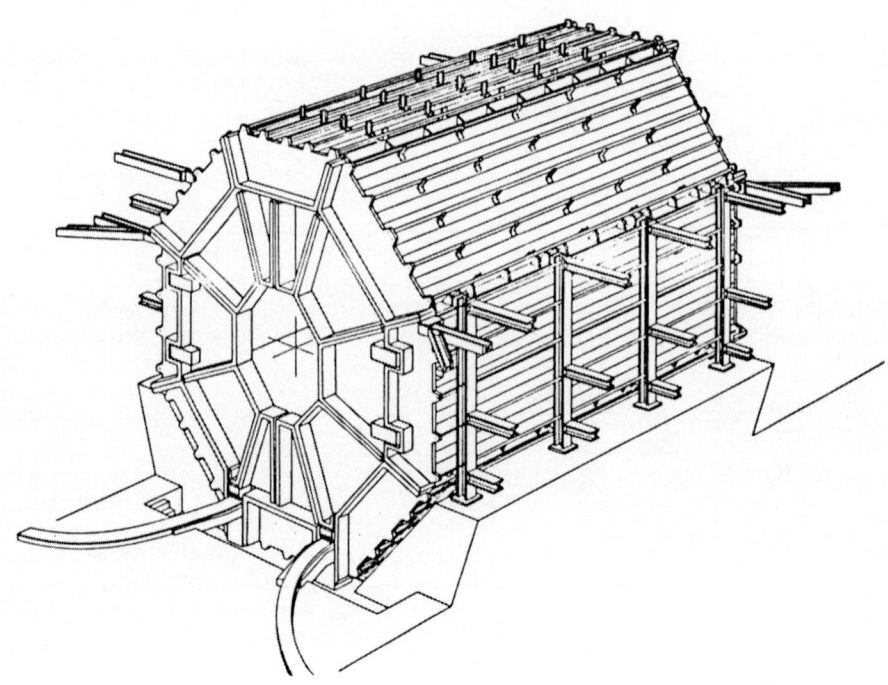

Fig. 6. The L3 magnet.

3.6 absorption lengths thick, and is backed up by a 5 layer brass muon filter of 0.9 absorption lengths. The 4×10^5 chamber wires are read out in 3.3×10^4 "minitowers", in a scheme which has been optimized for energy and spatial resolution [13]. Test beam results (see Fig. 12) indicate that the resolution of the calorimeter in the barrel region, with the BGO in front, will be $\sigma_E/E = 48\%/\sqrt{E} + 4\%$ — or probably better because of side energy leakage from the 100 crystal BGO matrix used in the tests. The tested resolution of the end caps is $75\%/\sqrt{E} + 3\%$.

All of the hadron calorimeter chambers were completed at the end of 1987. Completion of all calorimeter modules is scheduled for the early spring of 1988.

3.3.6 Muon Chambers

As shown in Fig. 5, the inner subdetectors of L3, out to the brass muon filter, are supported in a long 4.6 meter diameter tube. The main volume of the detector, outside the tube, is occupied by three layers of muon chambers arranged in 16 "half octants", each containing 5 precision chambers. A muon track is measured 16, 24 and 16 times in the three layers. Measurements in the non-bending direction, along the beam line, are provided by chambers with conventional accuracy (500 μm) mounted on the inner and outer precision chamber modules. The half-octant layout is shown schematically in Fig. 13.

The chamber modules in an octant are aligned at each end by an optical system consisting of an LED mounted on the inner chamber, a lens on the middle chamber, and a photodiode on the outer chamber to an accuracy of 10-15 μm. The alignment of the two ends is maintained by a rotating "laser beacon" that defines a reference plane. Internal precision is maintained by pushing the sets of 16 or 24 wires against precision glass edges, where the edges are aligned and spaced perpendicular to a zero thermal expansion carbon fiber rod at each end. UV lasers are used to ionize the gas along uninterrupted optical paths that pass through the chambers at several locations, where the laser beams may be directed by a precision set of mirrors. This allows for the simulation of an infinite momentum particle. Shifting the laser beam by up to a few centimeters, along some paths, also allows for the internal calibration of the drift velocity.

The first tests of completed octants showed agreement between the external optical alignment systems, and the internal UV laser system at the level of 50 μm. The sources of the differences are being understood, and this accuracy is being improved with time: we expect to reach the 20-25 μm level in all octants prior to installation. Given 16, 24, and 16 measurements along a track, and an individual coordinate accuracy of 180 μm obtained with cosmic rays and test beams, this will lead to a momentum resolution of 1.5% for 50 GeV muons, which corresponds to 1% mass resolution for $Z^0 \rightarrow \mu^+\mu^-$.

The results of a typical "laser shot test" are illustrated in Fig. 14.

Approximately two thirds of the muon chambers were completed by the end of 1987. All chambers, and all half-octant support frames will be completed by August 1988. Instrumenting, testing and aligning all octants will be done, and installation of the octants will begin, in December 1988.

4. PHYSICS AT LEP

LEP will open a new energy range for the study of e^+e^- interactions, and may bring with it a rich set of new physics processes to be explored. The higher luminosity at LEP, as compared to the SLC, will also provide the opportunity for precise tests of electroweak theory.

The breadth of the LEP program is illustrated in Table 6, which shows the list

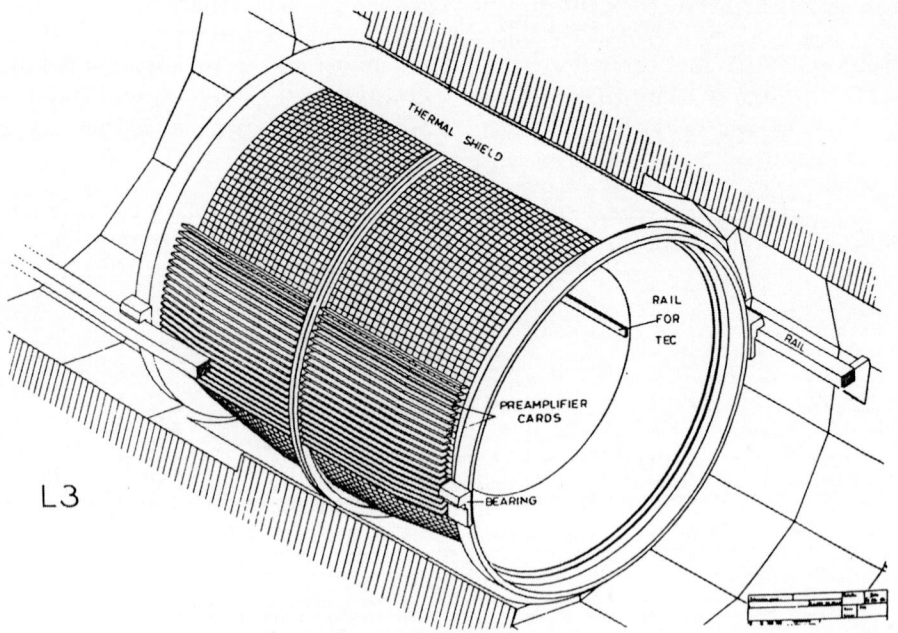

Fig. 7. The BGO barrel electromagnetic calorimeter, showing the carbon fiber support structure, and some of the preamplifier cards.

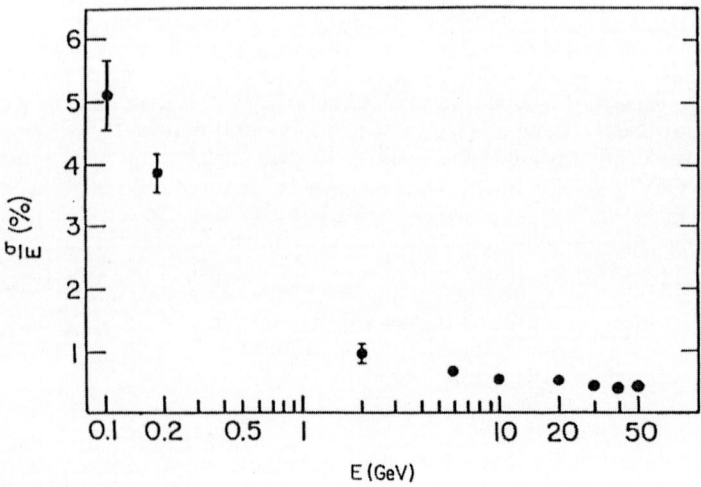

Fig. 8. BGO energy resolution for electrons, obtained in test beams at Cornell and CERN.

of event generators which are already available for simulations of electron-positron interactions at LEP (as of mid-1987). In addition to the known quarks and leptons, heavy quark and lepton, supersymmetric squark and slepton and gaugino, and Higgs production and decays are included in the list.

Some of the highlights of potential new physics and new experimental tests to be performed at LEP (necessarily using the perspectives of present-day theory) are summarized below, with emphasis on the L3 experiment. Some of the examples given are also contained, and discussed further, in the CERN Report which resulted from the 1986 LEP Physics Jamboree: "Physics at LEP" [14].

4.1 Confronting the Standard Model at LEP

A new era of studying the Z^0 will be opened at SLC in 1988, or by LEP in 1989. Using the achievable goals for the SLC quoted by Dorfan in mid-1987 [15], the comparison of expected Z^0 production rates, and muon pairs/day for a charge asymmetry measurement at SLC and at LEP is summarized in Table 7. Measurements using polarized beams are not considered in the table because they are not likely to play a significant role in the first years' running.[*]

A reasonable scenario [14] for the early months of running at LEP, focused on the precise determination of the Z^0 mass and width, is outlined in Table 8. Following a rough determination of the Z^0 mass by scanning in 2 GeV steps at 2 pb^{-1} per point, a high statistics sample of decays events would be obtained at and near the Z^0 peak.

The Z^0 mass, if measured with the highest possible precision, will become a new "reference point" at $Q^2 = M_Z^2$, and a starting point for a test of the internal consistency of the standard model. This is because of the two relations that relate the Z^0 and W masses to $\sin^2 \theta_W$ at the tree level:

$$\sin^2 \theta_W = 1 - \frac{M_W^2}{M_Z^2} \tag{1}$$

$$\sin^2 \hat{\theta}_W = \frac{\pi \alpha(m_e)}{\sqrt{2} G_F(m_\mu)} \frac{1}{M_W^2}. \tag{2}$$

Equation (2) is modified by radiative corrections, which are dominated by photon vacuum polarization in going from the low energy scale ($Q^2 \sim m_e^2$ or m_μ^2) to the gauge boson energy scale ($Q^2 \sim M_Z^2$). This factor is usually expressed as $1/(1 - \Delta r)$. Recent calculations based on three quark and lepton generations, $M_{Higgs} \sim M_Z$ and $M_{Top} \sim M_Z/2$ have found [14,16,17]:

$$\Delta r = 0.070 \pm 0.002. \tag{3}$$

It is clear from Eqs. (1) and (2) that a test of the internal consistency of the standard theory will involve three steps:

---

[*] One notable exception could be a resonant spin depolarization test measurement, using a laser polarimeter. This would be valuable in providing a precise (10 ppm) absolute machine energy calibration.

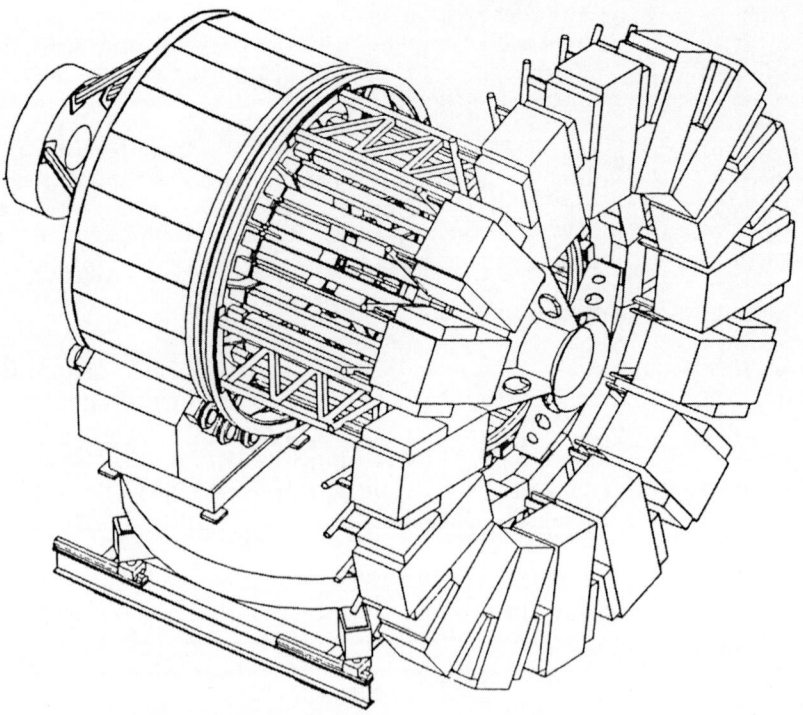

Fig. 9. The rotating test stand used to calibrate each of the 3840 crystals in a half-barrel of the BGO electromagnetic calorimeter.

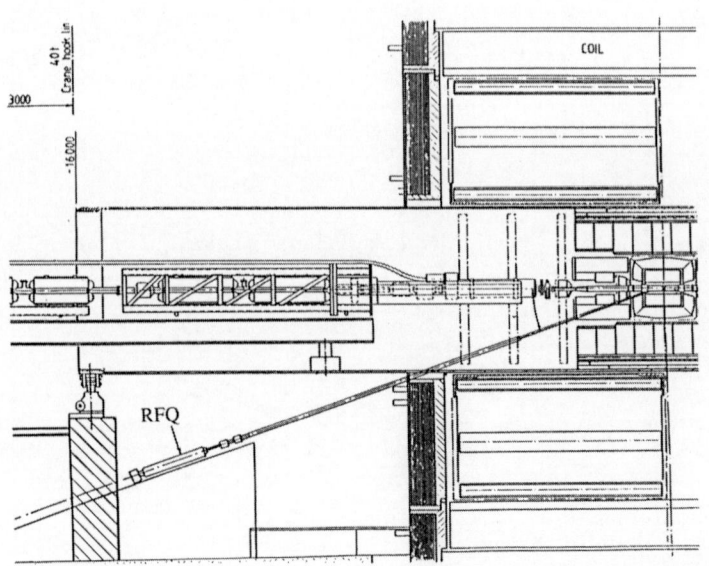

Fig. 10. The Radiofrequency Quadrupole (RFQ) accelerator situated below the LEP ring. A neutralized hydrogen beam will bombard a target next to the L3 vertex detector, producing 17 MeV radiative capture photons as a calibration source.

(1) Precise measurement of the Z^0 mass.

(2) Measurement of another quantity, such as the $\mu^+\mu^-$ forward backward charge asymmetry A_{FB}, which depends on $\sin^2\theta_W$.

(3) Measurements at and above the threshold for $e^+e^- \rightarrow W^+W^-$, including a precise determination of the W mass.

All three steps are within the scope of the LEP program: steps (1) and (2) at LEP Phase I, and step (3) at LEP Phase II. Following steps (1) and (2), and the application of radiative corrections, the self-consistency can first be checked by comparing the results for M_W resulting from Eqs. (1) and (2), using the world average for $\sin^2\hat\theta_W$ from experiments at lower energies. The real test will come at LEP II, where M_W will be determined directly, and precisely (see Section 6).

The measurement of the asymmetry over a large energy range will also provide additional information for the test. The accuracy required for a test down to the level which is sensitive to new physics at higher energies, is not attainable at the CERN or Fermilab $p\bar{p}$ Colliders.

Table 8 also emphasizes the need for precise acceptance. In practice this means that the acceptance for radiative muon-pair events must be well understood experimentally, on both sides of the Z^0 peak, if a systematic shift is to be avoided. Detailed tests of radiative corrections using muon pairs may best be done with L3, given its ability to measure the muon pair mass and the photon energy with sufficient precision directly. For a given set of acollinearity and photon energy cuts, a fit to the Z^0 line shape will have to be made to reduce the systematic error to a minimum.

The precise determination of the Z^0 mass will also rely on higher order radiative correction calculations. This is shown in Fig. 15, taken from the work of Altarelli and Martinelli [14]. The figure demonstrates the apparent peak shift in the Z^0 excitation curve as one includes first and second order corrections. The problem is seen in more detail in Table 9, which shows the mass shift resulting from the inclusion of first, second and third order leading log terms (LL1, LL2, and LL3, which represent powers of terms of order $\frac{\alpha}{\pi} \times (\ln(M_Z^2/m_e^2))$, along with terms ($F$) of order $\frac{\alpha}{\pi}$. In order to get to a theoretical accuracy below 20 MeV, which should be achievable experimentally, the leading logs will have to be resummed, and two loop contributions of order $\frac{\alpha^2}{\pi} \times \ln(M_Z^2/m_e^2)$ will have to be included.

The second step in confronting the standard model at LEP, especially with L3, will be the precise determination of the charge asymmetry A_{FB} in $e^+e^- \rightarrow \mu^+\mu^-$. The results expected from a year's running at LEP Phase I, with the design luminosity (200 pb^{-1}) at the Z^0 peak are shown in Table 10. On the basis of more than 10^5 accepted muon pairs, the charge asymmetry may be measured with a statistical accuracy of better than 0.3%. Systematic errors in the asymmetry should therefore be kept at a level below 0.2%; smaller if possible.

The L3 experiment, and the Collaboration, may achieve this accuracy by the following steps:

(1) Precise determination of the acceptance, through:

(a) Precise geometry of the muon chamber system.

(b) Muon tracks are nearly straight (4 mm sagitta for a 50 GeV track) and well measured ($\Delta p/p \sim 1\%$)

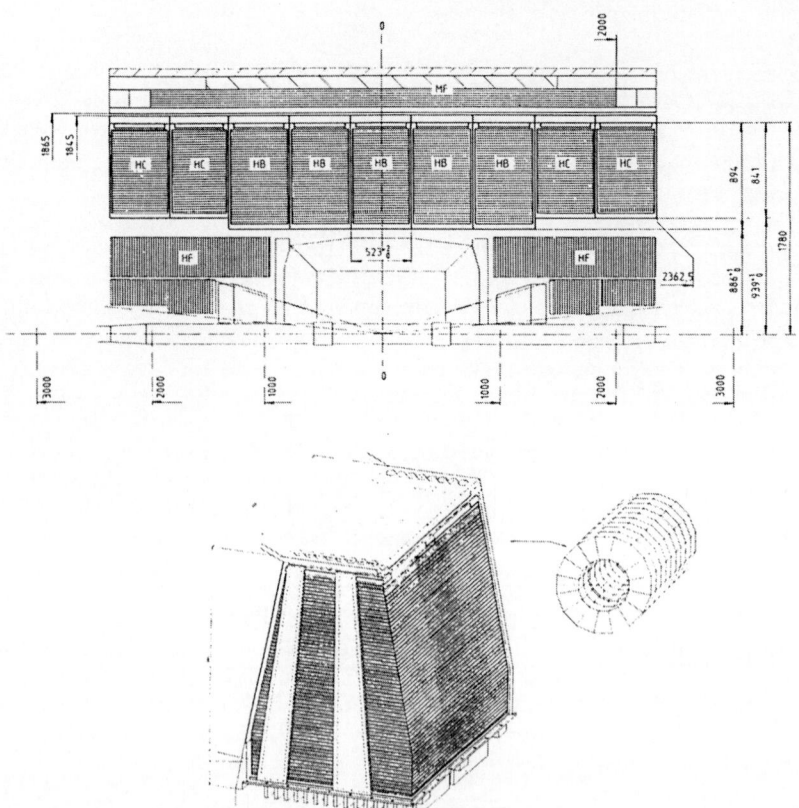

Fig. 11. Top: Cut view of the top half of the L3 Uranium-Copper hadron calorimeter, and brass muon filter.

Bottom: Perspective view of one barrel module with the cover removed. The assembly of 144 modules into 9 rings is shown on the right. Each ring weighs about 32 metric tons. The readout cards are mounted in the outer flange of the module (bottom side in this view).

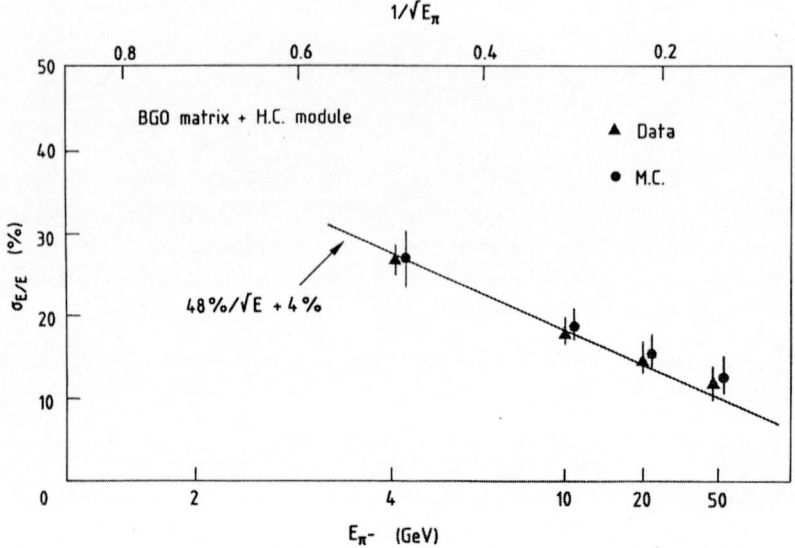

Fig. 12. Energy resolution of a BGO matrix and hadron calorimeter module, obtained at a CERN SPS test beam.

(c) Measurement of final state radiative photons, for cross-checks of radiative corrections.

(2) Precise cosmic ray tests, using the detector's extensive tracking capabilities.

(3) Cancellation of some systematics by running with each field polarity in the detector, for roughly 50% of the time.

Steps (2) and (3) have been used very effectively in the MARK J at PETRA. Application of these methods, and their extension to use with the more sophisticated L3 detector, should make the goal for the systematic error attainable.

An analysis of the relationship between $\sin^2 \theta_W$ and A_{FB} has been carried out by Baillon, Barbiellini and Treille in Reference 14. They find that there may be a significant contribution to the systematic error in $\sin^2 \theta_W$, from the non-reproducibility of the LEP beam energy. In the case where $(\delta E_{beam}/E_{beam}) = 2 \times 10^{-4}$, due to the short- and long-term non-reproducibilities in the LEP magnet lattice, $\delta \sin^2 \theta_W$ reaches 0.4% from this source alone. By controlling the overall LEP field integral to 10^{-4}, the systematic error in the A_{FB} measurement to 0.2%, and assuming a 0.5% statistical error in A_{FB}, the overall accuracy obtainable in $\sin^2 \theta_W$ will be approximately 0.6%.

As discussed in Reference 14 by Kleiss et $al.$, determination of $\sin^2 \theta_W$ corresponding to a given value of the charge asymmetry can only be obtained from the asymmetry measurement once higher order terms are included in the radiative correction calculations. As before, the acceptance used in the calculation must be a very accurate representation of the experimental situation.

4.2 A_{FB} Versus A_{LR} in the Determination of $\sin^2 \theta_W$

Much has been made of the advantage of the use of polarized beams in determining $\sin^2 \theta_W$ to greater accuracy. An analysis of the relationships between the accuracy of A_{FB} and $\sin^2 \theta_W$, and the left-right "polarization asymmetry" A_{LR} [14] shows that:

$$|\delta A_{FB}| \simeq 24\eta |\delta \sin^2 \theta_W| \tag{4}$$

$$|\delta A_{LR}| \simeq 8 P_e |\delta \sin^2 \theta_W|, \tag{5}$$

where P_e is the electron beam polarization, η is the familiar combination of the vector and axial vector neutral current couplings for the final state leptons $g_V = \frac{1}{2}(1 - 4\sin^2 \theta_W)$ and $g_A = -\frac{1}{2}$:

$$\eta = \frac{g_V g_A}{g_V^2 + g_A^2}, \tag{6}$$

and where A_{LR} is the difference of the cross sections for right- and left-handed circularly polarized electrons annihilating with unpolarized positrons to give muon pairs, normalized to the sum:

$$A_{LR} = \frac{\sigma(e^-(L)e^+ \to \mu^+\mu^-) - \sigma(e^-(R)e^+ \to \mu^+\mu^-)}{\sigma(e^-(L)e^+ \to \mu^+\mu^-) + \sigma(e^-(R)e^+ \to \mu^+\mu^-)}. \tag{7}$$

Assuming that $\sin^2 \theta_W \cong 0.23$ and that P_e is 50%, we obtain:

353

$$|\delta \sin^2 \theta_W| \cong \tfrac{1}{2}|\delta A_{FB}| \tag{8}$$

$$(\Delta = D\delta)$$

$$|\delta \sin^2 \theta_W| \cong \tfrac{1}{4}|\delta A_{LR}| . \tag{9}$$

To first order, the error in both A_{FB} and A_{LR} is just given by $1/\sqrt{N}$, where N is the number of accepted events (there is a factor of $\sqrt{1-A^2}$ which is quite unimportant at the Z^0 peak). There is therefore just a factor of two in accuracy, for a given number of events, between Eqs. (8) and (9). This has to be weighed against the fact that polarization will be difficult to obtain, control and measure with sufficient accuracy. Even if the physicists were willing to wait, PETRA experience has shown that the conditions for maximum beam polarization may not correspond to the highest luminosity or the lowest background conditions at the experiments. In the trade off, it is clear that the highest integrated luminosity should have top priority. This points the way to the use of the lepton forward backward asymmetry A_{FB}.

Experiments may also use the asymmetry in electron-positron pairs, assuming that the radiative corrections for this process are available to sufficient accuracy. The calculation is complicated by the presence of spacelike exchange diagrams in addition to the annihilation diagrams. The problem of experimental systematics will also be somewhat more difficult, as compared to the use of a clean muon pair sample, because of the presence of a large rate at small angles from the lower-Q^2 scattering process.

4.3 Radiative Muon Pairs at $E_{cm} - M_Z \sim 10$ GeV

A precise measurement of $e^+e^- \to \mu^+\mu^-(\gamma)$ at E_{cm} near 100 GeV has the following physics motivations:

(1) The precise shape of the Z^0 resonance can be measured.

(2) Sensitivity to the presence of new particles near the Z^0 resonance peak is increased.

(3) The effect of radiative corrections, which can reach a few hundred percent in this region, can be accurately measured.

(4) A cross check on A_{FB} as a function of the dimuon invariant mass can be obtained.

This is illustrated in Fig. 16, where the measured $\mu^+\mu^-$ mass is plotted against the muon acollinearity angle ξ for a center of mass energy 10 GeV above M_Z. If a radiated photon above 1 GeV is observed, it is included in the invariant mass. The corresponding asymmetries which may be obtained with an integrated luminosity of 400 pb^{-1} (approximately two years' running), as a function of the $\mu^+\mu^-(\gamma)$ invariant mass, are shown in Fig. 17.

In this way events with hard initial state and final state radiation can be identified separately, and the radiative corrections may be checked in detail. The presence of a new Z^0 or a $q\bar{q}$ resonance above M_Z would alter the distribution, and would tend to increase the relative number of events with final state, rather than initial state radiation. Another distinctive characteristic of the final state radiation sample is its large expected charge asymmetry, of approximately 45%, whereas the initial state radiation sample would have a small asymmetry. Thus the separation of the two samples may be enhanced. A deviation in the final state radiation asymmetry distribution, from the predictions for one Z^0, would provide a particularly sensitive

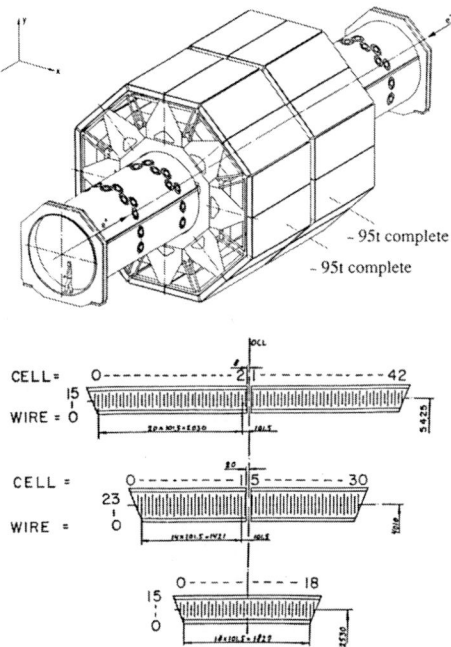

Fig. 13. Layout of the L3 muon chambers in "half-octants" surrounding the support tube which will contain the inner subdetectors, out to the brass muon filter. The layout of wire cells in each half-octant of five chambers is shown at the bottom.

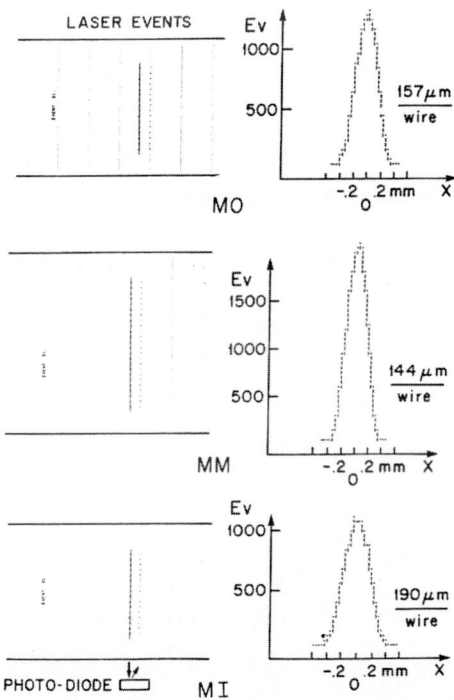

Fig. 14. Data for a laser shot through all three muon chambers. Beam is nearly vertical. Poorer resolution of MI is due to back-scattering of laser light.

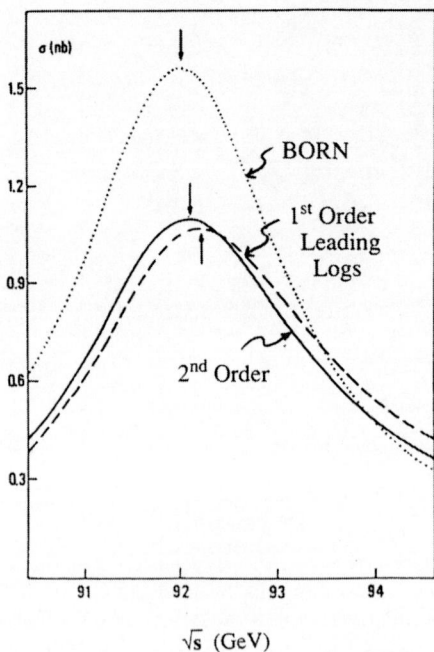

Fig. 15. Effect of radiative corrections on the peak position and shape of the Z^0 resonance. The importance of second order corrections for a precise determination is evident.

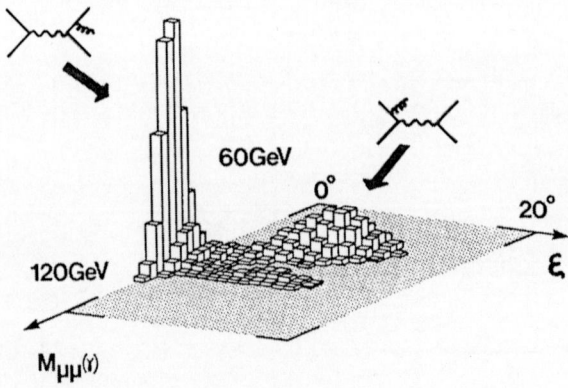

Fig. 16. Muon pair (plus photon) invariant mass versus acollinearity angle. The separation of samples resulting from initial state and final radiation of hard photons is indicated by the Feynman diagrams.

test for the existence of a second Z^0, even if its mass is much greater than the available center of mass energy.

4.4 Absence of the Top Quark, and Self Consistency of Electroweak Theory

Recent measurements of proton-antiproton events by UA1 at CERN indicate that the top quark mass M_t is above 45 GeV, using a choice of structure functions and a Q^2 scale that lead to the lowest limit [18]. Using what UA1 considers the best choice of structure functions, as embodied in the EUROJET Monte Carlo simulation of Ali et al.[19], they find a limit of 56 GeV at the 95% confidence level.

The continued absence of the long-awaited top quark would pose a problem for the internal consistency of the standard (GWS) model. Through vacuum polarization effects in the propagator, a very heavy top quark would make the values of $\sin^2 \theta_W$ (Eq. (1)) and $\sin^2 \hat{\theta}_W$ (Eq. (2)) significantly different. This is summarized in Table 11 and Fig. 18 [16]. The consistency of recent measurements of M_W and M_Z, and hence $\sin^2 \theta_W$ from UA1 and UA2 [18], with lower energy measurements, especially the ensemble of the world's neutrino-nucleon scattering experiments, may only be maintained for $M_t \lesssim 180$ GeV at the 90% confidence level [16,17]. The general lack of consistency of the measurements of $\sin^2 \hat{\theta}_W$ in low Q^2 experiments and $\sin^2 \theta_W$ in experiments at $Q^2 \sim M_Z^2$, for a top quark well above 200 GeV, is shown in Fig. 18.

One of the following physics scenarios is likely to be borne out by running at LEP:

(1) The top quark will be found at LEP, in the 100-200 GeV range.

(2) The top quark will be absent, even at LEP II. This will provide an (indirect) indication that there is "new physics", apart from the standard model with three quark and lepton families, in the energy region $\lesssim 1$ TeV.

(3) New physics may appear directly in the range covered by LEP II.

5. HIGGS DETECTION AT LEP

One of the cornerstones of electroweak gauge theory is spontaneous symmetry breaking, which is supposed to generate the heavy gauge boson masses. A key prediction of spontaneous symmetry breaking is the existence of at least one neutral scalar particle: the Higgs [20]. Finding the Higgs — the missing piece of the otherwise highly successful standard theory — is one of the highest priorities in the LEP program. This section reviews some of the most effective experimental methods that may be used in this search.

5.1 The (Standard) Higgs Mass, M_Z and A_{FB}

The effect of a standard, neutral Higgs on the relationship between A_{FB} on the Z^0 resonance peak, and M_Z is shown in Fig. 19, taken from Reference 21. At $M_Z = 93$ GeV, the difference between A_{FB} for a 10 GeV and 1 TeV Higgs is 0.5%, which is at the limit of detectability according to the discussion above. However, this is one of the few ways (along with a shift in the W mass, as discussed below) in which a 1 TeV Higgs could be "detected" prior to the succeeding generations of higher energy hadron colliders.

5.2 Direct Detection of Neutral "Higgs-Strahlung"

The Higgs may be detected directly by measurements of the missing mass recoiling against $\mu^+\mu^-$ and e^+e^- pairs at $E_{cm} = M_Z$. The neutral Higgs H^0 is produced in the "Bremsstrahlung" process shown in Fig. 20. The results of an analysis by L3, taking advantage of the lepton resolution, are shown in Fig. 21 for the e^+e^- case. It should be noted that the background calculation in the figure assumed a 30 GeV top quark mass. With the current limit of 45 GeV [18], top quark production at $\sqrt{s} = M_Z$ will be strongly suppressed or (more likely) absent. The background would then be much lower.

In LEP Phase II, the same technique may be used, with a better signal to noise ratio. This is illustrated in Figs. 22 and 23 for $\mu^+\mu^-$ and e^+e^- pairs at $E_{cm} = 165$ GeV. The background is cut away by requiring $M_{\mu^+\mu^-} > 30$ GeV (Fig. 22) and $M_{e^+e^-} > 60$ GeV (Fig. 23).

5.3 Toponium and the Higgs

The presence of a toponium Θ not too far above the Z^0 mass (keeping in mind Reference 18), would be of great help in finding the Higgs. The key is the "Wilczek Mechanism" [22]: $e^+e^- \to \Theta \to H^0 + \gamma$, which has a relatively large branching ratio at the peak of the toponium resonance. The events are separated from the background by the presence of the high energy monochromatic photon.

The rates expected for 1000 hours of running at LEP Phase I are shown, for various Higgs masses and center of mass energies, in Table 12, as computed by L3 for the Physics at LEP study [14]. The inclusive photon background originating from:

—— the γ and Z^0 contribution generated according to the Lund model, and

—— the contributions from competing Θ decays, weighted with their relative branching ratios,

was also computed. In each case a cut $|\cos\theta_\gamma| < 0.98$ was needed to eliminate photons from initial state radiation. The inclusive photon spectra expected from the background reactions were also analyzed. Examples of the results of this analysis are shown in Figs. 24 and 25 for running at the resonance peak, for toponium masses of 90 and 110 GeV. The distributions are arbitrarily normalized to 10^5 Lund hadronic events, corresponding to 2.35×10^4 and 5.25×10^3 produced toponia in the two cases. The energy, and the expected number of monochromatic photons per bin (refer to Table 12 for the actual number of events) are shown as heavy dots in the figure.

By summing up the contributions of all the backgrounds, as in Figs. 24 and 25, the integrated luminosity needed to find the Higgs with 3σ significance was obtained, as a function of the Higgs mass. The detector was assumed to have L3's BGO electromagnetic calorimeter (including end caps), which covers nearly 4π steradians, and which has a resolution which can be parametrized approximately as:

$$\frac{\sigma_E}{E} = \frac{1.65\%}{\sqrt{E}}, \qquad \text{for} \quad 0.02 < E < 2.7 \text{ GeV} \tag{10}$$

$$\sigma_E = 1\%, \qquad \text{for} \quad E > 2.7 \text{ GeV}. \tag{11}$$

The results (also see [14]) are shown in Fig. 26, for toponium masses of 90 and 110 GeV. The figure shows that the detection of inclusive photons with high resolution will make finding the Higgs up to mass values of 70% to 80% of the toponium mass possible.

Additional analysis, and cuts to select the $H^0 \to b\bar{b}$ two jet topology, would improve the signal to background ratio. These cuts will only be required if the Higgs is closer to the toponium mass, or if the toponium is near M_Z (since the inclusive photon background from hadron jets will be higher in this case). Overall, the results

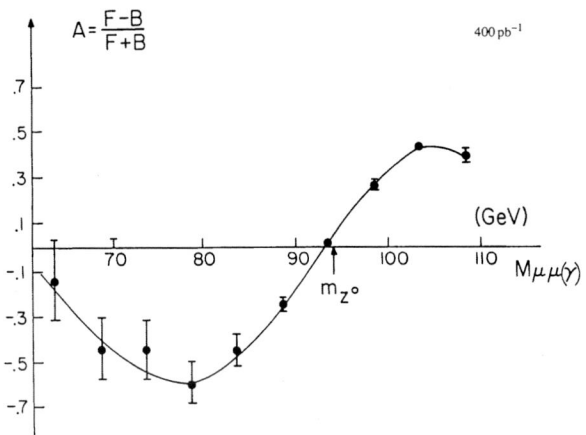

Fig. 17. The expected value of the forward-backward charge asymmetry A_{FB}, as a function of the measured muon pair (plus photon) invariant mass, grouped in 5 GeV bins. The entire spectrum could be obtained, as shown in this simulation, while taking data at the fixed center of mass energy of 104 GeV.

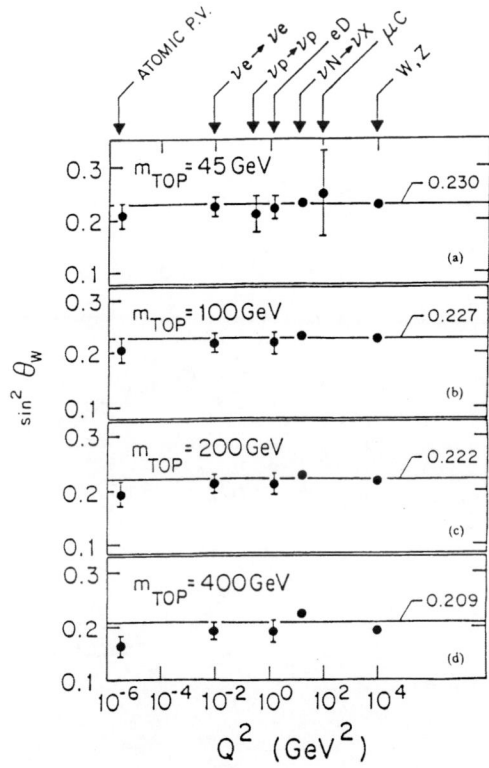

Fig. 18. (a) $\sin^2 \theta_W$ for various reactions as a function of the typical Q^2, determined for $m_t = 45$ GeV. The best fit line, $\sin^2 \theta_W = 0.230$, is also shown. (b-d) $\sin^2 \theta_W$ values determined for $m_t = 100, 200$ and 400 GeV.

359

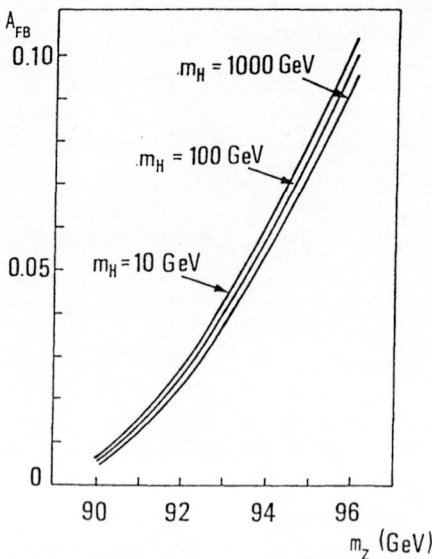

Fig. 19. The forward-backward charge asymmetry A_{FB} as a function of M_Z for $M_t = 30$ GeV and different values of M_H. The QED radiative corrections to the asymmetry have been subtracted.

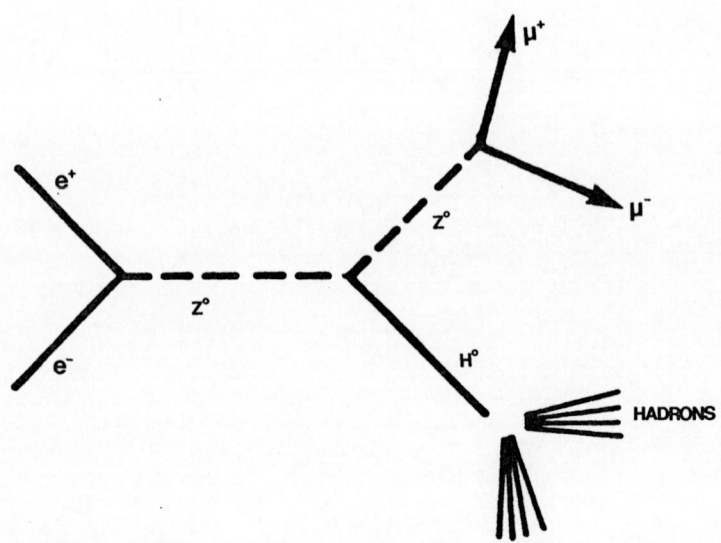

Fig. 20. Higgs production in $e^+e^- \to Z^0 \to e^+e^-H^0$, at $\sqrt{s} = M_Z$ at LEP I.

of Reference 14 indicate that with one year's running at LEP, the Higgs could be found up to 80 and 85 GeV, given the existence of a toponium of 90 and 100 GeV, respectively.

The precision on the H^0 mass per event, resulting from a 110 GeV toponium, is shown in Fig. 27. The solid line in the figure corresponds to a BGO detector as described above. The dashed line corresponds to a calorimeter with wire chamber readout, which would have an energy resolution typically five times worse.

6. W PAIR PRODUCTION AT LEP II

Following an exploration of the Z^0 region with high statistics, and the energy range above the Z^0, the energy of LEP will be raised above the threshold for $e^+e^- \rightarrow W^+W^-$. The design luminosity in this energy range is 10^{32} cm^{-2} sec^{-1}, corresponding to an attainable yearly integrated luminosity of 1 femtobarn^{-1} (see Section 2.6).

6.1 Gauge Cancellations: Testing the Non-Abelian Nature of the Theory

As introduced in Section 4.1, the third and perhaps crucial step in confronting the standard model consists of W-pair measurements, including a precise W mass determination. The shape of the W^+W^- cross section at threshold, as a function on energy, is sensitive to the presence of the three gauge couplings γW^+W^- and $Z^0W^+W^-$. This measurement therefore provides a direct test of the non-Abelian character of the weak interactions. LEP II provides, in fact, the only chance we may have to test the three gauge couplings before the year 2000 (see Reference 24).

The three lowest order diagrams contributing to W^+W^- production, and the familiar "gauge cancellations" needed to damp the cross section at asymptotically high energies are illustrated [24] in Fig. 28. The contribution of the ν exchange graph shown in the figure grows linearly with s (the square of the center of mass energy), while the standard model prediction including all three graphs decreases asymptotically as $\frac{1}{s} \ln\left(\frac{s}{M_W^2}\right)$.

6.2 W-Pair Sample Selection

A clean sample of $e^+e^- \rightarrow W^+W^-$ events can be obtained at LEP II by the selection of four jet events. Roughly half of the W pairs will produce a four jet (predominantly $q\bar{q}q\bar{q}$) final state, so that at $E_{beam} \simeq 100$ GeV, the resultant 4 jet signal $e^+e^- \rightarrow W^+W^- \rightarrow 4$ jets has a cross section of approximately 10 pb. The QCD 4 jet background, which has a production rate ~ 0.5 pb, may be cut by simple cuts on the jet angles with respect to the beam line and with respect to each other: e.g., $|\cos(\theta_{jet}) < 0.9|$ and $|\cos\theta_{ij}| < 0.9$ [24,25]. This leaves a signal with a net production rate, after cuts, of 5-6 pb above threshold.

6.3 Measurement of the W Mass

The result of a simulated measurement of the W pair threshold [24], corresponding to 100 pb^{-1}, is illustrated in Fig. 29. The cuts needed to obtain a clean four jet sample are not included in the figure: they would reduce the cross section shown by about 65%. Nevertheless, the measurement of the four jet data corresponding to the figure would give a statistical precision on the W mass of approximately 200 MeV. With 1 fb^{-1}, the statistical accuracy would fall to < 100 MeV, and the measurement could be dominated by systematics.

Other methods, such as a measurement of the electron energy spectrum in $W \rightarrow e\nu$ decays, or the dijet masses in four jet final states, may also be used to measure the W mass. These approaches are discussed in detail by Bijnens et al., in Reference 26.

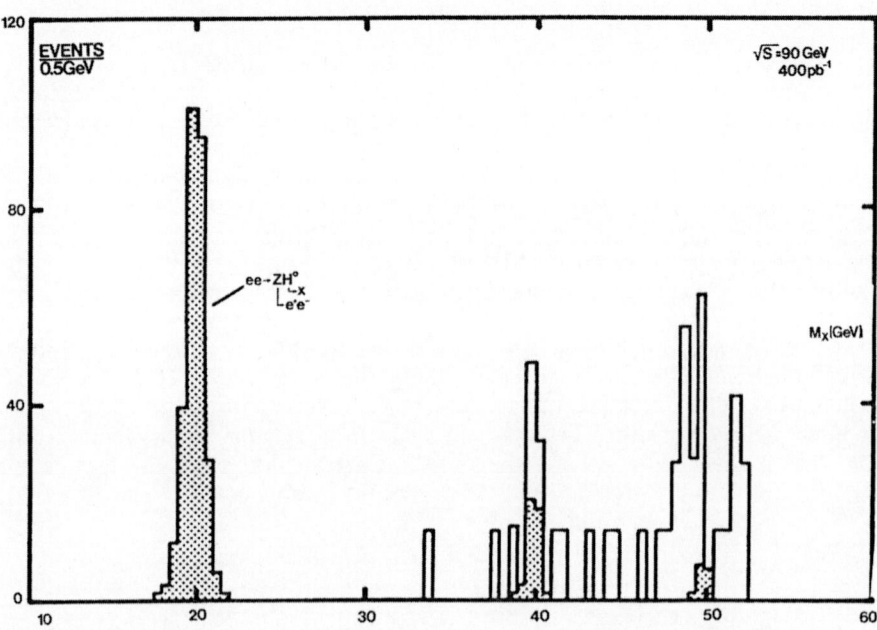

Fig. 21. Missing mass spectrum expected for Higgs of 20, 40 and 50 GeV, using the reaction in Fig. 20. The background should be much lower than shown, given the new limits on the top quark mass, as described in the text.

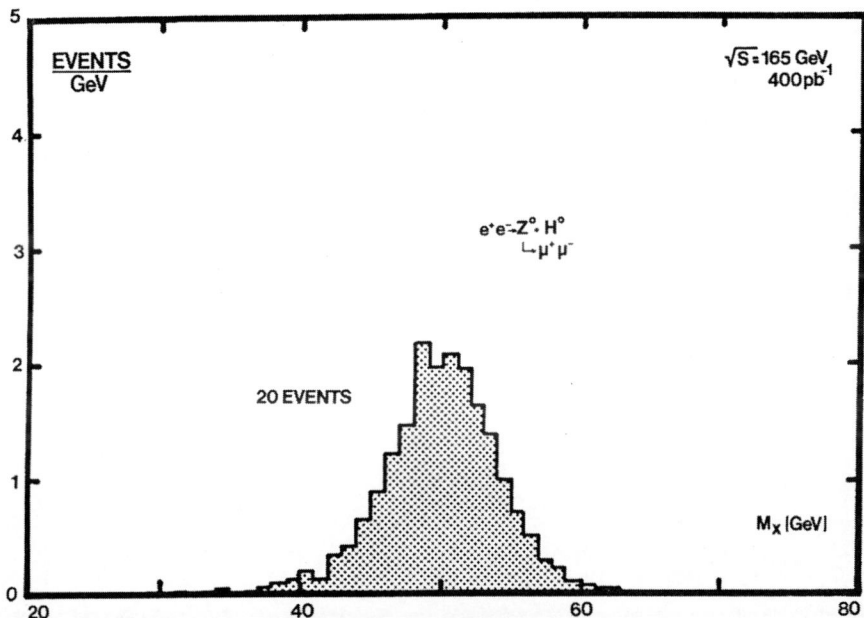

Fig. 22. Missing mass spectrum expected for the Higgs, measured by its recoil against $\mu^{+}\mu^{-}$ pairs from Z^{0}'s at $\sqrt{s} = 165$ GeV at LEP II.

363

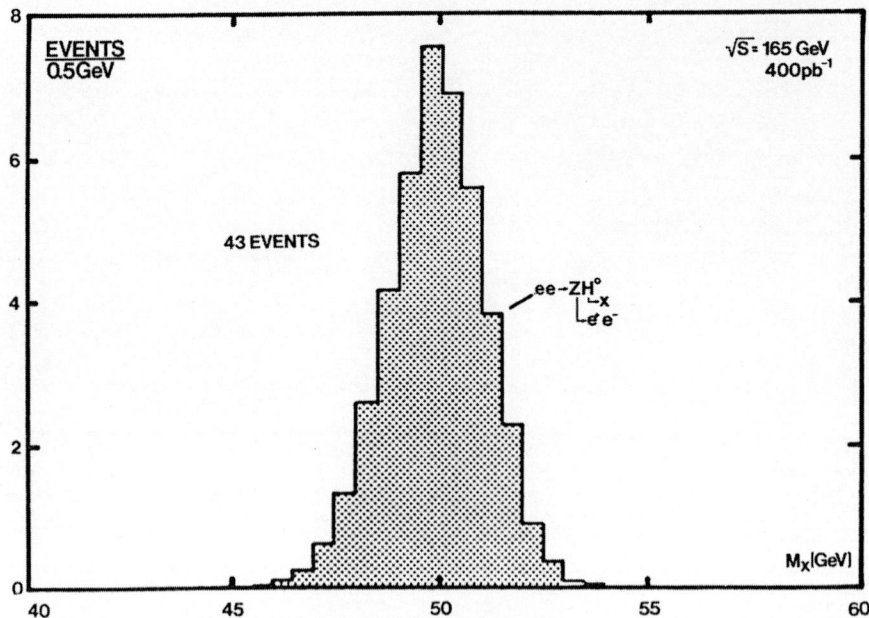

Fig. 23. Missing mass spectrum expected for Higgs, measured by its recoil against e^+e^- pairs from Z^0's, at $\sqrt{s} = 165$ GeV at LEP II, as measured in the BGO calorimeter.

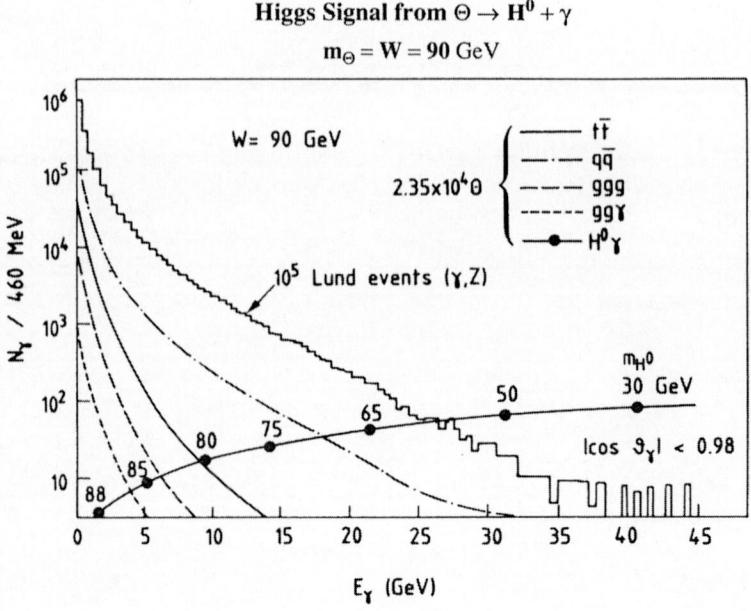

Fig. 24. Higgs signal expected from the Wilczek mechanism: $e^+e^- \to \Theta \to H^0 + \gamma$, for a toponium mass of 90 GeV.

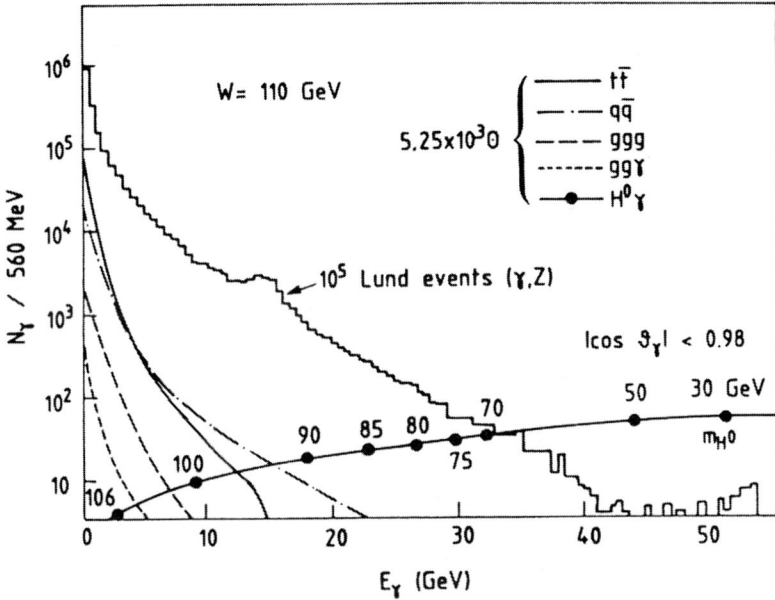

Fig. 25. Higgs signal expected from the Wilczek mechanism: $e^+e^- \to \Theta \to H^0 + \gamma$, for a toponium mass of 110 GeV.

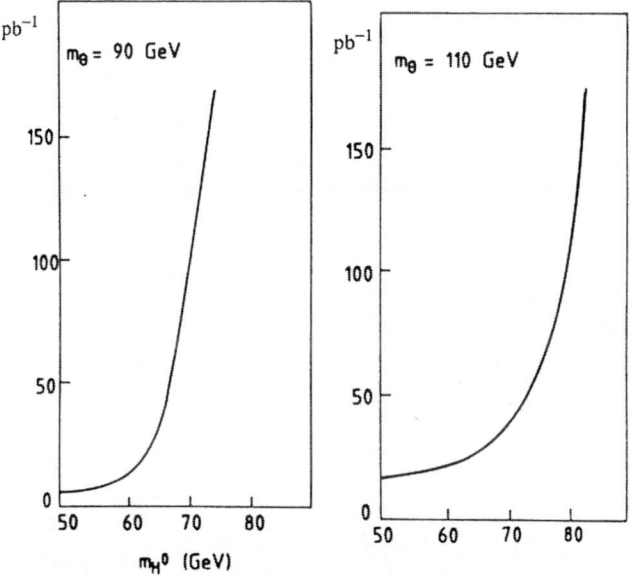

Fig. 26. Integrated luminosity needed to detect the Higgs, as a function of the Higgs mass, using the Wilczek mechanism. The cases where the toponium mass is 90 GeV and 110 GeV are shown.

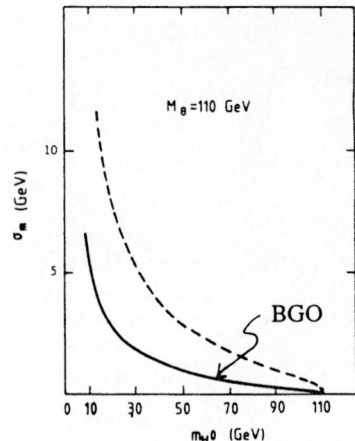

Fig. 27. The Higgs mass uncertainty per event, obtained in measurements of the monochromatic photons in $H^0 + \gamma$ final states. The solid curve corresponds to the resolution of the L3 BGO calorimeter, and the dashed line to the resolution expected for a typical electromagnetic calorimeter with wire chamber readout.

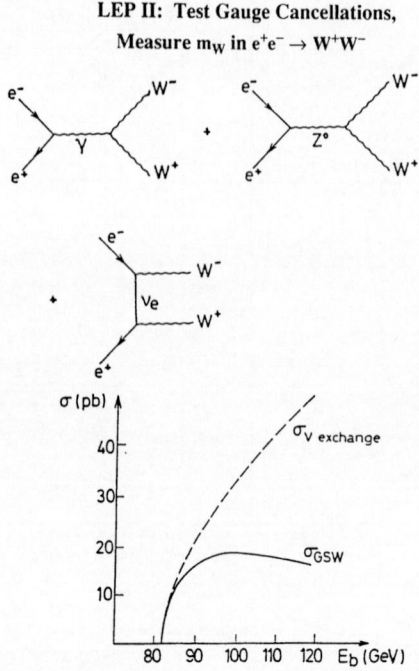

Fig. 28. Feynman diagrams and expected energy threshold behavior for W^+W^- production at LEP II (solid line). The dashed line shows the divergent result which would correspond to the neutrino exchange diagram alone.

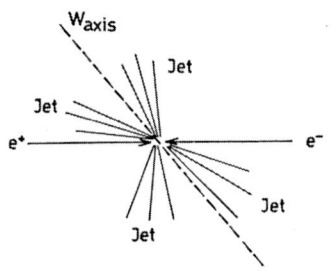

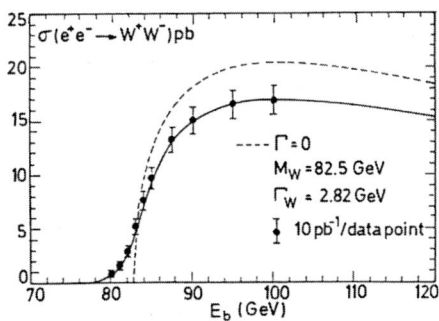

Fig. 29. Simulated measurement of the W pair threshold, using four jet events. The effect of the W width is shown (solid line) as opposed to the predictions with the width set to zero. Radiative corrections (not shown) must also be taken into account.

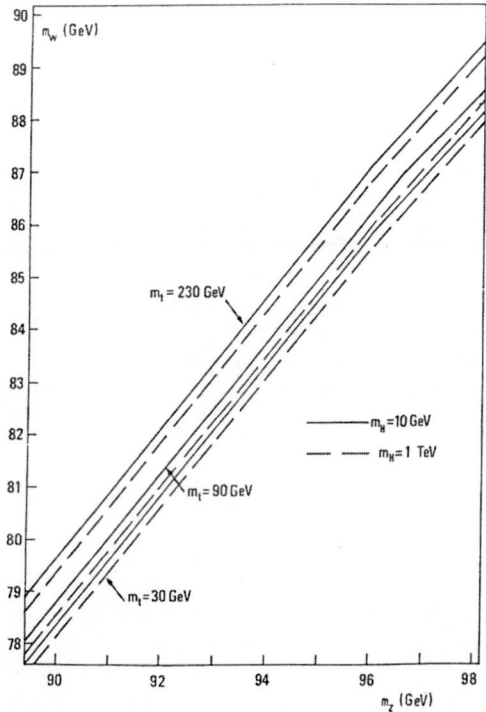

Fig. 30. M_W versus M_Z, for various top and Higgs masses, corresponding to $\sin^2 \theta_W = 0.23$. Very heavy top quarks, beyond the reach of LEP II, will have a measurable effect on the M_W/M_Z ratio.

Assuming 500 pb^{-1}, each of these techniques leads to statistical errors below 100 MeV. Systematic errors of 200-300 MeV, due to the nature of the detector and the analysis details, have to carefully corrected.

Overall, a precision of approximately 100 MeV in the W mass should be achievable, even at luminosities lower than the LEP II design values. As in the case of the precise determination of the Z^0 mass, higher order radiative corrections, and a thorough experimental study of the corrections will be needed to avoid significant systematic errors. For example, the leading order corrections in the W mass determination from the W-pair excitation curve cause an apparent shift of approximately 500 MeV [26].

6.4 The W Mass, Internal Consistency of the Standard Model, and New Physics

As discussed in Section 4, it should be possible to determine $\sin^2 \theta_W$ to 0.6%, and hence to predict the ratio M_W/M_Z to 0.3% using the charge asymmetry A_{FB}. M_Z itself will have been determined to $\sim 0.02\%$ at LEP I. The resulting accuracy in the difference between the value of M_W predicted in LEP Phase I, and M_W measured at LEP Phase II will be approximately 250 MeV.

The sensitivity to new physics which is contained in the precise comparison of M_W, M_Z and $\sin^2 \theta_W$ is shown in Fig. 30. The Figure [17] shows that the shift in M_W relative to M_Z, caused by the presence of a heavy top quark, should be detectable at LEP. The difference between the predictions for $M_t = 90$ GeV and $M_t = 230$ GeV, for example, would be measurable with 3σ significance.

The sensitivity to new physics, for various new physics scenarios, is also indicated in Table 13 [27]. Other scenarios, such as a non-minimal Higgs sector which could give a shift ~ 500 MeV, are discussed in Reference 26.

7. CONCLUSION

LEP I and LEP II will provide either the most precise tests of the standard electroweak theory, or a window on new physics in the range up to 1 TeV.

Maximizing the accuracy of the experimental measurements, and the sensitivity to new physics, often calls for the highest possible resolution in photon and lepton energy measurements, as in the determination of M_Z and A_{FB}. Mass reconstruction from leptons, photons and jets also plays an important part in many cases, as in the search for the Higgs or obtaining a clean W-pair sample. These specific considerations, and the historical importance of lepton and photon resolution in many recent physics discoveries, have been the guiding principles for the design of the L3 detector for LEP.

ACKNOWLEDGEMENTS

I would like to thank the members of the L3 Collaboration, and the participants in the "Physics at LEP" and the "ECFA Workshop at LEP 200" working groups for their excellent work in studying the physics of LEP, which made the preparation of this lecture possible. Discussions with many members of the CERN EP and LEP Divisions, especially S. C. C. Ting, U. Amaldi, W. Schlatter, A. Hoffman, L. Leistam and L. Walckiers are appreciated. Most of all, I would like to thank Prof. A. Zichichi, for his warm hospitality during this school, and for creating the Ettore Majorana Scientific Center as a scientific cultural resource for all of us.

TABLE 4

LEP I $\Rightarrow$ LEP II

APPROACHES to 250 GeV

- Add 192 S/C Cavities in Regions 4 and 8:

  32 MW Total: To E_{cm} = 220 GeV

- Magnet Lattice and Vacuum Chamber Limit:

  $E_{cm} \approx$ 250 GeV

- All 8 Regions with RF or 10 MV/m Needed[*]
  to Reach Limit: 64 MW to Beams

[*] With Niobium - Plated Copper Cavities,
Under Development at CERN

TABLE 5

SUMMARY OF THE NEW e^+e^- DETECTORS

	ALEPH (LEP)	DELPHI (LEP)	L3 (LEP)	OPAL (LEP)	Mark II (SLC)	SLD (SLC)
Vertex Detector	Chamber + μ-strips	Chamber + μ-strips	TEC	Chamber	Chamber + possible μ strips or CCD's	CCD's
Tracking Device	TPC 1 atm r = 1.8 m n = 300	TPC 1 atm r = 1.2 m + outer	---	DC 4 atm r = 1.6 m n = 160	DC 1 atm r = 1.5 m n = 72	DC 1 atm r = 1.0 m n = 80
Coil	s.c. 1.5 T	s.c. 1.2 T	warm 0.5 T	warm 0.4 T	warm 0.5 T	warm 0.6 T
Tracking Resolution	0.13% p	0.15% p	0.03% p μ only	0.15% p	0.12% p	0.13% p
Particle ID	dE/dx 4.5% σ	dE/dx + RICH	---	dE/dx 3.5% σ	dE/dx 7% σ	RICH
EM Calor.	Pb-gas	HDPC	BGO	Pb glass	Pb-LA	Pb-LA
EM Resolution	18%/$\sqrt{E}$	18-20%/$\sqrt{E}$	< 1%*	6.5%/$\sqrt{E}$	12%/$\sqrt{E}$	8%/$\sqrt{E}$
Hadron Cal. Resolution	100%/$\sqrt{E}$	100%/$\sqrt{E}$	48%/$\sqrt{E}$ +4% (Barrel)	100%/$\sqrt{E}$	---	55%/$\sqrt{E}$
Muon Det.	full	full	high res.	full	55%	full

Key: TEC - Time Expansion Chamber   RICH - Ring Imaging Cherenkov Detector
      CCD - Charged Coupled Device   HDPC - High Density Projection Chamber
      TPC - Time Projection Chamber   LA - Liquid Argon (p in GeV)
      DC - Drift Chamber

* Approximately 1.65% / $\sqrt{E}$ at low energies.

TABLE 6
GENERATORS

Process	Authors (* = L3)	Status
cosmic μ	E. Nagy*	available
beam-gas bremsstrahlung	X. DeBouard*	available
LUND	T. Sjöstrand *et al.*	available
$e^+e^- \to f^+f^-$	Berends, Kleiss	available
$e^+e^- \to e^+e^-$	Berends, Hollik, Kleiss	available
$e^+e^- \to \tau^+\tau^-$	Jadach *et al.*	soon
$e^+e^- \to \gamma\gamma$	Berends, Kleiss	available
$e^+e^- \to \gamma\gamma\gamma$	Berends, Kleiss	available
$e^+e^- \to e^+e^-e^+e^-$	Berends *et al.*	available
$e^+e^- \to e^+e^-\mu^+\mu^-$	Berends *et al.*	available
$e^+e^- \to e^+e^-L^+L^-$	Berends *et al.*	available
$e^+e^- \to e^+e^-q\bar{q}$	Berends *et al.*	available
$e^+e^- \to \mu^+\mu^-\mu^+\mu^-$	Berends *et al.*	available
$e^+e^- \to e^+e^-\gamma$	C. Mana*, M. Martinez	available
$e^+e^- \to \nu\bar{\nu}\gamma$	C. Mana*, M. Martinez	available
$e^+e^- \to W^+W^-$	Kleiss	available
$e^+e^- \to \mu^*\mu$	C. Mana*	available
$e^+e^- \to F^+F^-$	Jadach, Kuhn	available
$e^+e^- \to H^0Z^0$	Berends, Kleiss	soon
$e^+e^- \to H^+H^-$	Berends, Kleiss	soon
$e^+e^- \to H^0H^{0\prime}$	Berends, Kleiss	soon
$e^+e^- \to \theta \to H^0\gamma$		missing
$e^+e^- \to \tilde{W}_1\tilde{W}_2$	C. Dionisi*, F. Carminati*	available
$e^+e^- \to \bar{\gamma}\gamma\gamma$	C. Mana*, M. Martinez	available
$e^+e^- \to \bar{\nu}\nu\gamma$	C. Mana*, M. Martinez	available
$e^+e^- \to \bar{e}e\bar{\gamma}$	M. Martinez	available
$e^+e^- \to \tilde{W}e\bar{\nu}$	M. Chen*, M. Martinez	available
$e^+e^- \to \bar{e}\bar{e}$	C. Dionisi*, F. Carminati*	available
$e^+e^- \to \bar{\mu}\bar{\mu}$	C. Dionisi*, F. Carminati*	available
$e^+e^- \to \bar{\tau}\bar{\tau}$	C. Dionisi*, F. Carminati*	available
$e^+e^- \to \bar{q}\bar{q}$	C. Dionisi*, F. Carminati*	available
$e^+e^- \to \bar{Z}\bar{\gamma}$	C. Dionisi*, F. Carminati*	available

TABLE 7
SLC and LEP

	SLC (Z^0)	LEP (To 200 GeV)
START:	Spring 1988 (Est.)	Summer 1989
LUMINOSITY:		
MID-'88:	1.2×10^{28} (Est.)	---
MID-'89:	1.6×10^{29} (Hoped)	2×10^{31}
1990 +:	to 10^{30} ?	LEP II: 10^{32}
Z^0 SEEN PER DAY:		
MID-'88:	20	---
MID-'89:	270 ?	Fall '89: 28,000
MUON PAIRS/DAY for ASYMMETRY:		
By End of '89:	20 ?	950

TABLE 8

Z MASS and WIDTH: PRECISE MEASUREMENT
IN THE FIRST DAYS of LEP

- (1) SCAN: Two Days (2 pb^{-1}) Covering
  M_Z Range : ± 6 GeV

- (2) 100 DAYS NEAR Z PEAK: ≈ 100 pb^{-1}

- (3) L3 Only: RECONSTRUCT $\mu^+ \mu^- (\gamma)$ MASS:
  Reduce Systematics Due to Radiative Corrections

⇒ *STATISTICS ARE NOT THE PROBLEM.*

IN 10 DAYS OBTAIN:

- $(\delta M_Z)_{STAT} \leq \pm 10$ MeV (0.01%)
- $(\delta \Gamma_Z)_{STAT} \leq \pm 15$ MeV (0.5%)

- CONCLUDE: CONTROL SYSTEMATICS:
  - PRECISE ACCEPTANCE
  - CALIBRATE LEP to ± 10 MeV
  - THIRD ORDER RAD. CORR. (With CUTS)

.

TABLE 9

	Born	LL1	LL1+LL2	LL1+LL2+F	LL1+LL2+LL3+F
Max. (GeV)	92.02	92.22	92.11	92.10	92.12
Peak (nb)	1.59	1.09	1.15	1.12	1.11

$$LLn = O\left(\frac{\alpha}{\pi} \ln(M_Z^2/m_e^2)\right)^n$$

TABLE 10
STATISTICAL ACCURACY ON $\sin^2\theta_W$
FROM $A_{FB}(e^+e^- \Longrightarrow \mu^+\mu^-(\gamma))$

- AFTER 200 DAYS RUN (200 pb^{-1}):
  - 5.6×10^6 Visible Z^0's
  - 1.9×10^5 $\mu^+\mu^-(\gamma)$ for Asymmetry
  - $(\delta A_{FB})_{STAT} \approx 0.0023$
  - $(\delta \sin^2\theta_W)_{STAT} \approx 0.0012$ (0.5%)

ASYMMETRY SYSTEMATICS

- CONTROL TO $\leq 0.2\%$

- APPLY MARK J TECHNIQUES (BETTER in L3)

- PRECISE MUON AND PHOTON ENERGY CUTS
  - REDUCE SYSTEMATICS for RAD. CORR.

TABLE 11

Values of $\sin^2\theta_W$ determined by a global fit to all data for various m_t and M_H, the ratio $\sin^2\hat{\theta}_W(M_W)/sin^2\theta_W$, and the corresponding value of $\sin^2\hat{\theta}_W(M_W)$.

m_t (GeV)	M_H (GeV)	$\sin^2\theta_W$	$\dfrac{\sin^2\hat{\theta}_W(M_W)}{\sin^2\theta_W}$	$\sin^2\hat{\theta}_W(M_W)$
25	100	.229±.004	0.994	.227±.004
45	100	.230	0.991	.228
60	100	.230	0.991	.228
100	100	.227	1.01	.229
200	100	.222	1.05	.233
400	100	.209	1.18	.248±.005
45	10	.229	0.994	.228
45	1000	.231	0.982	.227

TABLE 12

Number of $\theta \rightarrow H^0 + \gamma$ Events per 1000

for Different Choices of m_{H^0}

m_{H^0} (GeV) m_θ (GeV) $[\int L\, dt \ (\text{pb}^{-1})]$	10	30	50	60	70	80	85	90	100
70 [32]	202	160	102	52					
90 [49]	107	95	75	60	43	22	12		
110 [65]	39	36	31	28	25	19	16	13	7

TABLE 13

NEW PHYSICS ABOVE THE Z^0

	δA_{FB}(LEP I)	δM_W (LEP II)
● HEAVY TOP QUARK $M_T \approx 180$ GeV:	0.0075	780 MeV
● HEAVY HIGGS $M_H \approx 1$ TeV:	− 0.0045	− 160 MeV
● HEAVY LEPTON $M_L \approx 250$ GeV	0.006	300 MeV
● HEAVY SQUARK (SUSY) $M_{SQ} \approx 250$ GeV	0.010	300 MeV
● HEAVY SLEPTON (SUSY) $M_{SL} \approx 250$ GeV	0.006	300 MeV
● TECHNICOLOR $SU_8 \otimes SU_8$:	− 0.018	− 500 MeV

$\Rightarrow$ LARGER EFFECTS for HEAVIER MASSES

$\Rightarrow$ *L3: AIM for* $\delta A_{FB} < 0.002$

REFERENCES

[1] G. Hanson *et al., Phys. Rev. Lett.* **30**, 1189 (1973); A. Litke *et al., Phys. Rev. Lett.* **32**, 432 (1974).

[2] M. Perl *et al., Phys. Rev. Lett.* **35**, 1489 (1975); M. Perl *et al., Phys. Lett.* **63B**, 366 (1976).

[3] W. Braunschweig *et al., Phys. Lett.* **57B**, 407 (1975); W. Tanenbaum *et al., Phys. Rev. Lett.* **35**, 1323 (1975); G. Feldman *et al., Phys. Rev. Lett.* **35**, 821 (1975).

[4] D. P. Barber *et al.*(MARK J Collaboration), *Phys. Rev. Lett.* **43**, 830 (1979); R. Brandelik *et al.*(TASSO Collaboration), *Phys. Lett.* **86B**, 243 (1979); Ch. Berger *et al.*(PLUTO Collaboration), *Phys. Lett.* **86B**, 418 (1979); W. Bartel *et al.*(JADE Collaboration), *Phys. Lett.* **91b**, 142 (1980).

[5] G. Arnison *et al.*(UA1 Collaboration), *Phys. Lett.* **122B**, 103 (1983); M. Banner *et al.*(UA2 Collaboration), *Phys. Lett.* **122B**, 476 (1983); G. Arnison *et al.*(UA1 Collaboration), *Phys. Lett.* **126B**, 398 (1983), and *Phys. Lett.* **129B**, 273 (1983); P. Bagnaia *et al.*(UA2 Collaboration), *Phys. Lett.* **129B**, 130 (1983).

[6] S. Glashow, *Nucl. Phys.* **22**, 579 (1961); S. Weinberg, *Phys. Rev. Lett.* **19**, 1264 (1967) and *Phys. Rev.* **D5**, 1412 (1972); A. Salam, in *Elementary Particle Theory*, Ed. N. Svartholm, Stockholm, p. 361, 1968.

[7] F. Bonaudi, "LEP – A Status Report", CERN/EF 87-5, June, 1987.

[8] J. P. Gourber *et al., Proceedings of the 1983 Santa Fe Particle Accelerator Conference, IEEE Trans. Nucl. Science* **NS-30**, No. 4, 3614 (1983).

[9] A. Hoffman, "LEP", Talk at the 1987 International Symposium on Lepton and Photon Interactions at High Energies, Hamburg (1987).

[10] A brief discussion of the experiments' characteristics, and the collaborating institutions, may be found in the annual report "Experiments at CERN". Further discussions are contained in the Technical Proposals and reports to the LEP Experiments Committee (LEPC).

[11] L3 Technical Proposal, L3 Collaboration, May, 1983.

[12] "The Construction of L3", Report by L3 to the Department of Energy, April, 1985; "The Progress of L3", Quarterly Reports to DOE by L3, 1985-1987.

[13] H. S. Chen *et al.*, "The Performance of an L3 Hadron Calorimeter Prototype Module with BGO", submitted to *Nucl. Instr. and Methods.*

[14] "Physics at LEP", CERN Yellow Report CERN/EP 86-02 (Vol. I and II), 1986.

[15] J. Dorfan, CERN Colloquium, July 1987.

[16] U. Amaldi *et al.*, "A Comprehensive Analysis of Data Pertaining to the Weak Neutral Current and Intermediate Vector Boson Masses", *Phys. Rev.* **D36**, 1385 (1987).

[17] G. Costa *et al.*, "Neutral Currents Within and Beyond the Standard Model", Reports CERN-TH 4675, MAD/TH 87-07, and LBL-23271 (1987).

[18] P. Jenni, "Weak Boson Production and Decay", in *Proceedings of the 1987 International Symposium on Lepton and Photon Interactions at High Energies*, Hamburg (1987), to be published; D. Froidevaux, "Physics at the CERN $p\bar{p}$ Collider: Some Recent Results and Prospects", Report CERN-EP-119 (1987).

[19] A. Ali, and B. Van Eijk, to be published; Also see B. Van Eijk, "Heavy Flavor Production and Heavy Flavor Mixing at the CERN Proton-Antiproton Collider", NIKHEF Thesis, 1986 (Unpublished).

[20] P. Higgs, *Phys. Lett.* **12**, 132 (1964); *Phys. Rev. Lett.* **13**, 508 (1964); *Phys. Rev.* **145**, 1156 (1966).

[21] B. Lynn and R.G. Stuart, *Nucl. Phys.* **B253**, 216 (1985); B. Lynn *et al.*, in Vol. I of Reference 14.

[22] F. Wilczek, *Phys. Rev. Lett.* **39**, 1304 (1977).

[23] H. Burkhardt *et al.*, ALEPH Note 87-59, August, 1987.

[24] G. Barbiellini *et al.*, "Physics at LEP at High Energies", in Vol. II of Reference 14.

[25] The simulation program for W^+W^- production was provided by R. Kleiss.

[26] J. Bijnens *et al.*, "Measurement of the W Mass at LEP 200", in Vol. I of the *Proceedings of the ECFA Workshop on LEP 200*, Ed. A. Boehm and W. Hoogland, Aachen, 1986. CERN Reports CERN 87-08; ECFA 87/108, June, 1987.

[27] Adapted from the work of B. Lynn, M. Peskin, and R. Stuart, in Reference 14.

DISCUSSION

- *Pitman:*

What are the various resolutions on missing energy or missing mass measurements?

- *Newman:*

Typically, in the Higgs search I showed, we do get into kinematic regions where the resolution on the missing mass is $\sim 1\%$, for relatively large Higgs masses.

- *Pitman:*

What percentage of the detectors are dead regions due to cables or supports?

- *Newman:*

In the couple of percent range.

- *Iacovacci:*

What are the possibilities at LEP to select different flavor or jets?

- *Newman:*

The only way that we would find a heavy flavor is through the semileptonic decay channels. Other experiments have, since the days of PETRA, hoped and, in some cases, even planned on tagging the flavors of jets. I think I cannot say for the future, but until this moment I do not know of a detector which has succeded in an efficient way to tag different flavors by looking, for example, at the highest momentum hadron in a jet. The one exception is $D^* \to D\pi$.

- *Iacovacci:*

I have another question: how often do you need to recalibrate your apparatus, on average?

- *Newman:*

As I said at the beginning, there is no compromise in trying to achieve the ultimate resolution for electrons, photons and muons. For example, the laser calibration system for the muon chambers, which monitor the chamber's alignment, are working all the time.

In case of the BGO electromagnetic calorimeter, I showed you a glimpse of the radio frequency quadrupole accelerator which produces photons (by radiative capture in a target) up to 18 MeV and which can calibrate all of the 12,000 crystals in an hour. The RFQ will run several times a day. It may also produce photons for short intervals during each data taking run.

– *Schmidt:*

How well can your detector distinguish between hard electrons and accompanying collinear soft photons?

– *Newman:*

The limitation on the accompaniment will be coordinate resolution of the BGO, which is extremely good. The position resolution for electrons or photons in the BGO above a couple hundred MeV gets down to one millimeter. So you have to consider that you have electromagnetic calorimeter of a radius of 50 cm. If there is an angular separation of a few degrees and the photon energy is not too small, we will separate the accompanying photon.

– *Jonker:*

How well do you think L3 can measure emission of neutrinos using single photons?

– *Newman:*

We expect 5000 events of the type photons plus "nothing" for each of the known neutrino types. We will get approximately 1,700 extra events of that type for each additional neutrino family. We expect to measure the number of neutrino types with a precision of 4% during LEP Phase I.

– *Jonker:*

Can you do neutrino family counting while sitting on the Z or do you have to make a special run?

– *Newman:*

The numbers which I gave you were for a special run which is 10 GeV above the mass of the Z. However, we will see inclusive photons of the type you just mentioned even scanning over the Z.

– *Ting:*

I want to make a comment. There was a question a few minutes ago about how well one can identify single electrons. That is a very interesting question. The proper way to ask this question is: you have e^+e^- collisions, you produce many jets, in these jets there are pions, electrons and photons. If you do a Monte

Carlo study given the coordinate resolution for the vertex chamber, and the energy resolution for the BGO crystals, the answer is one part in a thousand.

– *Wicklund:*

I can add one thing to a previous question which was about the possibilities of tagging flavors at LEP. I know that the general purpose detectors would tag, for example, charm by using the decay $D^* \to D^0\pi$, $D^0 \to K\pi$, and most detectors would tag bottom jets using high transverse momentum leptons and by looking for large impact parameters with vertex detectors. So those seem to be the flavors that can get tagged with some efficiency: charm and bottom.

– *Newman:*

Yes, I mentioned the role of semileptonic decays in tagging heavy flavours.

"GRAN SASSO PHYSICS" *

I.A. Pless

Massachusetts Institute of Technology
Cambridge, MA 02139, USA

A description is given of the Gran Sasso Laboratory and the approved physics program.
The four major detectors, LVD, ICARUS, MACRO and GALLEX are described. The programs of
neutrino physics, particle physics, astrophysics, cosmic ray physics, nuclear physics and
geophysics are outlined. The physics topics of neutrino oscillations, dark matter, solar
neutrinos, and proton decay are discussed in some detail.

I. INTRODUCTION

Underground physics has always been a subject of study throughout the world. In figure 1
there is a display of most of the underground laboratories.

The newest addition to the field of underground physics is Gran Sasso Laboratory, the
world's largest and most powerful laboratory.

The original idea of the Gran Sasso Laboratory was due to Professor Zichichi. His basic
concept was to build an underground laboratory where one could bring the forefront of our
advanced technology to bear on the important physics questions of our day. He recognized that
the Rome-Teramo highway, where it tunneled under the Gran Sasso mountain, presented an ideal
opportunity for the location of his laboratory. The largest trucks in Europe could enter directly
into the laboratory. As I will discuss, this site has much more to offer than just convenient
access.

A crucial characteristic of any underground laboratory is its depth. If the thickness of the
rock above a laboratory is too great, nothing (except neutrinos) gets through to your detector.
If the thickness is too small, the uninteresting surface cosmic radiation creates such a large
background that the interesting physics is buried and lost in the noise. The optimum thickness
is about 4,000 meters water equivalent. The Gran Sasso tunnel is at this depth and is one of the
major reasons Prof. Zichichi chose this site for the laboratory.

Another crucial aspect of underground laboratory is the amount of natural radioactivity in
the surrounding rock. Because of the special nature of the Gran Sasso rock (sedimentary rock
which has been folded upward), the radioactivity in this laboratory is at least an order of
magnitude less than in any other underground laboratory. Because of this Prof. Zichichi likes to
describe the laboratory as the "Cosmic Silence Lab." This "Cosmic Silence" was the final reason
why he chose Gran Sasso as the site for the laboratory.

The Superworld II
Edited by A. Zichichi
Plenum Press, New York, 1990

Figure 1. Location of most of the major underground laboratories.

Gran Sasso Laboratory is already world famous even before the first experiment has been installed. Physicists from Russia, China, the U.S.A. and major countries of Europe now have approved experiments in Gran Sasso Laboratory. In fact all available laboratory space has been allocated and there are plans to enlarge the Laboratory.

II. DESCRIPTION OF GRAN SASSO LABORATORY

Before describing the Gran Sasso Laboratory, it is useful to put its position with respect to the rest of the underground laboratories in some perspective. Figure 1 shows the locations of the major underground sites in the world.

Figure 2 displays the location of Gran Sasso Laboratory. It is about 120 kilometers east and slightly north of Rome.

Figure 3 is the plan view of the underground laboratory. For purposes of scale, Laboratory A is 100 meters long.

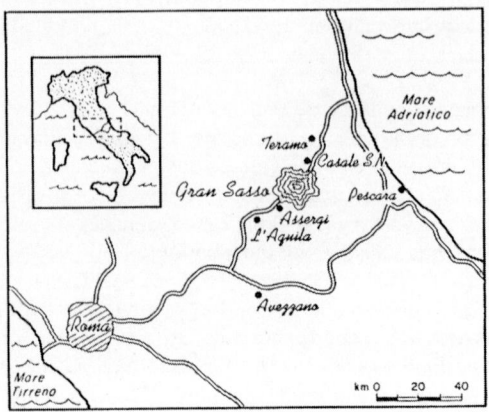

Figure 2. Location in Italy of the Gran Sasso Laboratory.

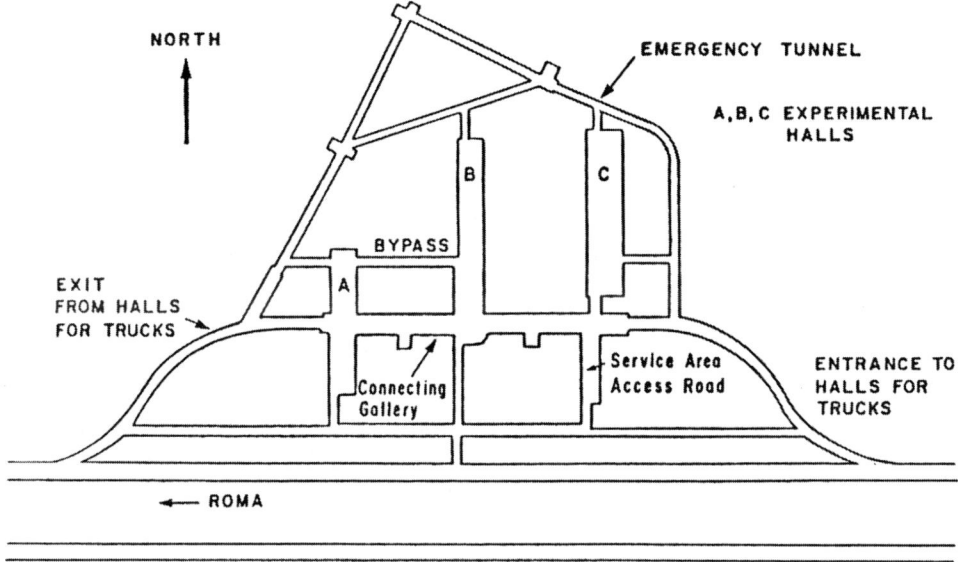

Figure 3. Plan view of the Gran Sasso Laboratory.

Figure 4. Gran Sasso surface laboratory.

381

PARAMETERS OF THE LABORATORY

Volume	60,000 M^3
Normal ventilation	20,000 M^3/Hour
Emergency ventilation	100,000 M^3/Hour
Lab. temperature	20° ± 1° C
Relative humidity	50% ± 5%
Electric power	5 megawatts (20 in future)

The laboratory also contains compressed air, water, computer facilities, safety systems, etc. Associated with the Gran Sasso Laboratory complex is an above ground laboratory. This laboratory houses offices, control rooms, conference rooms and other facilities. Figure 4 is a photograph of this above ground laboratory.

The construction of Gran Sasso Laboratory has been compared to building a submarine under water. The water pressure inside of Gran Sasso is 60 atmospheres. Anywhere you drill a hole, water pours out. Figure 5 shows the early stages of constructing a Gran Sasso Laboratory. The knee deep water is clearly apparent. The next figure (Figure 6) is a photograph of Hall A in the final stages of excavation. The size of this hall is impressive and again the water is everywhere. Figure 7 shows Hall A completely excavated and the water is now under control.

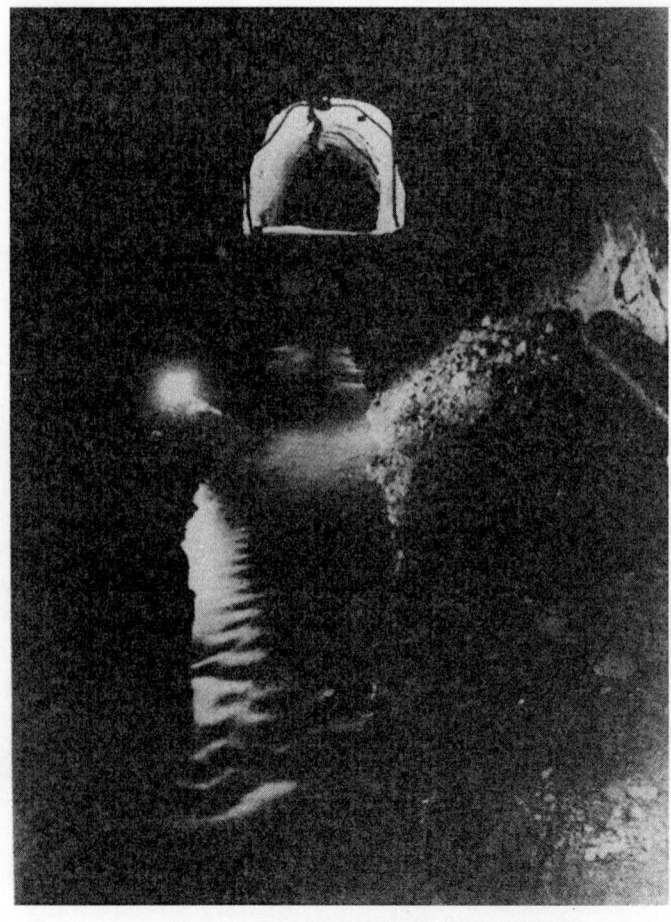

Figure 5:   Early construction stage of a laboratory.

Figure 6.   Final construction stages of Laboratory A.

Figure 7.   Hall A completely excavated.

Figure 8. Device for installing insulation inside Laboratory A.

Figure 9. Wall insulation for Gran Sasso Laboratory.

Figure 10.   Completed Laboratory A.

Figure 8 is a picture of the device of the which places the insulation on the surface of the experimental hall. This insulation is required to make the hall habitable. Without insulation, the temperature inside of the hall would be 6° C. Hence insulation is required. Figure 9 shows this insulation. Figure 10 is a recent photograph of Hall A complete with crane.

III. APPROVED GRAN SASSO EXPERIMENTS

The following experiments have been approved for Gran Sasso Laboratory:

° ICARUS
° MACRO
° GALLEX
° LVD
   DOUBLE BETA DECAY
   LASER INTERFEROMETER
   GRAVITATIONAL WAVE DETECTOR
   GFM
   CDS
   LAID

LVD, ICARUS. MACRO and GALLEX are the largest experiments approved for the Gran Sasso Laboratory. These detectors rival in size and complexity any of the major detectors being built for LEP, HERA, SLC or FERMILAB.

A. ICARUS (Imaging Cosmic and Rare Underground Signals)

ICARUS is a giant liquid argon drift chamber with two-dimensional space readout plus a drift time readout. Figure 11 is a drawing of ICARUS. [Note: Not yet funded by U.S.]

385

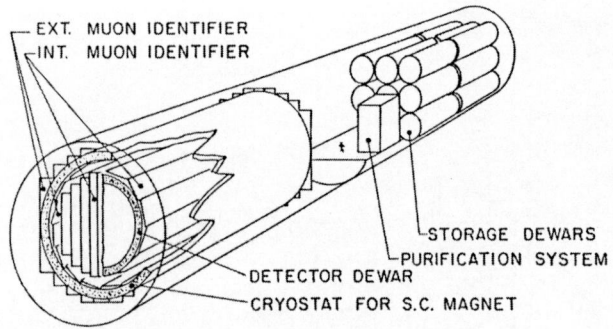

Figure 11.   Artist sketch ICARUS.

ICARUS PARAMETERS (1986)

Volume	4,600 M
Diameter	14.0 M
Length	30.0 M
Stereo angle	10 Deg.
Drift length	2,30 M
Drift field	1.0 Kv/cm
High voltage	230 Kv
Number channels	100,000
Spatial resolution	2 mm
Total memory	200 M Bytes

ICARUS is a very large, very ambitious project at the forefront of current technology.

B. MACRO (Monopole, Astrophysics and Cosmic Ray Observatory)

MACRO is a large detector consisting of three layers of scintillation counters, streamer tubes tracking chambers, plastic track ETCH detectors and concrete. Figure 12 is a sketch of the MACRO detector.

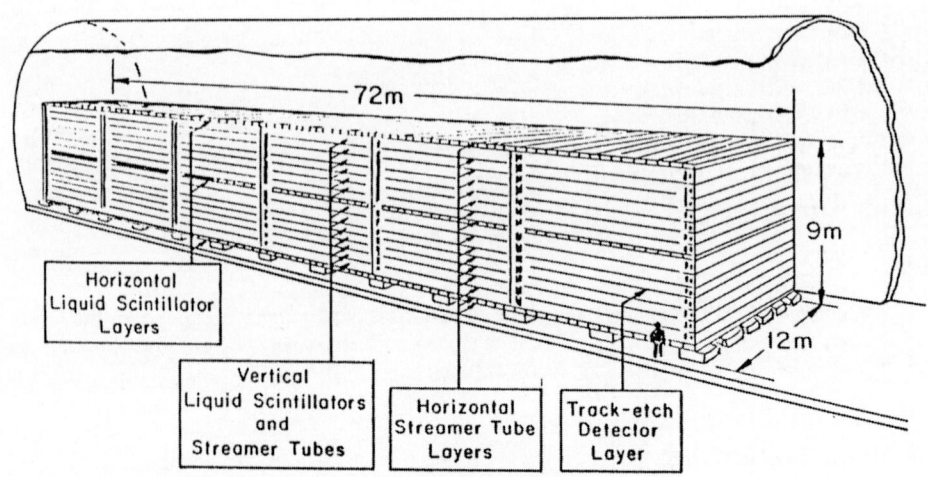

Figure 12.   Sketch of the MACRO detector.

MACRO PARAMETERS

Length	12 M
Width	12 M
Height	9 M
Mass of Scintillator	800 T
Analogue channels	242
Scintillator counters	484
Tracking channels	139,000

MACRO is a large, well understood device.

C. GALLEX (Gallium Experiment)

GALLEX is a solar neutrino experiment. It contains 30 tons of gallium dissolved in hydrochloric acid. The gallium is the active detector material. In some sense, GALLEX has a similarity to a large chemical plant. Figure 13 is a sketch of the GALLEX detector.

D. LVD (Large Volume Detector - the experiment of Prof. Zichichi)

. Since LVD is the experiment of Prof. Zichichi I will spend a little more time on this subject. LVD can study a wide spectrum of physics topics. The following are the LVD parameters.

LVD PARAMETERS

Length	42 M
Width	12 M
Height	13.2 M

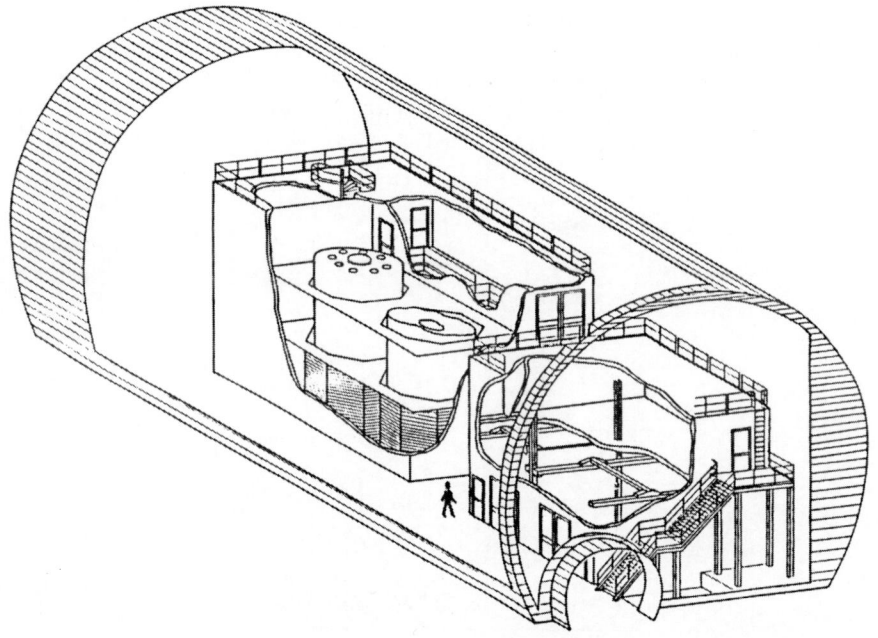

Figure 13.   Sketch of the GALLEX detector.

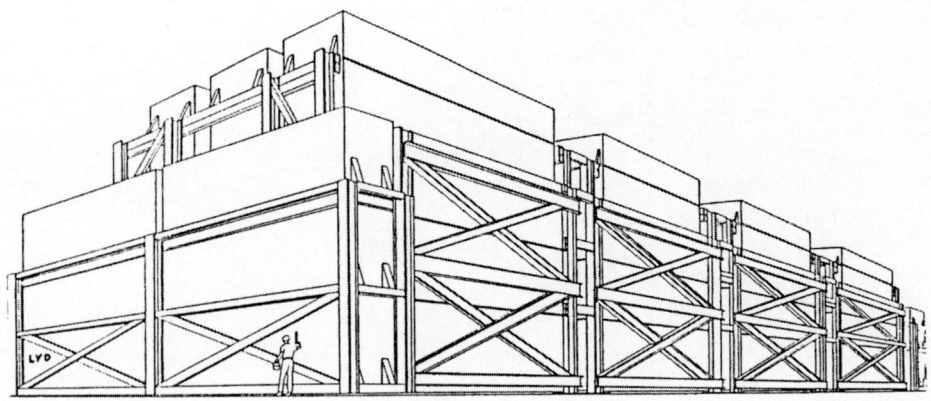

Figure 14.   Sketch of the LVD detector.

Scintillator counters	1,600
Photo tubes	6,400
Analogue channels	12,800
Tracking channels	100,000
Number of computers	7
Weight scintillator	1,800 T
Weight steel	1,800 T
Modules	200

Figure 14 is a sketch of the LVD detector. In this sketch the support structure and the 200 modules are clearly visible. Figure 15 displays the scintillation counter. The container for the liquid scintillator is 1.5 meter x 1 meter x 1 meter. The liquid is viewed by three 15 centimeter photo tubes for energy resolution and by one fast 5 centimeter phototube for time of flight determination.

Eight scintillation counters are grouped into sets of eight counters as shown in Figure 16. Figure 16 is a sketch of a porta tank which is the mechanical support structure of the module. Each porta tank has two panels of streamer tubes and readout strips attached to it. The streamer tubes are arranged into two staggered layers which gives a spatial resolution equivalent to two centimeters. This streamer tube arrangement is more clearly shown in Figure 17.

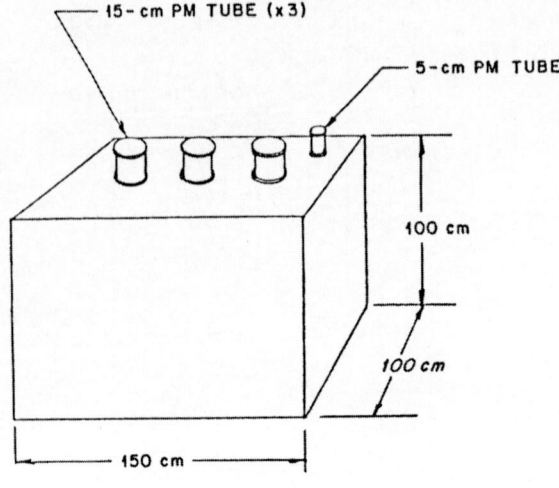

Figure 15.   LVD scintillation counter.

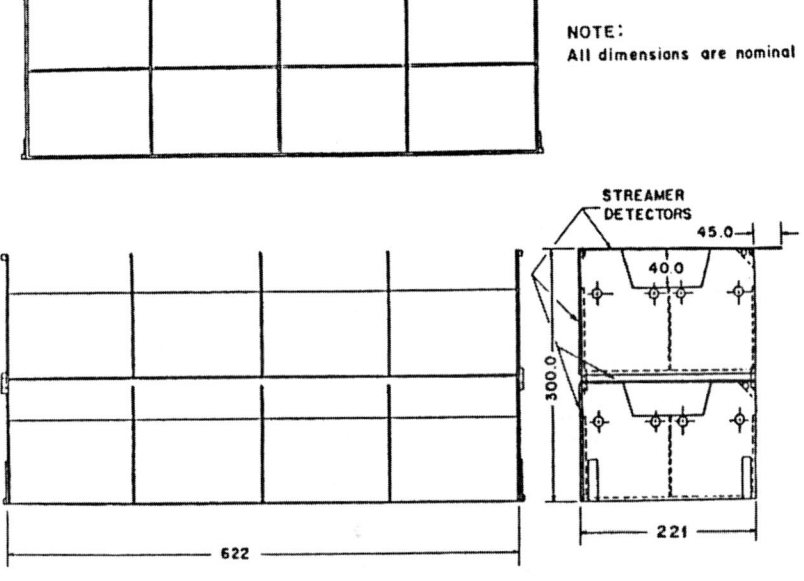

NOTE:
All dimensions are nominal

STREAMER
DETECTORS

45.0

40.0

300.0

221

622

Figure 16: Porta tank with streamer tube tracking detector.

The modular structure of LVD has two practical purposes. First, the modular design allows data taking to start as soon as the first module is installed and continue uninterrupted throughout the construction stage. Second, you can fill the laboratory with detector. Figure 18 illustrates this point. Figure 19 demonstrates how the small panels of streamer chambers attached to each porta tank can be connected so as to make large area tracking chambers.

IV. FRONTIER PHYSICS

The experiments approved for the Gran Sasso Laboratory propose to study the following physics topics:

°Neutrino oscillations
°Stellar collapse
°Solar neutrinos (Boron)

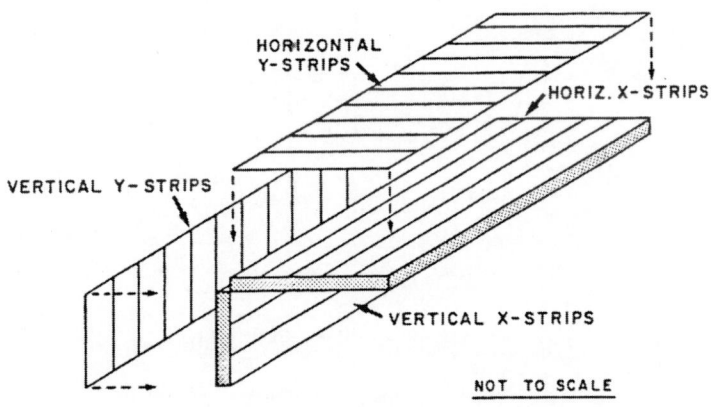

HORIZONTAL
Y-STRIPS

HORIZ. X-STRIPS

VERTICAL Y-STRIPS

VERTICAL X-STRIPS

NOT TO SCALE

Figure 17.   Sketch of streamer tube tracking detector.

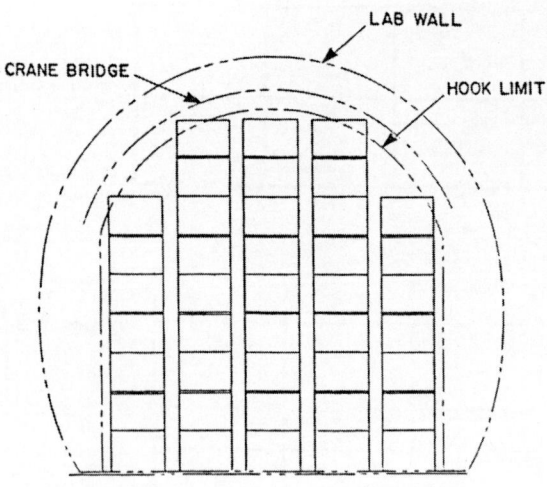

Figure 18.   LVD modules stacked inside Hall A.

°Dark matter
 Cosmic ray sources
 Neutrino astrophysics
 Monopoles
°Proton decay (super symmetric)
 Gravitational waves
 Double beta decay
 Cosmic ray element adundances
 Fault motion measurement
 Geophysical neutrinos
 Geophysical electric fields
 Geophysical magnetic fields

The °'s indicate those experiments for which LVD has been optimized.

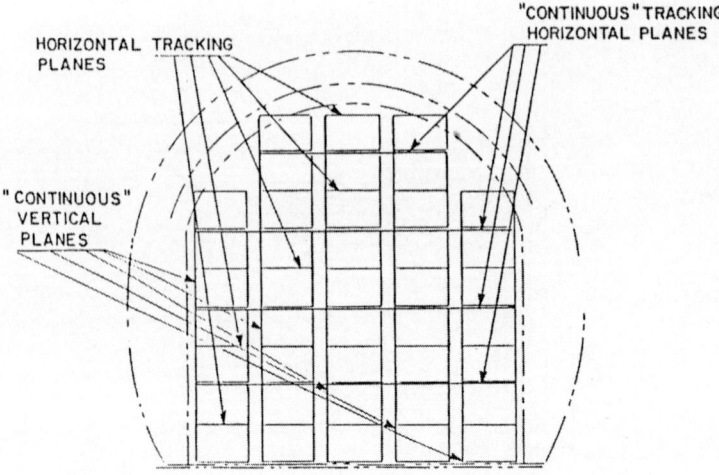

Figure 19.   LVD large area tracking planes.

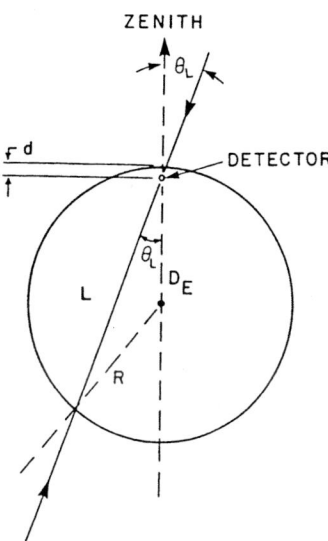

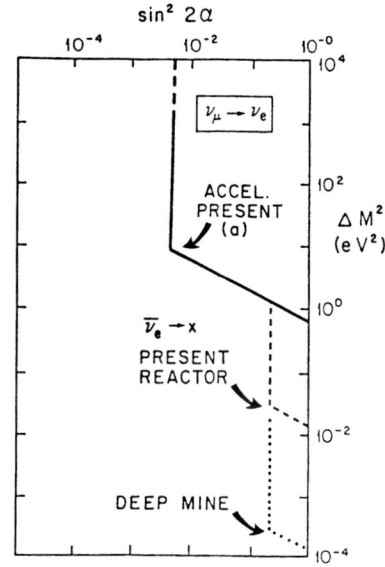

Figure 20. Single detector, two source arrangement for measuring neutrino oscillations.

Figure 21. Neutrino oscillation limits.

NEUTRINO OSCILLATIONS

There are at least two ways to measure neutrino oscillations. The first is to have one source of neutrinos and two "identical" detectors at different distances. The second is to have two "identical " sources at different distances and one detector. The measurement consists of detecting either a different number of neutrinos from the two "identical" detectors or a different number of neutrinos from the two "identical" sources.

In Gran Sasso, one uses the technique of two "identical" sources. Figure 20 shows this technique.

The crucial key to making this measurement is to detect charge current interactions occurring in the detector creating muons going upward as compared to charged current interactions in the detector creating muons going downward. The difference between the number of muons going upward and the number going downward "measures" neutrino oscillations.

Of all the detector in Gran Sasso, only LVD can make this measurement in this "clean" manner. Figure 21 illustrates the present limits on neutrino oscillations and the limits we expect to achieve with LVD.

STELLAR COLLAPSE

One of the most awesome events in the universe is the creation of a supernova or stellar collapse. It has now been demonstrated that terrestrial instruments can detect the neutrinos and antineutrinos that are emitted in the early stages of a stellar collapse. The surest way to identify a stellar collapse is to detect the antineutrinos. While some of the other detectors in Gran Sasso can detect neutrino and antineutrino interactions, only LVD can identify antineutrinos.

LVD identifies antineutrino interactions by the following two-step process:

$$I. \quad \nu + p \rightarrow N + e^+$$
$$e^+ + e^- \rightarrow 2\gamma$$

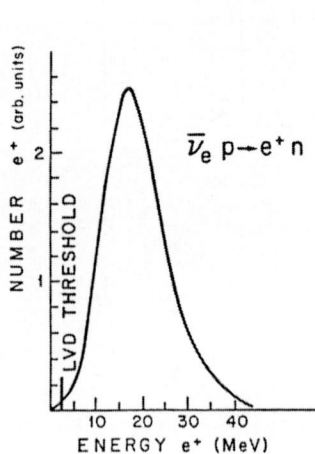

Figure 22. Energy spectrum of positron from inverse beta decay due to incident antineutrinos from a stellar collapse.

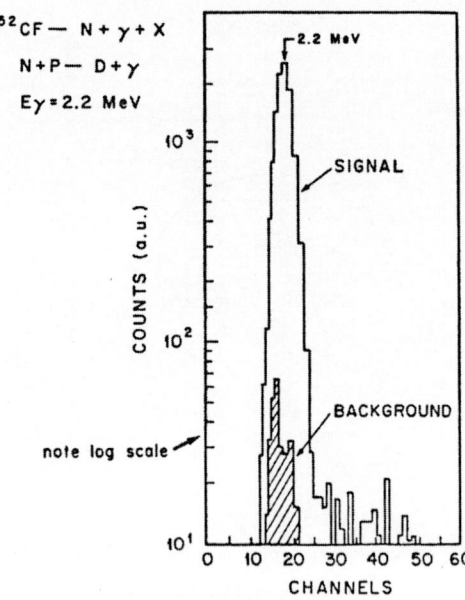

Figure 23. Gamma ray signal due to neutron proton fusion.

II. $N + p \rightarrow D + \gamma$ ($E\gamma = 2.2$ MeV)

Figure 22 shows the expected energy spectrum from stage I. Stage II produces a line spectrum (smeared by instrumental resolution). The moderation time of the neutron is expected to be about 170 micro seconds.

In Figure 23 we show the result of an experiment to measure the $N + p$ fusion gamma ray ($E\gamma = 2.2$ MeV). The experiment consists of placing a 252_{CF} source inside one of the counters. The decay of 252_{CF} is as follows:

$$252_{CF} \rightarrow N + \gamma + x$$

A solid state detector located next to the 252_{CF} source is triggered by the γ which then gates open the three 15 cm photo tubes of the counters for 500 micro seconds. The data in Figure 23 shows that the 2.2 MeV fusion line (smeared by our resolution) standing well above background. Figure 24 shows the arrival time of the 2.2 gamma ray after the decay of the 252_{CF}. As can be seen, this distribution fits a 170 microsecond neutron moderation time very well.

These data demonstrate that LVD can identify antineutrino interactions.

SOLAR NEUTRINOS

Figure 25 displays the theoretical spectrum for solar neutrinos. The only measurement to date is sensitive to the ^8B reaction which furnishes about 10^{-3} of all the solar neutrinos. This measurement was made by Davis and his collaborators at the Homestake mine during the period 1970 - 1984. The technique used involves the absorption of a solar neutrino in chlorine which changes chlorine to argon, and then detecting the decay of the argon back to chlorine.

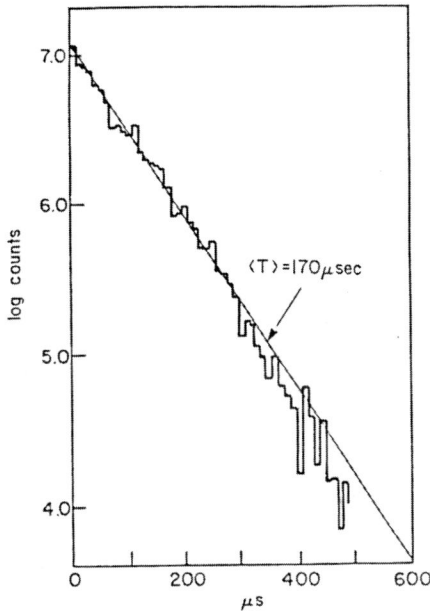

Figure 24.   Neutron moderation curve.

The reaction is:

$$V_e + {}^{37}CL \rightarrow {}^{37}AR + e^-$$

followed by:

$${}^{37}AR \rightarrow {}^{37}CL + e^+ \; (2.8 \; KeV)$$

The actual measurement is performed by detecting the 2.8 KeV positron.

The well known result is that only 1/3 of the expected solar neutrino flux was found. There are many theoretical explanations for these data, none of which are very satisfactory. What is clearly needed is a measurement of the major section of the solar neutrino spectrum, not the very small high energy tail.

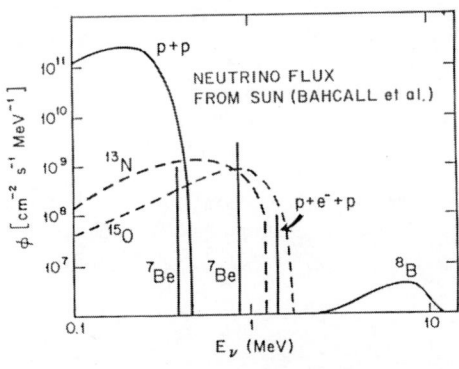

Figure 25:   Theoretical solar neutrino spectrum.

The GALLEX experiment proposes to do just that. This experiment also depends on a two-stage reaction:

$$\text{I. } V_e + {}^{71}\text{GA} \rightarrow {}^{71}\text{Ge} + e$$

$$\text{II. } {}^{71}\text{Ge} \rightarrow {}^{71}\text{GA} + \gamma$$

$$\gamma_K = 10.4 \text{ KeV}, \gamma_L = 1.2 \text{ KeV}$$

The ^{71}Ge is swept out of the mixture of gallium and hydro-chloric acid by bubbling helium through the detector liquid. After much chemistry, a gaseous compound is formed - ^{71}Ge ^{4}He. This gas is placed inside a proportional counter in which the K and L shell gamma rays are detected.

GALLEX expects to detect about 100 solar neutrinos per year, so that in three years they will have better than a 10% measurement on the flux of low energy neutrinos from the sun.

LVD and ICARUS will be able to detect about 2,000 solar neutrino events from the ^{8}B reaction per year by means of the following reactions:

$$\nu_e + e \rightarrow \nu_e + e$$

Each experiment will gather in one year more events than the Davis experiment gathered in 14 years. Gran Sasso will generate abundant data to compare to the standard model of the sun and, hence, perhaps resolve some of the present confusion. It should be noted that LVD is a proven technique with a well understood energy resolution, and hence must be considered the best detector to study the Boron solar neutrinos.

DARK MATTER AND EXOTIC PARTICLES

There exist good observational evidence that most of the mass of the universe is non-luminous, i. e. dark matter. From a theoretical point of view, it is difficult to identify this matter with protons, neutrons, electrons or massive neutrinos.

Recent studies indicate that massive supersymmetric particles might be good candidates for dark matter. It has been suggested that these massive particles and their antiparticles could be trapped by the sun's gravitational field and produce neutrinos when they annihilate. These neutrinos could be detected by the instruments in Gran Sasso and hence indirectly establish the existence of supersymmetric particles.

LVD has a large amount of scintillator and tracking chambers in a uniform matrix. Time of flight measurements will allow pointing back to the sun, even for contained events. This makes LVD unique. No other detector in Gran Sasso can do this.

At this meeting Prof. Dimitri Nanopoulos showed a slide which indicates that LVD could possibly detect the existence of the neutrinos from photino - antiphotino annihilations. The reaction quoted is:

$$\gamma + \gamma = T + T$$

His slide is shown in Figure 26. The sharp rise of the number of neutrinos at the position of 240 on this graph ($E_\gamma = 2.4$ GeV) is the signature of a 40 GeV top. These curves are the result of a detailed Monte Carlo calculation. The basic statement is that if the photino mass is greater

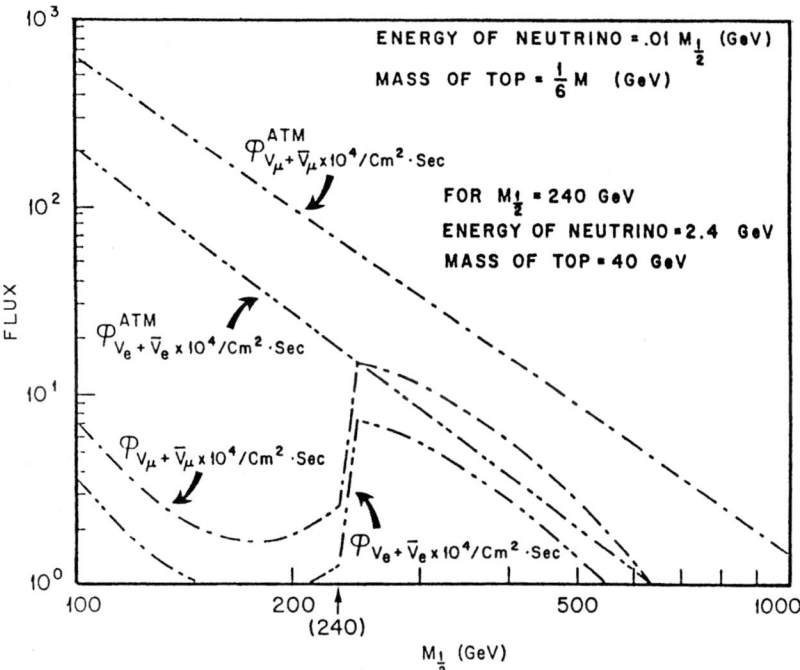

Figure 26: Theoretical neutrino spectrum from the reaction photino plus anti-photino going to top plus anti-top followed by the decay of the top and anti-top.

than the top mass, LVD has a chance to detect the photino and at the same time measure the top mass.

This is just one more example of how Gran Sasso might advance the knowledge of high energy physics.

PROTON DECAY

Supersymmetric models of proton decay predict the dominant decay of the proton to be:

$$P \rightarrow K^+ + \nu$$

$$\tau (p \rightarrow K^+ + \nu) > 1.5 \times 10^{31} \text{ years}$$

LVD is uniquely suited to detect this decay mode. The K^+ from the proton decay has 105 MeV kinetic energy. Fast electronics will see the K^+ from the proton decay, followed by the decay of K^+ to $\mu^+ + \nu$. The μ^+ finally decays into e^+. The scintillator electronics can record this whole decay chain. This is a cleaner signal than that available to water-Cherenkov systems.

Ten years of operation will yield a limit of 3×10^{32} years.

Finally, I would like to comment that although LVD has been optimized for a particular set of physics topics, it can do many other topics in a very competitive manner. Let me list some of those topics.

GENERAL PHYSICS (LVD is very competitive)

1. Neutrino point sources
   Cygnus X-3, Vela X-1, etc.

   - LVD has a large geometric factor $\sim 7000 \, M^2 \times ST$

   - LVD has a large projected area near the horizon

2. Single muon distribution

   - More than 10^6 events/year

3. Primary cosmic ray composition

   - Excellent muon bundles detection

4. Monopole studies

   - Eight layers of scintillators make an excellent monopole detector

V. SUMMARY

Gran Sasso is the unique World Underground Laboratory

1. It is the world's largest
2. It employs state-of-the-art technology
3. It is at the right depth
4. It has the lowest radioactive back ground

Gran Sasso has a broad physics program

1. Neutrino physics
2. Particle physics
3. Astrophysics
4. Cosmic ray physics
5. Nuclear physics
6. Geophysics

Prof. Zichichi is to be congratulated that he has put Europe (Italy) at the forefront of this exciting field of physics.

DISCUSSION

– *Villasante:*

Could you tell us something about the gravitational wave detector?

– *Pless:*

The heart of a gravitational wave detector is an aluminum cylinder, cooled below $1^0 K$ with supercooled helium, which can detect a vibration of the order of a small fraction of the proton radius. It is an acoustic device; when there is a gravitational occurrence, such as a supernova collapse, there is produced a gravitational field which hits the aluminum cylinder and causes it to vibrate. At these low temperatures they can detect this small vibration. They claim to be able to detect any supernova collapse in our galaxy. The detector is being built at CERN and will be tested first at CERN, and if it passes the surface tests, it will be moved to Gran Sasso. This experiment needs to be done underground because the cosmic ray radiation would release enough energy in the detector to vibrate it, which would produce what looks like false gravitational wave signals. The mountain shields the detector from most of the cosmic radiation. As Prof. Zichichi points out, it is possible to make a coincidence between this gravitational wave antenna and LVD, which also is an excellent detector for collapsing stars. The combination of a gravitational wave detector and LVD would separate real signals from false ones.

– *Miele:*

How is it possible to discriminate between the neutrinos which come from dark matter and those which come from other sources?

– *Pless:*

The neutrinos we are interested in come from photino-antiphotino annihilation in the center of the sun. The neutrinos which do not come from the center of the sun are background. LVD has the ability to distinguish the difference between neutrinos coming from the center of the sun and neutrinos coming from other sources, even if the neutrinos have the same energy, due to its excellent directional capability.

– Heusch:

Would you be good enough to explain the directionality you can achieve for cosmic rays or solar neutrinos and also give details about how the LVD will achieve 1 ns time resolution over such a large volume?

– Pless:

We have 13 layers of tracking chambers that each have at least count of 2 cm, and which gives a spatial resolution a little better than 1 cm. The height of the detector is 13 m. If you take the height of the detector and the spatial resolution of the tracking chambers, and if you calculate the angular resolution from those two numbers, one finds find that this matches the multiple scattering in the rock from muons with energy of the order of 10 TeV. Hence, we have designed the detector so the angular resolution of the detector matches the scattering in the rock.

– Heusch:

How do you find the directionality of solar neutrino events with a detector composed by streamer tubes and scintillators?

– Pless:

We have no directionality information at all about the solar neutrinos of the sun. But in one year we have more than enough events to measure a spectrum and compare it to the one expected from ^{8}B. However, if gamma-ray radiation from the rock creates background with energy greater than 7 MeV, this would hide the signal in the noise. We will be the first to test this background radiation from the rock using LVD as the detector.

– Heusch:

Can you tell me how you achieve the 1 ns time resolution you mentioned before?

– Pless:

With the tracking chambers we can determine the track path to better than 2 cm. Thus, we know the distance from the 5 cm phototube, which has a rise time of less than 1 ns. So knowing the track coordinates, the distance, the rise time of the phototubes, and the delay through the phototubes (which we have measured), the overall measurement we can make on the time-of-flight from a particular point on the track to that phototube is of the order of 1 ns. By using an offline program we can achieve a one nanosecond time resolution.

– Zichichi:

We demonstrated 20 years ago that this technique using plastic scintillators could achieve a fraction of a nanosecond.

– Liu:

How do you detect monopoles?

– Pless:

The way you detect monopoles depends on what you think the monopole characteristics are. The only kind of monopoles of current interest are very massive monopoles that travel very slowly and have very high energy.

There are two theories about what happens when such monopoles go through a liquid scintillator. One theory is that you get an enormous amount of energy (1 GeV/cm) deposited in the liquid scintillator. They travel so slowly that from the top of the detector to the bottom of the detector takes about 150 μs. So, each scintillator lights up one after the other in such a manner that it takes 150 μs from the top to the bottom. It is trivial to detect this in LVD because each of the scintillating tanks is self-triggering. Off-line you can look at these enormous pulses in each scintillator tank. Then, knowing the path and the time resolution in each tank, you can determine if the particle has uniform velocity and you can measure very accurately this velocity. The second theory is that instead of a single enormous pulse you get 20-30 smaller pulses from each scintillator tank in a period of 15 μs. Off-line you cannot miss such a situation and, as before, you can determine the velocity. By the way, we can put a very good limit on the monopole flux: less than 10% of the Parker bound in 1.5 years of running.

– Iacovacci:

What are the most important background processes at Gran Sasso and at which rate you will work at Gran Sasso?

– Pless:

There are two serious backgrounds which especially affect the solar neutrino studies.

One background we think we can control is the ^{60}Co in the steel in our equipment. The ^{60}Co can give two photons going at the same time in the same tank each having about 1.5 MeV energy adding up to 3 MeV. Given our resolution, this 3 MeV could be interpreted as 4-5 MeV, which falls in the ^{8}B region. So, we are carefully choosing steel with a very low ^{60}Co content.

The other background comes from the slow neutrinos from the rock which are captured by the nuclei of heavy metals in the liquid scintillators or steel containers, giving a gamma-ray cascade. This cascade can easily add up to 6-7-8 MeV released into perhaps one scintillator tank, again falling in the ^{8}B region. This background is so small that it can only be tested once the experiment is running.

To answer your second question: the average rate we expect is 1 count per hour per square meter, which gives in our detector about 1 count in one-tenth of a second. However, in the case of a supernova or of a monopole event, we could have a counting rate up to 1 KHz. Therefore the electronics must be able to handle at least 1 KHz. But, in fact, all of our equipment can handle up to 2 KHz.

– *Pitman:*

How sensitive are you to neutrino oscillations?

– *Pless:*

$$\Delta M^2 = 10^{-4} - 10^{-5} \ eV.$$

– *Pitman:*

Could you elaborate on the calculations of the incoming neutrino flux?

– *Pless:*

We do not have to calculate the neutrino flux. In the acceptance angle, the flux of neutrinos from below is equal to the flux of neutrinos from above within a 3% error due to the earth's magnetic field differences and to a mass effect, calculated by Bethe.

– *Pitman:*

What is the energy resolution on the kaon you hope to observe from proton decay?

– *Pless:*

For energies up to a few hundred MeV we have a resolution of $10\%/\sqrt{E}$, hence we can get 105 MeV energy very accurately. We have 1 ns time resolution on the decay of the K to the μ. This decay time is about 12 ns, so we get a 10% error measurement. Then we do the decay of the μ to the electron to a very high precision. We get extremely good measurements on the particle energies and on the decay times. This allows us to recognize unambiguously the process.

– *Nimai Singh:*

Unlike other experiment set-ups, in this case the background neutrinos do not affect the measurement of the proton decay life time. Is it right? If so, what would be the upper limit of the decay rate of the proton?

– Pless:

There is a careful study by our Russian colleagues that shows that the upper limit of the accidental coincidences of the different backgrounds is about 1 count a year. This is a factor 20 better than any other experiments which, unlike ours, are limited by neutrinos created in the athmosphere. We hope to have a limit of $10^{32} - 10^{33}$ years.

– Ito:

How do you measure the nuclear composition of primary cosmic rays from underground?

– Pless:

If the cosmic rays were composed entirely of protons, you see in LVD about one muon per event. If the cosmic rays were 100% iron, then each time you see in LVD a bundle of roughly 20-50 muons, one for roughly each nucleon in the iron nucleus. The number of muons in the bundle can tell you the kind of nucleus in the cosmic ray. Unfortunately the cosmic rays are a mixture of protons, iron, carbon and other nuclei, so we obtain a spectrum of multiple muons in each bundle. However, we can match this spectrum to a mixture of proton, iron and carbon nuclei with a Monte Carlo to determine the composition of the primary cosmic rays.

HERA

V. Soergel

DESY

Notkestrasse, 85, D-2000 Hamburg 52

INTRODUCTION

HERA is the electron-proton collider under construction now in
Hamburg at the DESY-laboratory. The name HERA stands for hadron electron
ring accelerator. In HERA electrons (or positrons) of 30 GeV will be
made to collide with protons of 820 GeV.

The HERA project has been authorized in April 1984, it is scheduled
to give first ep collisions in 1990. HERA is being built with international collaboration. The German Government made this a condition for
approving the project. The countries collaborating either by contributing components or by sending people are Canada, China, France, Italy,
Israel, USA, Poland, the Netherlands, the United Kingdom. Here notice-
ably Canada and Italy had declared their firm intention to make important contributions in components at an early stage, and in the process
to obtain final approval this was an essential element.

HERA will be the first storage ring facility to collide particles
of different masses, and it will make the colliding beam technique
accessible to the investigation of the lepton-nucleon scattering process, which so far was still a domain of fixed target experiments. Here,
HERA will extend the range of squared momentum transfer by two orders of
magnitude, when compared with the highest energy accelerators in operation today, like the SPS at CERN or the TEVATRON at Fermilab. HERA will
then allow to probe the lepton-quark interaction at very high center of
mass energies, going up to 300 GeV in the limit. Since leptons do not
interact with gluons, ep-scattering at high momentum transfer is mainly
pure electron-quark scattering. This will greatly simplify the interpretation of the results. HERA will allow to collide both, electrons and
positrons, with protons; moreover one aims for longtiduinal polarization
of the electrons at the interaction point, with either helicity. This
would be of particular interest for the investigation of parity violating effects in the weak interactions.

HERA will consist of two storage rings, one for electrons or positrons, and one for protons, which are mounted on top of each other in a

common ring tunnel of 6.3 km circumference, built 10-20 m below surface adjacent to the DESY site (Figs. 1 - 3). Four large underground halls to house the experiments are located symmetrically around the tunnel, in 360 m long straight sections. The proton ring will be built with superconducting magnets - a real technological challenge.

In what follows, the storage ring facility and the present status of its construction will be described, followed by a brief outline of some interesting physics problems to be studied with HERA, and a presentation of the two detectors now under preparation by large international collaborations, H1 and ZEUS.

THE HERA FACILITY

The main parameters of the two storage ring are given in Table 1. The two beams are made to collide initially in the three halls South, North and East, all located outside the DESY site (Fig. 2). The hall West, on the DESY site, will be left without collision for the running in phase in order to facilitate commissioning of the machine. It can later be transformed into an intersection, if the need arises.

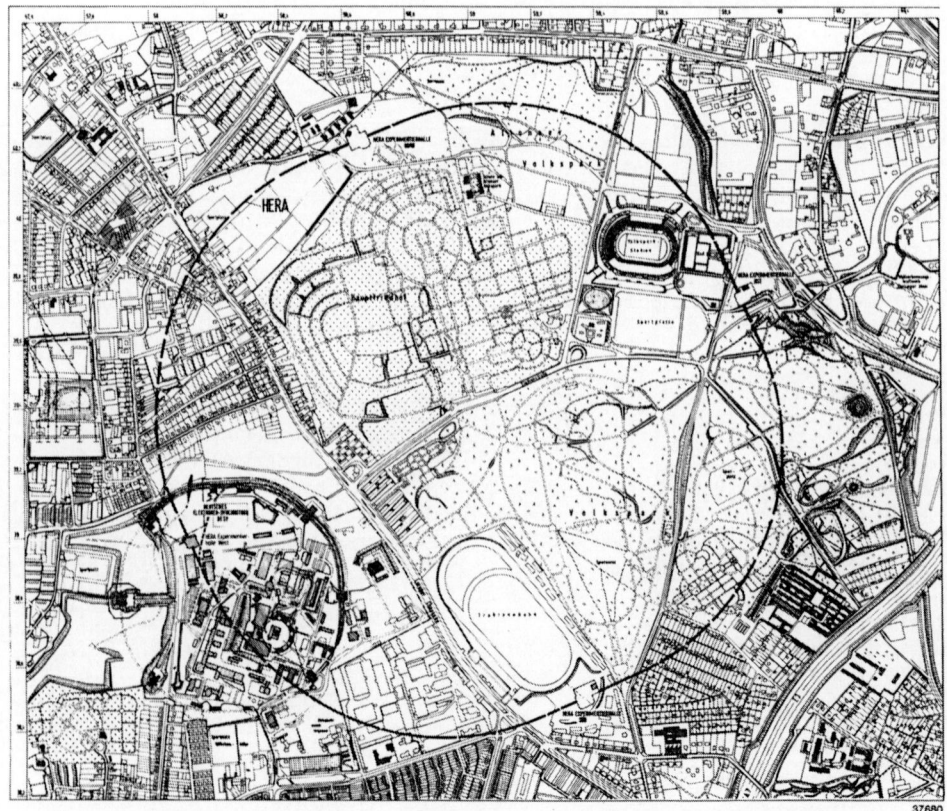

Fig. 1

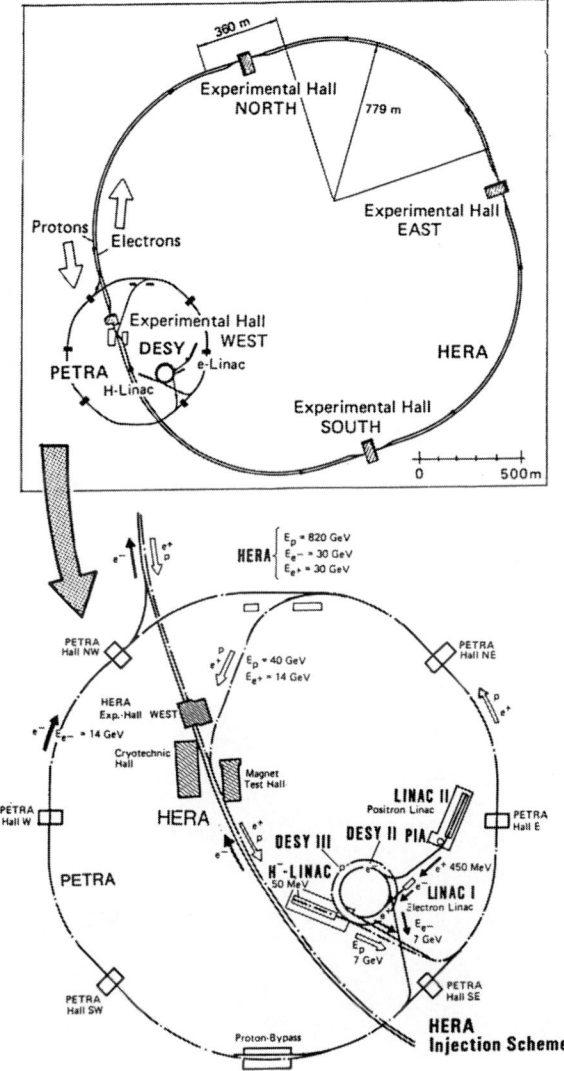

Fig. 2

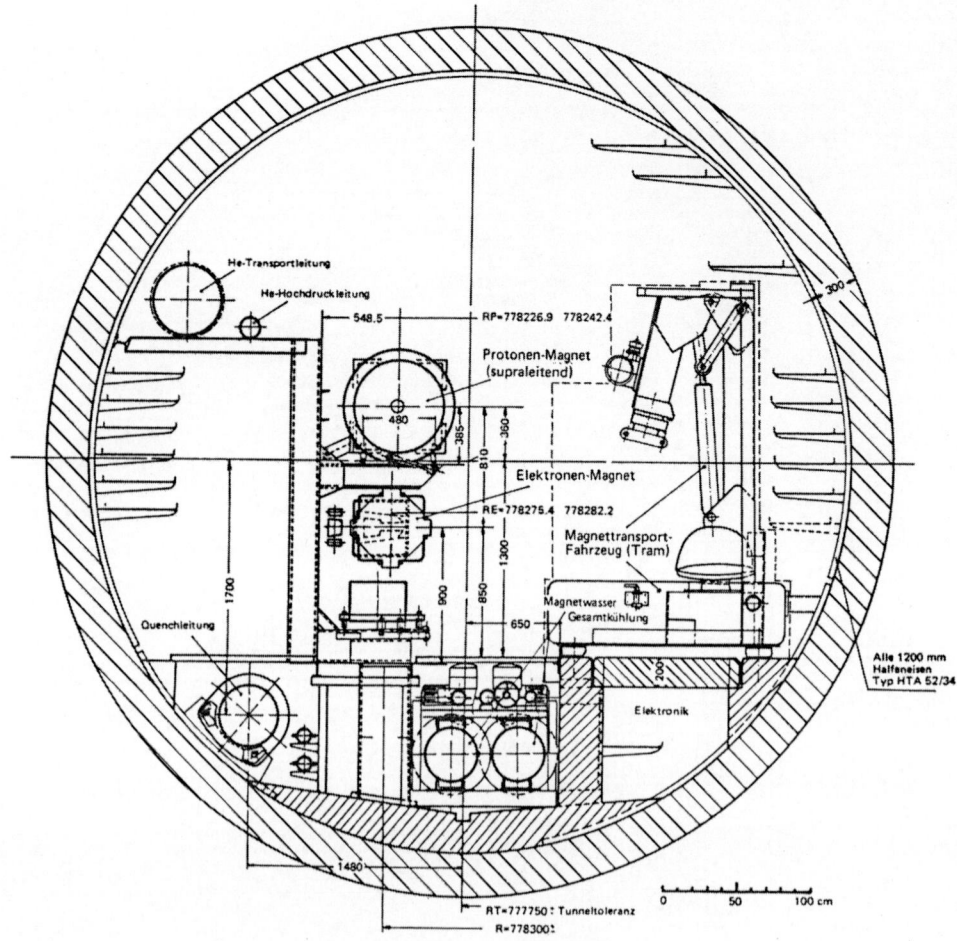

Fig. 3 HERA tunnel, cross-section in the arc

HERA Main Characteristics

	p	e$^{\pm}$		
energy	820	30	(35)	GeV
bending field	4.7	.18		Tesla
bending radius	604	541		m
number of bunches	200			
beam current	160	60		mA
beam size at crossing horizontal	.17	.18		mm
vertical	.055	.02		mm
bunch length	100	10		mm

head on collisions

	p	e$^{\pm}$	
polarization p_e (max) (spin rotators)		80 % (at 30 GeV) (85 % at 35 GeV)	
$E_{injection}$(PETRA)	40	14	GeV
luminosity	2×10^{31} cm^{-2}s^{-1}		

Buildings

The storage rings are mounted in a ring tunnel of 6.3 km circumference and 5.2 m open diameter. For somewhat more than 50 % of its circumference the tunnel is below the water table. The four arcs of the tunnel connect the four large underground experimental halls with dimensions 25 m (in beam direction) times 43 m for halls N,E,S, and 25x37 m for hall W. Two tunnels of 350 m and 100 m length between PETRA and HERA house the injection channels. In addition, there are two large halls on surface for the cryogenic plant and the magnet measuring facility for superconducting magnets (Fig. 2).

All these buildings are nearly completed including the tunnel. In most places, installation work is well under way.

Electron ring

Magnets

The magnetic elements of the electron ring are mounted in 11 and 12 m long modules (Fig. 4), of which 420 are needed. Each module holds on a common iron structure a 9 m long dipole, a quadrupole, a sextupole, and a correction dipole. The modules are mounted and prealigned on surface and lowered into the tunnel as units. The dipole magnet is excited by a single Al-bar, allowing for a very simple mounting. All magnets are in production, a large fraction is delivered. In one quadrant of the ring the modules are already installed. The vacuum chamber with a adjacent pumping channel housing distributed ion getter pumps, is made from an Cu-alloy so as to absorb a large fraction ($\sim$ 75%) of the synchrotron radiation already in the walls of the chamber. Further shielding is provided by lead profiles.

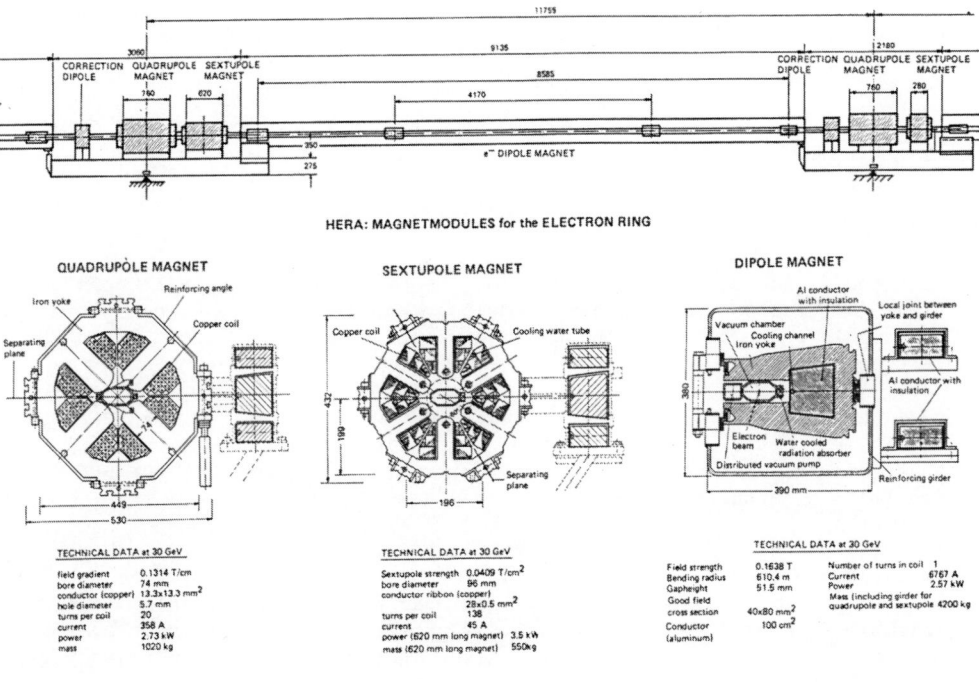

Fig. 4

Radio frequency systems

A 500 MHz RF-system is used, like in the other electron ring accelerators at DESY. Initially, clystrons and 85 Cu-cavities previously
used in PETRA will be installed, sufficient to reach 26 GeV.

For higher energies, the installation of superconducting Nb-cavities is
foreseen, which are now under development at DESY, jointly with industry. A 4 cell prototype cavity manufactured in industry has given
accelerating gradients of 6 MV/m. Depending on the result of a beam test

in PETRA, planned for November 1987, a pilot series of 8x2 of such
cavities will be ordered to be installed in HERA and bring the beam
energy to $\sim$32 GeV.

Polarization

Electrons and positrons in HERA will - hopefully-obtain a high degree of tranverse polarisation, $P_T \leq 92\%$), through the Sokolov-Ternow
effect in the emission of synchrotron radiation. By means of spin
rotators installed at the intersections, this will be turned into longitudinal polarization where both helicities can be chosen to allow the
investigation of parity violating phenomena (Fig. 5). The maximum value
of the longitudinal polarization which could be obtained if HERA was an
ideal machine, is $P_\ell = 80\%$ at $E_e = 30$ GeV, $P_\ell = 82\%$ at $E_e = 32$ GeV.
However, at this moment it is still uncertain whether in the "real" HERA
electron ring the tranverse polarization will reach values which make
experiments with polarized electrons attractive.

The build-up time of transversal polarization is proportional to
$1/E^5$; at 26 GeV, $\tau_p = 49$ min, at 32 GeV, $\tau_p = 17$ min. That's why the installation of the superconducting cavities is imperative to allow meaningful studies on polarization.

Initially, one pair of spin rotators will be installed in hall East
for polarization studies. If a sufficiently high degree of polarization
is achieved, rotators will also be provided for the intersections S
and N, were the detectros ZEUS and H1 will be located.

It is planned to have a stored electron beam in HERA by late summer
1988.

Proton ring

superconducting magnet system

The magnet system in the arcs of the HERA proton ring will consist
of 416 superconducting dipoles, 9 m long, and about 250 superconducting
quadrupoles. Two dipoles and one quadrupole form one half-cell of the
lattice.

The cold beam pipes of the dipoles and quadrupoles serve also as
cryo-pumps. Wound on the beam pipe of the dipoles are superconducting
quadrupole and sextupole correction coils, and a supraferric correction
dipole is mounted in the quadrupole cryostat.

The 9 m dipole (Fig. 6), jointly developed by DESY and BBC (Mannheim), has a 2 shell coil wound from Rutherford cable, which is clamped
by an Al-collar and surrounded by a cold iron return yoke. This whole
assembly is housed in a cryostat and supported at three points. The
development of this collared coil magnet is based on the pioneering work
at Fermilab.
With a design field of 4.7 Tesla for 820 GeV proton energy, five proto-

type magnets have reached about 6 Tesla without training, at 4.3 K, with good field quality. One magnet was cooled to 3.9K and reached 6.95 Tesla, as expected from the short sample behaviour of the conductor. The heat insulation achieved is also within specification.

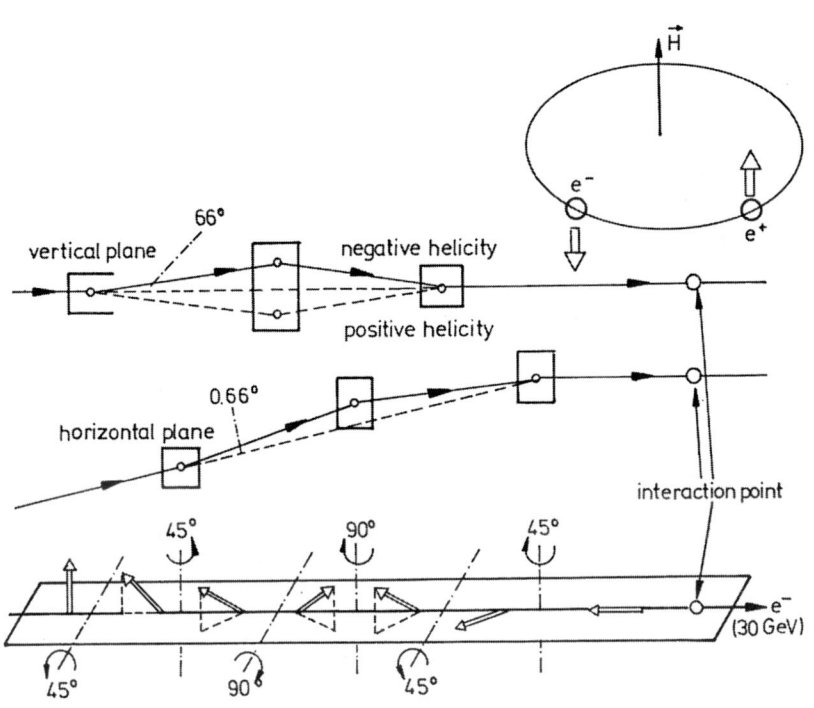

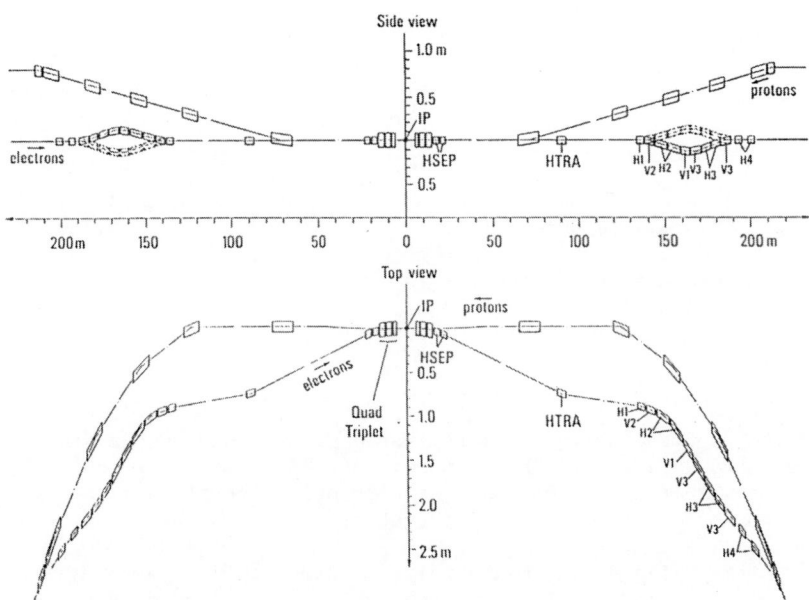

Fig. 5 Intersection region with spin rotator

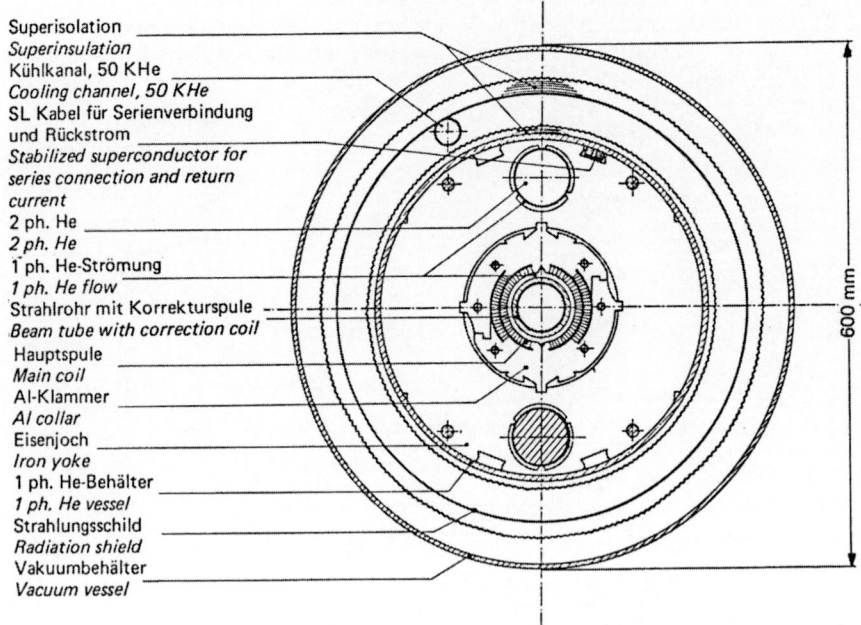

Superisolation
Superinsulation
Kühlkanal, 50 KHe
Cooling channel, 50 KHe
SL Kabel für Serienverbindung
und Rückstrom
Stabilized superconductor for
series connection and return
current
2 ph. He
2 ph. He
1 ph. He-Strömung
1 ph. He flow
Strahlrohr mit Korrekturspule
Beam tube with correction coil
Hauptspule
Main coil
Al-Klammer
Al collar
Eisenjoch
Iron yoke
1 ph. He-Behälter
1 ph. He vessel
Strahlungsschild
Radiation shield
Vakuumbehälter
Vacuum vessel

600 mm

Fig. 6 HERA superconducting dipole, cross-section

Orders have been placed by NFN in Italy with three firms, LMI, Ansaldo and Zanon, which jointly build 50 % of the magnets as Italian contribution to HERA, and in Germany with BBC, Mannheim, for the other 50 %. Preseries production is in progress, the first magnet from Italy to be delivered to DESY by November 1987.

The quadrupoles have been developed at Saclay, and two prototypes have shown excellent performance, exceeding the specified gradient by about 30 %. Half the quadrupoles will be manufactured by the French firm Alsthom as a French contribution, the other half by two German firms, KWU and Noell. The orders have been placed and preseries production is in progress.

The correction windings and supraferric correction dipoles are all manufactured by Dutch industry as Dutch contribution to HERA, and production is in progress.

All elements of the superconducting magnet system should be delivered to DESY by late summer 1989.

Each superconducting magnet will be cooled down and tested at DESY before being installed in the tunnel, and a large measuring facility has been set up which allows simultaneous cooling down and measuring of eight magnets.

System Test
Five superconducting magnets, three dipoles and two quadrupoles, were mounted in a row in the same sequence as later in the HERA proton ring, and with the inclination of the tunnel where it has its steepest slope, to perform a test of the magnet system. Important results of this system test were:
  - the cooling of the magnets is satisfactory in the row and with the inclination, and the overall heat leak does not exceed the sum of the individual heat leaks
  - the system could be excited to the quench limit of the

weakest magnet, and

- most important - a quench fired in one magnet was confined
  to that magnet and did not propagate into the neighbouring
  magnets.

These results are very encouraging for the later operation of the
magnets in the HERA proton ring.

Cryogenic Plant and Distribution System

A large cryogenic He-system, consisting of 14 screw compressors and
three coldboxes, each with a cooling power of 6.3 kW at 4.3 K and 20 kW
at 40 K, has been installed and is being commissioned at present. The
first coldbox, in operation since April 1987, reached the specification.
The cold Helium at both temperatures will be delivered around the tunnel
in a 4 pipe transfer line to supply the magnet system, the superconducting cavities and the superconducting solenoids of the experiments.

"warm" parts of the proton-ring

In the straight sections the proton ring comprises many warm magnets to steer the beam towards the collision points with a vacuum system
at room temperature. Orders for most of these parts have been placed.

RF-System

The proton ring will have two RF-systems, one at 52 MHz for capture
of the bunches at injection, built at Chalk River as a Canadian contribution, and one of 208 MHz for acceleration and storage, now under development at DESY.

All components of the HERA-proton ring should be installed in au-
tumn 1989 for commissioning.

Injection

The HERA injection scheme is shown in the lower part of Fig. 2.
The PETRA machine, after some modifications, will serve as injector
synchrotron for protons of 40 GeV and electrons and positrons of 14 GeV
into HERA through two newly built injection channels, which have been
tested with e^+ and e^--beams on April 1st and July 1st, 1987, respectively.

The electron injection into PETRA uses the existing linear accelerators, the e^+-accumulator PIA, and the new injector synchrotron DESY II
which first accelerated electrons to 7 GeV in February 1987.

For the protons, a completely new system has to be build. It starts
with an H^--source delivering ions of 18 keV, which are then accelerated
to 750 keV in a radiofrequency quadrupole. These two first stages are
already operational. The H^- ions will then be accelerated to 50 MeV in a
linear accelerator and transferred to a 7 GeV synchrotron DESY III,
where injection is achieved by stripping off the two electrons. The
7 GeV protons are then injected into PETRA.

The linac and the proton synchrotron DESY III are presently under
construction, the transfer channel between them has been provided by
Canada and is already installed.

The complete proton injection scheme should be in operation towards the end of 1988.

With the completion of the HERA proton ring in 1989, the commissioning of the whole HERA facility is planned for spring 1990.

PHYSICS AT HERA

The basic physics process to be studied at HERA is electron-quark scattering, mediated by neutral or charged currents. With its high center of mass energy, HERA is then a unique machine to study the weak ineractions in space like processes.

The event topology will be quite characteristic: In the final state, the lepton - presumably an electron or a neutrino - and the struck quark will balance the transverse momentum, and the two remaining quarks will carry little transverse momentum and remain in general undetected (Fig. 7). The struck quark will of course appear as a jet. The large difference in the beam energies gives the event a boost in the direction of the proton momentum. A typical electron-quark scattering event will then have a lepton side and a hadron-(jet) side, which are well separated - a characteristic feature of HERA-events when searching for new phenomena.

Important physics questions to be addressed with HERA include

- the properties of space-like currents, charged and neutral, in particular the search for vector bosons heavier than the Z^o and the $W^{\pm}$. The effect of an additional boson W2 on the charged current cross section is shown in Fig. 8, and it can be expected that HERA experiments will be sensitive to masses up to 800 GeV for charged and for neutral bosons.

- Search for right-handed currents. Here a good polarization of the electrons will be instrumental.

- Search for flavor changing currents giving rise to processes like $e^- + d \rightarrow \tau^- + b$, to be identified by an anomaly on the lepton side of the event.

- The structure of the nucleon and its constituents down to about 3×10^{-18} cm.

HERA will be a unique facility for these physics questions. Electron-proton collisions at high cm energies have also a good potential to search for SUSY-particles, heavy leptons, lepto-quarks, and heavy quarks (like top). At HERA energies one will here be sensitive typically for masses up to 100-150 GeV. Heavy quarks will mainly be produced in photoproduction through the so-called "photon-gluon" fusion process (Fig. 9).

Of course, with such a big step in the center of mass energy as provided by HERA for ep collisions, we hope to find something really new and unexpected, and to do in this sense truly new physics.

HERA experiments

Two detectors are presently being prepared for HERA by large international collaborations, named H1 (Fig. 10) and ZEUS (Fig. 11). Both experiments are similar in their general lay-out: A central tracking chamber, jet-chambers in both cases, is surrounded by a hermetic calorimeter with electromagnetic and hadronic sections. Axial magnetic fields are provided by superconducting solenoids, and the return yoke iron structures surrounding the whole detectors serve both as μ absorber and spectrometer and, instrumented with streamer tubes, as backup calorimeters. The forward boost of the events in direction of the proton momentum is taken care of by additional forward tracking chambers - with transition radiation detectors for particle identification, by corresponding asymmetric layout of the calorimeters, and by special forward μ-spectrometers.

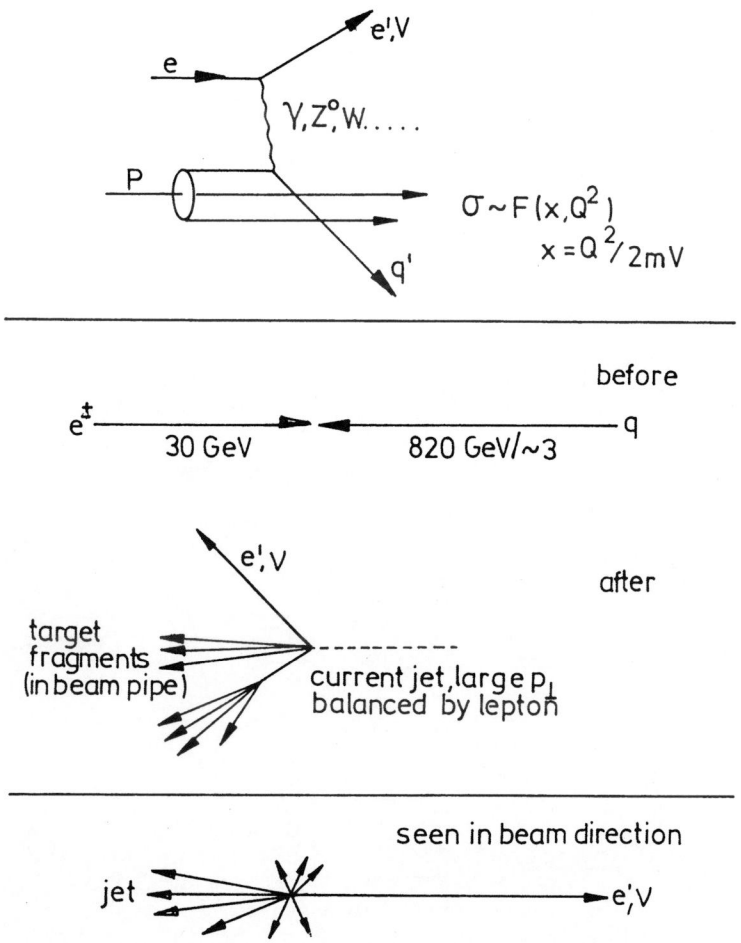

Fig. 7 Electron-quark scattering event

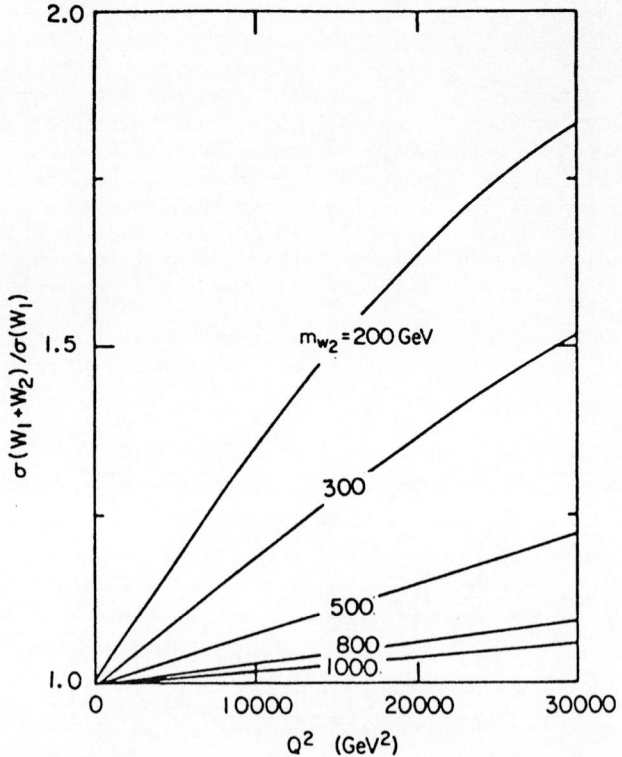

Fig. 8 The ratio of the cross-sections for two
and one W.

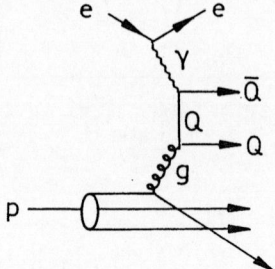

Fig. 9 Quark pair production by
photon gluon fusion.

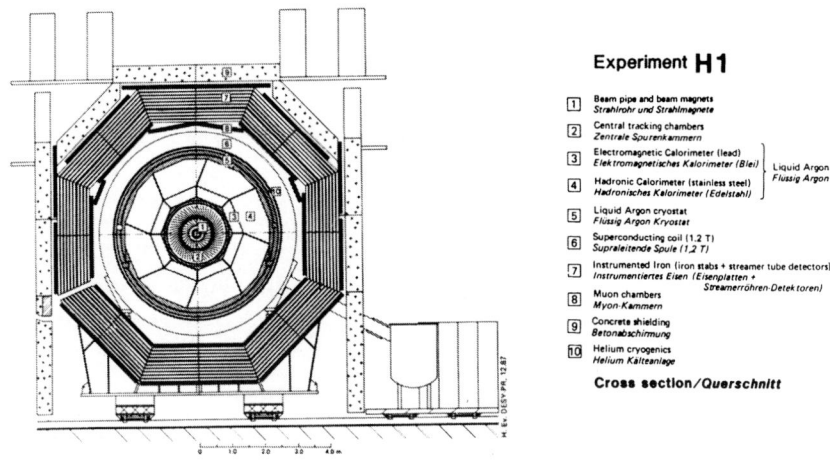

Experiment H1

1. Beam pipe and beam magnets
   Strahlrohr und Strahlmagnete
2. Central tracking chambers
   Zentrale Spurenkammern
3. Electromagnetic Calorimeter (lead) ⎫ Liquid Argon
   Elektromagnetisches Kalorimeter (Blei) ⎪ *Flüssig Argon*
4. Hadronic Calorimeter (stainless steel) ⎭
   Hadronisches Kalorimeter (Edelstahl)
5. Liquid Argon cryostat
   Flüssig Argon Kryostat
6. Superconducting coil (1.2 T)
   Supraleitende Spule (1.2 T)
7. Instrumented Iron (iron stabs + streamer tube detectors)
   Instrumentiertes Eisen (Eisenplatten + Streamerröhren-Detektoren)
8. Muon chambers
   Myon-Kammern
9. Concrete shielding
   Betonabschirmung
10. Helium cryogenics
    Helium Kälteanlage

Cross section/Querschnitt

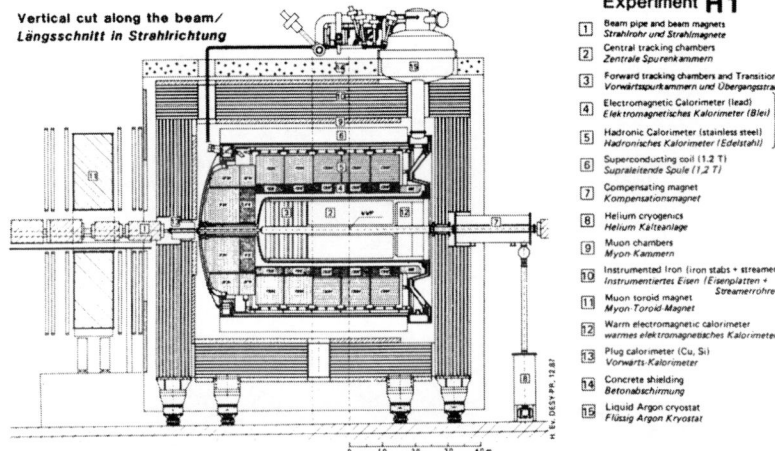

Vertical cut along the beam/
Längsschnitt in Strahlrichtung

Experiment H1

1. Beam pipe and beam magnets
   Strahlrohr und Strahlmagnete
2. Central tracking chambers
   Zentrale Spurenkammern
3. Forward tracking chambers and Transition radiators
   Vorwärtsspurkammern und Übergangsstrahlungsmodul
4. Electromagnetic Calorimeter (lead) ⎫ Liquid Argon
   Elektromagnetisches Kalorimeter (Blei) ⎪ *Flüssig Argon*
5. Hadronic Calorimeter (stainless steel) ⎭
   Hadronisches Kalorimeter (Edelstahl)
6. Superconducting coil (1.2 T)
   Supraleitende Spule (1.2 T)
7. Compensating magnet
   Kompensationsmagnet
8. Helium cryogenics
   Helium Kälteanlage
9. Muon chambers
   Myon-Kammern
10. Instrumented Iron (iron stabs + streamer tube detectors)
    Instrumentiertes Eisen (Eisenplatten + Streamerröhren-Detektoren)
11. Muon-toroid magnet
    Myon-Toroid-Magnet
12. Warm electromagnetic calorimeter
    warmes elektromagnetisches Kalorimeter
13. Plug calorimeter (Cu, Si)
    Vorwärts-Kalorimeter
14. Concrete shielding
    Betonabschirmung
15. Liquid Argon cryostat
    Flüssig Argon Kryostat

Fig. 10

The most important component in these detectors is the calorimeter, which should allow to measure the momentum of the hadronic jet with high precision - in fact, for charged current events the whole information is contained in this quantity! And the essential difference between the two detectors is in the calorimetry: ZEUS uses a compensating calorimeter with depleted uranium as absorber and plastic scintillator with wave length shifter read-out as detector. By proper choice of the thicknesses of the uranium plates and the scintillator plates, equal response for electrons and hadrons could be achieved, resulting in a good energy resolution in particular for hadrons. Over an energy range from 10-100 GeV, a line width of about 35 %/$\sqrt{E}$ (GeV) for hadrons and 18 %/$\sqrt{E}$ (GeV) for electrons has been achieved in test experiments. H1 uses the liquid argon technique with absorber plates from lead for the electromagnetic part and stainless steel for the hadronic part. Compensation for the π°'s in the hadronic shower will here be done off line by weighting the π°- spots identified throught the fine sampling of the calorimeter. In a test set-up a line width of 50-55 %/$\sqrt{E}$ (GeV) for hadrons was obtained, again for energies ranging from 10-100 GeV.

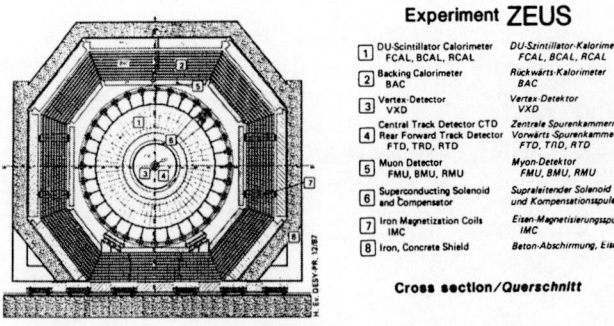

Experiment ZEUS

1	DU-Scintillator Calorimeter FCAL, BCAL, RCAL	DU-Szintillator-Kalorimeter FCAL, BCAL, RCAL
2	Backing Calorimeter BAC	Rückwärts-Kalorimeter BAC
3	Vertex-Detector VXD	Vertex-Detektor VXD
4	Central Track Detector CTD Rear Forward Track Detector FTD, TRD, RTD	Zentrale Spurenkammern CTD Vorwärts-Spurenkammern FTD, TRD, RTD
5	Muon Detector FMU, BMU, RMU	Myon-Detektor FMU, BMU, RMU
6	Superconducting Solenoid and Compensator	Supraleitender Solenoid und Kompensationsspule
7	Iron Magnetization Coils IMC	Eisen-Magnetisierungsspulen IMC
8	Iron, Concrete Shield	Beton-Abschirmung, Eisen

Cross section/Querschnitt

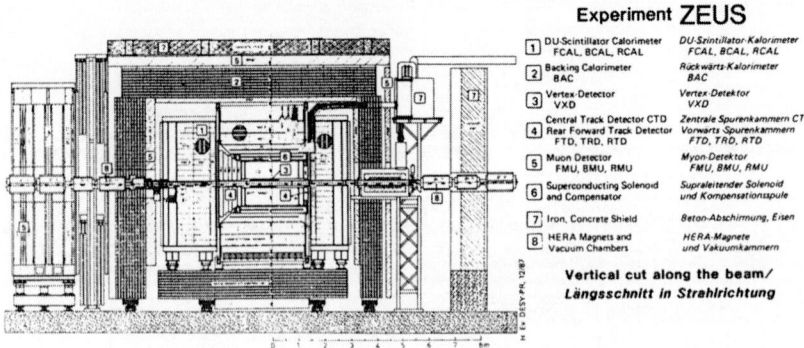

Experiment ZEUS

1	DU-Scintillator Calorimeter FCAL, BCAL, RCAL	DU-Szintillator-Kalorimeter FCAL, BCAL, RCAL
2	Backing Calorimeter BAC	Rückwärts-Kalorimeter BAC
3	Vertex-Detector VXD	Vertex-Detektor VXD
4	Central Track Detector CTD Rear Forward Track Detector FTD, TRD, RTD	Zentrale Spurenkammern CTD Vorwärts-Spurenkammern FTD, TRD, RTD
5	Muon Detector FMU, BMU, RMU	Myon-Detektor FMU, BMU, RMU
6	Superconducting Solenoid and Compensator	Supraleitender Solenoid und Kompensationsspule
7	Iron, Concrete Shield	Beton-Abschirmung, Eisen
8	HERA Magnets and Vacuum Chambers	HERA-Magnete und Vakuumkammern

Vertical cut along the beam/
Längsschnitt in Strahlrichtung

Fig. 11

In comparison, the two techniques of calorimetry are somehow complementary in their virtues and weaknesses: the depleted uranium calorimeter gives the best resolution for hadrons and has built-in hadron-electron compensation, however the calibration must be carefully monitored. The liquid argon calorimeter has built-in long term stability, the calibration being determined by the purity of the argon, it can however not achieve the good resolution of the uranium calorimeter, and the essential compensation must be done off-line.

Both collaborations are making good progress in setting up their detectors, having obtained the approval by DESY only in November 1986. The large components like supercnductng coils, iron structures, liquid argon vessel, have been ordered, and their installation in the HERA halls South (ZEUS) and North (H1) will begin in spring 1988. It is the goal of both collaborations to be ready with the first stage of their detectors for initial data taking at HERA in summer 1990.

With HERA we are looking forward to a new and interesting chapter in particle physics in the next decade.

DISCUSSION

– *Liu:*

(1) You did not mention the Higgs in your list of physics.

(2) Could you explain why, just briefly?

– *Soergel:*

(1) I don't think that HERA is a very good machine to discover the Higgs;

(2) Not briefly, no.

– *McBride:*

(1) Can you comment on the future of DORIS?

(2) And when will ARGUS be taking data?

– *Soergel:*

(1) DORIS will continue to run for high energy physics, which means for ARGUS, as long as there is an interesting physics program.

(2) ARGUS will take data again in 1988. In the year 1987, the proton-synchrotron for HERA injection is being installed in the same tunnel, which also houses the electron synchrotron needed as the injector into DORIS and PETRA. Therefore we cannot operate DORIS this year.

– *Pless:*

If you convert all R.F. cavities to superconducting, what is the maximum energy of the electron beam?

– *Soergel:*

50 GeV; it is the limit of the magnets, provided of course, that we succeed with the superconducting cavities.

– *Schmidt:*

What is the plan for the third experiment at HERA?

– *Soergel:*

There is no plan yet.

The Einstein Podolsky Rosen paradox:
might nature be more imaginative than us?

Oreste Piccioni Werner Mehlhop, and Brian Wright

University of California at San Diego

Physics Department, La Jolla California 92093 USA

1. Summary

We present arguments against the existence of the EPR state using only elementary fundamental properties of quantum mechanics. The popular arguments in favor of the EPR, such as the example given by Einstein et al, the superposition principle, and the conservation of angular momentum are discussed and found not compelling. With particular emphasis we argue that an action at a distance which would justify the EPR must be a message violating special relativity.

2. Introduction

We welcome the opportunity to include a second seminar on the Einstein Podolsky Rosen paradox (EPR) in the proceedings of this distinguished Majorana center. We hope it will stimulate in the students more attention on this subject which, though often misunderstood, is a very promising subject for elementary particles and their interactions. To avoid one such misunderstanding we emphasize that the basic assumption of this seminar is that quantum mechanics (QM) correctly and beautifully describes the phenomena of the constituents of nature and at present needs no correction, especially not in the fundamental concepts involved in the EPR. Thus we beg the reader not to interpret our considerations as inspired by classical mechanics (CM). This article is not a survey of the relevant studies made on the EPR. The reader is referred to the reviews of Clauser and Shimony, of Selleri and Tarozzi, and of Pipkin for such information [1]. Our purpose is to discuss the existence of the EPR purely from a theoretical point of view, debating the arguments pro and con the assumption of its existence, outlining only the schemes of the experiments. We describe the issue of the EPR in a very simple fashion, independently of the original presentation and of the historical discussions of the intervening 50 years. In particular, concepts and misconcepts of "reality" need not be mentioned at all. Also the justly famous Bell's inequalities can now be avoided, because they spoil the simplicity of the issue, with the consequence that what should be obvious might appear debatable. The discussion about the EPR reduces to the recognition that the elementary particles involved are inevitably well separated from each other, thus unable to interact with or "influence" each other. This is really the simple, key point of the issue, generally buried under descriptions of "extensive states". If we then accept the proved principle of QM that no machinery can prepare particles in eigenstates of two non-commutable variables, and reject the invention of an action at a distance (aaad), we immediately conclude that the EPR state, such as the singlet state of far apart fermions, would flagrantly contradict QM. Given the abundant proof that QM is valid we cannot assume the existence of that state. After describing the elementary steps which bring us to this conclusion, we examine the arguments against it. We note that no experiment, the one of Aspect et al [2] included, has proved the existence of the EPR or of the aaad. We also discuss the theoretical arguments (principle of superposition, conservation of angular momentum) supporting the thesis that QM demands the production of the EPR state. We show that they are not at all compelling unless we essentially assume at the start the existence of the EPR, or of an aaad. We argue that our fundamental knowledge of interactions would be destroyed by the existence of an aaad. For instance, the aaad needed for the EPR would clearly violate the Schroedinger equation, which does not imply an aaad. It would also violate the special theory of relativity, despite the popular opinion that it would not because humans cannot use the EPR to transmit messages. We show that the famous example of Einstein et al [3], which has always been considered a proof of the EPR based on purely basic principles, is not valid even if we correct it after the suggestion of Epstein [4]. That is important for us, because if the EPR state was really demanded by QM we should expect that a valid demonstration along the simple lines of Einstein et al would by now have been found.

Thus the existence of the EPR state has certainly not been demonstrated. What we probably have is a very interesting puzzle, namely that the experimental data show a correlation "as if" the aaad, with properties that cannot be taken seriously, indeed existed. Such a situation is not new in physics. Unfortunately, the belief in the aaad is a strong obstacle against a vigorous experimental program because it suggests that "everything is clear". Instead, while our analysis indicates that the solution of the puzzle does not exist now, our confidence in QM makes us believe that it will be found and it will be very rewarding.

3. The analysis of the singlet state of far apart fermions proves that the EPR contradicts QM

We will use the model of the very familiar singlet state of fermions, $ud - du$, though a direct experiment based on that state has never been done. That model, due to Bohm [5], is justly preferred because it enhances the QM property of the EPR state while the case of photon polarizations has too much of a classical flavor. In particular the fermions model easily allows to consider the singlet state of two particles bound together. The EPR linear polarization state xy-yx for two photons propagating in the two opposite directions of the axis z is totally analogous to the state ud-du, thus the reasoning for fermions is easily applicable to photons. Let us then suppose that we have a source which produces pairs of fermions, #1 and #2, not necessarily identical, which are emitted in opposite directions and, much after leaving the source, enter two Stern Gerlach devices, SG1 and SG2, so that their spin states can be analyzed. Let us then assume that the two particles, even when they are far from each other, are described by the state ud-du. We start with the SG's oriented along z.

The state $ud - du$ predicts that for an average of 50% of the times we will measure +1/2 for #1, and -1/2 for #2. The other 50% of the events will give us -1/2 for #1 and +1/2 for #2. Importantly, as we will discuss below, the fermions (as well as the photons in the most recent experiments) are confined to relatively small wave packets so that they cannot interact with each other nor with the source except at the time of production (see below).

When the packets are separated, the Schroedinger equation (which implies no aaad) guarantees that by measuring one particle we do not at the same time produce any change in the state of the other. A theory that implies the contrary statement inevitably implies the existence of an aaad, whether or not some types of direct experimental demonstrations are possible. Thus without an aaad the states of the particles must be prepared exclusively by the source.

Then the predicted correlation that when we measure +1/2 for #1 we will always observe -1/2 for #2 is tantamount to establishing that such result for #1 is a signal indicating that the source has prepared #2 in the eigenstate with the eigenvalue -1/2. This eigenstate is denoted as d. Making this simple consideration four times we are forced to conclude that the source is made in such a way as to produce a classical mixture of pairs of fermions which are of two types: A, made of eigenstates +1/2 for #1 and -1/2 for #2, namely ud, and B, -1/2 for #1 and +1/2 for #2, namely du, all states being described along z. It is hardly necessary to emphasize that this conclusion is only valid because it is based on an arbitrarily large number of measurements.

Our analysis has given a complete description of the source as far as the study of the spins of the fermions is concerned, because for spin 1/2 particles the description along any arbitrary direction is sufficient for the prediction of the result of any experiment on their spins, and is not at all a statement "contingent" on making the measurements with the SG's oriented along that direction. In fact we could substitute the source with one of our own design which would produce such A and B pairs, alternating type A and type B at random, and any experimental set-up would give the same result with the original as with the substitute source.

Moreover, the source will keep producing eigenstates in the z direction whether or not we observe the products with the SG's, because they are only passive devices where the fermions enter long after leaving the source. The idea that a macroscopic object such as a SG could be connected with another big object such as a source in virtue of an aaad has never been proposed.

So the analysis of the source with respect to z contradicts the assumption of the state ud-du at distance, because such state, being spherically symmetrical, does not describe that the fermions are prepared by the source in eigenstates along z or any other particular direction. As a practical consequence our analysis in z predicts that if we turn the SG's along x we must find no correlation among the spin states of #1 and #2, because particles in a given eigenstate of z are all perfectly identical and cannot be relabeled part in the eigenstate +1/2 (x) and part in -1/2 (x).

Instead, the spherically symmetric ud-du predicts the practical consequence that we will actually observe the same perfect correlation in x or y.

This is the simple, irremediable contradiction, which cannot be wiped out by all the semantics, metaphysics and mathematics written about the EPR.

In retrospect, even without the detailed steps of our demonstration it should be obvious that the state ud-du is impossible when the fermions are confined to clearly separated wave packets. Consider, for instance, that if all pairs are identically described by ud-du, which is a pure state with no privileged direction, the spin state of each fermion must be pure and without a polarization. However that is impossible for a beam of free, isolated fermions, because any state a(u) + b(d) is polarized in a direction determined by a and b.

It should be noted that in our presentation the elusive notion of "reality" has no role. We do not wish to play with the idea that humans can define a reality independent of their human minds, and we interpret that Einstein meant by reality nothing else but what is commonly meant by that word.

We also note that our reasoning does not use Bell's inequalities because we only want to display a contradiction with QM, not with hidden variables, which have never been shown to be a possible replacement for QM. We think it is very important to describe the EPR without Bell's inequalities because not only do they obscure the simplicity of the phenomenon but they also demand from the experiments a costly accuracy that is not needed. [14] Moreover, some experiments cannot give results in terms of the inequalities, though they can furnish very important information.

As mentioned before, the essence of the demonstration given above can be translated into the context of photon pairs instead of fermions.

4. The localization in space

In our proof of non existence of the EPR state, the space localization of the fermions (or the photons), which is the extent of their wave packets, is the essential basis for arguing that an action on one cannot provoke any change in the other without a very offensive violation of the Schroedinger equation. From that basis one must conclude that the source alone is what prepares the eigenstates in z. This point has not been sharply focused in the past literature, where the concept of "reality" was relied upon. As a consequence, some EPR analyses are still based on the notion of extended states and plane waves.

In Fig. 1 we show a histogram taken during an experiment of ours on the polarization properties of the photons from the annihilation of positronium. It shows the rate of detection of both photons versus their relative delay. The width of the histogram, 3.4 nanoseconds, (3.4 feet) is so exactly equal to the resolution of our electronic apparatus that we can conclude that with a more refined device the photons would indeed be localized along their direction of propagation within an inch or so. Transversely, the collimators limit the wave packets to a circle of two inches diameter. There is no room for fantasies that photons cannot be localized, despite the popular references to the "wholeness" of Bohr. Incidentally, we fully agree with the article of Ne'eman [6] that "to quote Pauli or Bohr in a physical argumentation, in the manner in which one quotes the Scripture or St. Augustine" is no substitute for a valid demonstration.

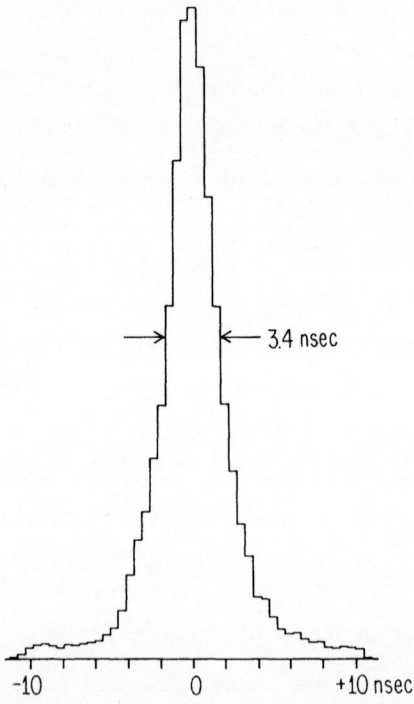

Fig. 1. Showing the localization of .51 MeV photons from the annihilation of positro nium. The histogram reproduces the experimental width (see text).

5. The experiments do not prove the existence of a superluminal aaad

Fig. 2 is the box diagram of the atomic cascade experiments of Aspect et al [2]. The previous works of Clauser [7], Fry [8] etc did not have the deflectors D, which alternate the paths of the photons produced by the source, between the analyzers A1 and B1 and between A2, B2. The alternation of #1 and #2 is done with different frequencies, so that in effect a random switching takes place.

According to Aspect et al, the inspiration for their experiment was contained in two published lines of Bell [9], who in turn quoted that Bohm and Aharanov [10] had noted that "the orientation of the measuring apparatus could quickly be changed, so that the spin of B would have to respond immediately." A and B are the two particles, and A is measured before B.

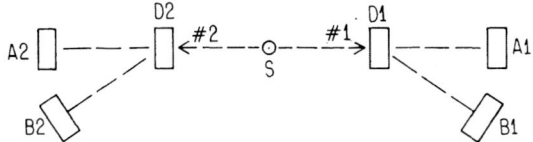

Fig. 2. Sketch of the apparatus of Aspect et al. [2].

Clearly the authors did not want to specify any mechanism, but we observe that the singlet state implies that it is only after A is measured that the message to B can be sent, to tell B which state it should assume. B does not make a choice by "looking" at its analyzer, thus no confusion can result from switching the mirrors, given that the electronics registers which analyzer is crossed by A or B. The point is that the mirrors could not have made a change in the outcome because as far as we can see the needed message is between the particles, not between the mirrors.

Nevertheless, trying new experimental directions is what the EPR needs, and we admire the Paris group for doing that experiment. If they had found an effect, no matter what the cause, it surely would have been a break through. Unfortunately, no effect was found. Thus the obvious conclusion is that the experiment, excellently done, gave no evidence for an aaad. We cannot take that null result as a proof not only that the aaad exists, but that it also has the magic property of being faster than light. Note that the authors did not write such conclusion.

6. The aaad related to the EPR indeed violates special relativity

It should have been clear even before the Paris experiment that if the aaad existed, it would indeed be a message which would contradict the special relativity. The often heard statement that it would not because humans could not use them to communicate is a gratuitous opinion.

To see that, let us first note that the popular discussion in terms of two observers is not needed: we can replace them with a recording device. Let us say that the source emits the two photons P1 and P2 (see figure 2) at the time zero. P1 arrives at analyzer A1 at the time t1, and chooses, completely at random, whether to register x or y. As noted, this choice cannot occur before P1 arrives at A1. Suppose it chooses x. Then P2 must register y when it arrives at A2 at t2. If both t1 and t2 are large with respect to the duration of the wave packets and are closely equal, the message from P1, if it only travels with the velocity of light, cannot arrive to P2 before P2 arrives at A2 and makes its own choice. Same for the message from P2 to P1. If the choices are independent of each other the correlation will disappear. A numerical analysis of the situation for the experiments of Clauser [7], of Fry [8] and especially of Wilson, Lowe and Butt [11] shows that the independent choice would have occurred with enough probability to spoil the correlation at a noticeable level. A speed faster than light by a factor ten would have been needed to make the correlation as good as shown by the data.

An equally important point is to recognize that the needed aaad is a full fledged message and does violate special relativity. This is so because, though we are only spectators, we know, if we accept QM, that P1, at the time t1, makes a random choice C1 between x and y, in the perfect sense of a random QM choice. In no way C1 exists before t1. Yet, at a large distance from P1, photon P2, at t2, makes a choice C2 which is strictly a consequence of C1. Thus C1 must be available at the time t2, at the position of P2. Thus C1 has been transmitted, because a significant information cannot pop up at a distance without being transmitted, just like energy cannot. Even if a human accompanied P1, the scientifically most reliable way of testing whether the messages to P2 are strictly obeyed would be to send random choices.

Thus if the aaad were the explanation of the EPR, for t2 closely equal to t1 we would be the spectators of a clear violation of special relativity. The restriction that humans must be the actors of the play is a conveniently invented law.

Furthermore, if the aaad did exist, we should really expect that it should also be possible to develop "usable" ultrafast communications. Indeed the fact that the aaad has been proposed to explain an elementary phenomenon does not limit the application of the aaad to single quanta. No connection with Plank's constant can be perceived in the aaad to indicate such a limitation, thus one should expect its application, in some form, to large numbers of quanta, which would allow "practical uses".

7. The original state of Einstein et al. is not a valid demonstration of the existence of the EPR

Were it only for its elegant simplicity, it is quite appropriate that the original article of Einstein et al [3] be still cited. However, it is generally cited in a wrong fashion. The authors wrote a state which was time independent and which they claimed satisfied the equations of QM. Yet, they claimed, it was physically impossible. We think the first claim was wrong, and the second, the more important, was right. On the contrary, even at the present time probably the majority of physicists expresses exactly the opposite assessment. See two important papers in the Joensuu 1985 proceeding [12,13].

The state in question (SEPR) contemplates two particles #1 and #2 moving in a one dimensional space. It is a delta function of their distance D:

$$(SEPR) = \int\limits_{-\infty}^{\infty} exp[iK(x1 - x2 - D)]dK$$

The position of each particle, $x1$ and $x2$ is completely indetermined but the distance between the particles is exactly D, invariant with time. Thus if #1 is in an eigenstate of x with the eigenvalue X1, #2 will be in the eigenstate with the eigenvalue X2=X1 - D. The state also shows that if #1 is an eigenstate of momentum with the eigenvalue k1, #2 will be in the eigenstate of eigenvalue -k1. Hence the famous sentence that we can know exactly the value of x or of k of #2, at our choice (but not both x and k for the same pair, of course) without disturbing #2 at all.

Since the ingredients of (SEPR) are unquestionably fundamental elements of QM, the conclusions based on the (SEPR) have always been given for granted by everyone (including Bohr [15]), and still are. We first noted [14] that the question whether (SEPR) satisfied the Schroedinger equation was not discussed in 1935.

It is easy to see that (SEPR) as it is does not satisfy the Schroedinger equation because the second derivative with respect to x of a delta function is not proportional to the delta function, and it is necessary to introduce in the equation a potential, with which the state does not represent two free particles but two particles connected by a rod of some sort. Measuring one certainly disturbs both so the contradiction with QM disappears.

Actually it is not obvious that the state (SEPR) can indeed be prepared, because it is the superposition of states of separate particles with a precise phase relation between the separate particles. Moreover we must recall that the theoretical superposition principle cannot prove the existence of a state "not physically realizable" (see below). Our reasoning above shows that without aaad, the EPR is not realizable.

Here we will leave this point undecided and assume that the state (SEPR) is physically realizable.

Epstein [4] suggested to modify (SEPR) introducing the time dependent factor representing the kinetic energy, which is a function of the momentum, but to our knowledge no one has published an analysis of (SEPR) so modified. To do that we studied the time development of the best suited initial state. We found that the product of the uncertainties in x2 and k2 from the measurement of x1 or k1, respectively, never violates the uncertainty principle. This is so because to measure k1 accurately (in order to predict k2 accurately) we must take a minimum amount of time, during which the uncertainty in x2 increases. (see appendix) Thus there is no paradox in the state proposed by Einstein Podolsky Rosen.

It is hard not to conclude that if QM really demanded the EPR an unquestionably viable example along the simple lines of (SEPR) would have been found.

8. The superposition principle

A strongly cited argument that the EPR is a "straight consequence" of QM is based on the superposition principle. It is especially tempting to use that principle to demonstrate the existence of the singlet state ud-du for distant particles, because ud or du are perfectly possible states. However, that principle "reigns supreme" only when the resultant state is "physically realizable" as it is emphasized in a famous article [16]. A linear combination of a neutron and a proton cannot exist as a free particle.

Our demonstration (above) shows that without assuming the aaad the singlet state is not possible for free particles, thus the superposition principle must be considered ruled out by a "superselection rule" [16].

9. The conservation of angular momentum and the dissociation of a bound singlet state

The argument that two particles, for example two atoms, initially bound in a singlet state then separated by a spin independent interaction must "remain in their states" is not viable. The singlet state does not imply any particular state for each of the particles, thus they cannot "remain" in the "state where they were" because there is no such a thing. The argument is popular because it appeals to a classical mechanics intuition, in which context is of course true.

On the other hand, saying that "it is the system of the pair that must remain in its state" is playing with words. Any subsystem must obey the Schroedinger equation with its proper hamiltonian, thus the analysis separates in two independent parts and we must again conclude that to remain mutually dependent the particles would need some yet unknown agent, like the aaad.

The argument that the fermions must remain in the singlet state because of conservation of angular momentum is more complex, but again not compelling. We start with two particles bound in a state of total spin zero, and we can indeed observe the state of the total spin of such bound state (at the price of being unable to observe the spin state of each individual particle because they disturb each other.) Instead, after the separation we cannot observe directly the total spin of the two particles but we can observe the spin state of each individual particle. However, the sum of the individual results cannot give us the state of the total spin unless the particles are each in an eigenstate, no matter which quantization axis we choose. Assuming that, would be assuming at the start that the two free fermions are in a singlet state. That assumption would give us consistency but not a proof.

The root of the trouble is of course that an object of spin zero has all components perfectly determined, like a classical object, but the two isolated fermions are each constrained by the uncertainty principle to have only one accurately known component.

Thus the argument in favor of the existence of the singlet state of two separate fermions, made on the basis of that conservation law, cannot be compelling.

The picture that we favor, if only by intuition, is related to the present impossibility to describe the measurement process in adequate QM terms. The separation of the two particles can be considered a measuring process, because, as we said, it allows us to measure each individual spin state without the interference from the other.

The initial state ud-du is rotationally invariant, thus it is equivalent to a sum of states each described respect to one of a large number of quantization axes distributed in the entire 4 p solid angle sphere. All axes are directed outward from the center of the sphere and for each axis there is one in the opposite direction. Since the term -du becomes +ud when rotated by 180 degrees, the state relative to each axis is just ud. This representation of ud-du makes it natural to expect that when the two particles fly apart (not when one is measured) the initial spin state should collapse into a state ud with respect to one of those axes chosen at random (Furry model) [17], just like when a particle is emitted in an S state, the S state eventually collapses into a state with a determined momentum vector.

To justify such a collapse, let us suppose that the bound singlet state is held together by an attractive potential Vs which is proportional to the spin dependent projection operator for the singlet state and is rapidly decreasing with distance. When Vs becomes less and less significant, any arbitrarily small, noise-like potential could allow the transition into the Furry state.

Finally, it is important to remember that the experiments do not identify with certainty the final state of the observed process, which may actually be more complex than ordinarily described, so as to provide some other way to satisfy all requirements of QM. Clearly the considerations made above do not solve the EPR issue, because at any rate the described Furry state would not, by itself alone, explain the observed perfect correlation [14], yet they show that the argument of the angular momentum, absent an adequate description of the whole picture, is not at all as compelling as our demonstration given above that the EPR state contradicts QM.

10. About objection to "contrafactual" statements

A popular objection to reasonings such as ours is an ad hoc made rule against "contrafactual" statements. It points out that when predicting the x spin component of #2 from the measurement of #1 we cannot, for the same pair, predict the z component of #2. Thus our conclusion cannot be proved "with facts". The only thing clear in that obscure idea is that in order for our argument about the EPR to be valid, we should prove that we can determine both the x and z components for #1. But in that case our QM would be totally destroyed. Instead, our reasoning does not kill QM, but simply suggests that we work harder at finding a solution within our proven QM.

Beside that, if that objection is distinguishable from stating that the aaad binds the particles forever, one wonders why should we assume that the source, which is a heavy piece of machinery, may change. Note that the most used method in experimental particle physics is based on analyzing a beam and then using it with the firm belief that the machine will not change. We could not determine the scattering cross section of antiprotons if we had to identify by annihilation each of them before they scatter.

11. Description of our experiment

Feynman narrates that at Los Alamos he tried to show to a police officer that a safe could be opened without knowing the proper code. The officer waited and waited, then asked "are you making any progress?" Feynman answered: "there is no progress at this game. Either I open it or I don't." We think the EPR puzzle is analogous to the safe. Nevertheless, we insert these two sections as examples of experimental efforts which should be done, hoping that eventually one will open the safe.

Our experiment was aimed at investigating whether the photons from positron annihilation responded as demanded by the Klein Nishina (KN) formula concerning the polarization acquired upon scattering off electrons. Such pairs of photons, as well known, are supposed to be in an EPR state, though the first authors of theoretical or experimental work [18,19,20] did not focus on the paradoxical aspect of the phenomenon, which was only emphasized by more recent authors [10,21]. On the other hand, if the EPR state does not exist, which we believe is the case, then one could try to explain the observed correlations by assuming that the EPR particles, in our case the annihilation photons, are not of the "ordinary type", but possess anomalous properties. In particular, one would expect that taken singly they should have a sharper correlation, between their direction of scattering and their linear polarization, than that provided by the Klein Nishina (KN) formula.

For instance, the photons could have spin more than one, and yet have only two substates as a relativistic consequence of their speed. Their phase space would thus still be the same as for ordinary photons and any consequence of the detailed balance principle would not be altered. This is an important difference from schemes of hidden variables, suggested without a physical proposal, which would alter the available phase space thus contrasting with well tested detailed balance relations.

We were also intrigued by the possibility that the photons of the EPR state of atomic beams could also be "anomalous" and, because of their higher spin, could have a detection efficiency (by a photomultiplier) less than that of ordinary photons. After crossing a polarizer, they could become ordinary photons with normal efficiency. Our "invention" is the reverse of the enhancement [22], which had appeared to us as very improbable, yet it would have the same beneficial effect of simulating a perfect correlation.

Briefly: the EPR state predicts that if # 1 registers x, # 2 always registers y, so that the counting rate is the same with or without the polarizer A2 in place and oriented along y. The Furry state (see above) does not predict such 100% correlation [14]. However, if, with the polarizer A2 removed, the photons, being anomalous, have less efficiency than with A2 in place, the correlation may appear to be 100% as the EPR demands. Our hypothesis could be numerically adjusted to reproduce the experimental correlation within the error.

In the hypothesis that EPR photons should be described by hidden variables as proposed by several authors [23] we should also observe a sharper scattering correlation than predicted by KN.

We want to make it clear that we certainly did not think it probable that photons with hidden variables or spin greater than one would exist. But when we consider that a popular candidate for the EPR is the miracle of aaad, we feel compelled to explore even suggestions which are "almost impossible" but not obviously absurd.

In the experiment one photon of the pair scattered by an angle Asc off a counter C1 and the polarization it acquired from that scattering was determined from the measurement of the azimuthal asymmetry of a second scattering by an angle of 90 degrees, at the azimuth Az. Figure 3 shows the schematic of the apparatus. The size of the counters, typically one inch cube, was large enough that a Monte Carlo (MC) computation was needed for comparison with the theory. The density matrix method of A. Wightman [24] was used. The polarization of the photon was followed by the MC through multiple scatterings and the amplitude distribution of the pulse in each counter was computed by the MC.

Instead of choosing patterns of scatterings with a probability given by the theoretical KN formula, our MC chose the patterns with uniform randomness and computed the theoretical probability of the events. We found this method preferable to the traditional one in order to maximize the MC accuracy. Since we had to impose acceptance limits to the pulse amplitudes in each counter, the MC should have reproduced the experimental amplitude spectra quite well.

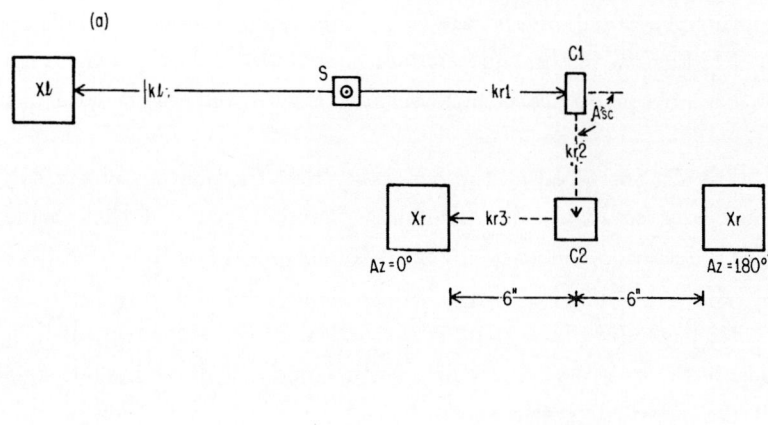

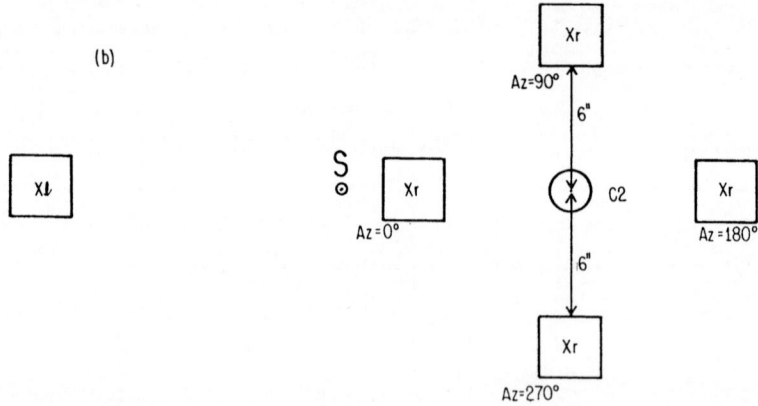

Fig. 3. Our apparatus to test the Klein–Nishina prediction for the polarization ac
quired by an unpolarized photon upon scattering.

Figures 4,5,6 show that the MC indeed reproduced such spectra. Note that the width of the MC spectra was not arbitrarily adjusted but it was determined from the known resolution of the respective counters. For the crystal, the MC was supposed to fit the shape of the peak only.

The experiment clearly shows an acquired polarization typical of ordinary .511 Mev photons obeying the KN formulas (Table 1). Since only a Lorentz transformation is needed to change the energy of a photon, the result implies that also the photons of the atomic cascades such as those of the experiments of Clauser, Fry and Aspect are the ordinary quanta of light.

We conclude that neither our model nor the models based on hidden variables are likely to be the explanation of the EPR. However, we cannot exclude that some other model of anomalous photons might be successful.

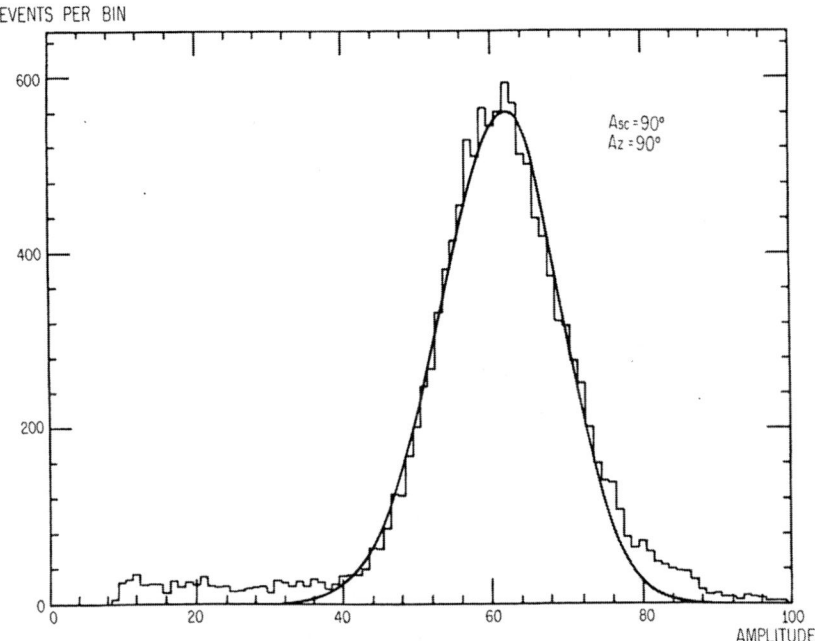

Fig. 4. Experiment–Monte Carlo comparison for the amplitude spectrum of counter C1 for Asc=90 degrees; Az=90 degrees.

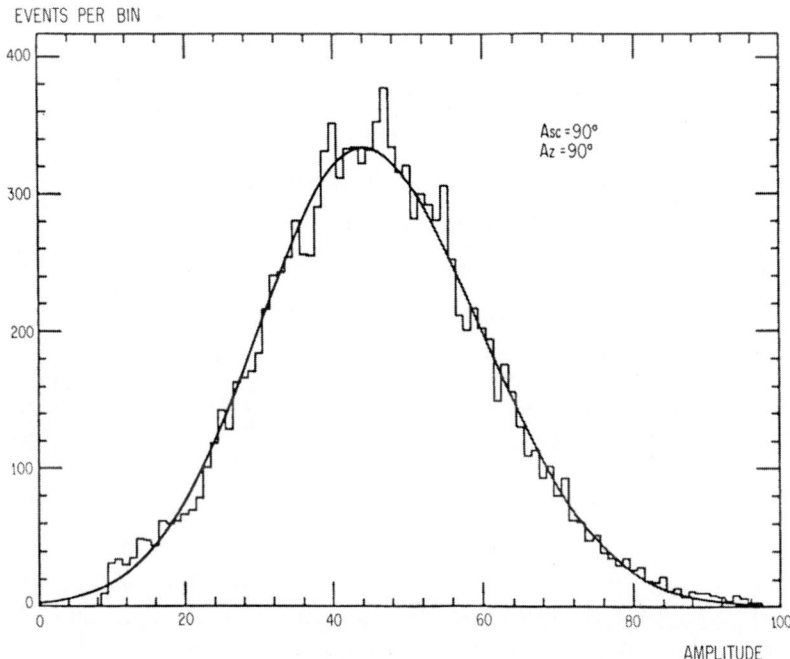

Fig. 5. Same as figure 4, for C2.

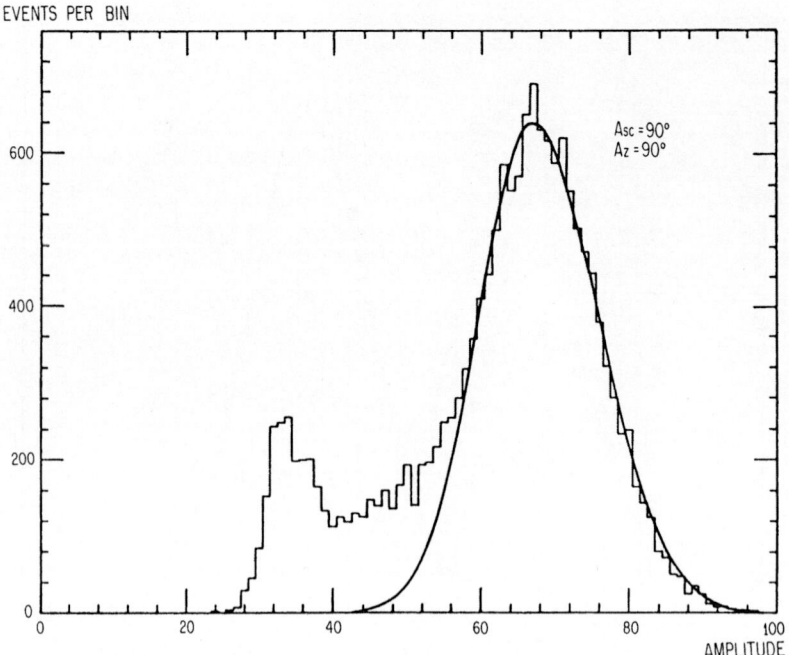

Fig. 6. Same as figure 4, for the crystal counter XR.

12. Discussion of other experiments

The suggestion from our work also comes from a beautiful experiment made at Stirling on the circular polarization of atomic photons, along the line of an experiment made by Clauser [25] in 1976. The Stirling authors [26] were seeking to display a violation of Bell's inequalities. However, with their set up Bell's inequalities were rather useless because they could not prove the EPR effect, on account that the apparatus could not distinguish a linear combination from a mixture (see our article [27] in the proceedings of the 1986 Tokyo conference).

This is an example of how important it is to have in mind a good physical picture beside good formulas. Indeed, while the Stirling's circular correlation by themselves could not prove the EPR effect, their comparison with the linear correlations previously measured by the same group shows that the degree of circular correlation observed is too much to be compatible with the high degree of observed linear correlation. Indeed, a machine which produces linear eigenstates and circular eigenstates at our choice is an example of the EPR effect. Models of hidden variables or of higher spins, invented to explain the correlations of the linear polarizations, cannot explain those of the circular polarizations without very unnatural inventions.

An example of overemphasis on formulas is also the fact that, despite its obvious importance, the circular correlation experiment has been overlooked for so long after Clauser's work. Moreover, it was essentially ignored in the reviews of Ref. 1. We dare to suggest that it has been so because of an undue conviction that the EPR could only be discussed on the basis of Bell's relations, which, as far as we know, are not usable for the circular case.

13. Conclusion

We face the fact that the most popular candidate for the explanation of the EPR, the aaad, must violate the Schroedinger equation, the principle of eigenstates and the theory of special relativity. Thus we have a puzzle to solve, and we suggest that the EPR might indicate a door, so far closed, to an interesting new knowledge within QM and without miraculous phenomena. Accepting that nature has more imagination than us, we should gladly accept the challenge, and look into the unfamiliar. An action conveniently propagating with infinite velocity is not unfamiliar: it is probably described in many science fiction books. An unfamiliar idea is the following: "By viewing the vacuum as a Lorentz-invariant medium, we would expect that many of its parameters, such as the color dielectric constant, the average values of some symmetry numbers,..., can be subjected to changes. That may lead us to other striking phenomena, hitherto completely unknown to us." (Seminar at Columbia University by T.D. Lee. The proposal was not made in the context of the EPR.)

We thank our secretaries Celeste Schmid and Arianna Gerace for their cooperation.

Appendix

To study the time development of the (SEPR), the state proposed by Einstein et al, it is obviously convenient to choose the center of mass frame of the two particles, which coincides with the laboratory frame where $k1+k2 = 0$ and to choose the relative variables

$$x = x1 - x2 - D; mr = (m1)(m2)/(m1 + m2)$$

and

$$k = ((m2)k1 - (m1)k2)/(m1 + m2).$$

The analysis can then be carried on as if we were studying the development of a single wave packet. Obviously, the uncertainty in x will cause an equal uncertainty error in the prediction of x2 from the x1 . However, accepting the (debatable) superposition of (SEPR), we have k1=-k2 independently of the time, so the error in the prediction of k2 depends only on the accuracy of the measurement of k1. Following Schiff [28], (QM section 12, page 63) the state of the relative particle can be expressed in terms of the uncertainty in x, (Ex,0) and of k, (Ek,0). The original (SEPR) has (Ex,0) equal zero and (Ek,0) equal infinity. Instead, we will leave both undetermined, but assume that their product has the minimum value of 1/2, valid when the uncertainties are RMS values.

Thus at t=0, the state is

$$(SEPTx,0) \; = \; (2\pi(Ex,0)^2)^{-\frac{1}{4}} \; exp \left(- \frac{(x- <x>)^2}{4(Ex,0)^2} \; + \; \frac{i<p>x}{\hbar} \right)$$

Since we refer our formula to the center of mass the average of the momentum p respect to the QM uncertainty is zero, and we can also assume the same for x.

The uncertainty in the prediction of k2 depends upon the time t taken to measure k1.This is so because since particle #1 is free, a measurement of k1 gives us the value of the energy W1, with an error

$$(eW1,t) \; \geq \; \frac{\hbar}{2t} \tag{1}$$

which is related to the uncertainty on k1 (ek1,t).We assume that the latter uncertainty, which affects the actual measurement of k1,is appreciably less that the initial uncertainty (Ek,0) .

For our purpose we can assume the particles to be non relativistic. Then

$$W1 \; = \; \frac{(\hbar k1)^2}{2m1} \qquad\qquad (eW1,t) \; = \; \frac{dW1}{dk1} \, (ek1,t)$$

So:

$$(ek1),t) \; = \; \frac{(eW1,t)m1}{\hbar^2 k1}$$

We can substitute for k1 its RMS value at the time t, which in turn is equal to that at t=0. We make the approximation that the latter is equal to the uncertainty of the relative quantity k, denoted by (Ek,0). Recalling (eq 1) we have

$$(ek1,t) \; \geq \; \frac{m1}{(2(Ek,0)\hbar t)} \tag{2}$$

where $m1$ is the actual mass of particle #1 . The prediction of x2 must of course be done also for the time t4. The prediction will have at least the uncertainty that x has at the time t. The square of the uncertainty in x at the time $t > 0$ is trivially derived from Schiff's equation 12.21 and is

$$(Ex,t)^2 \; = \; (Ex,0)^2 \; + \; \frac{(\hbar t)^2}{(4(mr)^2(Ex,0)^2)} \tag{3}$$

We indicate the uncertainty error in x and in k, due to the development of the wave packet until the time t as (Ex,t) and (Ek,t). However the uncertainty in k remains constant with time until the momentum of particle #1, k1, is measured, thus (Ek,t) = (Ek,0) . The mass mr is the "reduced mass".

Multiplying the square of (eq 2) side by side by (eq 3) we obtain

$$(Ex,t)^2(ek1,t)^2 \geq \left(\frac{(Ex,0)m1}{2(Ek,0)\hbar t}\right)^2 +$$

$$\left(\frac{\hbar t}{2(mr)(Ex,0)} \times \frac{m1}{2(Ek,0)\hbar t}\right)^2 \tag{4}$$

From which, noting that m1 is larger than mr, we have for the product of the errors in our prediction:

$$(Ex,t)(ek1,t) > \frac{1}{2}\frac{m1}{mr}$$

Equation 4 is of course not valid for t=0, because we focused on the case of interest, where the measurement of k1 is affected by an error less that (Ek,0). This is especially true for (SEPR) where (Ek,0) is infinite. No paradox remains after applying the constraint of the Schroedinger equation to (SEPR). The magic of the original state of Einstein et al apparently was entirely due to the overlooking that the state could not stay constant without a rod linking the particles.

Table 1 - Azimuthal asymmetries $(A_z = 180)/(A_z = 90)$

The error quoted is the statistical error

First Scattering Angle Asc(DEG)	Experiment	Monte Carlo	Exp/MC
60	2.15 ± .04	2.07 ± .02	1.04 ± .02
64	2.13 ± .04	2.11 ± .04	1.01 ± .03
90 Geometry 1	2.79 ± .02	2.87 ± .01	.97 ± .01
90 Geometry 2	2.61 ± .01	2.76 ± .01	.95 ± .01

REFERENCES

1. J. Clauser and A. Shimony, Rep. Progr. Phys. 41, 1881, 1978; F. Selleri and G. Tarozzi, Riv. Nuovo Cimento 4(2), 1, 1981; F. Pipkin, Adv. in Atomic and Molec. Phys. 14, 281, 1978.

2. A. Aspect, J. Dalibard and G. Roger, Phys. Rev. Lett. 49, 1804, 1982.

3. A. Einstein, B. Podolsky and N. Rosen, Phys. Rev. 47, 777, 1935.

4. Cited by M. Jammer, The Philosophy of Quantum Mechanics; Wiley, N.Y., 232, 1974.

5. D. Bohm, Quantum Theory; Prentice Hall, N.J., 614, 1951.

6. Y. Ne'eman, Symp. Foundations of Modern Physics, eds. P. Lahti and P. Mittelstaedt; World Scientific, Singapore, 481, 1985.

7. J. Clauser, Phys. Rev. Lett. 36, 1223, 1976.

8. E. Fry and R. Thompson, Phys. Rev. Lett. 37, 465, 1976.

9. J. Bell, Physics 1, 195, 1965.

10. D. Bohm and Y. Aharanov, Phys. Rev. 108, 1070, 1957.

11. A. Wilson, J. Lowe and D. Butt, J. Phys. G. 2, 613, 1976.

12. M. Jammer, Symp. Foundations of Modern Physics, loc. cit., p. 129.

13. N. Rosen, Symp. Foundations of Modern Physics, loc. cit., p. 17.

14. O. Piccioni, P. Bowles, C. Enscoe, R. Garland and W. Mehlhop, Open Questions in Quantum Physics, eds. G. Tarozzi and A. van der Merve; D. Reidel Publ. Comp., Dordrecht, Holland, 103, 1985.

15. N. Bohr, Phys. Rev. 48, 696, 1935.

16. G. Wick, A. Wightman and E. Wigner, Phys. Rev. 88, 101, 1952.

17. W. Furry, Phys. Rev. 49, 393, 1936.

18. C. Wu and I. Shaknov, Phys. Rev. 77, 136, 1950.

19. A. Wheeler, Ann. New York Academy of Sciences 48, 219, 1946.

20. C. Yang, Phys. Rev 77, 242, 1950.

21. L. Kasday, J. Ullman and C. Wu, Rendiconti Scuola Internazionale di Fisica, Course 49, ed. B. d'Espagnat; Academic Press, N.Y., 195, 1971.

22. J. Clauser and M. Horne, Phys. Rev. **D10**, 526, 1974.

23. T. Marshall, E. Santos and F. Selleri, Phys. Lett. **98A**, 5, 1983.

24. A. Wightman, Phys. Rev. **74**, 1813, 1948.

25. J. Clauser, Nuovo Cimento **33B**, 740, 1976.

26. A. Duncan, W. Perrie, H. Beyer and H. Kleinpoppen, Second European Conference on Atomic and Molecular Phys., Amsterdam, The Netherlands, 1985; H. Kleinpoppen, Proc. 2nd Int. Symp. Foundations of QM, eds. M. Namiki et al; The Physical Society of Japan, Tokyo, 59, 1987.

27. W. Mehlhop and O. Piccioni, Proc. 2nd Int. Symp. Foundations of QM. Tokyo, loc. cit., p. 72; O. Piccioni and W. Mehlhop, Symp. at Joensuu, Finland, P. Lahti and P. Mittelstaedt, eds. (World Scientific, Singapore) 197, 1985; O. Piccioni and W. Mehlhop, Conference on New Techniques and Ideas in Q. Measurement Theory, Ann. N.Y. Acad. of Sciences, **480**, 458, 1987.

28. L. Schiff, Quantum Mechanics, 3rd edition; McGraw-Hill, N.Y., 63, 1968.

29. (Reference added in proofs.) A beautiful new model of EPR apparatus was built by C.O. Alley and Y. H. Shih using laser technology without atomic beams. The authors successfully demonstrated the same full EPR correlation of linear polarizations as that present in the atomic beam experiments. C.O. Alley and Y.H. Shih, Proc. 2nd Int. Symp. Foundations of QM, Tokyo, loc. cit., p. 36.

DISCUSSION

– Nicrosini:

You say that EPR state does not exist, or better, it contradicts QM. Is there any general principle on this point? For instance the QM principle of superposition of far apart states.

– Piccioni:

The superposition principle does not guarantee at all that if you superpose two QM states you get a physical state. You can take a neutron and a proton, both satisfy the QM equations; but a linear superposition of them does not exist, the deuteron of course is not such. The famous article by Wick, Wightman and Wigner says essentially, that if the resulting state is not physically realizable, there is a superselection rule against the superposition principle.

– Cass:

The experiment by Aspect, et al., is generally taken to support Bohr's QM interpretation against that of Einstein, et al. You don't seem to agree with this. Could you say which QM interpretation you favour and suggest experiments to distinguish between different ones?

– Piccioni:

The point is really not what Bohr or Einstein said about QM fifty years ago. The point is what the QM equations say. About the Aspect experiment, many people believe that it proves the existence of actions at a distance, and of signals which propagate with superluminal velocity. Both of those statements are not correct. Essentially they are based on the wrong assumption that the EPR state must exist because of the superposition principle and because of the wrong but attractive argument that if you have two particles bounded in a singlet state, and you separate them gently, they will remain in the state in which they are. However none of them is individually in any defined state. At any rate my point is that QM cannot contemplate a transition into a non existent state.

– Quackenbush:

Question: When you say a "gentle" separation you imply that you are interfering in the evolution of the state instead of allowing it to evolve naturally. Is it a case where the experiment may effect the system?

– Piccioni:

Take for example, an atom of Mercury which decays in a cascade of two levels emitting photons which superficially one may say, are in a singlet state. True, the quantum numbers of the initial and final state of the atom would be satisfied by the emission of an object with the quantum numbers of the singlet state. But when the photons are 40 feet apart, are they really in a singlet state? The experiment gives only one parameter, the correlation, which is just one number because QM tells us that a spin one particle can only give an intensity proportional to a cos^2 or sin^2 so the shape of the correlation dependence on angle does not add any information. Beside, there may be two or three very low energy photons accompanying the pair of photons or they may be very special photons (my experiment says that such is not the case). Such inventions are repulsive but not as absurd as a miracle. The point is that you can't separate "gently" two photons in a singlet state leaving them in "their own state" because the single photons are not in a definite state to start with.

– Pitman:

You mentioned that the spin of the photon could be greater than one. Does not the standard model require its spin to be one? And, what experiments could be done to look for photons of higher spins?

– Piccioni:

We choose not to worry about the standard or any other model. We asked ourselves "What would work?". Take for example, the two photons which are emitted in an annihilation. A solution of the problem could be that those photons are photons which have a sharper correlation between their polarization and the direction in which they scatter. Any explanation is more plausible than a miracle.

– Liu:

What do you mean by spin two photon? Do you mean the particle which mediates the electromagnetic interaction?

– Piccioni:

If we start with an initial state which is a single state and we suppose that single states at distance don't exist, one might suppose that what comes out is a very rare channel in which quantum numbers are conserved not by the emission of common photons, but by nonconventional photons. I emphasize that in comparison with the absurd invention of an action at a distance, the fact that we contemplated the existence of a photon of spin more than one should not infuriate you. I believe that when the solution of the puzzle offered by the experimental data will be found, it will be very interesting but will not be a miracle nor incompatible with Quantum Mechanics nor with whatever is correctly demonstrated to be a consequence of it.

THE ELOISATRON PROJECT

A. Zichichi

CERN

Geneva, Switzerland

1. INTRODUCTION

In the last forty years, our understanding of Nature and its Laws has undergone a dramatic development. The "old" elementary particles (leptons, mesons, baryons) have been replaced by "new" elementary particles (quarks, gluons, leptons, $W^{\pm}, Z^0$).

Where can we expect to go? To the fundamental constituents of the elementary particles (leptonic quarks, preons, rishons) and to the supersymmetric particles. There is an immense amount of phenomena to be discovered. We know the list and it is a very long one.

The energy level has been and will continue to be a key factor in the progress of our knowledge towards the final objective of the unification of all the fundamental forces of Nature.

1.1 – The position of Italy

Italy is not a superpower. Instead of trying to do many things and not doing them very well it is considered far better to do a few things and do them excellently.

Authentic innovations stem from the discovery of new fundamental Laws of Nature. And this is a field where Italy has engaged its intellectual forces.

This is the reason why the unprecedented scientific development in this advanced field of Modern Science did not find the Country unprepared. Italy has nowadays reached an outstanding position in this basic field of Science and Technology.

1.2 – The main objectives

Following the above considerations, Italy, through the CIPE (Comitato Inter-ministeriale per la Programmazione Economica) has undertaken a preliminary step of financing the ELOISATRON Project in the 5–years INFN plan 1984-1988.The main objective being to promote the construction of the largest subnuclear machine in the world, in collaboration with those Countries which share its interest in a field so vital for scientific, cultural and technological progress.

Such a machine is identified as a (100+100) TeV proton proton Collider. The implementation of this project would in fact constitute a unique instrument for expanding the frontiers of our present knowledge.

Many fundamental scientific problems–like, for example, the family, the hierarchy, the proliferation and the compositeness–will open up new horizons.

Furthermore, the construction of the Collider would provide for technological development in several leading sectors of the so–called High Technology, i.e. the technology of the year 2000. These leading sectors are:

◊ cryogenics;
◊ superconductivity;
◊ powerful magnetic fields;
◊ high vacuum;
◊ ultra fast electronics;
◊ supercomputers;
◊ new particle detectors.

Finally, it should not be forgotten that ELOISATRON is a scientific project whose goals are Peace and civil progress. For this to be real, a new trend of Scientific and Technological solidarity between the NORTH and the SOUTH is needed. As shown by the list of Authors, the ELOISATRON Project has stimulated a tremendous interest among scientists from developing Countries. Only if peaceful high technology is made competitive, will victory over war technology become possible. Science for Peace can only become a reality through concrete projects.

1.3 – The logic

In order to build the largest accelerator in the world, it was felt that the most effective approach would be as follows.

First of all to develop the required technologies. Here the ELOISATRON Project has chosen to follow a totally new approach to the problem.

It was, in fact, recognized that amongst the various technologies needed for the construction and the experimental exploitation of such a machine, there are for example, accelerating techniques or particle detection techniques, which are best developed in Research Laboratories and others like, for example, the construction of superconducting magnets, which are much better pursued by Industry.

Accordingly, it was considered essential for the ELOISATRON Project to promote, at a very early stage, the interest of those industries where the most basic components of the machine could be constructed. Next, in order to establish the basis for such a project to become real, the creation of a large area of collaboration, the world over, was of vital importance. In fact the project cannot be isolated from the scientific and cultural environment. It must promote a stronger and wider interaction among all the scientists interested.

1.4 – The basic steps

The basic steps of the ELOISATRON Project are shown in Table 1.1.

So far the time schedule has been followed as expected. At present we are fully engaged in the R&D (LAA) project for new Detectors to be used at the MultiTeV Supercollider such as the 10% ELOISATRON model. The next step is to start the construction of the 10% model. The last step is at the level of conceptual design. Intensive R&D in fundamental technologies is going to be an integral part of this step.

All the steps towards ELOISATRON are discussed in more detail in the following pages.

Table 1.1 - Steps towards ELOISATRON

◇ 1979 CS Project (Superconducting Cyclotron) resumed
◇ 1981 Start of the CS construction (Ansaldo–LMI–Zanon)
         The HERA Project is presented to the INFN Council
◇ 1982 First Meeting of the HERA designers with Ansaldo–LMI–Zanon
◇ 1985 The construction in Italy of the superconducting magnets
         (prototypes) for HERA begins
◇ 1986 R&D for Detectors - LAA Project is approved and financed
◇ 1987 R&D for LSCM (Long SuperConducting Magnet) prototypes
         10% ELOISATRON Model–starts at Ansaldo
◇ 1988 The construction of the HERA magnets is expected to be completed
◇ 1989 Magnet construction for the 10% ELOISATRON Model
         is expected to start
         The Full Scale ELOISATRON PROJECT:
         1st–Conceptual Design
         2nd–R&D in fundamental technologies
         3rd–Construction of the Superconducting Long Magnets

1.4.1 – The Superconducting Cyclotron (CS)

In 1979 INFN took up again the project of building in Italy a Superconducting Cyclotron. Besides the scientific importance for the studies in Nuclear Physics, the aim of the CS Project was to develop around it a scientific undertaking of high technology affecting three Italian industries: one specialised in magnets (ANSALDO), another in superconducting cables (LMI) and a third one in cryogenics (ZANON).

This first step was crowned with success. In 1981 the construction of the CS began.

1.4.2 – The superconducting magnets for HERA

The following year the second step started with the participation of Italy in the construction of the most delicate part of the German Project HERA (Hadron–Electron Ring Accelerator): that is, the superconducting magnets for the 820 GeV proton ring at DESY, Hamburg. Also this step has been successfully completed. In fact Italy is building 50% of the superconducting magnets for HERA. The series production of about 300 dipoles has already started and will be completed by 1988.

1.4.3 – The 10% ELOISATRON Model

After completion of the CS and HERA phases a general test at the 10% level of the Full–scale ELOISATRON Project is needed.

This test must be made at the minimum cost and in the most efficient way using the best structures existing today.

It should be noticed that in order to construct this machine only moderate R&D of the magnets developed by the Italian industry for the HERA machine is needed.

1.4.4 – Research and Development for Detectors (LAA)

Experience with the LEP and HERA programmes has shown that the design and construction of the required complex apparatus needed in present accelerators is an undertaking of the same order of magnitude as the construction of the machine itself.

Moreover, considering the complexity of the event and the severe experimental conditions involved in a multi–TeV hadron Collider, it is recognized that with present–day techniques nobody would know how to perform an experiment at such a machine.

Notice that this situation is completely reversed for electron Colliders. In this case the detector doesn't present relevant difficulties, but the technology to build the machine is unknown.

In parallel to the development of the 10% ELOISATRON Model it was therefore deemed necessary to start the LAA Project: an intensive R&D programme for new experimental techniques. The goal is to prove, on the basis of prototypes, the feasibility of essential components for a detector capable of operating at the next generation of Colliders, the first one being the 10% ELOISATRON Model. Special attention must therefore be paid to radiation hardness, rate capability, momentum resolution and hermeticity of such a detector assembly.

Bearing this in mind, the LAA Project covers the following most important components:

◇ **High precision tracking devices**
◇ **Calorimetry**
◇ **Large area devices for muon detection**
◇ **Leading particle detection**
◇ **Data acquisition and analysis**
◇ **Theoretical QCD calculations and MonteCarlo simulations**
◇ **Very high magnetic fields**
◇ **Superconductivity at high temperature.**

At present a number of projects in the various fields listed above has been approved and financed by the LAA Scientific and Technical Advisory Board. They are under way.

1.4.5 – The Full–scale ELOISATRON Project

In addition to the problems related to the machine itself, two most basic problems had to be considered. Where to build ELOISATRON and how to build it.

The Civil Engineering problems have been studied in specialized Working Groups and possible solutions have been found. A detailed study of the machine lattice and of the full–scale project with its executive plans has not been attempted since it was felt that this must constitute an intrinsic part of the effort put in the 10% test model.

However, existing studies of projects at the (10–20)% scale (LHC,SSC) and the studies for the full–scale ELOISATRON performed so far, allow to conclude, as will be discussed later, that such a project has no intrinsic prohibitive features.

1.5 – The structure

According to the principles discussed above, the structure of the ELOISA-TRON Project is, at present, as follows:

The Project is subdivided in a number of Working Groups to which scientists from various Research Laboratories, Universities and Institutions participate together with several Italian industries.

Table 1.2 shows the Working Groups and the scientists, scientific institutions or industries in charge of organizing the activity in the various domains.

Table 1.2 - ELOISATRON Working Groups

1	**Superconducting Magnets**	*Ansaldo, LMI, Zanon*
2	**Refrigeration Power Plants**	*Zanon*
3	**Civil Engineering**	*L. Valeriani, Grandi Lavori*
4	**Geophysical Studies**	*E. Boschi, ING*
5	**Machine**	*K. Johnsen*
	Workshops (Puglisi, Torelli, Villa)	
6	**Experimental Areas**	*H. Wenninger*
7	**R&D for Detectors (LAA)**	*A. Zichichi*
	Workshops (G. Charpak)	
8	**Supercomputers**	*T.D. Lee*
	Workshops (R. Mount)	
9	**Physics – Experimental**	*S.C.C. Ting*
	Workshops (L. Cifarelli)	
10	**Physics – Theory**	*A. Salam*
	Workshops (A. Ali)	

The results of the studies performed are then presented and discussed by the scientific community in Specialized Workshops.

The required research and development of prototypes is carried on, either in Laboratories, in the framework of the LAA programme, or, in industries, in the framework of concrete projects apt to develop the appropriate technologies at the industrial level.

The crucial point in the ELOISATRON Project is that the location of the main Lab is where all detailed R&D is going on. And this means that ANSALDO, LMI, ZANON are basic labs of the Project.

2. – THE NEED FOR ELOISATRON

2.1 – A new era

A new era in modern scientific thought has opened up. New ideas, new concepts and new phenomena make physics of just twenty years ago seem as old as millennia.

Einstein's quadridimensional space-time seemed to be a conquest beyond which none would be able to go. This, however, seems to be an incredibly narrow outlook for two reasons, both fundamental: the <u>number</u> of dimensions and the <u>property</u> of those dimensions.

No one had thought, before the sixties, that there could exist space-time dimensions with "fermionic" properties. Those of Einstein are "bosonic". This is how the new concept of Superspace was born, and with it, Superparticles and Supermatter. In Table 2.1 a synthesis is given of the present state of our knowledge of Matter – quarks, leptons, photons, strong and weak gluons, Higgs particles – and Supermatter.

Table 2.1 – Present status of World and Superworld

MATTER $R=0$	SUPERMATTER $R \neq 0$		
SPIN 1 $\longrightarrow$ SPIN $\frac{1}{2}$		GAUGE PARTICLES	F O R C E S
GLUONS $\longrightarrow$	GLUINOS		
PHOTONS $\longrightarrow$	PHOTINOS		
($W^{\pm}$, Z^0) $\longrightarrow$	($W^{\pm}$, Z^0) INOS		
SPIN 1 $\longrightarrow$ SPIN ZERO		MATTER PARTICLES	M A T T E R
QUARKS $\longrightarrow$	SQUARKS		
LEPTONS $\longrightarrow$	SLEPTONS		
SPIN ZERO $\longrightarrow$ SPIN $\frac{1}{2}$		GAUGE BREAKINGS	S S B
HIGGS $\longrightarrow$	SHIGGS		

exception : NEUTRINO $\longrightarrow$ SNEUTRINO

SSB = Spontaneous Supersymmetry Breaking

The world in which we live and the matter we are made of could have their roots in a Superspace with many more than four "bosonic" dimensions plus the corresponding "fermionic" ones. And this is not all.

The concept of "point" that has held its position for centuries and centuries, falls by the wayside. In its place there is a Superstring or a membrane: i.e. either a unidimensional or a multidimensional entity with a non point-like structure in the "still to be definitely identified" pluridimensional Superspace. Fig. 2.1 shows the present status of our knowledge. In this extraordinary progress towards the unification of all the fundamental forces of Nature, the winning parameter has been so far, and will certainly remain, the energy level.

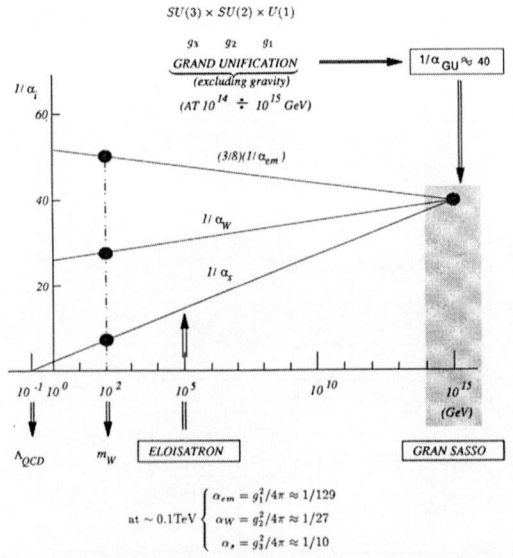

Fig. 2.1 – All the fundamental forces of nature should be generated by a unique force. The convergence of the three fundamental constants g_1, g_2, g_3, is the basis for GUT.

2.2 – Why start now to discuss a multi–TeV machine?

Given that by 1990 both LEP and HERA will be operating, and if, as many of us believe, Subnuclear Physics is to be pursued with vigour and without a break

450

of many years, then it is appropriate to start work now on a new hadronic machine in the 10^2 TeV range.

2.3 – From new ideas to reality

Some people might say that it would be better to wait for new ideas. But past experience has shown that to wait is not a wise choice.

In fact, a long time is needed to transform new ideas into reality. Two examples should suffice:

i) Superconducting high–field magnets, first proposed in 1961, became "reality" in 1986 (the Tevatron): 25 years were required.

ii) Collective field accelerators were proposed by Veksler, Budker and Fainberg in 1956: a third of a century later, there is still no practical design for a high–energy machine based on these ideas.

In this context it is important to notice that the ELOISATRON design is based on extrapolation from known facts and technologies. Nevertheless,

– new acceleration techniques,

– new superconductivity technologies (at high temperatures),

should be encouraged for the full–scale Project.

The 10% test model should not wait for these results. It should consolidate the validity of the extrapolation of present–day technologies.

2.4 – The desert and the lesson from the past

At the end of the seventies, a fashionable approach to the extreme energies was the "desert", i.e. nothing should exist in the energy range from a few 10^{-1} TeV up to about 10^{12} TeV.

The "desert" would be a serious obstacle if there were no problems whatsoever in this field of extreme energies. The high energy limit of our present knowledge has two frontiers.

One is in the domain of experimental physics. At present the known limits on the inverse radii of all known leptons (e, μ, τ) are of the order of 10^{-1} TeV. This means that point–like structures of the particles existing at present, are already in the multi–TeV range.

The other is in the theoretical domain. Here we have the family and the hierarchy problems, which, together with the proliferation of the Higgs sector and of its associated large number of parameters, make the multi–TeV range overcrowded with problems to be understood.

But even if everything looked "perfect", we should not forget the lesson we can learn from past experience. Other "deserts" were believed to be present in all the history of Physics and none of them has ever resulted to be there.

In the last 400 years, at least 5 times the following statement was made: *"There is nothing to discover, we have really understood everything"*. This happened after the discoveries of

1 – Galilei [Unification of Rectilinear and Circular motion]
2 – Newton [Universality of Gravitational forces:$F = k(m_1 \times m_2)/r_{12}^2$]
3 – Maxwell [Unification of Electricity, Magnetism and Optics]
4 – Planck, Einstein [Quantum ($h \neq 0$) and Relativistic ($c \neq \infty$) Physics]
5 – Fermi [Three "matter" particles (p, n, e) and three "glues" (π, γ, g)]

Five times a "desert" was predicted in the world of Physics and the lesson we can draw from the past is that this desert is our enemy if seriously considered as the truth. One example of desert not mentioned in the above list, in spite of being the most famous one, is the one predicted by Lord Kelvin in 1897: *"There is nothing to be discovered in Physics now, all that remains to be done is more precise measurements"*. A few months later J.J. Thomson discovered the electron.

The last example in the list is the most recent case: when Enrico Fermi was convinced that, with the discovery of the "nuclear glue" - the π meson – there was nothing more to be searched for, in order to understand the microscopic world. Since then a continuous confirmation of the lack of any predicted or unpredicted desert has characterized the development of modern physics. It is quite appropriate to review this in more detail.

◇ 30 GeV PS (CERN, BNL)

Original motivations for the construction of this accelerator:

$(\pi p)(pp)$ scattering, phase–shift analysis
test of isospin (I) and time reversal (T) invariances

Discoveries:

– new particle states $\Rightarrow$ SU(3) (Gell-Mann, Ne'eman)
– (ω, ϕ) mixing angle
– $(e^+ e^-), (\mu^+ \mu^-)$ production in hadronic interactions
– electromagnetic structure of the proton in the time–like region
– antideuteron
– $\nu_e \neq \nu_\mu$
– $\nu_\mu \neq \bar{\nu}_\mu$
– C, P, CP, T violation
– weak neutral current
– J particle state

◇ SLAC

Original aims:

- electromagnetic form factors of nucleons
- electromagnetic transition form factors ($N - N^*$)
- checks of Quantum ElectroDynamics (QED)

Discoveries:

- point–like structures inside the proton
- parity violation in purely electromagnetic interactions

◇ ADONE

Conventional motivations:

- checks of QED and radiative corrections
- (μe) electromagnetic equivalence
- study of the tails of the vector mesons

Unconventional proposals:

- search for heavy leptons via acoplanar (μe) pair techniques
- search for leptonic quarks

Unexpected discovery:

- the ratio (hadronic cross–section/muonic cross–section) greater than theoretically predicted

◇ SPEAR and DORIS

Discoveries:

- heavy lepton
- J/ψ family of new particles
- open charm states

◇ PETRA and PEP

Discoveries:

- evidence for gluons (expected)

◇ ISR

The ISR was a special case. In fact, both the choice of the energy range and the technical performances of the machine were optimal. However the Physics results obtained were not as outstanding as expected since the quality of the experimental setups used was not adequate. This points out the importance of a dedicated R&D programme for experimental setups, such as the present LAA Project, before the construction of any new machine is undertaken.

◇ 400 GeV SPS (CERN, FERMILAB)

Discoveries:

 – Υ states (unexpected)

◇ Sp$\bar{\text{p}}$S (540 GeV)

Discoveries:

 – $W^{\pm}, Z^0$ (expected)

◇◇◇ **TO CONCLUDE**, the lesson from the past can be expressed in the following terms:

1 – the same effort must be devoted to experimental set-ups and to machine construction (ISR docet);

2 – watch the energy gaps:

      ADONE   SPEAR   PETRA

      ‖———|————‖ ‖—‖

          3 GeV   9 10 GeV

          (J/Ψ) (Υ)

3 – when unexpected discoveries are made, the expected ones appear as a DESERT OF IMAGINATION.

This is why at present, the need for a multi–TeV machine and the effort to build such a machine are justified.

3 – FROM THEORY TO "DOWN–TO–HEART" PHYSICS

3.1 – The great problems of MultiTeV Physics

The theoretical goals outlined above can be expressed in terms of the following basic questions:

◇ do new heavier quarks and leptons exist?

◇ do supersymmetric partners exist?

◇ are there other intermediate bosons?

◇ do the Higgs bosons exist?

◇ are quarks and leptons composite?

◇ would some unexpected exotic process occur?

In addition to this impressive list of new physics, we should not forget the straightforward extrapolation expected from present knowledge.

3.2 – Extrapolation from present knowledge

In order to go from theory to "down-to-earth physics", i.e. in order to be able to make experiments, it is necessary to study which are the expected observable features of events produced by physics at multi–TeV energies and derive from these the main characteristics of the experimental apparatus. Rare phenomena, for instance, suggest that the highest luminosity should be aimed at. Let us consider, for example, $10^{33} cm^{-2} s^{-1}$. Furthermore, as discussed later, if only one event per bunch crossing is desired, a short separation time between bunches (<100 ns) is needed. This implies specific performances from the experimental setups (see Section 4.2).

Let us then review some of the expected physics features in the multi-Tev range.

The present QCD knowledge on hard parton interactions predicts that at multi–TeV energies the machine would essentially behave as a "broad band gluon-gluon collider". By extrapolating CERN ISR results ($\sqrt{s} = 62$ GeV), it was possible to predict the average charged particle multiplicity and the multiplicity distribution at the CERN $p\bar{p}$ Collider ($\sqrt{s} = 540$ GeV). These predictions are in good agreement, as shown in Fig.s 3.1 and 3.2, with experimental results. This agreement shows the relevance of the leading subtraction method. And this means that it is necessary to consider not the nominal energy $\sqrt{s}$, but the effective energy available for particle production once the "leading" particle effect - see Fig. 3.3 - is taken into account, i.e. once the energy carried away by the "leading" particles is subtracted from $\sqrt{s}$. The same procedure at $\sqrt{s} = 20$ TeV gives an average charged multiplicity of 100 with important fluctuations up to multiplicities greater than 200, as shown in Fig. 3.4.

Hard parton interactions will provide quarks and gluons which fragment into jets. They will also presumably produce new massive states decaying into intermediate bosons, leptons, quarks and gluons, therefore predominantly into jets. Jets and leptons are expected to be fundamental instruments to extract interesting rare events from the standard parton scattering processes. Moreover, "missing jets" – i.e. missing energy/momentum – would provide signatures for events with undetected particles such as neutrinos or photinos, or other "inos".

It should be noticed that high p_T jets must not only be considered as a test for QCD, or as a background to cope with, when hunting for more interesting new phenomena: they could reveal if quarks are composite. The cross–section behaviour with p_T should in fact be modified by the effect of a deeper contact interaction caused by a superstrong binding force.

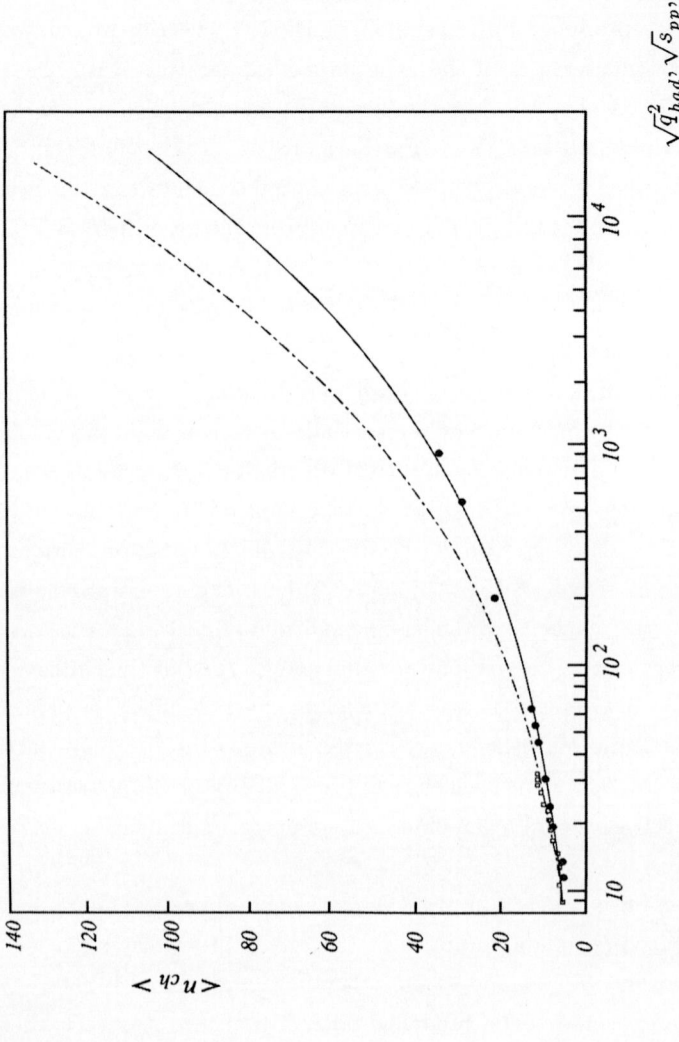

Fig. 3.1 – The average particle multiplicities as a function of $\sqrt{q_{had}^2}$ (dashed line) and $\sqrt{s}$ (full line). The dependence of the average particle multiplicity on $\sqrt{s}$ is derived from its dependence on $\sqrt{q_{had}^2}$.

$\sqrt{q_{had}^2}, \sqrt{s_{pp}}, \sqrt{s_{p\bar{p}}}$ (GeV)

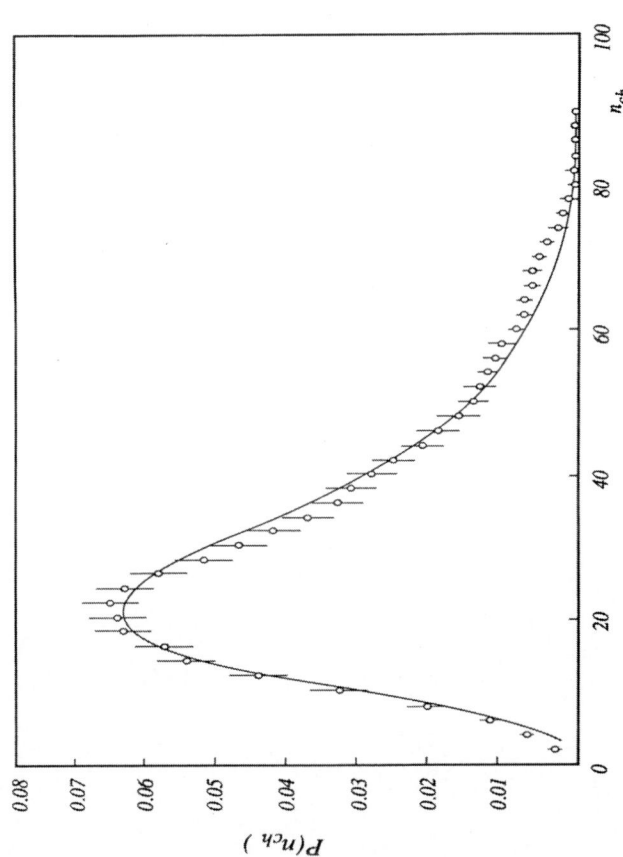

Fig. 3.2 – The charged particle multiplicity distribution measured by the UA5 collaboration at $\sqrt{s} = 540\ GeV$. The curve is the prediction with the BCF method of taking into account the leading effect.

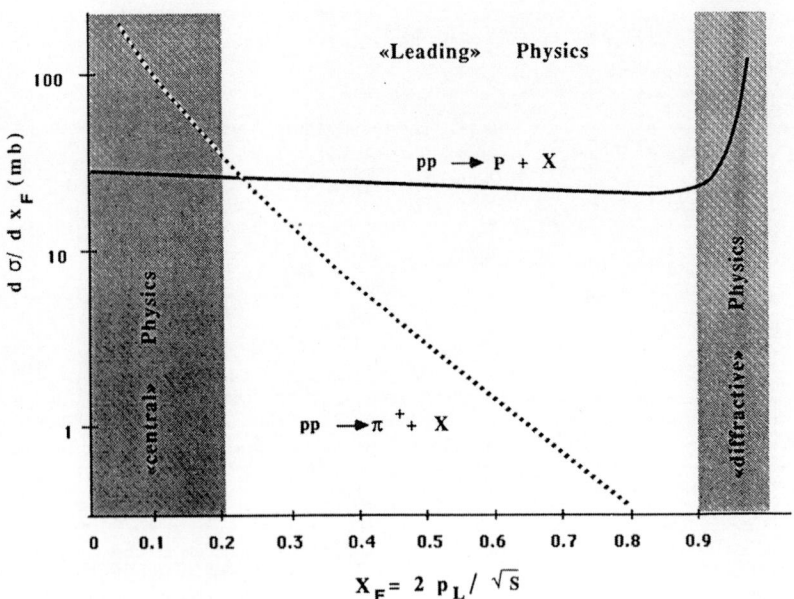

Fig. 3.3 – Protons and pions x_F distributions in (pp) interactions. Central, leading and diffractive regions are also shown.

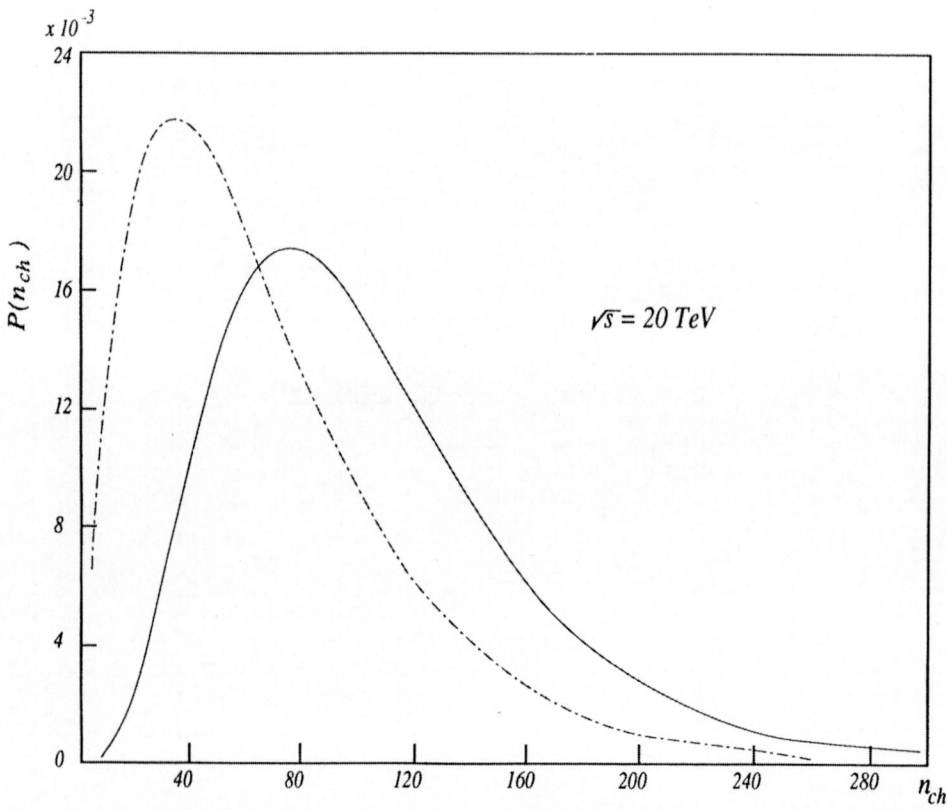

Fig. 3.4 – The charged particle multiplicity distribution predicted at $\sqrt{s} = 20\ TeV$ following the UA5 method (dashed line) and the BCF method (full line).

Concerning heavy flavour production, it is not unfair to predict that a large hadron Collider will be a good old and new quark factory. The calculations, mostly based on the fusion mechanism, show that at $\sqrt{s} = 20$ TeV heavy quarks with masses up to 400 GeV could be observed. It must also be recalled that the fusion mechanism underestimated the rate of c and b production at the CERN ISR and failed to reproduce the shape of the longitudinal momentum distributions.

Therefore the cross–section evaluations based on this model have to be taken as lower limits. In fact, the "leading" production observed at the ISR for Λ_c^+ and Λ_b^0 should be taken into account. In addition, the "leading" mechanism could be much more efficient in the detection of very heavy quarks. In fact, the semileptonic decay of baryonic states carrying the heavy flavour is expected to produce a distinctive pattern: the sign and the p_T dependences of the lepton charge are directly correlated to the "up–like" or "down–like" nature of the new heavy quarks and to their masses, respectively.

In general, the signatures for new heavy flavour production would consist of multi–jet events or, if their semileptonic decay is considered, of jets + lepton (within or nearby a jet, with high p_T with respect to the jet axis) + missing energy. Multi–lepton events should also be relevant in this kind of search. Substantial rates up to a few hundred GeV leptonic energy are expected from c, b and t decays. Leptons from heavier quarks will be even more energetic.

SUSY production, for instance $pp \rightarrow \tilde{g}\tilde{g}X$, with $\tilde{g} \rightarrow q\bar{q}\gamma$, could be investigated via a huge missing p_T signature due to the non–interacting photino. For $m_{\tilde{g}} < 1$ TeV and $\sqrt{s} = 20$ TeV, the cross–section could exceed 0.1 pb, giving 10 events/day at $10^{33} cm^{-2}s^{-1}$ luminosity.

Higgs production through WW fusion could occur at the 1 pb level for $m_H = 400$ GeV and $\sqrt{s} = 20$ TeV. Assuming $H \rightarrow WW$ decay, the 4–jet invariant mass could be a valuable tool to detect the signal.

Another feature of present-day knowledge to be extrapolated in the field of "flavour physics", is the structure of the multi–quark states, still to be discovered. There are hundreds of these "particles" as shown in Fig.s 3.5 - 3.8.

3.3 – The observable effects

The impressive list of phenomena to be searched for, in terms of "down–to–earth Physics", corresponds to the following experimental effects to be studied:

- *Short–lived hadrons* should be tagged by secondary vertex identification. Adequate tracking power is needed. The ability to define points in space will be of considerable advantage.
- *Long-lived hadrons* must be distinguished from leptons and photons and their energy precisely measured. The high multiplicity, jet–like events produced

in multi–TeV collisions call for high granularity detectors, capable of reconstructing the jet directions.

- *Electrons and photons* should be identified, even if produced inside jets. Their energy should be measured with the highest possible resolution.
- *Neutrinos and other non-interacting particles* can be "identified" as energy unbalance in the fully reconstructed event. Hermeticity will be an essential attribute of any apparatus, since most of the expected "new" physics will produce missing energy.
- *Muons* should be identified in the largest possible solid angle. Their charge should be measured up to the highest possible momentum, in order to detect charge asymmetries. Therefore, high precision, large area, low cost devices are in order.
- *Leading protons* can be identified by their momentum. Energies up to the maximum attainable should be measured with good precision, in the smallest possible angle with respect to the beam direction.

On the basis of past experience discussed in Section 2.4, we have to add to this impressive list the other one: "the unexpected". So far, as we have already said, "when unexpected discoveries are made, the expected ones appear as a DESERT OF IMAGINATION".

The ELOISATRON Project is before us with an unprecedented challenge: can we think of something to be discovered, which is more exciting than all the items in the above list? Needless to say that this list is really impressive. Could someone imagine a "new" breakthrough which is not part of our supervast horizon? This is the goal for multi–TeV physics.

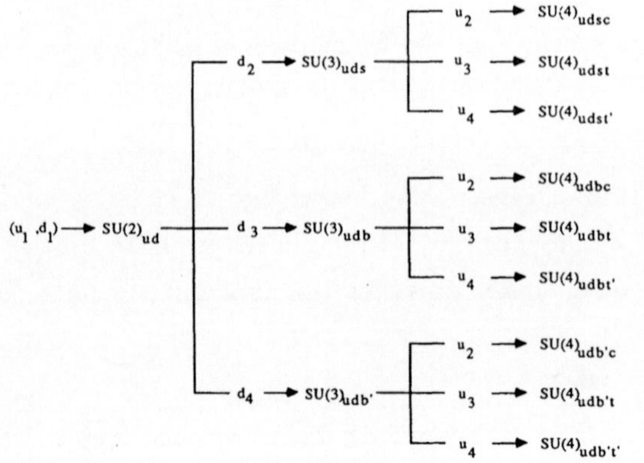

$$\begin{pmatrix} u & c & t & t' \\ d & s & b & b' \end{pmatrix} \equiv \begin{pmatrix} u_1 & u_2 & u_3 & u_4 \\ d_1 & d_2 & d_3 & d_4 \end{pmatrix}$$

Fig. 3.5 – $SU(3)_{uds}$ repeated with b and b' replacing s, and $SU(4)_{udsc}$ repeated with t and t' replacing c. Notice the above equivalence between old and new notation for flavours.

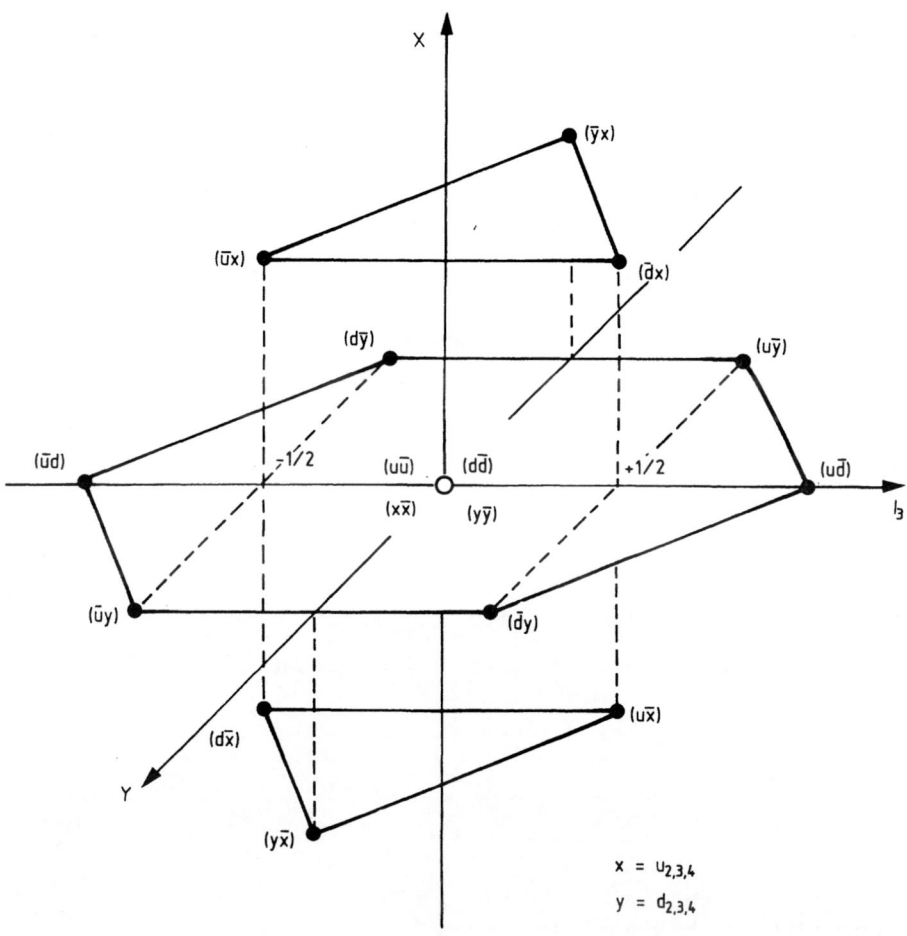

Fig. 3.6 – The $SU(4)_{udxy}$ structure for $J^P = 0^-$ and $J^P = 1^-$ meson states.

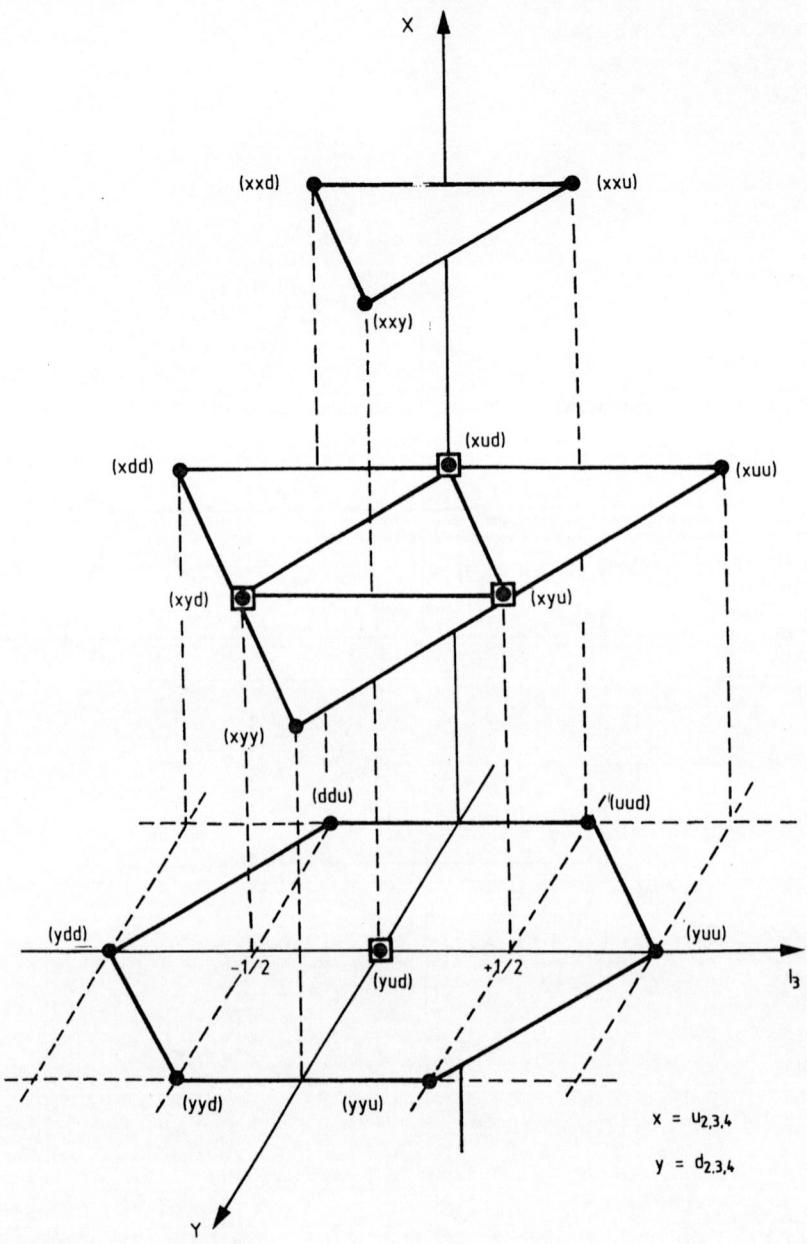

Fig. 3.7 -- The $SU(4)_{udxy}$ structure for the $J^P = (1/2)^+$ baryon states.

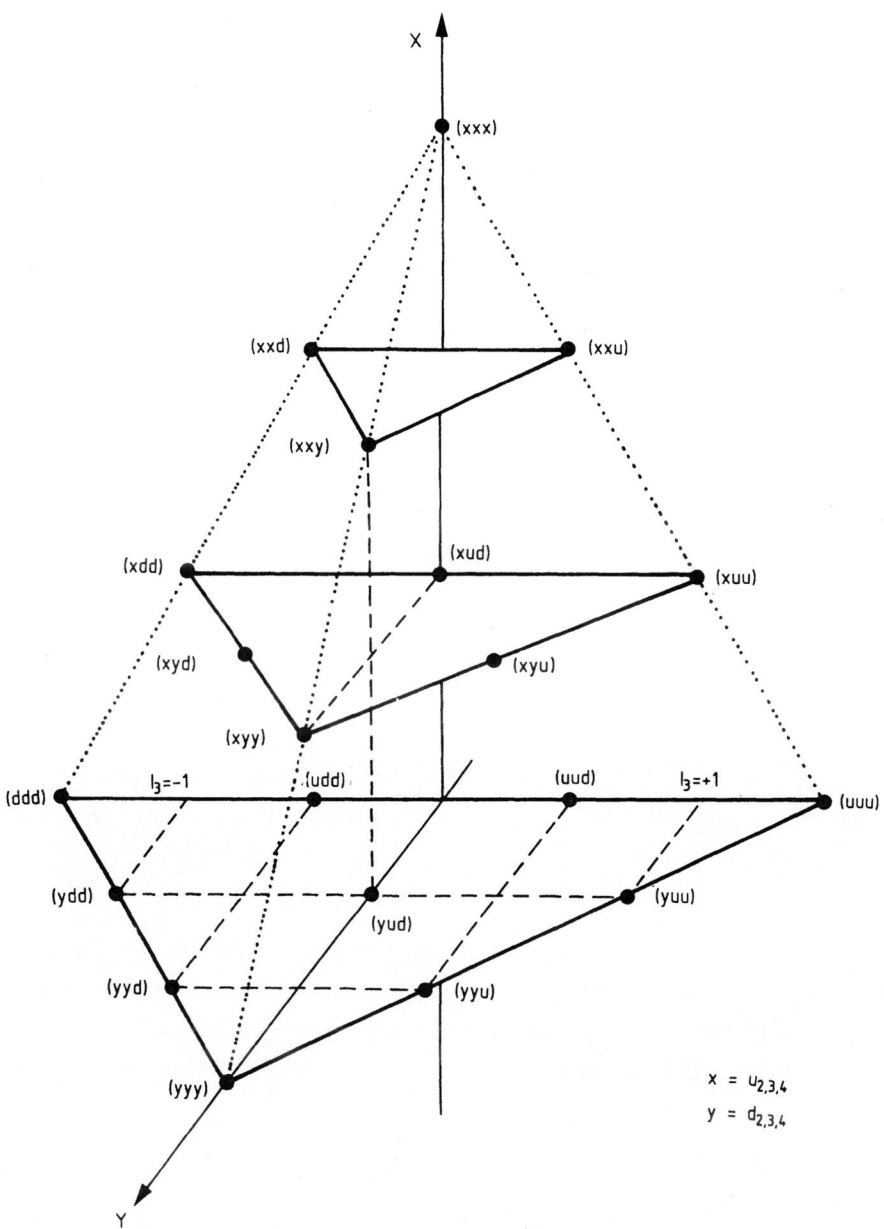

Fig. 3.8 – The $SU(4)_{udxy}$ structure for the $J^P = (3/2)^+$ baryon states.

4 – The Eloisatron Lab

Since the very beginning, the conception of the ELOISATRON Project was based on a new way of tackling the problem of big projects. This new way is based on the following points:

1) Do not start to build a new Laboratory if the work needed can be carried out within existing structures.

2) Involve the Industry in the implementation of the project. And this also means that the R&D work should be done in the laboratories of the industries in close connection with the physicists.

4.1 – Hardware for the machine: ANSALDO, LMI, ZANON

At present the ELOISATRON Lab, as far as Machine Components are concerned, is housed in three places:

- at Genoa (ANSALDO)
- at Florence (LMI)
- at Vicenza (ZANON)

In these three places R&D work is being undertaken:

1) at ANSALDO (Genoa): Superconducting technology to build powerful and high precision magnets is being studied.

2) at LMI (Florence): Superconducting cables technology is going on.

3) at ZANON (Schio-Vicenza): Powerful cryogenic plants and specific cryogenic problems are being studied.

Instead of having hundreds of people working in a new Laboratory, the ELOI–SATRON Project has hundreds of people working in these industries, each one highly specialized for the key work needed to build ELOISATRON.

Note that these industries are not only engaged in pure R&D work. They have been and are fully engaged in the implementation of the various phases of the ELOISATRON main steps: at present their major involvement is in the construction of the HERA superconducting dipoles and in the collaboration with the LAA Project.

In the following paragraphs we will report briefly on the work being done by ANSALDO, LMI and ZANON, in order to give a closer view of the status of this ELOISATRON Lab.

As mentioned before, one of the main points of the ELOISATRON Project has been to stimulate the Italian industry to acquire the technologies neeeded for the implementation of the Project. One of these technologies, which has a special key function, is the series production of superconducting magnets.

Ansaldo has specialized in the design and contruction of the magnets, LMI (La Metalli Industriale) in the production of the superconducting cables and Zanon in the cryogenics.

The main characteristics of the superconducting dipoles that these industries are at present building are those of the 820 GeV proton ring of HERA. For completeness and in order to give to the reader a close view with the real performance of these ELOISATRON Labs we report in Table 4.1 the effective data for these dipoles.

Table 4.1 – HERA Superconducting Dipole Magnets

DESIGNED FIELD	6 Tesla
DESIGNED STORED ENERGY	756 kJ (without yoke)
TYPE OF WINDING	2 Dipole coils
DESIGN NOMINAL CURRENT	4992 A
CABLE INSULATION	1 overlapped (58%) taping of kapton tape (0.025 mm thickness) + glass fibre tape (0.13 mm thickness) impregnated with B–stage epoxy resin wrapped around the cable with a 3 mm gap between adjacent turns.
CONDUCTOR	Rutherford–type superconducting cable built up of 24 individual twisted wires of 0.84 mm diameter; total dimension $10 \times (1.28; 1.67)$ mm; trapezoidal shape.
TYPE OF COOLING	Helium bath
COIL WEIGHT	250 kg approx. (without collars)

The production capability is 250 dipoles per year. This is just a starting point. The production capability could be easily doubled or increased even more, if needed.

The total length of cable produced for the HERA magnets is about 500 km with a production rate of 2 km per day. The critical current measured on a short sample, taken from the first 20 km produced, is 9010 Amps at 5.5 T, 4.6 ^0K.

The **conclusive remark** is that this part of the **ELOISATRON Lab** is **fully operative** and **very efficient**. And, as discussed in the following Section, this is not all.

4.2 – R&D for Detectors (LAA Project)

If a Supercollider were built, no physicist would know how to do experiments. A multi-TeV accelerator with luminosities in the $10^{32} - 10^{33} cm^{-2}s^{-1}$ range opens a totally unexplored field in detector technology. An intensive R&D programme for new experimental techniques is needed in order to proceed along the ELOISATRON line, and this is the main goal of the LAA Project. In the following pages we report briefly on the main features of this programme.

In order to study the new and rare phenomena of multi–TeV physics it is necessary to aim at the highest possible luminosity ($10^{33} cm^{-2}s^{-1}$). As far as the detectors are concerned, the limiting luminositiy is:

$$L \; =<n> /(t_b \times \sigma_{pp})$$

where

- $<n>$ is the maximum average number of events per bunch crossing that the experiment can tolerate;
- t_b is the minimum time between bunch crossings;
- σ_{pp} is the total pp cross–section.

The average number of events per bunch crossing $<n>$ must be one if the missing energy is to be used as a signature in event selection and analysis. Note however that $<n>= 1$ still corresponds to a probability of 0.26 to have more than one event per crossing. The total (pp) cross–section at the 10% ELOISATRON Model energies can be derived from theoretical extrapolations to be at the level of $10^{-25} cm^2$ (100 mb).

With this cross–section value and $<n>= 1$, a luminosity of $10^{32} cm^{-2}s^{-1}$ can only be reached by having a very short bunch spacing, in the 100 $ns(= 10^{-7}s)$ range. The lower limit on t_b is determined by the time response of the detectors used by the experiment and by the occupancy of the detector elements. Correspondingly, the interaction rate would be of the order of 10^7 Hz.

A higher luminosity, $10^{33} cm^{-2}s^{-1}$, with many events per bunch crossing, could be dedicated to rare phenomena with cross–sections

$$\sigma = 10^{-33} cm^2 = 1 \; nb \quad \Rightarrow \text{production rate} = 1 \text{ event/s}$$
$$\sigma = 10^{-37} cm^2 = 0.1 \; pb \quad \Rightarrow \text{production rate} = 1 \text{ event/day}$$

Complex trigger configurations are needed to take data both with an acceptable rate and with a high efficiency.

Hard parton interactions will provide quarks and gluons which fragment into jets. They will also presumably produce new massive states decaying into intermediate bosons, leptons, quarks and gluons, therefore predominantly into jets.

The formidable task of particle identification in a high rate, high energy and high multiplicity environment, where precision measurements are needed, both in time and space, call for a dedicated R&D in order to solve the many problems still ahead of us.

The presently–available technologies for particle detection have to be pushed to their limits, new applications of old technologies have to be invented, and new technologies have to be developed.

For these purposes to be achieved the LAA Project must cover the following items:

- high–precision tracking;
- calorimetry;
- large area devices for muon detection;
- leading particle detection;
- application specific integrated circuits (ASIC);
- data acquisition and analysis;
- supercomputers and Monte-Carlo simulations;
- very high magnetic fields;
- superconductivity at high temperature.

Let us briefly review each one separately.

4.2.1 – High precision tracking

i) Scintillating fibre microtracker.

The goal is to investigate and improve the properties of narrow scintillating fibres (light yield and attenuation, radiation hardness), and to design the micro-tracker itself (assembly of multi–fibre bundles in shell structures), together with high–speed optoelectronics read-out chain.

ii) Vertex detector based on multidrift gas proportional tubes.

The idea here is to develop, build and test a fast, triggerable and modular detector made of hexagonal thin carbon fibre tubes (30 mm diameter, 70 signal wires per tube), together with its dedicated Large Scale Integrated (LSI) circuits for the on–tube electronics.

iii) GaAs microstrip detector.

The properties of GaAs in terms of performances and radiation hardness up to the Mrad level will be investigated in order to prove the feasibility of GaAs microstrips with integrated front–end electronics to be used, instead of Silicon microstrips, in future high–precision tracking devices.

4.2.2 – Calorimetry

i) Electromagnetic calorimeter based on BaF_2 scintillator.

A prototype of calorimeter with BaF_2 scintillators and photosensitive wire chambers for shower containment up to 100 GeV will be built. An exceptional time resolution, obtainable by detecting the fast component emitted from the BaF_2 crystals in the UV region, and very good position accuracy, due to the application of photosensitive wire chambers, will be tested.

ii) "Spaghetti" hadron calorimeter.

The goal is to build a prototype of hadron calorimeter made of "spaghetti", i.e. scintillating plastic fibres, as active material and lead absorber. This calorimeter is of particular interest for its compactness and hermeticity. Its uniformity of response and expected compensation capability ($e/h = 1$) will be tested.

4.2.3 – Large area devices

i) Comprehensive study of muon detection in multi–TeV hadron colliders.

Particular emphasis will be devoted to the construction problems. The goals are:

- construction of plastic chambers (LST) with new design read–out strips for higher accuracy and higher rate capability;
- study of alternative solutions such as plastic tubes filled with liquid scintillator and viewed by photomultiplier tubes;
- study of high–precision drift chambers;
- feasibility study of high field magnets of large dimensions.

ii) High precision alignment study.

The development of an alignment and calibration system for large drift chamber detectors will be carried out. The goal is the construction of precision tools to measure angles, lengths, straight lines and planarity with a precision of 20 μm over 5 m (UV Lasers). Test drift chambers, using "cool" gases and newly developed electronics, will be constructed.

4.2.4 – Leading particle detection

The idea is to integrate a detector design with an advanced accelerator design in order to detect and measure particles produced at very low angles with respect to the beams. The problems of producing radiation hard electronics and detectors will be investigated, together with mechanical problems. The goal is to produce a prototype microstrip system small enough to contain its associated electronics within 30 mm around the detectors. The required radiation resistance is a few Mrad.

4.2.5 – Application specific integrated circuits

The development of front-end and read-out electronics will be carried out.

Microelectronics, consisting of highly integrated circuits in which detection, amplification and data processing are done, will be developed. The goals are:

- low noise amplifier design;
- fast signal comparator design;
- fast analogue pipelining, using CCD;
- fast on-chip digitization, using Sigma–Delta modulation;
- process compatibility development: electronics in Silicon detectors;
- evaluation and improvement of radiation hardness of electronics.

4.2.6 – Data acquisition and analysis

Data acquisition components and new data acquisition architectures in both hardware and software, via several small and coordinate projects, will be developed. Particular emphasis will be put on the problems of second level trigger and experiment monitoring.

4.2.7 – Supercomputers and Monte-Carlo simulations

The idea here is to develop and build a supercomputer for QCD dynamics lattice calculations. This supercomputer would be more powerful than the presently available CRAY. A programme chain, using our best knowledge on QCD, with particular emphasis on Heavy Flavour Physics in future high energy hadron colliders, will also be implemented.

4.2.8 – Very High Magnetic Fields

The final dimensions of an LAA–like Detector to be operated in a multi–TeV Collider will depend critically on the value of the Magnetic Field in the vertex volume.

At present, if no progress is made in this area, the dimensions of the LAA–like Detector would be of the order of 30 metres along the beam and 20 metres transverse, for a total weight of about 30 ktons.

To gain an order of magnitude in the vertex, Magnetic Field is of great value so as to reduce these dimensions. This is why the study of large volume Magnetic Fields is an essential part of the LAA Project.

To set up at CERN a laboratory for these studies would be extremely interesting, but very expensive.

However, such a laboratory exists in one of the most powerful and advanced Italian industries: ANSALDO. We are in collaboration with this industry and therefore this item of the LAA Project will be developed along this line of collaboration with ANSALDO.

It should be noticed that progress in the maximum value of B to be achieved for the vertex volume, implies progress in precision measurements in the Vertex Detector. The two progresses are strongly coupled and this is a further reason to have R&D for High Magnetic Fields in Large Volumes as an integral part of the LAA Project.

4.2.9 – Superconductivity at High Temperature

Superconductivity at high temperature has two possible applications in the field of High Energy Physics. One is in the area of detector technologies; the other is in the production of superpowerful magnetic fields.

To be in close contact with those centres where such research is pursued is the purpose of the LAA Project. On the other hand, an Italian industry, LMI, is at the forefront of "classical" superconductivity, and is now engaged in the field of Superconductivity at High Temperature.

Superconductivity at High Temperature will be followed by the LAA Project, with the purpose of establishing a series of contacts with the various research centres in the World and to stimulate close collaboration between European experts and CERN.

4.3 – Geophysical studies

The problem of finding a site for such a large machine as ELOISATRON with appropriate geophysical characteristics has been investigated by ING (Istituto Nazionale di Geofisica) in a 1985 study. The results of this study are summarized in Fig. 4.1. At least seven sites, in Italy, are suitable for the ELOISATRON ring.

4.4 – Civil Engineering studies

The civil engineering problems have been investigated in a specialized Working Group. Depending on the flatness of the ground, the machine could be built partly as prefabricated tract (Fig. 4.2) and partly in a tunnel (Fig. 4.3). The cost of this solution has been estimated to be not higher than other possible solutions, such as the pipe–line, once all needed items are included. Fig. 4.4 shows the schematic plan view of an experimental hall.

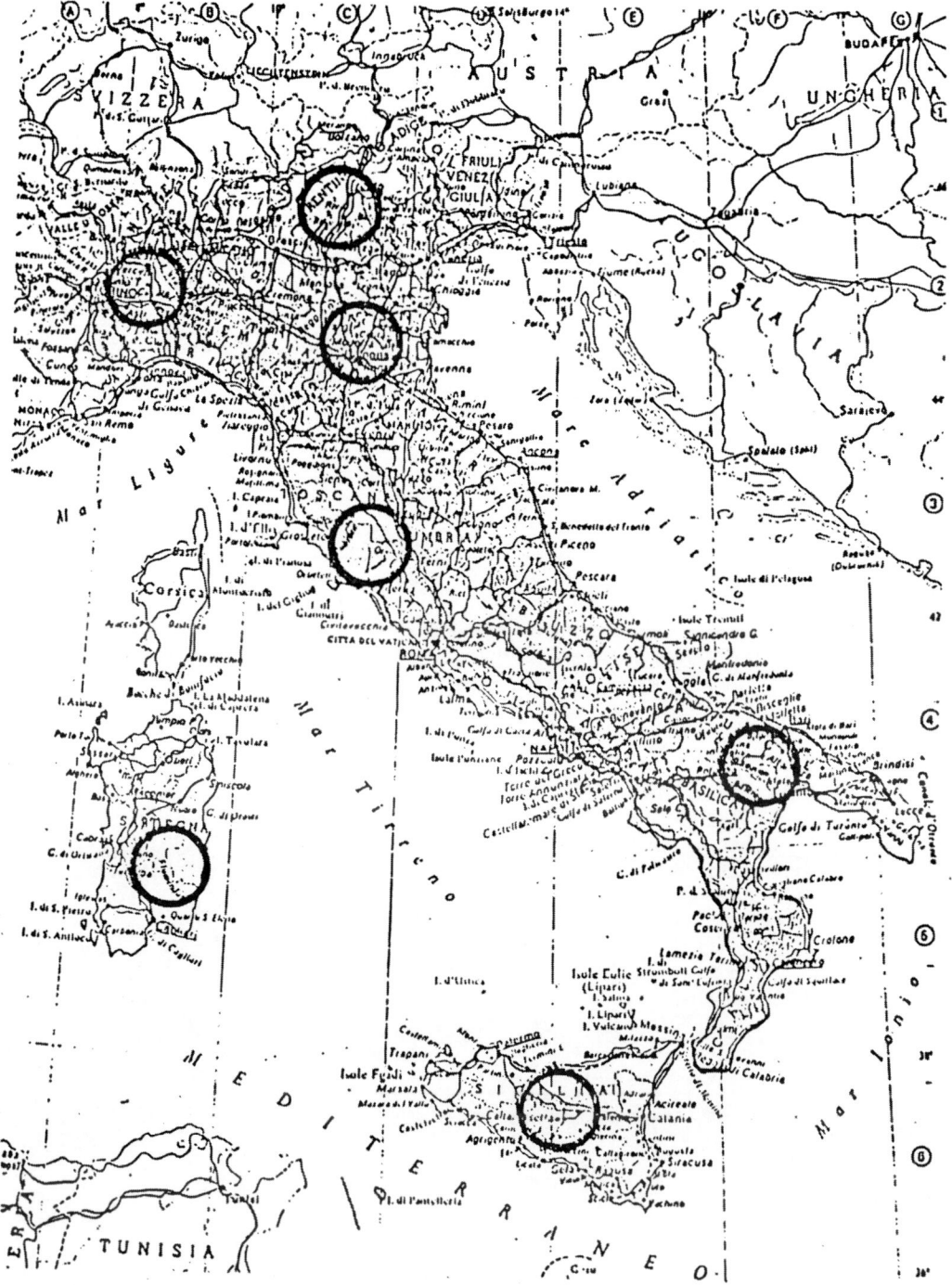

Fig. 4.1 – Possible sites for the large ELOISATRON ring. [Study made by the Italian National Institute for Geophysics (ING), 1985].

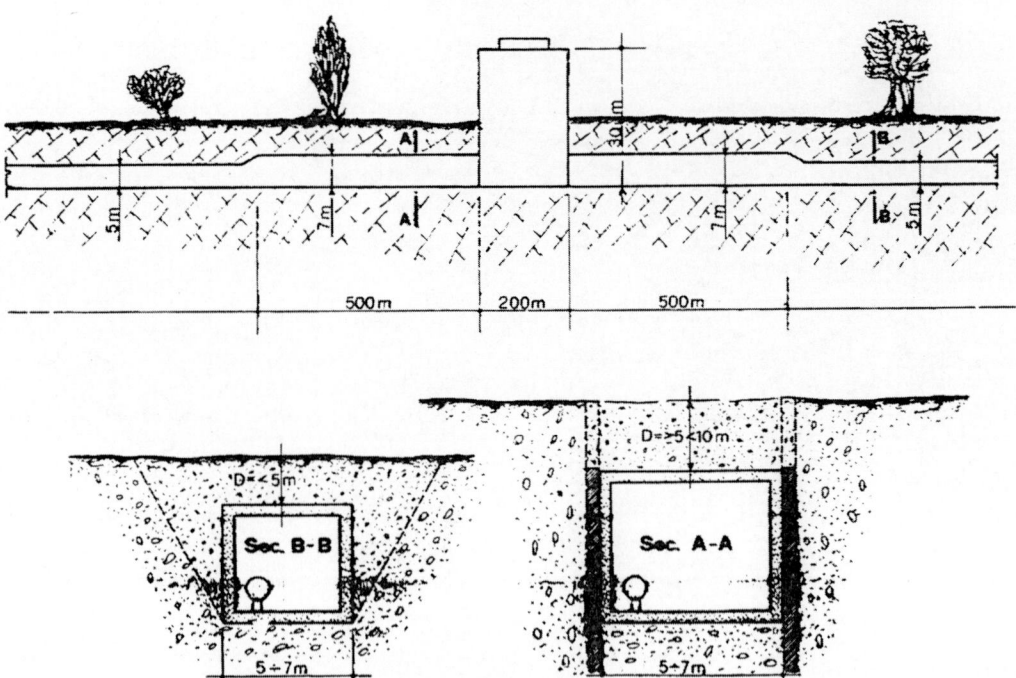

Fig. 4.2 – The schematic longitudinal section of an experimental hall and typical cross-sections for prefabricated tracts.

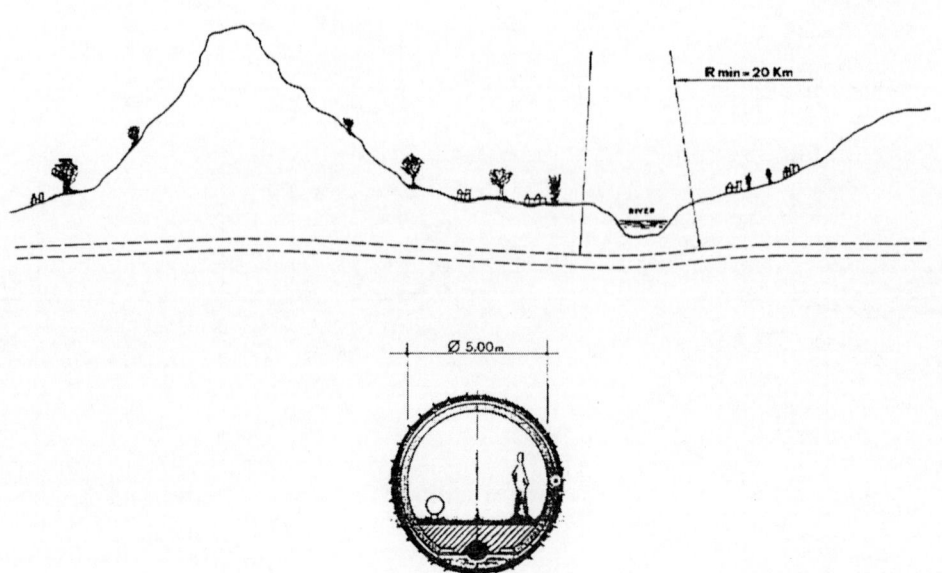

Fig. 4.3 – A typical longitudinal cross-section of the ELOISATRON tunnel.

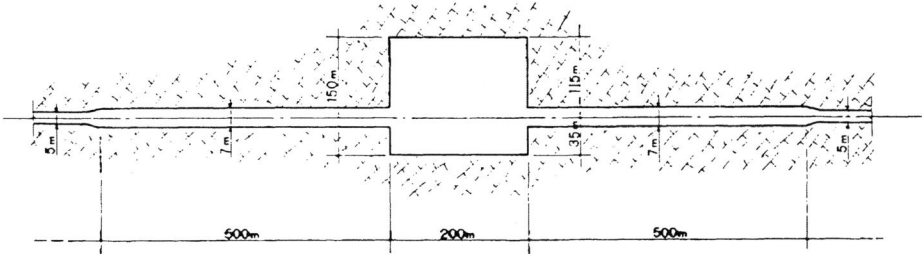

Fig. 4.4 — The schematic plan view of an experimental hall.

4.5 – Workshops

Four Workshops were held at the Ettore Majorana Centre for Scientific Culture in Erice in the years 1986-1987 as listed below:

◇ Seminar on New Techniques for Future Accelerators;

◇ Workshop on Vertex Detectors: State of the Art and Perspectives;

◇ Workshop on New Techniques for Future Accelerators: RF and Microwave Systems;

◇ Workshop on Very High Energy Proton-Proton Physics.

Other workshops are scheduled.

4.6 – The 10% Model

The 10% Model is a necessary step towards ELOISATRON. Its main data are:

◇ 1600 Dipoles; length 20 m, B ≈ 8 Tesla

◇ 400 Quadrupoles

◇ Estimated cost: 1000 GLiras

Notice that in order to build this machine only a modest development of the technologies presently available to Italian industry is required. In fact, as mentioned above, series production of 10 m long, 7T dipoles is already possible.

If the 10% ELOISATRON Model is to be ready without a long break between its start up and the end of the LEP and HERA operation, its implementation has to be done in such a way that the most efficient structures and facilities must be incorporated, wherever they are.

One obvious possibility, but by no means the only one, would be to build the 10% ELOISATRON Model in the LEP–tunnel at CERN. Fig. 4.5 shows that there would be no conflict between the LEP programme and the 10% ELOISATRON Model. Other possibilities should be investigated because the crucial point is that, no matter where the 10% ELOISATRON Model is going to be built, its implementation has to be linked with the full scale project.

473

5.– CONCEPTUAL DESIGN OF THE FULL SCALE ELOISATRON PROJECT

5.1 – Introduction

The main part of the ELOISATRON Project is a 2×100 TeV proton–proton collider of the maximum possible luminosity. With a circumference of 300 km the bending field of the collider will have to be about 10 T. The total bending length will in that case be 209.6 km/ring. The rest of the circumference will be used for focusing quadrupoles, correcting magnets and other auxiliary magnets, acceleration

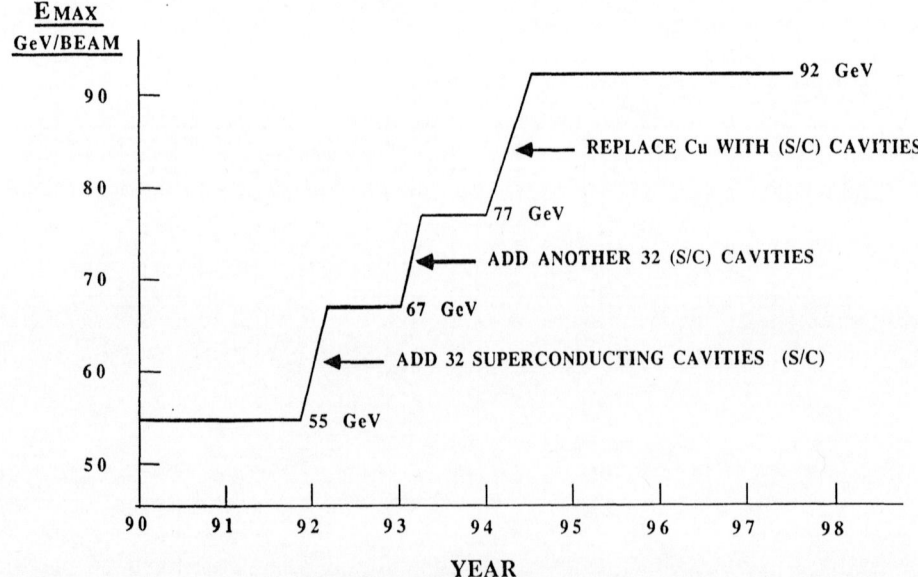

Fig. 4.5 – Possible time evolution of the LEP programme.

cavities, beam instrumentation, injection and extraction systems, and, most important of all, the insertions for the interaction regions. It is proposed to house these in two major groups, each group occupying about 15 km. The facility will thus have the shape of a race track similar to the arrangement already proposed during the LSR study about 10 years ago and also adopted in the SSC design. However, the long sides will contain some bending, so they are not quite straight. Within each group it could be envisaged to have three interaction points, as indicated in Fig. 5.1. However, other solutions should be analysed as well, for instance having many interaction points like a string of pearls with many detectors almost continuously along the 15 km.

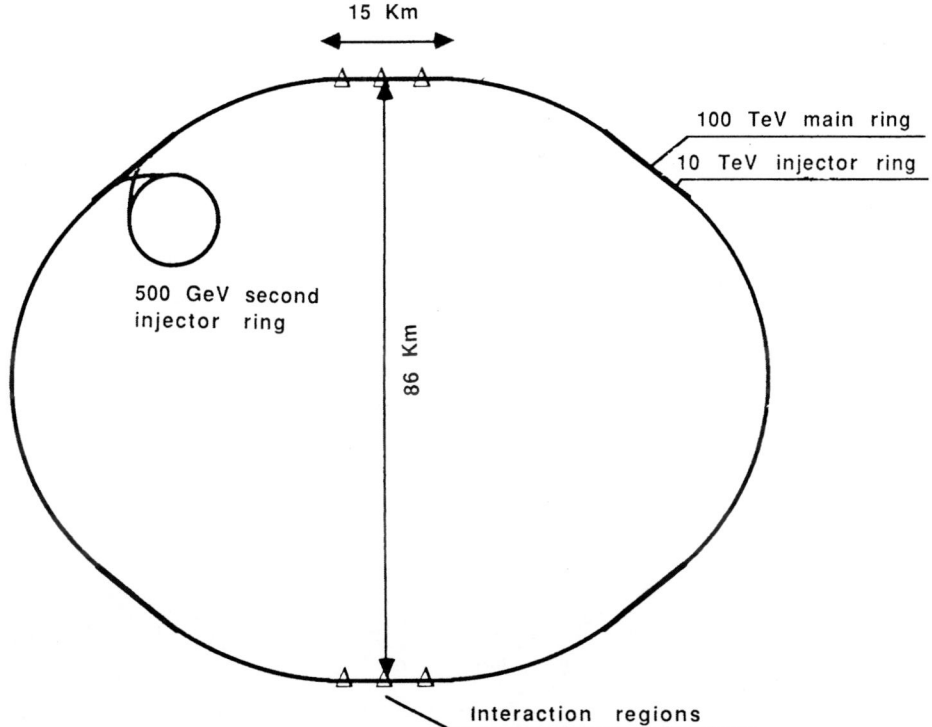

Fig. 5.1 – Possible ELOISATRON layout.

The main rings will be fed from a cascade of synchrotrons (most likely three in succession) which in turn will be fed from a linear accelerator. This is also shown schematically in Fig. 5.1. It should be noted that the last of the injector synchrotrons is, in this sketch, assumed to be placed in the same tunnel as the main rings, and therefore does not appear as a separate ring. Other arrangements are possible as will be mentioned in Section 5.4.

In the following are some comments on the main subsystems.

5.2 – Insertions

The main purpose of the experimental insertions is to transform the beam properties such that the crossing regions become very intense event sources and still maintain accessibility for detectors and otherwise make the facility flexible and convenient to use.

These are often contradictory requirements, and compromises must be found, perhaps leading to different solutions for different interaction regions. The final arrangement will depend much on detector developments over the next decade or so, for instance, on how beam–focusing systems can be integrated into detector components. It seems likely however, that under any circumstance interaction regions of very high luminosity will be needed, leading to the requirement of the strongest possible focusing near the interaction points. For instance, a β of about 1 m or less will be required to get luminosities above $10^{33} cm^{-2} s^{-1}$.

In order to get an idea whether this is feasible and practicable, simple scaling from the corresponding LHC arrangement can be performed. Only the triplets at either side of the interaction point will be considered since these are the most critical elements. The scaling gives a 75 m long triplet consisting of four 15 m long quadrupoles with 5 m between each. The gradient of the quadrupoles becomes 350 T/m. The free space from the interaction point to the first quadrupole is 30 m and the system is symmetric with respect to the interaction point. With such an arrangement the β at the crossing point can be made 1.25 m and the maximum β in the triplet becomes 8250 m.

This illustrates that a β of the order of 1 m is realistic for the interaction point, but much study is needed, both on the machine side and on the detector side, to arrive at optimum insertion designs.

5.3 – Main rings

5.3.1 – Lattice

As stated in the introduction the main rings should provide two colliding proton beams of 100 TeV each with the highest feasible luminosity in the interaction regions. A total circumference of 300 km will require a bending field of about 10 T. This seems a safe projection of what can be obtained on an industrial scale in some years, and 10 T is therefore the assumed figure for the present study. If in the future higher fields are proven feasible and practical, the total circumference can be correspondingly smaller.

Considerable further analysis and computation are needed to arrive at an optimum lattice for the machine. However, simple scaling from existing designs of similar smaller projects gives a good starting point for such analysis. An example of such scaling is given in Table 5.1. The numbers are for one ring only.

In this example the number of dipoles per period is fairly high, and if the construction techniques allow about 20 m long magnets, the more convenient number of 8 dipoles per period could be adopted.

TABLE 5.1 – LATTICE PARAMETERS

Length of period	200 m
Phase advance per period	$\pi/3$
Betatron wavelength	1200 m
Bending angle per normal period	4.7 mrad
Number of normal periods	< 1332
Number of quads per period	2
Effective length of each quad	13.6 m
Total number of quadrupoles	2664
Maximum dipole field	10 T
Number of dipoles per normal period	12
Effective dipole length	13.1 m
Total number of dipoles	15984
Bending radius	33356 m

5.3.2 – Performance evaluation and performance-related parameters

Once the energy has been fixed, the next most important parameter for the ELOISATRON Project is the luminosity given by

$$L = N_p^2 f_b/(4\pi\sigma^2) = N_p^2 f_b \gamma/(\beta^* \varepsilon)$$

Here N_p is the number of particles per bunch, f_b is the average bunch frequency, σ is the r.m.s. beam radius at the crossing points, and ε is the normalized emittance defined by

$$\varepsilon = 4\pi\gamma\sigma^2/\beta^*$$

where β^* is the beta value at the crossing point.

These equations illustrate where to put emphasis to obtain high luminosities. There are, however, a few limitations; the most important ones for the design can be listed as follows:

i) The detectors have difficulty in discriminating between events if the bunches cross more frequently than once every 25 nsec. This is not a hard limit, and may decrease. However, for the time being this is taken as giving the following limit on the bunch frequency

$$f_b \leq (25 \times 10^{-9} s)^{-1} = 40\ MHz$$

or 7.5 m between bunches.

ii) The beam–beam tune shift parameter must be kept smaller than a given number. This parameter is given by

$$\xi = N_p r_p / \varepsilon$$

where r_p is the classical proton radius. For pp interactions the total tune shift over all interaction regions should be less than 0.01, which gives

$$\xi < 0.01/6 = 0.00167$$

in the case of six interaction regions.

iii) It is in practice very difficult to make β^* very small at very high energies. For SSC at 20 TeV it is assumed that β^* can be made 0.5 m. For 100 TeV we assume optimistically that the smallest we can make it is $\beta^* \sim 1$ m.

iv) Synchrotron radiation at 100 TeV becomes a serious load on the cryogenic system of the magnets. It does not seem reasonable to assume more than say 2 W/m of the vacuum pipe, and even in that case it will be very desirable to try to trap much of this on a separately cooled screen inside the vacuum chamber.

v) The normalized emittance from present–day proton accelerators is $5\pi \times 10^{-6}$ or larger. However, since iv) above will be imposing a limit on the circulating current far below the capability of an accelerator, it is assumed that the emittance can be correspondingly reduced to say

$$\varepsilon \sim 0.75\pi \times 10^{-6}\ m$$

It is difficult to put all these conditions into simple mathematical formulae. It should also be remembered that some of the conditions are rather flexible and may change as the various technical solutions are being studied. Examples, however, have been worked out and some iterations have resulted in tentative numbers presented in Table 5.2.

The most important parameter is, as already stated, the luminosity, and one might speculate a little on how this can be increased without violating the other constraints. One simple way would be to by-pass two out of the six interaction regions during the highest luminosity runs. This would permit a reduction of emittance (if possible) to $0.5\pi \times 10^{-6}$ m and an increase in luminosity by a factor 1.5

TABLE 5.2 - SOME TENTATIVE PERFORMANCE PARAMETERS
FOR ELOISATRON

Energy per beam	100 TeV
Number of bunches	39600 per beam
β-value at interaction point	1.25 m
Normalized emittance	$0.75\,\pi \times 10^{-6}$ m
r.m.s. beam radius at inter. point	1.25×10^{-6} m
Circulating current	16.43 mA
Particles per bunch	2.56×10^9
Beam-beam tune shift	1.67×10^{-3}
(with 6 active crossings)	
Bunch spacing	25×10^{-9} s
Stored beam energy	1.623×10^9 J
Luminosity	$0.91 \times 10^{33} cm^{-2} sec^{-1}$
Energy loss per turn due to synchrotron radiation	23.34 MeV
Radiated power (per beam)	385 kW
Power per unit length of one beam	1.89 W/m
Transverse em. damping time	1.2 h

to 1.4×10^{33} $cm^{-2} sec^{-1}$. One might also speculate about increasing the circulating current, but this can only be done after the handling of the synchrotron radiation is better known (perhaps as a later improvement programme). Further, how to dump 100 TeV protons with stored energies in the gigajoules range must be studied.

5.3.3 – Aperture

The physical aperture of an accelerator constitutes a hard limit on the beam in this respect that it cannot be changed once the machine has been built. For this reason the aperture requirement must be estimated on the basis of conservative assumptions.

In the next chapter arguments will be given for an injection energy of 10 TeV. However, since it is equally likely that the final choice may be about 5 TeV the latter figure will be assumed for the aperture estimates.

In section 5.3.2 a normalized emittance of $0.75\pi \times 10^{-6}$ m was assumed for performance estimates. This may, however, be difficult to achieve, in particular in the early operational phase. For aperture determinations we therefore assume the more commonly chosen figure of $5\pi \times 10^{-6}$ m.

If it is assumed that no physics is done at energies below 2×20 TeV, the beam should not need more room than 4σ at injection. This would give $8\,\sigma$ at 20 TeV, and therefore good beam conditions from this energy upwards.

The closed–orbit deviations will depend on the quality of the magnets, the beam observation and the correction system. In the lack of more detailed information LEP values will be scaled. Results from this kind of evaluation are given in Table 5.3.

TABLE 5.3 – APERTURE EVALUATIONS

Max. beam radius (4σ) at injection	2.59 mm
Corrected closed-orbit error	$\pm$ 4 mm
Radial extent of bucket	$\pm$ 0.17 mm
Needed vertical aperture	±6.6 mm
Needed horizontal aperture	±6.8 mm
Aperture of vacuum chamber	±7.5 mm
Aperture of coil	±12.5 mm

Note should be taken of the present big uncertainty in the space needed between the vacuum chamber and the coil. This space will be needed for vacuum chamber wall, correction windings, and possibly insulation etc. to keep the vacuum chamber at $4.6\,^0K$ while the coil must be at $1.8\,^0K$ in case niobium–titanium coils are chosen.

If we want to have space for 8σ of this beam in the insertion triplets, their aperture must be made larger than the rest of the machine. However, a more realistic scenario is to assume that the low-β insertion will only be fully activated when an emittance of $0.75\,\pi \times 10^{-6}$ m has been achieved. This will then give $8\sigma = 2.5$ mm in the maximum β point at 20 TeV, i.e. the same aperture can be assumed in the insertion triplets as in the rest of the machine.

5.3.4 – Comments on some of the technical components

5.3.4.1 – Magnets

Some remarks can be made about various possibilities:

i) One can use niobium–titanium as winding material. By going to $1.8\,^0K$ one can probably reach 8–10 T. However, the highest possible field may not correspond to an optimum cost, and this will eventually have to be studied.

ii) If Nb_3Sn is chosen, 10–12 T may be reached at normal liquid helium temperature. Much development on metallurgical processes for wire production and on the thermal treatment of the insulated windings is needed to arrive at a satisfactory design. Such work goes on at many places in the world, Europe included.

iii) With the recent dramatic development of "warm" superconducting materials, there is hope that such materials will become available for accelerator magnets in the future. This must be followed closely. It must however also be remembered that such solutions will introduce new problems. For instance, the advantage of cryogenic pumping is lost and distributed pumping of very small transverse dimensions will have to be developed.

iv) Two possibilities exist for the magnetic circuits of the two rings. The conventional approach is to make the two rings magnetically independent, i.e. each magnet has its own return yoke. Another approach is to let each magnet gap take the return flux from its neighbour. This is called the "two-in-one" approach. This has two advantages. The first one is that space is saved. The LHC has adopted this version in order to arrive at a design that will fit in the LEP tunnel. The second advantage is that this solution leads to some savings in superconducting material, steel and in the cryostat.

The main disadvantage of the "two-in-one" solution is the restriction it imposes on operational flexibility. It becomes difficult to tune and operate the rings separately. This has led the SSC Design Group to choose the independent rings. The same is recommended for the ELOISATRON, in particular, since the savings mentioned above are very small compared to the cost of the whole project.

However, it should be mentioned that it may still be desirable to put the two magnets in the same cryostat. This will introduce a little operational dependance, but much less than if they were magnetically coupled.

5.3.4.2 - Radiofrequency systems

The radiofrequency system is needed for acceleration, replacement of synchrotron radiation losses, and keeping the beam stable and longitudinally bunched. Very likely a 200 MHz system based on the HERA development will be appropriate.

5.4 – Injection

In order to obtain an estimate for the aperture requirements an injection energy of 5 TeV was assumed. There are, however, reasons to investigate the possibility of a higher injection energy, for instance 10 TeV. The main advantage comes from the effect of the persistent currents in the magnet windings of the main ring. These effects become more serious the larger the ratio between maximum and minimum field. A ratio of twenty is known to be bothersome. A ratio of ten will ease the situation.

The injector will be a very low duty–cycle machine. Since it seems wasteful to keep such a machine at low temperature all the time, it is suggested to make the injector, or rather the whole injector chain, with normal magnets. With such a choice it is natural to study the possibility of having the injector in the same tunnel as the main ring, as this would require about 1 T bending magnets, a natural field for conventional magnets. Since this machine cannot be permitted to go through the interaction regions, by–passes will have to be provided. The savings in the civil engineering may nevertheless be substantial and thus justify this approach. This accelerator needs not have a fast–cycling capability.

The next ring in the injection chain can have an energy of about one twentieth of the 10 TeV injector, i.e.about 500 GeV. The ring may therefore be rather similar to the SPS at CERN. Its injector, at 25 GeV, may be rather similar to the CERN PS or the Brookhaven AGS, but the whole chain would be realized with more modern designs. The injection chain would be realized as sketched in Fig. 5.2.

The above remarks about the injection system are possibly based on unduly conservative assumptions. If the development of "warm" superconductors leads to feasible and economical magnet designs for temperatures at or above liquid nitrogen the remarks about the relative advantages of normal conducting magnets for the injector chain will no longer be valid. In such a case the 5–10 TeV injector ring will be taken out of the main-ring tunnel and made as a separate, relatively small, injector ring. The other lower energy injectors will also be made with superconducting magnets. However, even the most superficial design of such a system is premature with present-day knowledge. This may, however, change fast.

5.5 – Further Comments

Two comments are needed in connection with the problem of the full scale ELOISATRON project.

i) R&D for superconductivity at high temperature

As mentioned before, superconductivity at high temperature presents possible applications of extreme interest for High Energy Physics both in the area of detector technologies and in the production of superpowerful magnetic fields.

An intensive R&D programme in this field is therefore needed to ascertain the practical relevance of high temperature superconductivity for the full–scale ELOISATRON Project.

ii) Executive design of the full–scale ELOISATRON

A detailed study of the full-scale ELOISATRON has yet to be done. This will be better done, in fact, once the possible implications of the high temperature

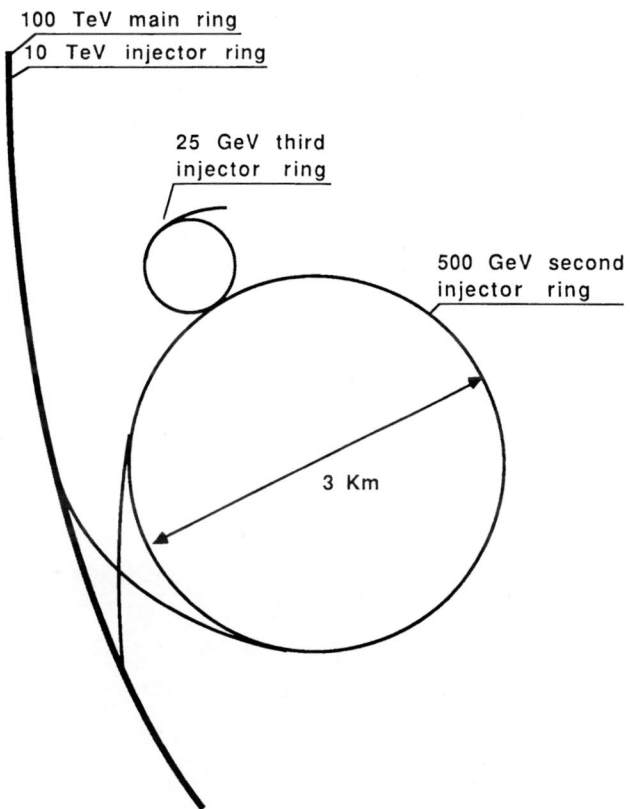

Fig. 5.2 – ELOISATRON injection.

superconductivity on the full-scale design are more fully understood.

However, it is important to notice that all the studies performed so far at the conceptual design level, point to the fact that the full-scale Eloisatron is feasible by extrapolating present technologies with intensive R&D.

Therefore, there is no point in spending a large amount of money (10 GLiras) for theoretical studies of the full-scale Project now. The R&D needed for it must be, on the contrary, an intrinsic part of the effort put in the 10% ELOISATRON Model.

6 – WHERE THE ELOISATRON PROJECT HAS BEEN PRESENTED AND DISCUSSED

The ELOISATRON Project and its different phases of development have been presented and discussed in many national and international forums, the world over.

A relevant factor in this context is the presentation of the Project to the Italian Government via the CIPE (Interministerial Committee for Economical Planning), which has undertaken a preliminary step of financing the ELOISATRON Project in the 5–years INFN plan (1984–1988) with the sum of 50 GL ($\sim$ 75 MUS$). In the following the scientific presentations are listed:

Versailles Summit Meeting (France)   June 1982

Sanremo (Italy)   **2 May 1983**
Nobel and Galileo Celebrations

Rome (Italy)   **9 May 1983**
Nobel and Galileo Celebrations

Castiglione della Pescaia (Italy)   **2 June 1983**
2nd INFN Meeting on Advanced Detectors

Erice (Italy)   **8 August 1983**
Ettore Majorana Centre for Scientific Culture
International School of Subnuclear Physics
21st Course

Como (Italy)   **2 September 1983**
3rd International Conference on
Physics in Collisions

Washington (DC,USA)   **3 October 1983**
National Science Foundation
Working Group Meeting on
High Energy Physics

Salerno (Italy)   **4 December 1984**
University of Salerno

Potenza (Italy)   **3 March 1985**
University of Potenza

Bologna (Italy)   **14 March 1985**
University of Bologna

Livermore (California, USA)   **9 April 1985**
Lawrence Livermore National Laboratory

Boulder (Colorado,USA)   **11 April 1985**
University of Colorado
Conference on World Affairs

Castelgandolfo (Italy)   **10 July 1985**
INFN National Conference

Erice (Italy)   **13 August 1985**
Ettore Majorana Centre for Scientific Culture
International School of Subnuclear Physics
23rd Course

Erice (Italy)   **3 September 1985**
Ettore Majorana Centre for Scientific Culture
International Conference on Advanced Solid
Earth Geophysics

Trieste (Italy)   **8 October 1985**
21st Congress of the SIF
(Italian Physical Society)

Geneva (Switzerland)   **11 November 1985**
CERN - Senior Staff Consultative Committee

Strasbourg (France)   **20 November 1985**
Council of Europe
Parliamentary Assembly Committee
on Science and Technology

Bologna (Italy)   **15 December 1985**
Academy of Sciences
Invited Plenary Address for
the opening of the academic year

Moscow (USSR)   **29 January 1986**
Presidium of the USSR Academy of Sciences

Cagliari (Italy)   **15 March 1986**
University of Cagliari

Geneva (Switzerland)   **19 June 1986**
CERN - RECFA

Erice (Italy)   **12 August 1986**
Ettore Majorana Centre for Scientific Culture
International School of Subnuclear Physics
24th Course

Moscow (USSR)   **15 January 1987**
Presidium of the USSR Academy of Sciences

Rome (Italy)   **9 June 1987**
INFN Symposium

7 – CONCLUSIONS

Past experience and present knowledge call for a big jump in the Energy level to be achieved in Subnuclear Physics.

The most effective approach towards this goal is the ELOISATRON Project which, as shown in table 1.1, has already realized a number of important steps.

Let us summarize a few basic achievements:

  i) a powerful collaboration between Industry and Research Laboratories;
 ii) the first series production (at an industrial level) of technically highly developed superconducting magnets;
iii) an intensive R&D programme for detectors (LAA) to be used in MultiTeV Colliders.

The next step is the 10% ELOISATRON Model, whose final goal is the most powerful superaccelerator in the world: (100+100) TeV.

The activity being pursued by the Italian Industries (Ansaldo, LMI, Zanon, Laben), in close collaboration with the most advanced Research Centres, will permit the involvement of Italy in the next big step towards the MultiTeV SuperCollider and its Physics achievements. In this context it should be emphasized that a few more steps in the superconducting magnet are required in order to achieve the best performances with present technologies. In fact we have been able to reach a level of technological accomplishment which is higher than the one aimed at. In other words, the 10% ELOISATRON model can be started practically with our present level of technological goals, already achieved (i.e.superconducting magnets 10 metres long with 8 Tesla).

During the construction of the 10% model an intensive R&D programme should be conducted so as to follow the spectacular developments of high temperature Superconductivity. However, even if this new, high technologically advanced field, does not produce effective results in the next five years, the R&D using present–day technologies would grant the feasibility of the full–scale ELOISATRON Project. In fact, 10 Tesla magnets are certainly within reach if the R&D programme in standard Superconductivity is carried out in the next few years.

The ELOISATRON Project, realized along the outlined programme of collaboration between Research Centres and Industries together with the effective participation of Third World Scientists, represents the greatest challenge for Subnuclear Physics to push forward our present knowledge in Science and Technology with a highly qualified component of International Solidarity. This is very much needed in order to halt the present trend of a continuously increasing technological gap between the NORTH and the SOUTH.

8 – THE AUTHORS

The authors of the work reported are:

K.S. Adhao, K.B. Aglawe, C. Aglietta, P.B. Agrawal, Z. Ahsan, K. Alberini, A. Ali, M. Ali, D. Allasia, B. Alpat, Q. An, F. Anghinolfi, F. Anselmo, D. Antreasyan, G. Anzivino, A. Arefiev, M. Arneodo, F. Artemi, F. Arzarello, A.V. Asole, B.H. Aurade, P.V.K.S. Baba, V.B. Badhe, G. Badino, S.D. Bagchi, M.K. Bagde, K.R. Bagree, D.R. Bahekar, Bai Fengning, Bai Jingzhi, U.D.N. Bajpai, J.S. Bakare, B.A. Bambah, A.M. Band, G. Bandyopadhyay, S. Bandyopadhyay, H. Banerjee, S.K. Banerjee, A.N. Bannore, G. Barbagli, E. Barberio, G. Bari, T. Barillari, M. Basile, M. Bassetti, B. Basu, D. Basu, R. Battiston, A. Bazzani, U. Becker, S.G. Bedi, S.Y. Bedore, L. Bellagamba, R.B. Belorkar, N.G. Belsare, J Berbiers, J. Berdugo, B.V. Berezinski F. Bergsma, R. Bertin, A. Bertin, V.S. Bhagde, D.R. Bhakre, P.D. Bhalerao, K.S. Bhamra, M.K. Bhan, N.C. Bhattacharya, T. Bhattacharya, S.N. Bhongle, S.S. Bijwe, R.K. Bock, S.M. Borikar, M.S. Borkar, D. Boscherini, E. Boschi, R. Bouclier, E. Braun, G. Bruni, P. Bruni, R. Bruzzese, J. Burger, D.K. Burghate, N. Cabibbo, Cai Yixing, Cai Zhiguo, Cai Xu Dong, U. Camerini, M. Campbell, Cao Wei, L. Caputi, G. Cara Romeo, R. Casaccia, C. Castagnoli, A. Castellina, G. Castellini, A. Castelvetri, H. Castro, S.M. Chandekar, G. Charpak, U.K. Chaturvedi, Chen Bofei, Chen Chaoqiang, Chen Huijun, Chen Jianpu, Chen Limin, Chen Pengri, Chen Senyu, Chen Siyu, M. Cherian, M. Chiarini, S.P. Chiney, N.H. Christ, L. Cifarelli, F. Cindolo, G. Cini, F. Ciralli, M. Civita, A. Clare, Coa Zang, E. Colavita, A. Contin, S. Cooper, M. Costa, Cui Huachuam, Cui Ruyu, Cui Shiangzong,

Cui X, Cui Xiao-Ying, Dai Guiliang, V.L. Dadykin, G. D'Alì, C. D'Ambrosio, S. D'Auria, C.K. Damle, D.N. Darane, M. Dardo, T.K. Das, M. Dasgupta, B. Datta, B.K. Datta, R. Datta, U.R. Datta, S. Datta Majumdar, D. Davasumony, M. De Felici, S. De Gennaro, S. De Pasquale, C. Del Papa, Deng Shusen, Deng Yanping, S.L. Deoghare, V.S. Deogoankar, V. de Sabbata, R. De Salvo, G. Di Sciascio, D.W. Deshkar, A.A. Deshmukh, B.T. Deshmukh, D.T. Deshmukh, K.C Deshmukh, K.G. Deshmukh, R.M. Deshmukh, V.R. Deshmukh, A.M. Deshpande, G.T. Deshpande, M.V. Deshpande, P.D. Deshpande, J.A. Dhanaraj, S.B. Dhannasare, G.B. Dhanokar, R.V. Dharaskar, M.V. Dhargave, M.T. Dharmadhikari, T.N. Dhobe, I. Ding, P. Diodati, O. Di Rosa, Y. Dong, M. Dutta, Du Wenfu, Du Xijiu, K. Eckhard, T. Ekelöf, R.B Elgira, M. Enorini, J.P. Fabre, F.L. Fabbri, Fang Cheng, Fang Shouxian, R.N. Faustov, S. Ferrara, L. Ferrario, M.I. Ferrero, F. Fiori, E. Fishbach, D. Fong, P. Ford, F. Frasconi, M. Fukushima, W. Fulgione, P.M. Gadkari, V.J. Gaikwad, Y. Galaktionov, S. Galassini, P. Galeotti, H. Galvez, B. Gao, Gao Cuishan, Gao Wei, Gao Wenxiu, M. Gasperini, J. Gaudaen, K.N. Gawande, S.G. Gawande, S.W. Gawande, S. Gerardi, M.H. Ghate, S.M. Ghatole, G.A. Ghike, S.G. Ghosh, S.N. Ghosh, G.T. Gillies, P. Giusti, K. Goebel, V.G. Gokhale, V.M. Gokhale, R. Goldoni, Gong Zhu Fang, N. Gopalkrishnan, Y. Gorodkov, M. Gourdin, C. Grinnel, M. Guanziroli, Gu Mengping, Gu Shudi, Gu Songhua, Guang Fengqi, Guo Weixin, Guo Yanan, Guo Zhiyuan, V.D. Gupta, T. Gys, E.S. Hafen, Han Qian, P. Haridas, E. Heijne, He Mengjia, S. Hellman, A.M. Henriques, G. Herten, W.B. Hirurkar, Hu Jialiu, Hu Jiawei, Hu Lidong, Huang Kaixi, G. Hu, Huang Li, Huang Nan, Huang Yinzhi, P.P. Huddar, S.I. Hussain, G. Iacobucci, M.F. Ingalgi, A.G. Ingle, B.G. Ingle, R.P. Ingole, M. Italiani, E. Jagel, J. Jain, S.K. Jain, R.K. Jambhorka, P. Jarron, B. Jeckelmann, P. Jenni, A.S. Joalikar, L. Jones, Jia Yanwu, Jiang Chunhua, Jiang Yanling, Jin Qingshou, Jin Youkai, D.N. Joharapurkar, K. Johnsen, V.B. Johri, S.V. Joshi, D.K. Kadu, J.M. Kale, M.G. Kale, Y. Kamyshkov, T.M. Karadei, U. Kasper, D.M. Kataria, S.A.H. Kbatil, Ke Songbai, D.G. Kelkar, G.M. Kesharwani, G.S. Khadekar, M.T. Khadse, G.A.N. Khan, M.Q. Khan, R.A. Khan, S. Khan, A.T. Khandare, M.L. Khandelwal, R.B. Kharade, T.M. Kharade, R.B. Kharat, U.U. Kharat, S.M. Khodragade, N.W. Khodragade, J. Kirkby, M.S. Kishirsagar, T. Kitamura, G.E. Kocharov, M.S. Korde, E.V. Korolkova, P.V. Kortchaguin, V.B. Kortchaguin, A.P. Kowale, B.P. Kowale, W. Krischer, Kuang Benlin, V.A. Kudryavtsev, A.B. Kulkarni, D.K. Kulkarni, R.S. Kulkarni, V.G. Kumar, G.N. Kumbhare, I. Laakso, G.R. Labde, G. La Commare, S.A. Ladhake, G. Landi, Lang Pengfei, R.B. Lanjewar, H. Larsen, G. Laurenti, T.D. Lee, A.R. Leo, M. Leo, H. Leuz, G. Levi, V. Liakhovets, Li Binfang, Li Bomin, Li Dashi, Li Dezhong, Li Guanglin, Li Jia, Li Jiacai, Li Jiang Feng, Li Jianguo, Li Jin, Li Qun, Li Youmeng, Lin Qitang, Ling Dachun, L. Linnsen, L. Lisowski, Liu Dekang, Liu Diankui, Liu Shiyao, Liu Rui, Liu Tai, Liu Yabo, Liu Yucheng, C. Liverani, G.C. Lokre,

L. Lone, Long Shida, Lu Changguo, Lu Jihua, Lu Weida, G. Luches, D. Luckey, H.S Lunge, Luo Yingxiong, Luo Zihua, Ma Jimao, G. Maccarrone, J.K. Madhugiri, R.K.W. Mahalle, S.N. Mahapatra, M.A. Mahure, C.K. Majumdar, M. Malavasi, B.N. Mandal, S.V. Manmode, J.L. Manputra, Mao Huishun, P. Marchesini, A. Margotti, A. Marino, M. Marino, A.S. Markov, S. Marmi, S. Maselli, T. Massam, C.G. Mathe, T. Matsuda, D. Mattern, B.R. Maurya, P. Mazzanti, G. Meddler, C.A. Mehare, K.H. Meier, V.N. Melnikov, D. Mencarini, V.R. Metkar, R. Meunier, Mi Jianlin, Mi Rongsheng, Mi Yong, K.D. Mishra, R.B. Mishra, Y. Mir, I. Mirza, B. Mitra, R.S. Mishra, S.B. Mishra, V.G. Miskin, S.G. Modak, A.K. Modi, G. Mohanty, S.C. Moholkar, C. Morello, K. Morey, J. Moromisato, M. Morpurgo, F. Motta, R. Mount, A. Mukherjee, S. Mukherjee, S. Mukherjee, K.V. Muley, V.D. Muley, A.M. Mundhada, S. Muthiah, A.B. Nadange, P.T. Naiekar, S.Y. Nandanwar, R. Nania, R.Y. Narkhede, S.G. Nasare, V. Nassisi, L. Natrajan, G. Navarra, K.A.N. Nerkar, H. Newman, M. Niaz, B.R. Nikhade, F.M. Nirwan, V. O'Shea, H.P. Paar, R.S. Pagrut, J.P. Pal, M. Pal, L. Palermo, S. Palermo, E. Pallante, F. Palmonari, Pan Huibao, N.K. Pande, S.N. Pandey, T.N. Pandey, U.S. Pandey, Pang Jiabiao, K.G. Pangarkar, G. Papini, S.C. Paranjape, Y.R. Paranjape, D.V. Parwate, G. Pasotti, C.B. Patil, M.B. Patil, P.D. Patil, S.P. Patil, B.A. Patki, A. Paul, K.V. Pawar, S.P. Pawar, P. Pelfer, A.W. Pendharkar, L. Periale, C. Peroni, E. Perotto, V. Peskov, A. Pitas, V. Pjidaev, I.A. Pless, V. Plyaskin, G. Pocci, R.B. Pode, H.M. Poharkar, M.E. Pol, C.N. Potdar, N.B. Potdar, V.K. Pratapwar, G. Prisco, M. Puglisi, H.S. Pundkar, I. Qazi, Qian Zhuming, J.M. Qian, S. Qian, Qiao Jimin, Qin Jiu, Qiu Hong, N. Qureshi, V.D. Raghatate, S.B. Raichur, V.B. Rajurkar, M. Rama Mohan Mao, A.D. Rangari, S.P. Rant, V.A. Ratate, G.D. Rathod, A.D. Raut, D.R. Raut, Ren Wenbin, Ren Zhong Liang, M. Ricci, G. Rinaldi, F. Rivera, H.A. Rizvi, C. Rizzuto, F. Rohrbach, P. Rotelli, R. Roy, Ruan Tongze, V.G. Ryassny, O.G. Ryazhskaya, H. Rykaszewski, O. Saavedra, N. Sacchetti, E. Saletan, P. Salvadori, D.L. Samudralwar, G. Sartorelli, S. Sarwar, N.N. Saste, R.V. Satpute, F. Sauli, W. Scandale, M. Schioppa, J. Schipper, H. Schöenbacher, R. Scigoki, M. Scioni, R. Scrimaglio, J. Seguinot, J. Seixas, S.H. Selukar, D. Sen, M. Sen Gupta, N.D. Sen Gupta, G. Servizi, F. Sgamma, Sha Hanying, A. Shah, D.I Shahare, D. Shambroom, V. Sharan, V.P. Shelest, Shen Boahua, Shen Junpeng, Shen Tianji, V. Shevchenko, Shi Yinsheng, Z. Shi, V.S. Shiramwar, S.P. Shrinkahnde, Shu Weisan, Shu Yuede, J.B. Shukla, E. Shumilov, Shyampati, S. Siboni, M.D. Singh, R.D. Singh, T. Singh, S.K. Singh, J. Singh Rana, K.P. Sinha, P.V. Soadekar, V.S. Soitkar, G. Soliani, D.N. Soni, M. Spadoni, L. Sportelli, R.S.L. Srivastav, A. Staiano, A. Starobinsky, M. Steuer, Sun Songlan, Sun Yuelin, S.N. Supe, G. Susinno, S.P. Swami,

A.A. Syed, V.A. Tabhane, Y. Takano, V.K. Tale, U.P. Talockin, R.K. Talwekar, T.M. Tamhane, H. Tanaka, Tang Baoping, Tang Cheng, Tang Shuming, M.S. Tapi, E. Tarkowski, S. Tazzari, M.T. Teli, T.J. Telranhe, B.R. Tembhurne, Teng Kejian, A.E. Terraneo, V.A. Thakhare, S.C. Thaker, A.W. Thakre, R.V. Thakre, V.M. Thatte, E. Thomas, Tian Fang, S.D. Tikar, R. Timellini, S.C.C. Ting, B.P. Tiwari, S.A. Tiwari, G. Torelli, G.C. Trinchero, V.S. Tumram, G. Turchetti, D.U. Umak, S. Valenti, P. Vallania, V. Vasileyev, G. Venturi, S. Verasani, S. Vernetto, I. Vetliski, K. Vidya Gaikwad, P. Vikas, F. Villa, A. Vishnoi, A. Vitale, E. Von Goeler, L. Votano, D.G. Wadke, S.D. Wadke, R.V. Waghamare, W.U. Waghmode, S.H. Walkey, A.B. Walkhade, Wang Bosi, Wang Dianchen, Wang Feng, Wang Hengfeng, Wan Hengjiu, Wang Jin, Wang Linlin, Wang Linzhou, Wang Man, Wang Shuhong, Wang Shuqin, Wang Taijie, Wang Yunyong, Z. Wang, Z.M. Wang, P.B Wani, A.M. Wankhede, P.C. Wankhede, T.V. Warhekar, T.S. Wasnik, Wei Kaiyu, Wei Zhuangzi, R. Weinstein, T. Wenaus, H. Wenninger, C. Werner, M. White, M. Wigdoff, R. Wigmans, C. Williams, M. Willutzky, Wu Baizhi, Wu Chuanchou, Wu Mian, Wu Shou Xiang, Wu Shu Lan, Wu Weimin, Wu Wentai, Wu Yingzhi, Wu Zhendong, B. Wyslouch, Xi Beixing, Xi Jiwei, Xiao Liangrong, Xie Jialin, Xie Peipei, Xie Xiaoxi, Xie Yenyi, Xu Jianming, Xu Shaowang, Xu Xiaokang, Xu Zhongxiong, Xue Jingxuan, Xue Shengtian, R. Yagannathan, V.F. Yakushev, Yan Binshan, Yan Jie, Yan Taixuan, Yan Wuguang, Yang Dajiang, Yang Guang, Yang Yun, Yao Xiaoguang, C.H. Ye, Q.H. Ye, S.H. Yeh, A.K. Yelne, S.V. Yenkar, J.M. You, T. Ypsilantis, Yin Zhaosheng, Yu Qinchang, Yu Qingfu, Yu Zhenze, Yu Zhongqian, Yue Wen, T.M. Zachariah, A. Zallo, G. Zanetti, G.T. Zatsepin, M. Zeng, Zhang Boachang, Zhang Chuang, Zhang Huashun, Zhang Pinghua, Zhang Yan, Zhang Yingping, Zhang Zhenjiu, Zhao Bin, Zhao Duo, Zhao Fuguang, Zhao Jijiu, Zhao Jingbao, Zhao Weiren, Zhao Yongjie, Zheng Guoqing, Zheng Linsheng, Zheng Shuchen, Zheng Youchun, Zheng Zhipeng, Zhong Shicai, Zhou Jikang, Zhou Shu, Zhou Xiaoguang, Zhou Yuehua, Zhu Fuquan, Zhu Liangsheng, Zhu Renquan, Zhu Ren-Yaun, Zhu Shangen, Zhu Yucan, A. Zia, A. Zichichi and K. Zographos.

DISCUSSION

– *Jonker:*

(1) It was not clear to me what you meant by the limit of "one event per bunch crossing". Does this also include the very forward minimum bias events? And where are the cutoff in p_t or angle?

(2) What are the advantages in doing missing energy physics instead of missing p_t physics? Actually I am worried that you will be swamped by many elastic events that will deposit large amounts of energy in the forward region.

– *Zichichi:*

(1) The key point is that the average number of events per bunch crossing, $< n >$, must be chosen to be <u>ONE</u>, if the MISSING ENERGY is to be used as a signature to identify "new physics". In fact the future of our field is going to depend on our ability to be sure that when a "missing particle" is observed, the "missing" signature is genuine. Not fake.

The limit of $< n > = 1$ is coming from the luminosity, the time between the bunch crossings and the total cross-section. This is why the time between bunch crossings has to be as short as possible. In fact the total cross-section is given by God, and the Luminosity we want it to be high. Note: $< n > = 1$ corresponds to a probability of 26% to have more than one event per crossing.

(2) In answer to your second question, <u>First:</u> missing energy and missing p_T are two basic quantities with different physical implications. <u>Second:</u> if you want to study multiTeV phenomenology you cannot ignore any range of phase space. The very forward range - where the leading effects dominate - must be under severe control. It is perfectly clear to me that the rate of events and the radiation level are tremendous. This means that the challenge we are confronted with is great.

– *Pitman:*

(1) You plan to use one device for vertex detection as well as tracking, which you plan to collapse in size. Does this pose limitations if you intend to use the device for other measurements such as dE/dx or momentum measurements? What sort of dimensions do you expect to shrink the detector to?

(2) There is probably some research being done on replacing completely drift chamber technology. Can you comment on this?

– *Zichichi:*

(1) The vertex detector technology is based on the following requirements:

  (a) unambiguous and precise determination of primary and secondary vertices;
  (b) tracking back to the vertex;
  (c) measuring the charge and multiplicities;
  (d) defining the topologies of the events;
  (e) very high spatial accuracy, double track resolution and redundancy because of large multiplicities, large number of particles in jets, and high momenta;
  (f) particle identification.

Superconducting, high–field magnets should be used in conjunction with a very high precision vertex detector. This is essential if we want to collapse the dimensions. Particle multiplicities are expected to be of the order of a few hundred per event at 200 TeV.

(2) Research and development is currently under way in the following areas:

  (a) high resolution scintillating fibres;
  (b) Gallium Arsenide and Silicon microstrip detectors;
  (c) multidrift gas proportional tubes;
  (d) superconducting magnets of order 6 Tesla (though the Ansaldo people in Genova are aiming at 12 Tesla).

There is another idea developed by Hans Sens, to use a superconducting current on the beam pipe which will allow a free space with a toroidal field of 10 Tesla. At present no one knows if a very powerful magnetic field can be achieved. It is only if such a field can become real that the problem of "dimensional collapse" in the detector becomes effective. The real answer will come in 5 years. In the meantime we will be constructing prototypes. I invite you to either join our group or to come in 5 years and get the answer.

– *Iacovacci:*

(1) Could you say a little more about the LAA project needed for flavour tagging in jets?

(2) What is the expected ratio of gluon and quark jets produced? This ratio should be dominated by gluon jets. You showed a plot of the charged multiplicity distributions which had two curves. Do these represent the quark and gluon fragmentation functions?

– Zichichi:

(1) The answer to your question will come from Monte Carlo studies based on the present knowledge of QCD. A group of physicists is at work within the LAA project. All predictions thus far are based on approximations which may or not be correct. The correct answer could come only from lattice-QCD calculations for dynamical processes. This would however require dedicated supercomputers which do not currently exist. No theorist on the planet could be taken seriously if he were to make predictions in the multi-TeV range based on present knowledge.

(2) The two curves you refer to are a dotted one based on scaling from data without leading effect and a solid curve showing the predictions based on leading particle effect. The data include those observed at the CERN Collider by UA5 at the so far highest possible energies, and data from the ISR, assuming the leading particle effect. Our group is the only one which has introduced in these studies the leading particle effect. The physics of the hadronic collisions is not given by the nominal energy. Each interaction must be studied individually because of the continuous range of energies available. This is shown in the curve. If the leading effect is correctly accounted for, the charged multiplicity in pp collisions does not deviate from that of e^+e^- collisions. We have demonstrated that e^+e^- physics is degenerate with proton-proton physics provided that you pick up in each interaction the right energy available. The distributions presented are based on these considerations. I expect that our predictions will be right at higher energies, where so far no data are available.

– Blazey:

Hermeticity at these energies is difficult. Given that many discoveries have been made with high resolution fixed target experiments and that you can minimize the effects or hermeticity in these experiments, are there any plans to include fixed target experiments?

– Zichichi:

The answer is no. I disagree with you. In the past the detectors have not been planned for a comprehensive physics programme. For example, UA1 was optimised for the discovery of the W and Z^0, not for hermeticity. It is a miracle that some hermeticity considerations could come out, given that there is no particle detection below 20^0. And this is the most hermetic gadget constructed so far. UA2 is clearly far from being hermetic. For the first time a detector will be built, such that all the details will be considered, not leaving holes for particles to escape. The *R&D* for this detector will be based on the best possible way to measure energy, momentum, particle identity and everything imaginable. The detector should be planned with the machine. Although I am not a machine physicist, this time I will follow the design of the machine closely. The answer to your question is that hermeticity has to be taken seriously, by far more than ever before.

– Heusch:

The preceeding remark is very important and this question of hermeticity was considered recently at an SSC workshop at Berkeley on detector possibilities. The problem of up to which rapidities you can use calorimetry, an important consideration for the study of missing energy and momentum physics, is technically challenging. You are right in pushing $R\&D$ in this area at this time. Up to which rapidities do you expect to cover with the hermetic detector?

– Zichichi:

The answer is 10 Mrad, not rapidity. The limiting factor from the detector development is determined by 10 Mrad, which the detectors and the associated microelectronics must withstand. As you know, 1 Mrad is equivalent to a year of running with luminosity 10^{32} at 10 cm distance from the beam.

– Heusch:

But the problem is that right now, even for the 10% Eloisatron, proposals must be made to obtain financial support. The SSC workshop concluded that a rapidity larger than 3.5–4, is technically unfeasible at this time. This makes the study of missing p_T physics difficult as a rapidity of at least 4.5 is required.

– Zichichi:

The LAA project is interested in pushing the limits. You should not be asking the classical questions to the LAA project. Instead you should be interested in the hardware side, to see what kind of instruments can be developed for the extreme limits. There are many interesting ideas floating around. It is not a question of meetings, workshops or seminars, but of $R\&D$ activity to overcome the present difficulties. They are not in terms of theoretical predictions. They are in terms of Radiation Resistance.

– Heusch:

If the chairman permits me, there is another parameter which must be thought in a quantitative way, namely the tradeoff between energy and luminosity. Energy is not the whole thing.

– Zichichi:

I disagree fully with you. Energy is the most important parameter.

– Heusch:

Many things we want to measure depend upon whether we see 1 or 20 events. Quantitative extrapolations of what may be observed in the future have been made by Eischten, Hincliffe, Lane and Quigg, in a massive opus for the SSC. For many effects it is just as well to have 10 × the luminosity as 10 × the energy. There you can see the tradeoff between energy and luminosity. And there is the question of what your detector can do at very small angles. I am not challenging what you say but one has to look at both these parameters, the energy and the luminosity.

– Zichichi:

You should change your approach. These are well known and old remarks. The future of subnuclear physics is to push the Energy frontier. The higher the better.

– Bolton:

How far are you away from achieving the 10-100 times improvement you need in radiation hardness that will be needed in order to operate a detector in the high luminosity environment of the Eloisatron?

– Zichichi:

The current state of microelectronics is that they can withstand 0.1 Mrad. For the detectors this limit is 1 Mrad. Our aim is to better this by at least one order of magnitude. This requires significant *R&D*.

– Miele:

Could you give a few details about the 10% E-model, the time scale involved and the physics it could produce?

– Zichichi:

The political leaders will ultimately determine whether the project will be realised. From our physics point of view the 10% project should be done as soon as possible. The physics programme we have is very exciting. Physicists should not waste time arguing about details such as energy versus luminosity. This energy step is needed before achieving the real goal of 100 TeV on 100 TeV. The LAA project should produce in the next five years many interesting results in the right direction.

– Reynoldson:

(1) Who made the study for choosing the possible sites for the Eloisatron and what aspects were considered in choosing particular sites?

(2) Will the machine run under populated areas?

– Zichichi:

(1) The Institute of Geophysics, the most competent organisation for determining the geophysical structure of the country, has come up with seven possible sites for the installation of the 300 Km ring.
(2) The seven sites selected do not pose problems to populated areas.

– Windey:

You only talked about radically new detectors. Why not about radically new acceleration techniques?

– Zichichi:

This is a very interesting remark. A new acceleration technique has recently been considered. This project which has been financed, would aim for 1 GeV/m. This would be fantastic. One of the aims of the Eloisatron is to encourage research into new frontiers. The 10% model however cannot wait for these developments and should be started as soon as possible. The past has shown, for instance in the case of superconductivity, that it takes the order of 20 years before a new idea becomes a working technique. We know how to construct the 10% model and industry is prepared to do it. It only depends on the agreement of a group of countries, no matter where the site will be. Physics is translational invariant.

– Morelli:

I think that there are two ways of performing fundamental high energy physics:
(1) searching for new particles and phenomena at very high energies, and
(2) to study known phenomena with high precision to be sensitive to the fine structure of the physics.

Can the Eloisatron give information about the second point or is there the danger that the project can fall into the category of what you called "non-approved" or "not completely financed projects"?

– Zichichi:

I have done high precision experiments like the g–2 measurement of the muon, QED at Frascati and high energy experiments. I agree with you that both paths should be pursued. Both are important though if a choice has to be made, there is no question that energy is the most important parameter. However, you must be careful not to leave any gaps in the energy range. The *J* could have been observed 10 years earlier at ADONE, if the energy had been pushed few percent higher than

3 GeV. At the time, I advocated the importance of going to higher energies.

– Liu:

Where do you get your financial support from and in particular I am interested in whether the Chinese government will support the Eloisatron?

– Zichichi:

Where we get the money from is a problem which does not concern me. The problem is in the hands of the political leaders who support the project. I can guarantee that there are some very good ones who support the project.

- Ferrara:

If the 10% Eloisatron is going to be built at CERN, will it interfere with the future upgrading of LEP?

– Zichichi:

The answer is no; it does not interfere. I can show you the relevant graph concerning the time schedule, which was presented here at Erice in a previous workshop by John Thresher. The graph shows that there are many periods in which the tunnel is free for access. This does not imply that the project will be done at CERN but this shows that there is no incompatibility. The 10% model can be built anywhere and we are open to collaboration.

–Ganchev:

Are there any eastern European institutes and laboratories in the Eloisatron project?

– Zichichi:

This question is open and as mentioned earlier is something to be considered by political leaders, not by physicists. We have been trying to attract the interest of many countries for this project. Eloisatron should be a world-wide project, open to all countries that want to contribute intellectually and financially. Even if we start with a few countries, others could join later, as the project is not closed to any contribution. So far we have a large contingent from China, India and the Soviet Union. The task of the physicists is to promote the project, demonstrate that it can be done and to form a large international scientific group to participate. So if you want to join, you're welcome.

– Pitman:

Do you have some idea about the bunch separation, which you would hope to achieve?

– Zichichi:

No. This is an important part of the machine development studies. However I have convinced the machine physicists that by no means should they stop at luminosities of 10^{32} and at 100 ns bunch separation. The design of the machine should not be based on the present capabilities of experimental detectors. Rather, the machine should be designed with the highest possible luminosity and energy. The only parameter of course is money, not the impossibility of doing experiments. The crucial point is *R&D* for Detectors. As an example, the pion discrimination technique using the so-called preshower method, developed by my group 25 years ago and called more correctly "the early shower" method, is still being used. New ideas are needed. Many of these new ideas are being implemented in the LAA project. Accelerator physicists should be encouraged not to listen to experimentalists, who think in the old "classical" way.

– Schuler:

My question concerns the theory. You will probably get a very high background from QCD which has to be calculated. Are there any new ideas for how the calculation could be done? I doubt that the cascade Monte Carlo based on the parton model will be sufficient.

– Zichichi:

The question of theoretical QCD predictions is linked to the power of the supercomputers available. The dream is to be able to do Lattice-QCD calculation for dynamical processes, as I have already mentioned. At present Lattice-QCD is able to produce "static" quantities only. And the QCD predictions are based on "perturbative" calculations. The theoretical component of LAA has two parts. One is the attempt to make the dream real. The other, waiting for the dream to become real, is to perform the best possible, the most powerful QCD perturbative approach.

THE GLORIOUS DAYS OF PHYSICS

A COMMEMORATIVE LECTURE SERIES IN HONOUR OF

Professor Martin Deutsch and Professor Herman Feshbach

of M.I.T., USA

Chien–Shiung Wu

Columbia University

NY, USA

I. Martin Deutsch: Pioneer in Nuclear Physics, Discoverer of Positronium

Summary: Martin Deutsch discovered Positronium, the fundamental "Atom" of quantum electrodynamics in 1951. He initiated and developed the study of the magnetic lens–type spectrometer in 1941, one of the standard instruments of nuclear physics. He promoted the uses of scintillation detectors for nuclear radiations; thus greatly improving the timing and efficiency of the measurements of nuclear radiations (see fig. 1 and fig. 2).

Starting in 1947, he was one of the pioneers to measure "an angular correlation between successive gamma–rays". Also, he was the first to measure the direction-polarization correlation of such radiations in 1950. Martin Deutsch was the first

The Superworld II
Edited by A. Zichichi
Plenum Press, New York, 1990

Fig. 1. Martin Deutsch in his lab (M.I.T. 1952). Discovery of the Positronium.

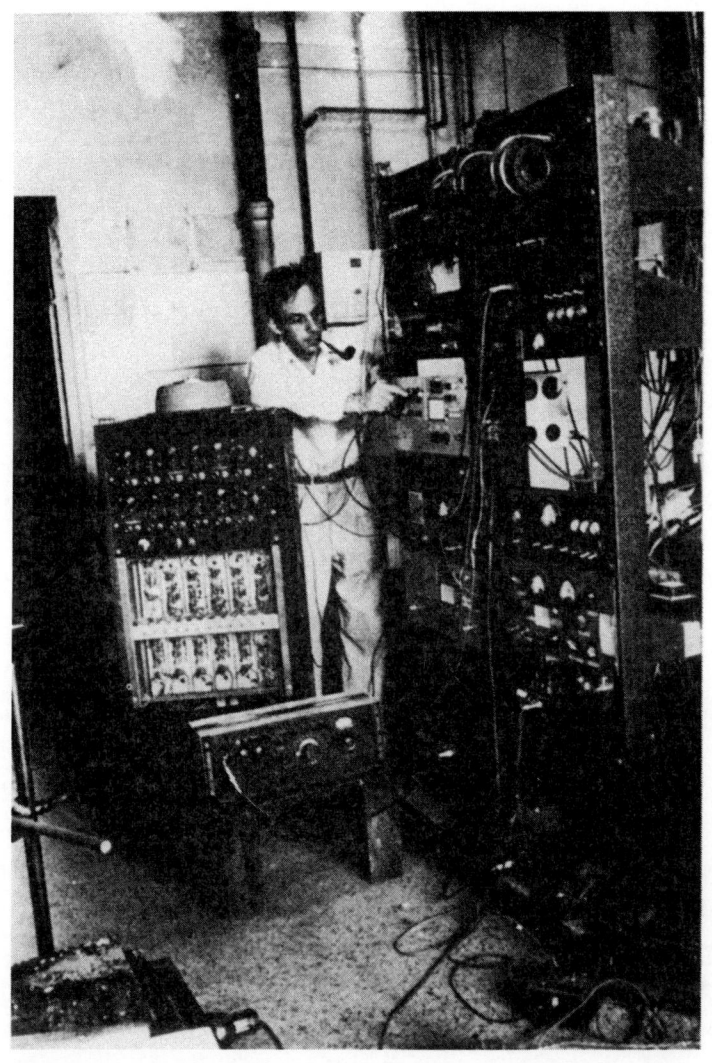

Fig. 2. Martin Deutsch checks out electronics on positronium (M.I.T. 1951)

to point out the Larmor precession of those correlations which could be used to determine the magnetic moments of nuclear states.

Through the use of all of these techniques on a number of the most significant radioactive nuclei, Martin Deutsch established many of the primary methods for physical determinations of the decay schemes of nuclear energy levels.

Martin has also made significant contributions in Elementary Particle Physics as in the study of the Compton Scattering of Very High Energies and the Search of the Identity and Characteristics of the Solar Neutrinos.

I am deeply honored to be invited here to participate in this celebration of the 25th anniversary of the world renowned and unique learning centre: "The International School of Subnuclear Physics in Erice". I wish to take this opportunity to thank the Director, Professor A. Zichichi, his colleagues and the friendly people of the town of Erice who have made all these intimate and tranquil scientific gatherings possible. We also wish the International Centre "Many Happy Returns of the Day!"

In this lecture session, we are commemorating first, the glorious days of Nuclear Physics developed in the past half century, primarily from those magnificent contributions by Professors Martin Deutsch and Herman Feshbach of M.I.T. in America. Professor Deutsch and Feshbach are not only known for their outstading contributions in research and discovery, but they are also known as the most admired and beloved teachers and comrades in the physics community for their warm and personable friendship.

Martin Deutsch was born in Vienna in 1917. Both his parents were physicians actively engaged in research and teaching. He received seven years of secondary education in a gymnasium (the equivalent to high school in America) which was very academically oriented. Professor Viki Weisskopf also graduated from that school but on the year that Martin Deutsch entered. However, because of the difference in their ages they never met back in Vienna. Not until almost twenty years later, when Martin first met Viki at Los Alamos National Laboratory, Viki produced a photograph of the teacher in the Viennese high school who had told each of them that they would end up in the gutter!

In February 1934, Martin followed his family when they moved to Zurich, Switzerland, rather abruptly as he had participated in the resistance movement to the Fascist seizure of power in Austria. He acquired a federal matriculation, the Swiss secondary school diploma. In October of 1935, Martin accompanied his mother to the United States and entered M.I.T. with permission to try for a bachelor's degree in two years. This he did in 1937 and then he continued on to obtain his Ph.D. in 1941, at the age of 24.

When Martin entered M.I.T. in the mid–thirties, the strength of its physics department was in theory and was led by Professor John Slater. The strongest experimental group, according to Martin, was in optics and was led by George Harrison and his spectroscopy laboratory. Martin still remembers very well that it was Harrison's junior course in atomic physics that had the greatest influence on him. Harrison's course was not very well organized; it consisted of a series of scientific anecdotes and vignettes; a style of teaching that somehow usually greatly kindles a young man's fancy and enthusiasm.

Futhermore, at that time the field of Nuclear Physics was in its relative infancy and there weren't many solidly established theories to study or experimental techniques to learn; but there was indeed much to initiate and do! Martin had always been a hard worker and also exhibited lots of creativity and organizing power. Among graduate students at M.I.T., there were several unusually talented co-workers such as Sanborn C. Brown, experts on electrical discharge in gases which was useful knowledge for fabricating gas counters. Arthur Roberts, who was an electronic circuit expert familiar with fast coincidence techniques. At that time, Dr. Stanley Livingston, who has just gotten his Ph.D. from the University of California at Berkeley, also joined M.I.T. and built a new cyclotron to produce radioactive sources. Then Martin himself plunged into his thesis work in demonstration of a lens–type magnetic spectrometer for β–ray spectroscopy in his own laboratory. All these ambitious plans were made ready to begin unravelling radioactive decay schemes on an assembly line basis. However, from 1941 on, with the threat of a second world war on, many of his original co–workers at M.I.T. were gradually called away to work on urgent defense work. Martin happened to be an enemy alien (as Austria was under German occupation) and therefore was only asked to work on some less urgent war works. From 1944–46, he was appointed as a scientist to work on the Manhattan Project at Los Alamos with some of the most famous nuclear physicists such as N. Bohr, R. Oppenheimer, E. Fermi and E. Segré. To work closely with E. Segré on nuclear fission physics was undoubtedly a very exciting experience for Martin.

In 1946, when Martin returned to M.I.T. after the war, he immediately resumed his ambitious work in Pure Nuclear Physics. He also was promoted to the rank of assistant professor of physics and became an associate professor in 1949. In 1953 he was appointed to full professor when he was only thirty–six years old and was also made Chairman of the Directing Committee for the Nuclear Physics Lab. His research was focussed on the properties and detection of gamma–radiations and also on the gamma–ray angular distributions and angular correlations.

In 1947, Deutsch and his students were the first ones to work on angular correlation of successive gamma–ray quanta (see fig. 3 and fig. 4).

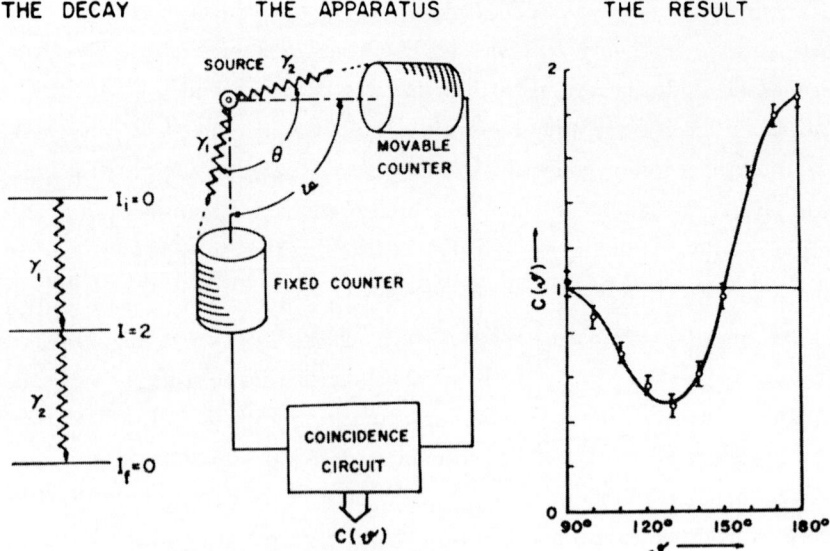

Fig. 3. Angular Correlation Arrangement.

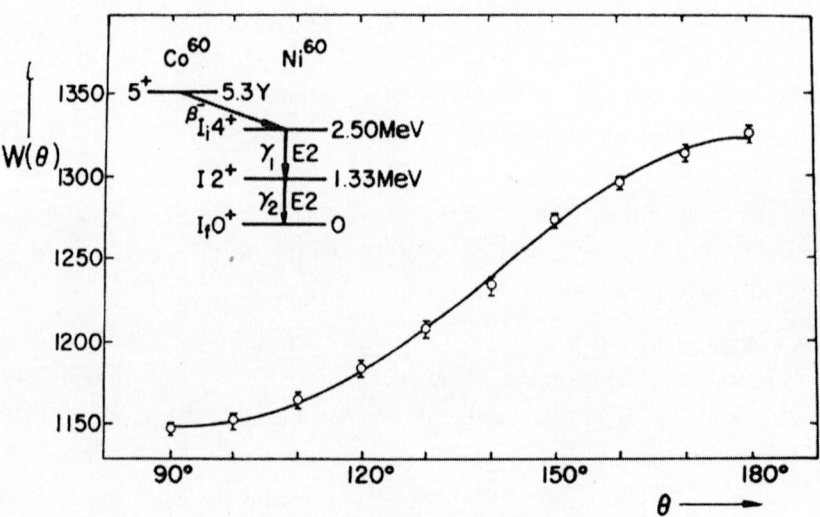

Fig. 4. Angular Correlation of ^{60}CO.

In 1948, he introduced to the physicists of the United States the scintillation counter of Kallman. Kallman was the first one to develop these fast and high resolution scintillation counters in Germany during the second world war. Martin learned the news through returning veterans but he was able to show the unusual power of the new counter by carrying out a series of hither–to almost impossible correlation and high energy resolution measurements (see fig. 1 and fig. 2). The immediate acceptance of this major tool of modern physics was due in no small part to Deutsch's powerful promotion.

In 1948, Deutsch also showed that the parity of nuclear states could be determined by angular correlation–polarization measurements (see fig. 5 and fig. 6). The tremendous vitality then aroused in the field of nuclear spectroscopy must be credited in a large part to Deutsch's innovations and dedication. Figure 7 by Alburger shows a level scheme of Pb^{206} by nuclear spectroscopy.

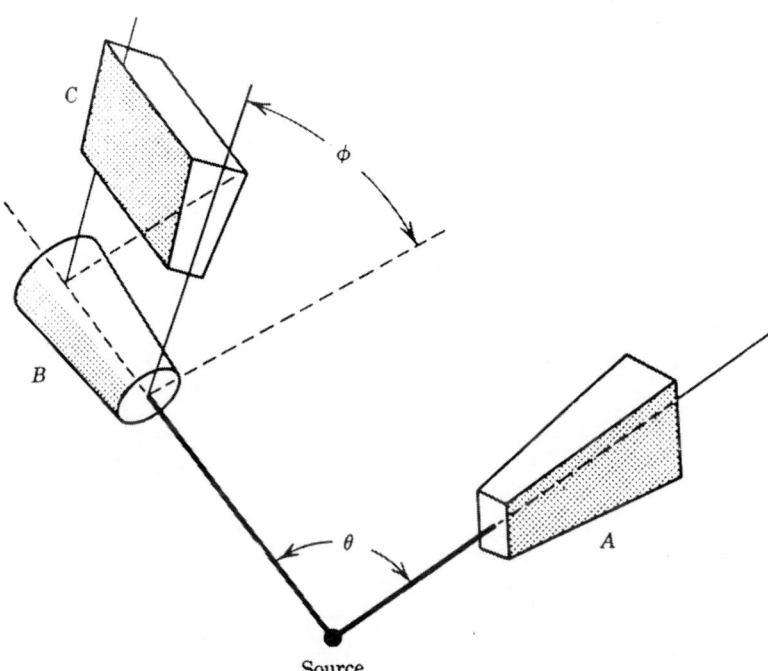

Fig. 5. Schematic diagram of a polarimeter. A, B, C are Scintillation Crystals.

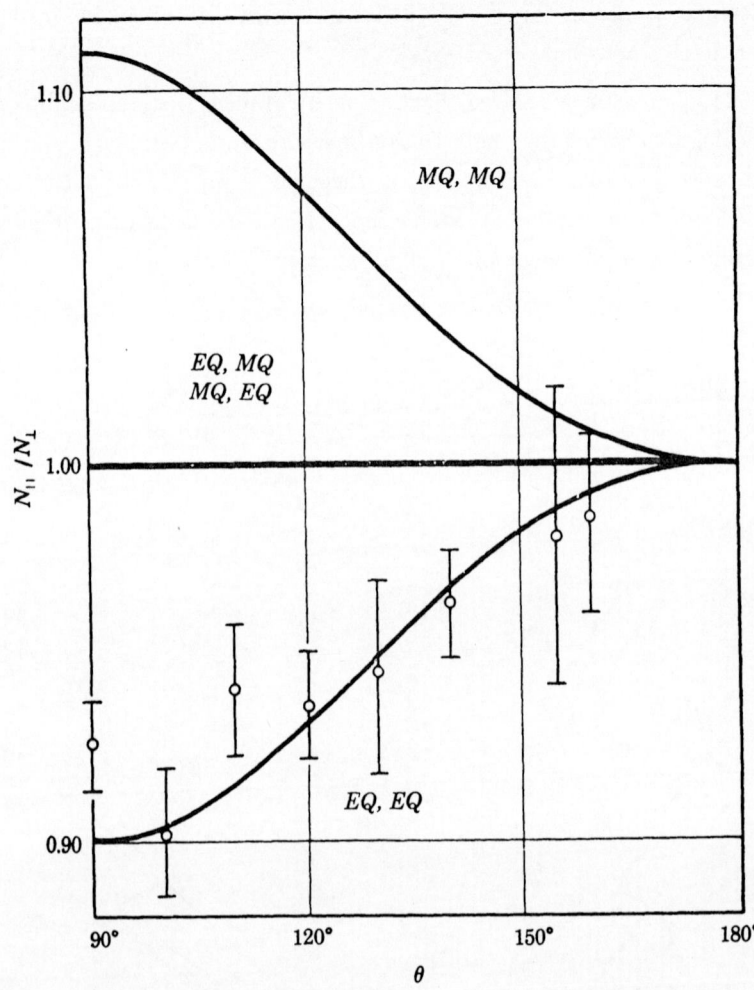

Fig. 6. The ratio $N_\parallel/N_\perp$ of the triple coincidence rates in the two extreme positions $\phi = 0$ and $\phi = \frac{\pi}{2}$ of figure 5 as a function of θ for the 1.12 MeV, 0.80 MeV cascade of Ti^{46} following beta–decay of Sc^{46}.

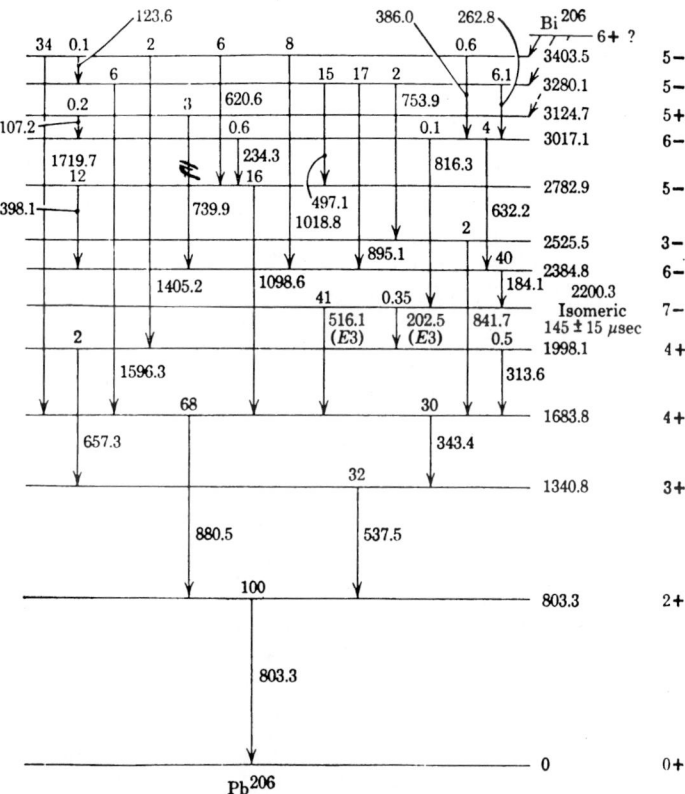

Fig. 7. Level Scheme of Pb^{206} following K-capture of Bi^{206} by Alburger.

In 1951, Martin Deutsch discovered positronium, the fundamental atom of quantum electrodynamics. During the few years, he carried out pioneering experiments on its properties (see figs. 8, 9 and 10).

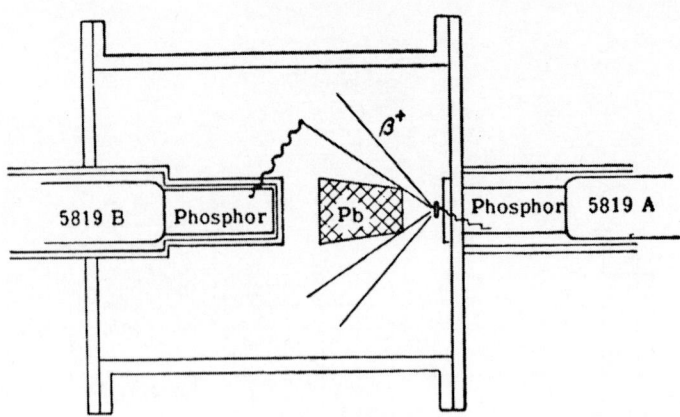

Fig. 8. Apparatus to measure the lifetime of positrons in gases (Deutsch M., Progress in Nuclear Physics 3, 141. Academic Press, New York, N.Y. 1953).

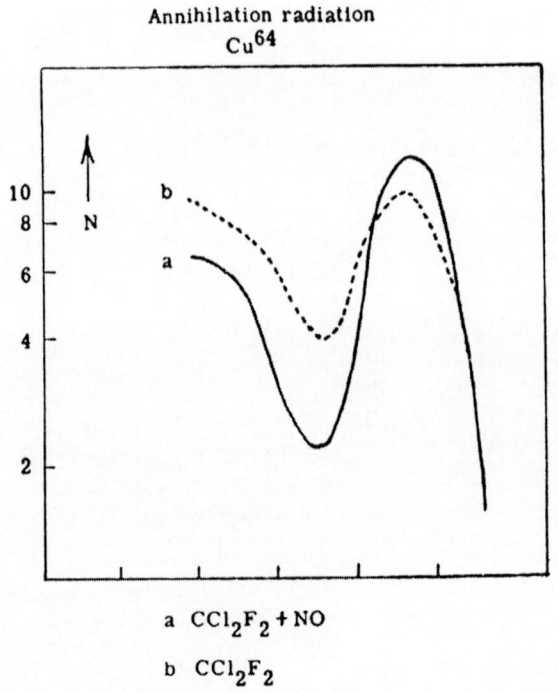

a $CCl_2F_2 + NO$

b CCl_2F_2

Fig. 9. Annihilation Spectrum for positrons, stopping in CCl_2F_2 (Freon) with and without the addition of NO showing how one can distinguish singlet and triplet annihilation [Deutsch, M. and Dulit, E., Phys. Rev. 84, 601 (1951)].

And these experiments, confirmed and extended during the past thirty years constitute <u>the strongest proof of the validity of quantum electrodynamics</u>. We are so glad today to have Professor Deutsch himself here to tell us the exciting story of positronium. The investigation of positronium by Martin was so thorough that it constitutes almost a complete chapter in Modern Physics. It is an almost unprecedent achievement in recent time for a single investigator to so fully unravel such a detailed and significant field during an important period. It is no wonder that Martin Deutsch was awarded the most prestigious "Rumford Prize in Physics" in 1985 at the centennial commemorative celebration of Niels Bohr's 100th birthday at the American Academy of Arts and Sciences in Boston, MA.

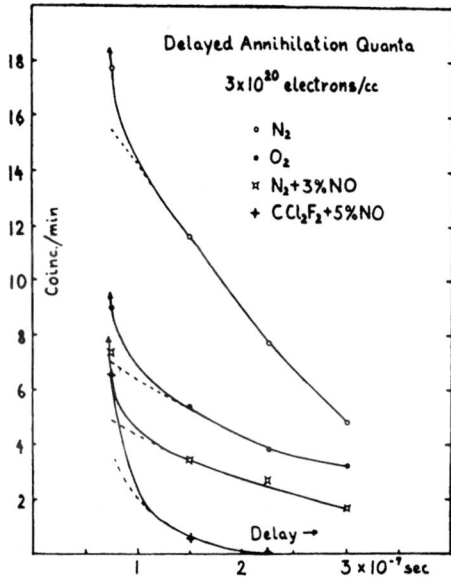

Fig. 10. Decay Curves of Positrons in several gases. The dotted lines are corrected for time resolution of the instrument.

It is important to point out that Martin's pioneering experimental investigation was actually carried out ahead of the appearance of the theoretical understanding of the formation of positronium. It was by the study of the formation of positronium in various gases which led Martin to realize that there are two different lifetimes of positronium. One is the singlet state of S=0 which is the short life state known as para–positronium: $(e^- \uparrow\downarrow e^+)$ decaying by two photons and with a short lifetime. The other one is the S=1 state known as the ortho–positronium $(e^- \uparrow\uparrow e^+)$ decaying by the three photon process with a long life time (see fig. 10).

Professor Deutsch also directed a very active and productive high energy nuclear and particle physics research laboratory at M.I.T. for several decades.

<u>He is truly a man of all seasons.</u>

II. Herman Feshbach

Professor Hans A. Bethe once quoted Victor Weisskopf's saying in a public talk at CERN, Geneva in 1964 and again in a slightly different form in his essay in the volume of "The Significance of Science" that is:

"Human Existence is based upon two Pillars;
Compassion and Knowledge
Compassion without knowledge is ineffective
knowledge without compassion is inhuman."

Of course, Viki Weisskopf is a great example of the combination of knowledge and compassion. However, now we would like to introduce and celebrate that 70th birthday of one of Viki's closest collaborators in nuclear physics at M.I.T., Professor Herman Feshbach. Herman joined M.I.T.'s famous theoretical group and sought and contributed to knowledge on one of the nuclear models described as a transparent and partially absorbing sphere. It is well known as "The cloudy–crystal–ball" model.

This model reproduced many properties of the nucleus and the agreements were very remarkable. At present, Professor Feshbach is the Institute Professor of M.I.T. and also served as the President of the American Academy of Arts and Sciences from 1982–1986. He had also served as the Chairman of the Division of Nuclear Physics of the APS; the Head of the Physics Department at M.I.T. and the Director of the Center for Theoretical Physics for ten years. In Chinese, we have a saying that "a person who is exceptionally capable, usually bears excessive responsibilities in all ways".

While Professor Feshbach was shouldering all these major responsibilities, he was also appointed to be the Editor of numerous major journals of the APS. Additionally, he found the time to co–author two monumental theoretical treatises or textbooks. One was the well known Methods of Theoretical Physics with Philip M. Morse. The other was the widely read and praised Theoretical Nuclear Physics with Amos de Shalit. The latter book was described as a fine achievement for presenting the nuclear research of the past half a century in a concise, systematic and lucid form. The second volume is still anxiously anticipated.

Professor Feshbach has also devoted a great deal of his energy and time to the problems of international security and human rights solutions.

The advent of the nuclear age has profoundly changed the nature of warfare and strategic relationship of the superpowers. The scientific developments that produced this revolution in warfare also created a new and special relationship between the scientific community and governments. Thus scientists became major participants in the formulation of military and foreign policies reflecting the new

514

technology. Over the years, many United States scientists have served in important governmental positions and as influential advisors on these matters.

In recent years, the National Academy of Sciences in the United States has been deeply concerned with the problem of international security and arms control. A committee on International Security and Arms Control was organized to reduce the threat of nuclear war. The members who are appointed to this committee are as expert a group of individuals as one could assemble to consider these critical problems. I am glad to mention here that Professor Feshbach has been appointed to this committee as he has devoted so much tought and work to these issues.

There have been some improvements in the international rights problems lately. I was very happy when I returned to the States from the Far East in May 1987 to read in the April issue of "Physics Today" and the July issue of the "Bulletin of the Academy of Arts and Sciences" that Professor Feshbach and a small group of presidents of private colleges in the States have had a very pleasant and meaningful meeting with Dr. Andrei Sakharov and his family in his apartment in Moscow (see fig. 11).

The encouraging news is that Dr. Sakharov looked very well in spite of the many indignities he had suffered in Gorky and that he is vigorously pursuing human rights issues as well as recent rapid developments in particle physics and cosmology. While detained in Gorky, he published no less than six scientific reports and also kept track of human right cases. Here is a recent picture (fig. 11) of Herman with Sakharov's family in Sakharov's apartment (1986).

Fig. 11. A recent picture of Herman with Sakharov's family in Sakharov's Moscow apartment (Winter of 1986).

THE DISCOVERY OF POSITRONIUM

Martin Deutsch

Massachusetts Institute of Technology
Cambridge, Mass. U.S.A.

INTRODUCTION

These are my personal recollections of a sequence of events which might prove nostalgic for the old or inspiring for the young. In parts my tale may even be cautionary. I do not guarantee rigorous precision of historical detail but a reasonable effort has been made at verification from contemporary records. Nor will I expand on the impact of the ideas and results which grew from the events in my story during the intervening forty years. For those who would like to learn the physics involved, I recommend the collection of contributions in "Adventures in Experimental Physics" vol. 4, Bogdan Maglich, Ed. For more recent reviews of positronium see, e.g. Berko (1980) and Rich (1981).

It would be logical to assume that a search for positronium was stimulated by the discovery and interpretation of the Lamb shift in 1947, but that was not true, at least in my case. The importance of positronium for quantum electrodynamics did not strike me until the events chronicled in this lecture had taken place and we started to plan the serious measurement of the fine structure splitting. My journal for 1948, 1949,1950, and the first months of 1951 contains only brief, dry references to the "positron experiment", while many other projects get detailed attention. Enrico Fermi, who, with his graduate student, Richard Garwin was searching for positronium at that time probably saw it in a clearer perspective than I.

When I returned from the war time Manhattan Project to academic work in 1946, I undertook to develop a program in radioactive spectroscopy at M.I.T. which I had started some years earlier. Many of the methods which became routine tools and remained so for decades had to be invented or at least implemented for the first time: Scintillation counters for the detection of beta and gamma rays; gamma ray angular correlation and polarisation measurements for determining quantum numbers of nuclear states; the perturbation of these correlations by atomic or externally applied fields to measure nuclear moments; the delayed-coincidence technique to measure short lifetimes. Those are some of the methods to which our group made significant contributions. Others, e.g. nuclear resonance scattering of gamma rays, we attempted, but failed.

Of course we tried to apply our newly acquired skills to experiments which promised to throw light on some important problems of the day, such as models of nuclear structure or the theory of beta decay but the direction of our work was largely driven by the possibilities of the new techniques and some of it was simply part of the cottage industry of characterising radioactive decay schemes.

BACKGROUND

In a brief note concerning a search for weak, high energy gamma rays from the decay of ^{64}Cu, published in the fall of 1947 (Deutsch 1947), I reported observing the annihilation of positrons in flight as a source of background radiation and wrote "Further experiments in this connection are planned." Thus, when a student, James W. Shearer approached me in the spring of 1948 and asked me for a first problem "involving a fundamental process", I suggested the rate of positron annihilation. There was no reason to doubt the validity of the straight-forward calculation given, e.g. in Heitler's book (Heitler 1944). To lowest order in α, the cross section of an electron for positron annihilation into two photons is given by

$$(1) \quad \sigma = \pi r_0^2 ((\gamma^2+4\gamma+1)\ln(\gamma+(\gamma^2-1)^{1/2})/(\gamma^2-1)-(\gamma+3)/(\gamma^2-1)^{-1/2})$$

Since two photon annihilation is only allowed in singlet collisions, (1) contains a factor of 0.25 from spin averaging. The equation is valid in the plane wave approximation and is applicable to measurements of σ performed at energies where distortions due to the atomic coulomb field are negligible. Shearer performed such experiments as his second project with ambiguous results (Shearer 1951). Our techniques were not yet ready. As far as I know, the first significant measurement of σ was performed later by my student Henry Kendall (1956). In 1948 my intention was to involve Shearer in the development of our new techniques for measuring sub-microsecond lifetimes using scintillation counters. I suggested that he attempt to measure the lifetime of positrons in matter, rather than the cross section.

At low energies, (1) becomes

$$(1') \quad \sigma = \pi r_0^2 c/v$$

where v is the relative velocity of positron and electron. The decay rate λ of positrons in matter should therefore be independent of their velocities and depend only on the average electron density n_e:

$$(2) \quad \lambda = \pi r_0^2 c n_e$$

In reality, we expect the great majority of the positrons to be slowed to thermal energy or at least to atomic energies and the wave functions are surely not plane waves. The coulomb attraction of the electrons increases the rate over that in (2) but the repulsion by the nucleus renders the inner electrons essentially inaccesible for annihilation. We could readily measure life times of the order of 10 nanoseconds, corresponding to the electron in a gas at atmospheric pressure. I expected that the experiments would throw light on the slowing down mechanism for positrons and on what now become known as positron chemistry. Information concerning the annihilation process itself could come only if we encountered unexpected simplifications - which we did ,in the end.

Note that this program is an example of an extremely bad research proposal: The quantity to be measured is buried in complex corrections and it can be calculated with certainty much better than the experimental precision even if a value could be extracted from the very murky experiment. Note also, however, that in that golden age I did not have to write a proposal.

THE FIRST EXPERIMENTS

Jim Shearer and I began by constructing our apparatus, which was to
become the prototype for most experiments in this field for almost a
quarter of a century. It consisted of a ^{22}Na positron source mounted in a
gas tight vessel, about 50cm long and two scintillation detectors for
gamma rays. Scintillation counters for gamma rays were a fairly new
technique. There were no commercial sources of scintillating plastics. The
best scintillator with sufficiently fast response for our purpose was
anthracene which had been introduced by P.R. Bell. Although I cannot find
any record, I believe that it was the material used by us. ^{22}Na emits a
positron accompanied by the emission of a 1.3 Mev gamma ray. One
scintillation counter placed very near the source, recorded the nuclear
gamma ray, providing a time signal for the birth of the positron. The
positron is stopped in the gas and the annihilation radiation is detected
in a second scintillation counter. A lead cylinder blocks gamma rays
reaching the annihilation detector from the source or its support. The
time interval between the pulses from the two counters is the life time of
the positron involved.

In those primitive times there were no time analysers to be bought;
no time-to-height converters; no multichannel pulse height analysers. A
commercial single discriminator was a vacuum tube circuit with its own
power supply, of a size comparable to a whole NIM crate. Amplifiers and
pulse shapers were home made. The time distributions had to be measured
one point at a time by changing the length of signal cable in one arm of
the circuit.

Our first results confirmed that the mean life of positrons in a gas
was indeed in the expected range. In order to establish that we understood
the situation, we proceded to show that the decay rate is proportional to
gas pressure, chosing air as the cheapest and safest gas available. To our
surprise, this proved not to be true. The life time at high pressures was
longer than expected. In addition, the decay did not seem to follow a
simple exponential, although this was difficult to establish with our
limited instrumentation. My first thought was that we were losing
positrons by diffusion to the walls of the vessel. This effect would
shorten the life time at lower pressures. We changed the size of the
vessel. At one point I replaced it with a meteorological balloon which
provided a convenient moveable boundary. The result was negative. At that
point I pushed the positron problem out of my mental focus, exhibiting
another example of bad judgement.

In November, 1948 I left M.I.T. for a two month stay in Stockholm, to
work with my friend Kai Siegbahn on radioactive decay schemes. It was my
first post-war return to Europe; my first as a professional physicist or,
for that matter, as an adult. To put the pace of work in those days into
perspective, note that we published at least three papers reporting
different results obtained during my stay in Stockholm. I gave seminar
talks, in Stockholm, in Copenhagen and in Zurich. In Zurich I spoke in the
lecture hall where, a few years earlier, I had listened to my first
University physics course, while the world of my childhood was
disintegrating. My lectures dealt with our new methods in nuclear
spectroscopy. They were well received. Even Wolfgang Pauli, sitting in the
first row, said nothing more insulting than "Das wissen wir auch"- this we
know,too. There was nothing to cause my attention to return to the murky
subject of positron annihilation.

I returned to Cambridge in January, 1949. There had been no progress
in my absence. I could think of only two possible explanations of our
observation that the measured decay rate was not proportional to gas
pressure:

If thermalisation of positrons was much slower than we had estimated, then the annihilation rate would decrease with time because fewer electrons are accessible to the slower positrons. The decay curve would not be exponential. Our measurement was confined to a relatively short range of time intervals and we were sampling different parts of the non - exponential curve at different pressures. Thus our observation could be explained.

The other possible explanation assumed that positrons could be captured in some stable atomic states in which the electron density available for annihilation is much lower than the average in the gas. We were aware of John Wheeler's prize-winning 1946 paper on bound positron-electron systems (the name "positronium" was suggested by A.E. Ruark in 1945). We did not know about the more detailed theoretical work concerning the formation and the annihilation rate of the several states being carried out at the time of our first experiments. I considered it very unlikely that free positronium would remain undisturbed in a gas at atmospheric pressure and believed that we were dealing with a more complicated system bound to a gas atom. We presented our results in a ten minute paper at the February,1949 New York meeting of the American Physical Society (Shearer 1949). The abstract states: " The tendency for the annihilation cross section to decrease wwith increasing pressure may be connected with the slowing down mechanism .... or perhaps with the formation of positronium."

It is not clear why I left our puzzle for eighteen months. Probably it was simply a matter of sloth. There were so many interesting, easy experiments to do and I did not want to spend months on a problem which I expected to have an uninteresting solution. I told Jim Shearer that we had to wait for new ideas or better techniques to break the deadlock and suggested that he abandon the low energy approach. He undertook a measurement of the annihilation cross section for energetic positrons, as mentioned earlier. In the event, new ideas and new techniques developed and matters progressed rapidly when the time was ripe.

SECOND ATTEMPT

In October,1950 Karl Darrow, secretary of the Physical Society asked me to give an invited paper on a subject of my choosing at the February meeting in New York. I submitted "Positron Annihilation" as the proposed title of my paper. In the months since we suspended the experiments some new ideas had developed giving me the courage for a commitment that forced me to solve the riddle in three months.

I had read the paper by Ore and Powell (1949) on the decay of the lowest triplet state of positronium which is formed in three quarters of the cases. The same paper also presents some valuable ideas concerning the formation and stability of positronium in gases. I now understood that the formation should normally occur in less than 10^{-10} sec, during the slowing-down process. It was also clear that positronium should be be stable against perturbations in most gases. The decay rate of the lowest triplet state, entirely due to three-photon emission, turned out to be comparable to the rate for two-photon annihilation of free positrons in a gas at atmospheric pressure. This made it hard to separate the two processes since the useful pressure range in our apparatus was limited by considerations involving the geometry and the positron range.

The lowest singlet state is formed in one quarter of the cases. Its decay rate into two photons was known to be of the order of 10^{10} sec^{-1}, far too high to be resolved by our circuits. We had always observed an initial "prompt" component in our decay curves but we ascribed it to annihilation in the walls of our apparatus. I had not been satisfied with that explanation since the phenomenon persisted even in my meteorological balloon, described earlier. In addition, the intensity of this component did not decrease with increasing gas pressure as fast as one should expect from the limited range of the ^{22}Na positrons. I began to suspect that we might be dealing with singlet positronium.

We had not yet seen Ore's thesis on the theory of the formation of positronium in gases but it seemed plausible that, like a chemical process it would depend on the structure of the gas molecules. This difference was even greater when I compared nitrogen with pure oxygen. In oxygen the observed number of delayed coincidences is only about half of the number observed in nitrogen at the same pressure. This meant that there is a mechanism inducing rapid annihilation of half of the positrons in oxygen but not in nitrogen. At this point my time ran out. I felt sure that I was looking at the answer without recognising it, but it was time to leave for New York to give my paper.

At that time going to New York from Boston did not mean taking the Air Shuttle at Logan Airport every hour, on the hour. It meant boarding a sleeper on the "Owl" of the New York, New Haven and Hartford Railroad at South Station in the evening and arriving in the morning at Grand Central Station. This gave me a night of interrupted sleep to search my subconscious for the repressed clue of which I was so confident. It did not turn up at the end of a series of syllogisms but rather like an event remembered or a pattern suddenly recognised: Suddenly the comprehension is there and one does not understand how he could ever have failed to see it. I had been trying to explain my observations in terms of factors like the energy dependence of the cross sections for annihilation and for positronium formation or break-up. The key was, instead the process of triplet to singlet conversion. The production of positronium is roughly equally probable in nitrogen and in oxygen and the triplet is formed three times as often as the singlet. Oxygen is one of the very few gas molecules with a non-zero electron spin in the ground state. In fact the ground state is a triplet and is paramagnetic. This can lead to collisions in which the electron spin in positronium is flipped, transforming the triplet into the singlet which very quickly decays into two photons. This "collision quenching" has a larger cross section than the annihilation cross section for free positrons. Hence, in oxygen all of the positronium disappears quite early and we observed the annihilation in collisions of the remaining free positrons which had too little energy to form positronium. The rate of this process is proportional to gas density, as expected. In nitrogen, on the other hand, the delayed counts are due to a mixture of triplet decays and free positron annihilation. The reduction in the number of delayed coincidences in oxygen compared with nitrogen indicates that the fraction of positrons forming positronium in oxygen is approximately 0.5.

That picture emerged with a feeling of absolute certainty and familiarity. My audience in New York was appreciative but not as convinced as I was. The most valuable suggestion I have ever received in my work came from the nuclear chemist Anthony L. Turkevich who was in the audience. He pointed out to me that if my interpretation of the oxygen data was correct, then the gas nitric oxide, NO, should be even more effective since its molecule contains an odd number of electrons. Thus spin-flip by exchange should be a very probable process in NO. Acquiring a cylinder of NO was not an entirely trivial matter. Nitric oxide (not to be confused with nitrous oxide - laughing gas) is very toxic. Not only had provision to be made for the safe handling of the gas but it had to be backed by appropriate paperwork. All of that could well take several weeks, but I was lucky. In 1946, when we were starting the first successful experiments on gamma ray angular correlations, I also wanted to study the effects of interatomic fields in perturbing these correlations.

My ideas on this subject were so confused that they must have been original. I thought that these perturbations should become manifest if we could study the same nucleus in a crystalline and a gaseous environment. Since we were working with ^{60}Co, we needed to produce a gaseous cobalt compound. My co-author in this work was a chemistry graduate student Edward Brady, who synthesized cobalt nitrosyl carbonyl, one of the very few gaseous cobalt compounds (Brady 1947). This synthesis required the use of NO. Thus <u>I actually already owned a cylinder of NO</u>!

THE DECISIVE RESULTS

Tony Turkevich's conjecture was immediately verified. The addition of a few percent of NO to nitrogen produced a decay pattern very similar to that in pure oxygen. The delayed component is reduced by a factor of three with the addition of NO indicating that more positronium is formed in the mixture than in pure oxygen, where the reduction is only a factor 2. An even more spectacular effect was observed in the gas known as Freon 12. I was using this gas because I expected that its high density would provide a short lifetime for free positron annihilation at conveniently low pressure. In fact, addition of five percent NO to freon at atmospheric pressure eliminated practically all decays beyond about 100 nsec.

One important new technique which had become available since the days of Jim Shearer's experiments was the use of NaI(Tl) - sodium iodide - scintillators introduced by Robert Hofstadter. This provided the possibility for very high efficiency, albeit low resolution, gamma ray spectroscopy. Replacing the anthracene in our annihilation counter by a sodium iodide crystal, I recorded the change in the gamma ray spectrum when NO was added to the gas. Although the resolution of these early measurements was poor, they established beyond doubt the increase in two-photon decay upon addition of NO. Nitric oxide continued for many years to be used in this manner to test for the presence of triplet positronium. My failure to acknowledge Turkevich's contribution in the original publication was an oversight of which I am ashamed.

Six weeks after the New York meeting I sent a Letter to Physical Review with the title "Evidence for the Formation of Positronium in Gases" (Deutsch 1951), summarizing my results to date. In June I could finally publish some results which could be quantitatively compared with a theory. In pure freon 12 abundant delayed decays were observed and the decay showed two distinct components. At pressures below about half atmospheric we could observe an initial decay with a rate approximately proportional to the gas pressure. It remained unaffected by the addition of NO. This was clearly the decay of the free positrons. At higher pressures this component disappeared too rapidly to be resolved by our apparatus. The remainng delayed component could be suppressed by NO. Its decay rate was nearly independent of pressure and was surely due to the three photon decay of triplet positronium. In pure oxygen the situation is reversed: The positronium decays rapidly, due to spin - flip quenching. The remaining delayed component shows a decay rate proportional to pressure as expected for free positron annihilation. My best fit to the freon12 data gave $\lambda = ((6.8\pm0.7)+0.3p)\cdot10^6$ .sec^{-1}, where p is the gas pressure in atmospheres.. The annihiltion rate of the positronium triplet state into three photons had been calculated in papers by Ore and Powell, by Lifshitz and by Ivanenko and Sokolov. The three results differ from each other by as much as a factor fifteen. Ore and Powell's value of $\lambda = 7.2 \times 10^6$ is in excellent agreement with the experimental result. The most recent published data (Westbrook 87) were obtained by essentially the same method and the authors claim the result

$\lambda = ((7.0516\pm0.0013)+0.2p)\cdot10^6$ sec^{-1}, obtained in nitrogen.

The theoretical value has been refined to $\lambda = (7.03830\pm0.00007)\times10^6$.

The good agreement of the life time with the predicted value showed that the positronium atom was not seriously perturbed by the gas. At that point I finally realized that a serious study of its properties could yield results of some importance. It has always been my conviction that the first thing to do when trying to get insight into a complex system is to apply a magnetic field to it. This has worked in the case of the Zeeman effect, in superconductivity, in medical Magnetic Resonance Imaging and many other cases. Fortunately, Henry Primakoff was spending a sabbatical year at M.I.T. Thanks to his patience and wisdom, I quickly understood the peculiar Zeeman effect of positronium. Today the concepts involved are familiar to every third year student. They were, of course equally elementary, but less familiar, in 1951. Once I had understood the Breit-Rabi formula and realized that the magnetic quenching experiment was closely analogous to the Stark quenching in the Lamb - Retherford experiment (Lamb 1950), I could carry out the necessary theoretical calculations. It seems strange that it took me several weeks to do it.

The splitting between the triplet and singlet levels of the ground state, ΔW had been calculated in lowest order to be $8.5\ 10^{-4}$ eV by Pirenne (1946). Berestetskii and Landau (1949) also calculated the Zeeman effect. Obviously no unperturbed state of positronium can have a magnetic moment, at least if parity and charge conjugation symmetry are conserved. In a magnetic field, the $m = \pm1$ substates of the triplet remain unperturbed but the singlet state and the $m = 0$ triplet are mixed, giving rise to a quadratic Zeeman effect. The displacement of each of these levels in a magnetic field B is

(3) $\Delta E = (\Delta W/2)((1 + x^2)^{1/2}-1)$

$x = ehB/(\pi mc\Delta W)$
The admixed amplitude is given by $a = x/2(1+x^2/4)^{1/2}$

Since the singlet component admixed to the triplet can decay rapidly by two photon annihilation the observed intensity of three photon decay in a field B should be given by

(4) $I(B)/I(0) = \lambda_t(1-a^2)/(\lambda_t(1-a^2)+\lambda_s a^2)$

where λ_s, λ_t are the decay rates of the two unpertubed states.

With the help of a new graduate student, Everett Dulit we assembled the necessary apparatus. A positron source of ^{64}Cu, which emits practically no nuclear gamma rays was placed inside a gas vessel of dimensions similar to those uised in the life-time experiment. This vessel in turn was located in the gap of a large electromagnet in Francis Bitter's laboratory. Note that no committee was needed to assign this state - of - the - art magnet to us. A much improved sodium iodide scintillation counter viewed the cavity at 90° to the field. The relative intensities of two- and three - photon annihilation were derived from the spectrum seen in this counter. The results were in excellent agreement with (4). From these data we derived a value of $\Delta W = 9.4\ 10^{-4}$ eV, with a very conservatively estimated error of 15 %. I took this good agreement with theory as proof that we were indeed dealing with relatively unperturbed positronium. We published our result under the title "Short Range Interaction of Electrons and Fine Structure of Positronium" by Deutsch and Dulit (1951).

Some time later, we realized that the theoretical curve with which we found such good agreement, was in error. The two annihilation photons from the mixed m = 0 state are not emitted isotropically. This changes the detection probability in our counter, located at perpendicular to the field. My conservative error estimate covered this and, in the event, we did not tell a lie. I do not recall whether we had considered the angular distribution. If we did, I surely made the intuitive argument that the decaying state has zero angular momentum and must therefore decay isotropiclly. Such are the limits to intuitive hand-waving.

There is another postscript to this experiment. Ev Dulit was a member of a fine jazz combo and it seemed to me that the relative importance he attached to it, compared to what I considered very exciting physics boded ill for his carreer. I advised him of this and he wisely changed profession. He became a successful psychiatrist and his combo, called "3 Finks and a Shrink" was still going strong when I last spoke with him.

This ends the story of the discovery of positronium. I was rather proud of our determination of the fine structure splitting, but when I told my pre-war mentor and collaborator Arthur Roberts about it, he seemed unimpressed. He startled me by claiming that we should be able to measure the splitting to better than one part in ten thousand. He did not, however, suggest how to do it. It was this challenge which made me invent the method and carry out the experiment a few months later. That also was an amusing story, but it leads beyond the topic of this lecture. The same is true of the many experiments concerning positronium chemistry in gases which followed. No presentation of the later work is possible without properly crediting and discussing the work of many others, especially the Yale-Columbia group, who cleaned up many of the sloppy loose ends that I left behind.

BIBLIOGRAPHY

Berestetskii, V.B. and Landau, L.D., 1949, J.Exp.Theor.Phys. 19:673,1130
Berko, S. and Pendleton, H.N., 1980, Ann. Rev. Nucl. Part. Sci. 30:543
Brady, E.L. and Deutsch M., 1947, Phys.Rev 72:870
Deutsch, M., 1947, Phys.Rev. 72:729
Deutsch, M., 1951, Phys.Rev. 82:455
Deutsch, M., 1951, Phys.Rev. 83:866
Deutsch, M., and Dulit, E., 1951, Phys.Rev. 84:601
Heitler, W., The Quantum Theory of Radiation Oxford University Press, London, (1944)
Kendall, H.W. and Deutsch, M., 1956, Phys.Rev. 101:20,
Lamb., W.E., and Retherford, R.C. 1950, Phys.Rev. 79:549
Ore, A. and Powell, J.L., 1949, Phys.Rev. 75:1696
Rich, A., 1981, Rev. Mod. Phys. 53:127
Shearer, J.W., and Deutsch, M., 1949, Phys.Rev. 76:421
Shearer, J.W. and Deutsch, M., 1951, Phys.Rev. 82:336
Westbrook, C.I. et al., 1987, Phys.Rev.Lett. 58:1328

FIFTY YEARS OF NUCLEAR PHYSICS

Herman Feshbach

Center for Theoretical Physics
Laboratory for Nuclear Science and Department of Physics
Massachusetts Institute of Technology
Cambridge, Massachusetts 02139 U.S.A.

Fifty years takes us back to 1937. I had just graduated from the City College of New York, where I received the Ward Medal of Physics, and I was engaged to Sylvia Harris. We subsequently married in 1940. I also had a job, which was remarkable. It was still the middle of the depression and very few CCNY 1937 graduates were employed. I was to teach at the City College for one year with a salary of $1,600, sixteen contact hours was the standard load at that time for everyone. With this salary and the money I earned teaching summer school, I was able to go to graduate school in the fall of 1938. I went to M.I.T.

I had become interested in physics in my senior year in high school mostly because of an excellent physics teacher, Irving Mosbacher, who introduced me to relativity and quantum mechanics, at the popular level of course. In college, I tried to learn quantum mechanics (I was a second year student) by reading Dirac's book, first edition, several times. Years later, I had to opportunity to tell Dirac just that. His reply – "Only book available at the time I suppose." Actually, he was not correct. There was an excellent book by Frenkel, called *Elementary Wave Mechanics* which I also studied. In my senior year I took special courses called "honors". In the first semester, Mark Zemansky asked me to read a *Handbook du Physik* article on the rigorous theory of diffraction developed by Sommerfeld. In the second semester under Henry Semat I reviewed the theory of the Dirac electron and the Fermi theory of β-decay.

New York in those days was an exciting place. Rabi and Breit ran a weekly evening seminar. Important slow neutron experiments were being done by Goldsmith. My friends included Gerjuoy, Hammermesh, Primakoff, Holstein and Schwinger. I was in awe of Schwinger. It was not that he had already published, but that he possessed such power and understanding of physics. Yet he was a pleasure to be with; quiet, unassuming, thoughtful and broadly knowledgeable. There were also a number of brilliant mathematicians whom we got to know because we sat in on Courant's courses at NYU. Courant was a superb teacher. The courses were at night so that we could attend. It was also a time of enormous concern because of the world political situation. The Spanish civil war, the persecution of the Jews by the Nazis, Italian fascism, the Moscow trials, the Japanese atrocities in China, and the economic depression formed a distressing background to our joy in doing physics as well as enjoying the music, art and theater of New York.

During my year 1937 – 1938 teaching at CCNY, I took some courses at NYU and did a paper with Clarence Zener, my first. Zener had just come to join the CCNY faculty. Some very brilliant physicists, such as Zener, were willing to take positions at CCNY in the depression years in spite of the crushing teaching load and the absence of any research facilities.

But on to graduate school where Phil Morse took me on for research. Morse had done a number of papers on the deuteron and nucleon-nucleon scattering with Fisk and Schiff. But when I arrived, his attention was focussed on acoustics, a field on which he had a major impact. In comparison with my fellow graduate students, Al Clogston, R. Lichtenstein, P. Rubenstein and D. Gelatis, and a couple precocious undergraduates, R. Feynman and T. Welton, I was badly educated, especially in classical physics. I spent much of the first year catching up. But I also was interested in research. In particular, I noticed that experimental results of the elastic scattering of electrons performed in cloud chambers using β-radioactive sources differed enormously from that predicted by Mott. Barber and Champion obtained 1/5 times the Mott cross-section in the scattering of electrons by Hg while Scherrer and Zunti obtained 20 times the theoretical value when the target was N. (Not the last time I was mislead by experimentalists.) Of course, there was no explanation for these results and eventually they were disproved in a series of experiments in which Van de Graaf, Buechner and I co-authored. I was the house theorist. It was a very good experience for me to watch and help to carry out and analyze an experiment (which from then on I believe is a miraculous concatenation of circumstances.) During World War II, I spend part of my time with the Van de Graaf group, building compact X-ray accelerators for the Navy and working on radiographic applications. After World War II, I continued my interest in electron scattering until the Stanford theorists, under the influence of Hofstadter's experiments, began to play leading roles in this area. William McKinley and I found a mistake in Mott's expansion of the cross-section in powers of $Z/137$ in the very first correction term and we went on to find closed expression for the next few terms. Also with L. Acheson I was able to show rigorously that the low energy electron scattering experiments measured the mean square charge radius of the nucleus. Hofstadter's experiments began what is now a most important branch of nuclear physics in which the electron accelerator becomes a giant electron microscope examining the structure of the nucleus. This audience will be interested to known that experiments at the Bates linear accelerator using polarized electrons will make important tests of the electro-weak theory.

I shall digress a few moments to talk about Van der Graaf. One doesn't read much about him these days in spite of the fact that much of experimental nuclear physics depends critically upon his invention. He was a wonderful man to work with; gentle and thoughtful. Some mistook his careful Southern speech as an indication of a slowness of the mind. They were wrong. He was, in fact, a brilliant experimentalist. Indeed in the last scientific conversation we had, he brought up the possibility and feasibility of $U - U$ collisions and what one could learn from such experiments. Much of the success of the post-war nuclear physics at MIT was a consequence of his close collaboration with W. W. Buechner. He died comparatively early of complications from the hepatitis he contracted as a consequence of a blood transfusion.

By 1937 several important advances in the understanding of the nucleus, starting with the discovery of the neutron, had been made. The separation of deuterium had made possible the study of the deuteron, its photodisintegration and the inverse process of capture. Neutron resonances had been described by the Breit-Wigner formula. Artificial radioactivity has been discovered. The 1934 Fermi theory of β-decay had completed the neutron-proton model of the nucleus. Nucleon-nucleon scattering had been observed and the concept of isospin and charge independence had been formulated. For the very first time use was made of an internal symmetry, $SU(2)$, not connected with spatial or temporal properties. It was found that the nuclear forces measured in terms of the electromagnetic forces were strong and short ranged, a result corroborated by the existence of neutron resonances. In analogy with the atomic case, it was hoped, naively as it turned out, that once the two-body force was known, it would be possible by solving the many-body Schrödinger equation, to predict the properties of nuclei.

Fortunately for beginners like myself, all of this had been described in considerable detail by Bethe in two marvelous *Review of Modern Physics* articles written with Bacher (1936) and with Livingston (1937), respectively. I still have these articles with my notes and queries in the margins. I particularly studied the two- and three-body problems and the theory of β-decay. Amusingly, considering my later interests, I found the material on nuclear reactions difficult — too many subscripts! Bethe and Bacher contained a review of the work of Feenberg and Wigner, a forerunner of the shell model. Wigner's $SU(4)$ model of nuclear structure was published in 1937. The effective range theory for the two-body system did not its appearance until after World War II. The Weisskopf, Landau, Frenkel evaporation model of nuclear reactions played a central role in the Bethe–Livingston review. That review did not contain Bethe's carbon cycle model for the production of stellar energy which he developed a few years later. It was the beginning of a new field, nuclear astrophysics which includes stellar energy and element production, stellar evolution, the properties of neutron stars and supernova. These studies rely on our understanding of nuclei and in turn provide information on nuclear matter and weak interactions.

My thesis was on ^{3}H, its β-decay and its binding energy. This had become an interesting problem because of the discovery of the quadrupole moment of the deuteron which in turn implied the existence of tensor forces. W. Rarita and Schwinger had thoroughly investigated the consequences for the two-body system. The three-body system was obviously next. I won't report on the results, or of calculations I carried out with W. Rarita and Robert Pease. These were the days before the discovery of the repulsive core in the nucleon-nucleon force which rendered these calculations invalid. We did learn that the D-state in ^{3}H made a smaller relative contribution to the binding energy of ^{3}H than the D-state in ^{2}H made for B.E. of ^{2}H. Sol Rubinow and I devised an equivalent two-body method which has found some use. Nowadays the three-body problem occupies a special place in nuclear physics. Its formulation by Faddeev has made possible a more effective treatment. It is in this system that exchange currents predicted by the boson exchange models of nuclear forces were unmistakably observed. Electron scattering of ^{3}H and ^{3}He now under study will provide further tests of important properties of nuclear forces.

I finished my thesis just about the time of Pearl Harbor. Many of the faculty were already involved in war work and a number of us were appointed as instructors in the fall of 1941. We then taught half-time and "half-time" went to the war effort. For the first few years I worked with the Van de Graaf group and then joined the Underwater Sound Group which was concerned with anti-submarine warfare. I taught many courses during those years but the most important was Morse's *Methods of Theoretical Physics*. I rewrote the class notes, expanding these considerably. It became the basis of the book that Morse and I wrote and published in 1953. It is still in print and selling. Once I finish the book I am writing now, I shall revise *Methods.* Morse was wonderful as a collaborator and he was responsible for the three-dimensional figures in the book. The war needs put a tremendous emphasis on classical physics, particularly wave propagation. Perturbation of boundary conditions, that is solutions of problems in which the shape of the boundary is perturbed or the conditions which are to be obeyed on the boundary surface are modified, had been a problem of interest to Richard Bolt, Albert Clogston and myself prior to World War II and we had published some papers on the subject. But during this war period I had hit upon a more general solution. Schwinger, at that time at the MIT Radiation Laboratory (which was responsible for radar research and development), had also studied this problem and our results were similar though we had not discussed the problem. Later, various extensions of the method were applied to problems in room acoustics and duct problems, in collaboration with Cyril Harris. I may as well add at this point, work with Albert Heins who taught me how to use the Wiener–Hopf method to solve a class of acoustic duct problems.

From a personal point of view perhaps the most important thing I gained from working in the Underwater Sound Group was my friendship and collaboration with Melvin Lax. Lax was and is an extraordinarily talented physicist who has that marvelous dedication and ability to "run" with an idea. Lax and I did several papers in acoustics and on photoproduction of pseudoscalar and vector mesons. We did not know at the time that the pions were pseudoscalar. This last research was stimulated by the construction of a 300 MeV electron accelerator at M.I.T.

Once the war was over, there were important changes at MIT as it acquired a whole group of nuclear physicists who had been at Los Alamos. The most important for me were Viki Weisskopf and Jerrold Zacharias. I had met Viki twice before, once when I was interviewed for an assistantship at Rochester (I didn't get it) and once at a Physical Society meeting at Brown to which Richard Present took me as I was looking for a position. This was before I was offered an instructorship at MIT. Viki became my dear friend. Over the years we faced many problems together. We found a profound concordance in attitude toward the issues of the day, scientific or otherwise. I learned from Viki that only when I could explain what I was doing in simple terms, using at most back-of-the-envelope calculations, did I truly understand my results. But as we shall see, there were many other ways in which Weisskopf made an important impact on my research. At this time I had the privilege of attending the Shelter Island Conference, followed in successive years by meetings in the Poconos and at Oldstone on the Hudson. These have been described by S. Schweber in a recent *Review of Modern Physics* article.

Following the publication of *Methods,* my research in classical physics dwindled rapidly to zero; although Felix Villars and I did do some studies of the properties of the ionosphere. Most of my research was dedicated to two problems in nuclear physics, the nucleon-nucleon problem and the theory of nuclear reactions. There was one paper on Nuclear Structure. I should include a number of papers on Elementary Particle Theory of which the *Reviews of Modern Physics* article with Felix Villars had some impact. Earle Lomon was my principal collaborator on the nucleon-nucleon problem. The nuclear reaction problem involved Weisskopf, Kerman, Koonin, Lemmer. A. Gal., J. Hüfner, W. Friedman, J. Lamarsh, V. Newton, L. Zamick and F. Iachello. The latter's thesis however was on the Interacting Boson Model in which the nucleus under consideration was ^{16}O, the p and f bosons were the well-known 1^- and 3^- states of ^{16}O. The model was quite successful but has not been followed up.

It differs from the Iachello–Arima model of the same name whose range of application is the low lying states of the vibrational and rotational nuclei, making extensive use of group theory. It was Iachello who found examples of supersymmetry in nuclear structure.

To the research effort I should add the book on which Amos de Shalit and I collaborated entitled *Nuclear Theory*. This was started in 1961 and didn't get published until nearly a dozen years went by, interrupted by de Shalit's untimely death. The long gestation period was primarily because we were involved in many other efforts. The second volume on reactions is almost done but will take an even longer period in the writing. Amos was a wonderful person to work with. It was an adventure as he was intellectually daring, incredibly quick, saw the essence of an argument instantaneously and was able to present it understandably. Somehow he was completely *"au courant"*.

The nucleon-nucleon model that Lomon and I developed gave the nucleons a finite size, in the sense that we assumed that for nucleon-nucleon separations less than $r_0 = 0.7$ fm ($1/2$ pion Compton wavelength), as it turned out empirically, the wavefunction was independent of the incident kinetic energy. The thought behind this was that at close distances the interaction was very strong. The consequence of that assumption is the boundary condition model in which the wavefunction in each partial wave satisfied a logarithmic derivative boundary condition at r_0. This model has been originally used by Gregory Breit and his colleagues as a phenomenological device to describe low energy S scattering. We generalized it to take account of tensor coupling and higher partial waves. The potential for $r > r_0$ was taken from the current meson theory of nuclear forces. This theory was quite successful but was never adopted by the community. The currently fashionable Paris potential makes use of more modern methods including dispersion theory to obtain the nuclear force for r large. They find an empirical potential for the short range interaction which is just the boundary condition model, although it is not expressed formally as such. Lomon later used this model to obtain quite good results for the structure of finite nuclei. And, of course, the quark model of the nucleon implies a structure similar to the one Lomon and I used. There is an important difference as Lomon and I did not and of course had no reason to include the effects of the degrees of freedom in the region $r < r_0$.

Calculations of the saturation energy and density of nuclear matter demonstrate that a two body nucleon-nucleon interaction will not reproduce observation. Three body forces are necessary. The early dream that once one knew the two-body nuclear force it would be possible to predict the properties of nuclei turns out to be entirely too simplistic.

By the beginning of World War II, it was commonly believed that nuclear forces were strong as demonstrated directly by nucleon-nucleon scattering. This was reinforced by the observation of resonances in nuclear reactions, the success of the compound nucleus theory of nuclear reactions, and the Bohr independence hypothesis. The last indicated long interaction times with a consequent loss of information. One, for example, could not from the spherical angular distribution of the evaporation reaction deduce the direction of the incident beam. I should mention at this point the paper with Hauser which extended the Weisskopf statistical model so as to include the effects of angular momentum and parity conservation. The results of this paper have been and are still being widely used for reactions involving compound nucleus formation. Corrections of importance when the nuclear levels excited are few in number have been found by Moldauer. Weidenmüller and his colleagues have obtained exact solutions for a problem in which it is assumed that the matrix elements and level spacings are random with distributions first given by Wigner and Porter and Thomas. Weidenmüller, French, and Bohigas and Giannini have speculated that systems which follow these laws are quantum examples of chaos.

This reaction mechanism failed completely to describe the (d, p) reaction or the elastic scattering of neutrons, which were observed once the post-war experimental investigations began to give results. In the first case, the angular distributions were sharply peaked forward and varied slowly with incident energy. This was clearly a reaction with an interaction time which was relatively short; too short for the formation of the compound nucleus. The angular distribution of elastically scattered neutrons was found to vary in a very regular fashion with mass number of the target nuclei. Porter, Weisskopf and I were able to show that this regularity followed if one assumed that the neutron moved in a complex potential well. This became known as the optical model. That model in which the structure of the composite nucleus seems to be of no importance is at first blush not consistent with the notion of the strong nucleon-nucleon interaction.

I had written two other papers with Weisskopf, the first with Peaslee was essentially a single channel version of the Wigner R-matrix theory. From my point of view it was a boundary condition model which was suggested to me by the work I had done with Lax on the scattering of acoustic waves by spheres whose surfaced had been treated. A similar intuition was involved in the boundary condition model of nuclear forces. I have often found that my understanding of acoustic wave phenomenon to be extremely helpful in thinking about nuclear reactions.

The picture of the nucleon-nucleus interaction presented by the optical model was in substantial agreement with that provided by the shell model of the nucleus, which had been discovered a few years earlier. It is difficult to describe the great excitement that accompanied that discovery or the revolution in the understanding of the nucleus that it generated. And of course the discovery of the optical model was an important corroboration. The question of how to make these empirical results consistent with the strong nuclear forces was common to both. But an additional dilemma was associated with the optical model. How could the optical model be consistent with the existence of resonances? The answer was given in the paper with Porter and Weisskopf. The elastic scattering data was taken with relatively poor resolution. One then measures an energy average and the fine structure of the resonances disappear. Or equivalently, one can say that poor energy resolution implies, via the Heisenberg uncertainty relation, that only the prompt part of the reaction was observed. The part of the reaction that is delayed because of the formation of the compound nucleus is therefore not part of the prompt reaction amplitude.

But this left a formal problem: the simultaneous description of prompt and delayed reactions. Somehow one formalism should be able to describe both compound nuclear resonances, the (d, p) reaction and the optical model. This was accomplished in a paper in 1958 and a sequel in 1962. My work on this problem started with the suggestion by Weisskopf that there should be an analog of the Kramers-Kronig dispersion relation for the optical index of refraction. That indeed proved to be the case as I was able to derive a dispersion relation for the nuclear optical potential. But much more came out; namely a solution to the problem posed at the beginning of this paragraph. The 1962 paper rephrased the 1958 results in terms of projection operators which turns out to be a most economical procedure providing an easy mechanism for formulating arguments based on intuition. A second feature of the 1962 article was the solution of the problem posed by the Pauli exclusion principle when the projectile and target are composed of identical particles. There is a second more formal problem of overcompleteness and non-orthogonality which automatically are also resolved. W. Friedman showed how the formalism could be used for several classes of experiments. L. Dohnert applied the results to the stripping reaction on ^{16}O. More recently, an easy generalization makes it possible to apply this method to heavy ion reactions. The 1962 paper also contains a better derivation of the optical model following a suggestion by M. Baranger.

As I have emphasized, the projection operator method created a way of thinking which led to a number of connected contributions. In summary, it was learned that a nuclear reaction proceeds in a set of steps. Each step leads to a stage involving a more complex class of excitations. The simplest stage is the entrance channel with the target nucleus and projectile in their ground states. The interaction time increases with the number of steps, that is with the complexity of each stage. The sequence of steps may be terminated at any stage by emission into the exit channel, that is, the final state. In a statistical theory the observed cross-section will be a sum of the cross-sections for the emission from each stage. The character of the observed cross-section will depend upon which stage or set of stages dominate that sum. That in turn will depend on the nature of the observables measured. For example, if the underlying single step cross-section is sharply peaked in the forward direction, multi-step processes will make the major contribution to the large angle cross-section. If we are concerned with large interaction times, compound nucleus formation will dominate. If small interaction times are selected by an appropriate choice of the energy resolution, the optical model will be the appropriate mode of description. If the energy resolution is improved one observes isolated doorway states, such as the giant resonances or the isobar analog resonances. If an isobar analog resonance is measured with "poor" resolution, one observes a beautiful single resonance shape. However, if the resolution is improved sufficiently it breaks up into many narrower resonances. The term "doorway state" was introduced in 1962 in a paper with B. Block which applied the statistical theory of such states to certain properties of the neutron strength function. These states are the first stage after the entrance channel, and must be "passed through" if the more complex configurations such as the compound nucleus is to be formed. If there are many such states in the energy range of interest, the probability for formation of the compound nucleus and therefore the neutron strength function would be large. On the other hand, if there are few it will be small. Suffice it to say this theory of multi-step processes has been very successful in its application to nucleon induced reactions. Collaborators on these efforts include Kerman, Lemmer, Koonin and Block. Comparisons with experiment have been carried out by L. Colli-Milazzo, R. Bonetti, P. Hodgson and their collaborators.

As was mentioned earlier, the shell and optical models had a common problem: How to reconcile the apparent non-interacting motion of the nucleon in the nucleus and the strong short-range nucleon-nucleon forces. One's first thought is that the interaction felt by the nucleon is the sum of its interactions with the individual nucleons making up the rest of the nucleus. However, that sum is very rapidly varying both in space and time. The mean field of the optical and shell models is obtained by taking the slowly varying component of the sum, averaging over the more rapidly varying component. The execution of this program, in the presence of strong and even singular forces is a major accomplishment of nuclear theorists. The mean field found in this way is dynamical, that it, it does depend through its parameters such as the nuclear radius on the nature of the nuclear configuration. It can rotate and vibrate, it can be deformed. From this mean field approach there follows not only the properties of the shell and optical model but also such special nuclei as the rotational deformed nuclei, the vibrational nuclei and such special states as the giant electric and magnetic (Gamow-Teller) resonances. By breaking the symmetry associated with baryon conservation one can develop a superfluid theory of nuclei. Bands consisting of the ground state of nuclei whose neutron number differs by two units are obtained. Observation of enhanced cross-section in (p, t) and (t, p) reactions verified this theory. Recently a super-deformed band has been found in ^{152}Dy predicted by the mean field theory. A "small vibration" correction to the mean field, referred to as the RPA method, has also been derived. Bohr and Mottelson and their colleagues, particularly Nilsson, took a leading role in these developments which are discussed in their book.

There are of course corrections to the mean field. The nuclear wavefunctions should not just be composed of independent single particle wavefunctions. There should be correlations. Correlations do appear in an average way in the calculation of the binding energy of nuclei. But it was clearly desirable to observe them experimentally. Curiously enough no clear cut measurement has ever been obtained. I should be careful, there are two types of correlations for which we have good evidence. One are the correlations induced by the Pauli exclusion principle which have a range of roughly $(1/k_F)$ where $\hbar k_F$ is the Fermi momentum, or about 0.8 fm. Another is referred to as the center of mass correlation. This is simply the statement that the coordinates of the A particles making up the nucleus are not independent since the states of the nucleus cannot depend upon the center of mass. This introduces a correlation length of the order of the nuclear radius. The dynamic correlations of interest are those arising from the action of nuclear forces. And the question is whether experiment could confront theory with respect to these correlations. Roughly, the corresponding correlation length is of the nuclear forces which unfortunately does not differ greatly from the Pauli correlation length.

Methods of measuring correlation lengths are well-known from condensed matter physics. Essentially they rely on double scattering measurements which depend upon the positions of the two scatterers. The intermediate propagator between the two can be off or on the energy shell. The interesting term is the off-shell component. A multiple scattering theory had been worked out by Kerman, McManus, and Thaler for high energy nucleon-nucleus scattering. The first term is a familiar one, familiar in elementary form to Lord Rayleigh, and given in its modern and more complete form by Melvin Lax. The second term is the correlation term which however at the time would be difficult to evaluate quantitatively. Gal, Hüfner and I and later Ullo, Lambert and Parmentola did find a good practical approximation valid for both space and spin dependent correlations. But Nature was not kind, as it turned out, that, except for the possibility of relatively long-range spin-spin correlations, correlation effects are not strong enough to be observable by this method.

Over these last fifty years, nuclear physics has greatly extended the range of experimental parameters available for the study of nuclear structure and reactions. This has been made possible by technical advances which permit increasing the energy of the projectiles, expanding the number of probes enormously and finally greatly improving the quality of the detectors. Fifty years ago the only probes were neutrons, gamma rays and relatively low-energy protons and alphas. Not only has the energy of these particles been raised to the many GeV range but a whole new family of projectiles, namely the heavy ions formed from the light nuclei, e.g. ^{12}C, through nuclei of medium weight, on to the heaviest ^{238}U. Energies extend into the relativistic range so that experiments are now on progress at BNL with laboratory energies of 15 GeV/A and at CERN with energies of 200 GeV/A. It is anticipated that a heavy ion collider (RHIC) 100 GeV/A on 100 GeV/A, will be build in the next several years. Electron accelerators providing pulsed electron beams with several hundred MeV energy have been available for several years. Experiments have been done at SLAC with a few GeV electrons while construction of a 4 GeV (hopefully greater than 4) CW electron accelerator (CEBAF) is underway in the U.S. And there is the low-energy anti-proton beam at CERN. Turning to unstable particles, nuclear physics facilities have included copious beams of pions, and muons. Low-energy kaon beams have been used for the production of hypernuclei, nuclei in which one of the baryons is a hyperon such as the Λ and Σ. Unfortunately, the facilities devoted to this effort have with one major exception been shut down.

This expansion of facilities has as a parallel the expansion of nuclear theory to include QCD physics. The discovery that the nucleon has structure and is of finite size (actually this was known very early, from the existence of the Δ) has presented a number of important questions to the nuclear theorist. First, one had to reconcile that fact with the traditional theory of nuclear forces based on boson exchanges between point nucleons. Note that the bosons also have structure. Secondly, how should one revise our successful understanding of nuclear structure based on point nucleons interacting via primarily (but not completely) two-body nuclear forces. And as a corollary, how is one to understand pion exchange currents in nuclei? A great impetus to these studies has been made by the discovery of the EMC effect. The experiments to be done at CEBAF at the higher energy should tell us a great deal about the structure of the nucleon the nuclear environment. The experiments to be performed at RHIC, daring experiments I should add, will hopefully produce another form of matter, a quark-gluon plasma, a most exciting prospect. In any event, nuclear theory now includes QCD nuclear physics!

In 1966, Kerman and I proposed a recoilless method of producing hypernuclei in a (K^-, π^-) reaction. (The same suggestion was made by Porodogetsky somewhat earlier). The suggestion was taken up by groups at CERN and BNL, and a number of hypernuclei, mostly in the p shell, have been produced, energy levels measured and γ-ray transitions observed. One of the notable results of these experiments is that the strength of the spin-orbit Λ-nucleon interaction is much smaller than that of the corresponding nucleon-nucleon spin interaction. The experimental observation of Σ-hypernuclei is at the present moment problematic. But the indications are that their width is much narrower than one would estimate from the strong Σ-Λ conversion and the greater Σ mass. To explain this phenomenon, Dover and I have proposed that the interaction in the baryon octet is to a good approximation SU(3) symmetric. In exact SU(3) symmetry the states would have zero width, with width arising from symmetry breaking. The elementary $\Sigma - p$ reaction is not in contradiction with this speculation but unfortunately there is essentially only one experiment with which to compare. If true, this would be an additional important fact to be understood in terms of the quark model. Interestingly, the interaction used by Jaffe in his discussion of the dibaryon when applied to the baryon-baryon interaction is SU(3) symmetric.

As a last topic let me turn to heavy ion physics. This is a vast subject as is attested to by the recently published seven volume work on heavy ion science edited by D. A. Bromley. Under study are the various ways angular momentum, energy, mass and charge are transferred from one of the colliding nuclei to the other. When the encounter is a distant one, electromagnetic excitation via the nuclear Coulomb fields is the principal reaction. Because these fields are large, multiple excitations become significant. For shorter distances, the direct reactions, called quasi-elastic, with short interaction time can occur. As the interaction time becomes longer it becomes possible to exchange large amounts of mass and energy. In this regime, referred to as deep inelastic scattering, the kinetic energy of the two colliding nuclei is completely converted into internal energy. The result came as a surprise. Finally when the interaction is much longer the two nuclei will fuse to form a compound nucleus which may decay by particle and gamma emission and by fusion. It was in a fission reaction that the superdeformed ^{152}Dy was formed.

I will touch on two phenomena to whose understanding I contributed. One is the fragmentation in the collision of relativistic heavy ions as observed at the Berkeley BEVALAC. There were speculations from that group that a new fundamental momentum scale had been discovered. However, Kerson Huang, M. Zabel and I showed that fragmentation is really a low-energy phenomena and could be understood in terms of standard nuclear physics. Central to the quantitative predictions was the development of a nuclear Weiszäcker–Williams method appropriate for peripheral high energy collisions.

Resonances have been found in the collision of certain light ions such as the ^{12}C+^{12}C reaction. I had been asked to review this subject and I was putting together my talk to be delivered the next day with the aid of a great deal of information which D.A. Bromley, one of the discoverers of this phenomena, had furnished when a qualitative understanding occurred to me. It involved three stages: An angular momentum window in the entrance channel. The formation of a doorway state and finally the development of fine structure. The doorway states were responsible for the observed resonances. More significant perhaps is the realization that clusters formed as a consequence of the strong interactions may nevertheless join together to form, in this case, long-lived "molecular" configurations.

This brings me to the present day. The paper with Dover is submitted for publication and I am being busy finishing the second volume on nuclear theory devoted to nuclear reactions. Many ideas remain to be developed and I must do a revision of *Methods*. Fortunately, I have been relieved of most, but not all, of my administrative responsibilities so that more time to do all that has become available. I have also not discussed my involvement with planning for and support of nuclear physics at a national and international level which began nearly 30 years ago. I served, for example, for four years on the Nuclear Science Advisory Committee, three years as Chairman. We produced the first comprehensive planning document for the field in 1979.

I feel extremely lucky that I chose to devote my energies to nuclear physics. It is an extraordinarily rich subject, and one which grows with time presenting engaging new problems whose solutions have a generality of importance to all of physics.

An important reward has been to be a part of the nuclear physics family, to have friends around the globe. Of course, it is with the members of the MIT Center for Theoretical Physics that I have had the closest and warmest relationships extending over many, many years. And the younger people we have brought in have been a real pleasure. M.I.T. has provided a supporting atmosphere. It has attracted superb students whose stimulation and capability were most important for the research just discussed. Not all have been mentioned since for some their research was not in nuclear physics. Saul Epstein, Marvin Mittelman, Father Nichols, E. Lerner and Jane Pease are examples.

We live in a world much of whose activities are driven by technology derived from the advances of science. It is our duty as scientists to provide an understanding of what science can or cannot do to the public, the politicians and the leaders of our communities and nations. Professor Zichichi has taken a leading role in this endeavor. For my own part, I have been most recently concerned with the rights of scientists and with the problem of nuclear weapons for many years. It may interest you to know that a declaration of scientist's rights was adopted by the International Council of Scientific Unions (ICSU) many years ago. Violation of scientists' rights have occurred in many countries (recent improvements in Argentina, the Phillipines and the Soviet Union should be noted). The Soviet authorities have made a good start and one hopes that the encouraging move made at the beginning of this year with regard to refuseniks and dissidents will be continued and amplified. I might add that this was a major concern of Andrei Sahkarov with whom I met in January. Another source of problems has focused on the Israelis as there are countries, India and China for example, which resist issuing visas to Israeli scientists for attendance at scientific meetings. In recent years Indian physicists have been successful in relaxing restrictions initially imposed by the Indian bureaucracy.

But of course the major problem brought on by technological development is the threat of nuclear weapons. Today's situation is bizarre. Each side has roughly 25,000 warheads, none of which would be used by any rational person. Yet we see in the negotiations now going on arguments are advanced, comparisons are made as if nuclear war were possible. These are totally meaningless debates, redolent of medieval scholasticism.

Underlying is the concept that a technical resolution is possible. The ultimate technical resolution is the unfortunate S.D.I. And here we can find fault with ourselves because much of the general public believe in one form or another that scientists can do everything or anything. There is no understanding of constraints imposed by physics and costs on what can or cannot be done.

I have been to Erice many times. Each visit was very productive in terms of the physics I learned and the ideas for research which were generated. I am most happy to be part of the celebration of the 25th Anniversary of the International School of Subnuclear Physics. Its director, Nino Zichichi, has made a most important contribution to physics through his creation of the Ettore Majorana Center and his perceptive direction of its activities.

Needless to say I am deeply appreciative of the remarks made by Professors Wu and Ting. It is amusing to recall that just after World War II, Professor Wu, Deutsch and myself gave invited papers at a session at a meeting of the America Physical Society. One of the positive aspects (they were not *all* positive) of my tenure as head of the M.I.T. Department of Physics is that I got to know Sam and Susan and to become very fond of both. I have been very fortunate in my friends.

CLOSING

CLOSING LECTURE

The Great LEP Forward

Sheldon Lee Glashow[†]

Lyman Laboratory of Physics
Harvard University
Cambridge, MA 02138

1. Introduction

1987 has been truly a super year. We have seen the first nearby supernova in 383 years, not only in light but in antineutrinos, and more recently, in X-rays and gamma rays. The repercussions of this discovery are still coming: What is the mysterious companion of SN1987a seen in speckle photometry? What limits can we place on neutrino masses lifetimes, and interactions? How are axion theories constrained by the fact that the explosion took place more-or-less according to the theoretical script? Should we believe the curiously premature Mont Blanc observations? Since it was a blue giant that blew up, should we revise upwards our earlier color-blind estimates of supernova frequency? Will our community be fully prepared for the next supernova?

Researchers in Texas and elsewhere have achieved superconductivity at super-high temperatures. Can we reconcile this with a theory of superconductivity that virtually forbids such a happening? What wonderful new technologies will emerge? The discovery of the first of the new ceramic superconductors earned Nobel Prizes for Alex Müller and Georg Bednorz of the IBM laboratories in Rüschlikon. Does the current uninterrupted series of eight European Nobel Laureates in Physics over the past four years signal a European resurgence or an American decline in basic science?

The Superconducting Super-Collider has been formally approved by the President of the United States. Thirty-five site proposals in twenty-five states survived initial scrutiny and a short list of potential winners, has been prepared. Will the SSC soon be funded by a deficit-ridden Congress? Will impatient Europeans press forward with their own large hadron collider such as the 10% Eloisatron or the LHC? Perhaps the time is finally come for international collaborations even more inclusive than CERN. Or, will Wall Street's super-crash lead to world depression and put an end to all of our grandiose ambitions and to dreams of future discoveries at the high-energy frontier?

Finally, 1987 is the year in which the high-flying superstring has come down to earth. We are told that the universe doesn't really have to be ten dimensional after all. Maybe it's as four dimensional as it looks. Superstring theories are now a dime

[†] supported in part by the National Science Foundation under Grant No. Phys 82-15349

a dozen and the dream of a unique and predictive Theory of Everything remains unfulfilled. Nonetheless, stringers enthusiastically pursue their fascination with ever purer mathematics, while some survivors grope towards the baroque to the beat of their superdrums. Perhaps we unstrung and unsung dinosaurs will have the last laugh after all.

I shall argue that LEP is a machine with enormous potential for spectacular discovery. LEP should accumulate tens of millions of analyzed Z decays, and will reveal the existence of almost any new particle sitting below 50 GeV: a light Higgs, the elusive top quark, a fourth family, or things even stranger. The neutrino-counting experiment will tell us a lot: Are there really only three fundamental fermion families? Are there triplet majorons, which will count as two additional neutrinos? Is the fourth neutrino heavy enough to be phase-space suppressed, thus yielding a *fractional* neutrino count? In this talk, I develop three minor variations of the standard model on which I, and perforce my students, have been working over the past year: Chiral color, a modification of the strong force; neutrinos at their limit, an implementation of the singlet majoron model; and a twist of the Higgs sector making top quarks hard to find. To some extent, this work is an elaboration of ideas presented at La Thuile [1].

2. Chiral Color

The most standard model is based upon the gauge group $SU(3) \times SU(2) \times U(1)$. Possibly, the low-energy gauge group is larger than this. Left-right theories involve a second $SU(2)$ factor whose existence is seriously constrained by experiment. The string inspired speak of a second $U(1)$ which may indeed be lurking at energies soon to be explored at LEP. Rather than play with the structure of the electroweak theory, we examine a minimal modification of the underlying group of the strong interactions. Explicitly, we embed unbroken color $SU(3)$ as the unbroken diagonal subgroup of spontaneously broken $SU(3)_L \times SU(3)_R$ [2]. Of course, if chiral color breaks well beyond TeV energies it's going to be hard to find evidence for it. Our hope is that the mass scale associated with chiral color is comparable with the electroweak mass scale.

An immediate implication of chiral color is the necessary existence of a color octet of massive axial-vector gauge bosons with couplings to quarks comparable with those of gluons. These axigluons are unstable and decay quickly into quark pairs. Cuypers and Frampton [3] point out that triplet quarkonium can decay into gluon plus axigluon. Since such decay modes of upsilon have not been detected, we may safely conclude that axigluons cannot be lighter than about 9 GeV, an admittedly unlikely possibility in any case. The story can be different for the still undetected toponium state, whose decays may be dominated by the gluon-axigluon mode, and whose width thereby may be dramatically enhanced. We shall see. Another way to find relatively light axigluons is in the decay of Z_0 if axigluon decay channels are open. Rizzo has computed the bremsstrahlung decay into quark, antiquark, axigluon [4], while Carlson, Jenkins and I computed the gluon-axigluon decay mode [5]. The former process dominates for axigluons less than half as massive as Z_0, the latter otherwise. We obtain a branching ratio of 10^{-3} for 60 GeV axigluons, and of 10^{-4} for 80 GeV. The detection signal is a 3-jet event where two of the jets peak at the axigluon mass. It is a difficult, but not an impossible, search for LEP to carry out.

Hadron colliders are ideal instruments with which to search for effects of heavier axigluons. There could be bumps in the two-jet mass distribution arising from axigluon production and decay. Remember, however, that W and Z have barely been seen as two-jet enhancements. So far, there are no good axigluon constraints from such data. Another possibility, recently explored by Bagger et al. [6], is the search for Jacobian peaks in one-jet distributions. From available Cern Collider data, they are able to exclude axigluons in a wide window extending from 150 GeV to 275 GeV. The window will soon be extended upwards (or axigluons will be found!) by experiments soon to be done at the Tevatron.

A respectable chiral color scheme must not endanger the very successful theory we already have. There are several additional sources of anomalies to threaten the renormalizability of the theory: those cubic or quadratic in one of the two $SU(3)s$. The fermion structure must be such as to lead to a cancellation of these anomalies in the spirit of Bouchiat, Iliopoulos and Meyer. Moreover, there is my persistent professional bugbear: the possibility of flavor-changing neutral currents. The theory must be concocted to give a natural explanation for the absence of such effects, whether mediated by $Z's$, by axigluons, or by Higgs bosons. These two constraints are not easily met. Aside from trivial models, or models with contrived and unnatural quark charges, Frampton and I have come up with only one plausible possibility [7]. We may not have a Theory of Everything, but at least our chiral color model is pretty much unique. And, it predicts the number of fundamental fermion families to be precisely five.

By a quark we mean a chiral fermion field transforming trivially under one $SU(3)$ factor and as a triplet under the other. Two axioms lead us to our chosen theory: (1) All left-handed quarks are triplets under one and the same $SU(3)$ factor, call it $SU(3)_L$. All right-handed quarks are triplets under the other. This is the gist of what we mean by chiral color: not just a second $SU(3)$ group, but a separate group for each quark chirality. Other $SU(3) \times SU(3)$ models can be constructed which do not respect this axiom, but in general they suffer from the appearance of flavor-changing neutral-current couplings of the axigluons. True chiral color models do not present this problem.

According to axiom (2), all left-handed quarks form weak doublets with charges $\frac{2}{3}$ and $-\frac{1}{3}$, while all right-handed quarks are weak singlets with corresponding charges. This assumption has two purposes. It excludes the possible existence of quarks with large and exotic charge, and it guarantees that there are no flavor-changing neutral currents coupled to Z_0. It also makes possible a choice of Higgs couplings that does not yield flavor changing effects in the Higgs sector. Models that evade this axiom are possible, but they are contrived and without evident appeal.

Frampton and I have identified the fermion content of the simplest model embodying our axioms. It involves five conventional fermion families, each containing a pair of quarks, and a conventional pair of leptons. However, one of the five neutrinos, as we shall see, develops mass and couplings that distinguish it from the remaining four (presumably massless) neutrinos. Thus, our chiral color theory offers the unambiguous prediction that there are four neutrino species to be counted at LEP, and four to affect early universe nucleosynthesis. A rich Higgs structure is needed to give mass to the five families: two distinct weak doublets with $(\bar{3}, 3)$ transformation properties under chiral color give mass to the quarks, and an additional weak doublet gives mass to the charged leptons. Many of these scalar mesons should survive at low energy as observable particles, including, for example, two color-octet charged mesons. Some parts of the rich scalar sector could be accessible to LEPI, and certainly will be accessible to LEPII.

The dichromatic $(\bar{3}, 3)$ of fermions is neutral under $SU(2) \times U(1)$, and serves to cancel the cubic chiral color anomaly. It obtains its mass from an isomorphic multiplet of scalar mesons which breaks chiral color but not electroweak symmetry.

Conversely, the Higgs scalar responsible for lepton masses breaks electroweak symmetry but not chiral color. Were these scalars absent, the axigluon mass would be determined to be about 300 GeV. In reality, it could be either larger or smaller.

The color singlet *quone* develops twice the mass of the color octet *queight*, and decays readily to queight plus axigluon. The queight itself decays, by Higgs-mediated couplings, to quark, antiquark, and lepton. Neither queight nor quone is produced copiously at LEP, which is a factory for producing only those particles with electroweak couplings. These particles, should they exist, will be tough to find.

The Higgs scalar responsible for down-quark masses can link the quone state to a linear combination of the five neutrino states, generating mass mixing of one of the neutrinos with the quone. Both states will become heavy and unstable. Thereby, the neutrino count for this chiral color model is reduced to just four. There must be a failure of lepton universality somewhere down the line, since one of the five neutrinos has grown into a heavy hadron.

The last of our exotic fermions is the *quix*, and it may turn out to be the most intriguing. Unless we introduce a heavyweight $(6, \overline{6})$ of scalars, the quix must get its mass through loop diagrams involving hypothetical scalar quarks. Its mass arises at second order in electroweak symmetry-breaking, so that the quix may be relatively light even if the chiral-color symmetry breaking scale is high. It should be one craziest of the new particles that may show up at LEP. Below the Z_0, it would be produced in pairs at $\frac{2}{3}$ of a unit of R. In Z_0 decay, its branching ratio (not including phase-space suppression) should be 2%. What does it do? The quix (more precisely, quixotic hadrons) are *stable* in the low energy theory. They may decay virtually into quark plus scalar quark. The scalar quark can decay to quark plus neutrino, but only at first order in the unification mass scale. As a crude guess, I would put the lifetime of a 30 GeV quix at the order of an hour. Here is a truly exciting challenge to LEP: to discover a new and essentially stable heavy hadron that is produced in pairs.

This chiral color model is unifiable. That is, the low-energy gauge group,

$$SU(3)_L \times SU(3)_R \times SU(2) \times U(1),$$

can be embedded [7] within a larger group characterized by a single gauge coupling constant. The unifying group is $[SU(4)]^6$, together with a discrete group which interchanges the factor groups. The 92 chiral fermions corresponding to the five families, the quix, the quone, and the queight are the low-energy survivors of a 252-dimensional representation of the unifying group. This representation, portrayed in moose notation, bears a striking resemblance to the chemists' picture of the benzene molecule.

3. Enriching Mr Higgs

It seems unlikely that the Higgs sector should consist of a single solitary Higgs boson. Strong CP arguments dictate a minimal enrichment involving two Higgs doublets and an invisibilized axion. But two doublets imply the survival of a charged scalar meson which, if it is light enough, can alter the nature of top-quark decay ensuring that the top quark would not have been detected at hadron colliders.

A recent paper with E. Jenkins explores this possibility [8]. If the top quark is heavier than the putative charged scalar, its predominant decay is into such a scalar and a lighter uplike quark. The scalar, in turn, decays about $\frac{2}{3}$ of the time into $c\overline{s}$, and $\frac{1}{3}$ of the time into tau plus neutrino. Conventional weak leptonic decay modes, the channels of discovery for hadron colliders, cannot compete with this semi-weak decay mode. Given the possible existence of a light enough charged scalar, no limit on the top quark mass has yet been obtained beyond those from PETRA or, perhaps, from TRISTAN.

If top is accessible in Z_0 decay, it will show up clearly at LEP, for example, as decay modes involving both tau(s) and hadrons. Pairs of charged scalars could also emerge from Z_0 decay, but with a branching ratio of 10^{-3} at best. In addition, the properties of toponium would be modified by the existence of a rapid (semiweak) decay mode for top. Decays involving charged scalars would dominate toponium decay, and assure that all of its states have widths well in excess of one MeV. Of course, this would not interfere with the potential discovery of top at TRISTAN through a step in R consisting of high sphericity events.

The charged scalars provide us with an unexpected bonus. Box diagrams involving charged scalars can generate the recently observed and surprisingly high level of $B_0 - \overline{B}_0$ mixing without the need for a heavy top quark and without upsetting the $K_0 - \overline{K}_0$ applecart. Thus, both the theoretical and experimental case for a heavy top quark are undone.

If there are indeed two (or more) Higgs multiplets which are separately responsible for up and down quark masses, then there is no real theoretical lower limit on the mass of the lightest one. Indeed, even in the minimal standard theory, the familiar Lindé-Weinberg mass limit can be evaded by a carefully chosen top quark mass. Furthermore, in a multi-Higgs model, the couplings of the lightest Higgs to fermions can be dramatically larger than in the minimal theory. Thus, it seems worthwhile to ask how a light scalar Higgs component might show up in the laboratory. We have computed the process electron plus positron into tau pair plus Higgs [9] for which there are two distinct contributions: the bremssstrahlung of a Higgs in $\tau\bar{\tau}$ production, and the appearance of a Higgs in τ decay. A barely detectable branching ratio of order 10^{-5} is obtained for standard model couplings of a Higgs at 200 MeV, with larger results at smaller Higgs masses. In multi-Higgs models, the branching ratio can be orders of magnitude larger. The Higgs yield is not very sensitive to beam energy, so that the search for tau pairs associated with a light Higgs (itself decaying into an e^+e^- pair) should best be carried out at lower energy machines like DORIS or CESR. To my knowledge, the experimenters have not yet provided us with any limit at all on this process.

Uri Sarid has considered the amusing possibility that each type of fermion has a Higgs of its own: one for uplike quarks, one for downlike, and one for charged leptons. There could be a fourth doublet as well contributing to neutrino masses through a see-saw mechanism: the more the merrier from the point of view of the preceding paragraph. The Peccei-Quinn symmetry becomes $U(1) \times U(1)$, or more. One of the potential Goldstone bosons gets a small mass from the $SU(3)$ anomaly to become the invisible axion. The other state remains essentially massless. Both particles must be invisibilized. Interestingly, a powerful limit on the mass scale associated with the true Goldstone boson follows from serendipitous observations of supernova SN1987a.

Suppose that a fraction f of the supernova energy is emitted as $\sim$ MeV Goldstone bosons. This corresponds to a flux on earth of $\sim 10^{11} f \, \mathrm{cm}^{-2}$. Some of these bosons can oscillate into photons during their transgalactic voyage. The probability for conversion is given by $P \simeq (BL/2G)^2$, where B $\sim 3\mu$ gauss is the intergalactic magnetic field, L is its effective coherence length ($\sim$ kpc), and G^{-1} is the coupling strength of the Goldstone boson to two photons. Note that $P \simeq 1$ for $G \simeq 10^{10}$ GeV.

It turns out that the Solar Maximum Satellite, even though it wasn't looking in quite the right direction, detected no gamma ray excess from the supernova at the time of the observed neutrino burst [10]. This puts a lower limit on G of order of 10^{13} GeV, which is considerably stronger than any known limit on the mass scale of the axion. Thus, we have both constructed and constrained our very own exotic multi-Higgs model—an example of the sort of onanistic physics we must do as we await exciting new data from the big machines we don't yet have.

4. Fat Neutrinos

Direct limits on neutrino masses are not very impressive. The electron neutrino may not have a Majorana mass of more than a few eV lest it lead to observable neutrinoless double beta decay. The muon neutrino weighs less than 250 KeV, and the tau neutrino less than 70 MeV. Yet, many of us have been taken in by our asteroid colleagues who argue that the sum of all neutrino masses cannot exceed 50 eV, or so. Not true! The argument only applies to *stable* neutrinos. For unstable neutrinos, the limit is given by

$$\sum n_i (\tau_i/T)^{\frac{1}{2}} \leq 25\,\text{eV},$$

where n_i and τ_i is the mass and lifetime of the ith neutrino, and T is the age of the universe. I have examined the possibility that neutrinos obtain their masses through a singlet majoron model, and that these masses, perhaps perversely, are at or near their direct upper limits [11].

An N-family model involves N weak-doublet neutrino states and N heavy singlets which get their masses from the vev of a scalar field. Dirac masses link the doublets to the singlets. For want of a better idea, we take the Dirac masses to equal the masses of the corresponding charged leptons. Since this is as questionable a notion as putting the top and bottom quark masses equal, read on with a generous measure of salt.

The light, mostly doublet, neutrinos ride the see-saw mechanism with the heavy neutrinos of mass M_i, to obtain their masses, n_i. Neglecting mixing effects, we obtain the approximate results:

$$n_e = m_e^2/M_e \leq 5\,\text{eV},$$
$$n_\mu = m_\mu^2/M_\mu \leq 250\,\text{KeV},$$
$$n_\tau = m_\tau^2/M_\tau \leq 70\,\text{MeV}.$$

In each of the three cases, we obtain a similar constraint on the mass of the heavy neutrinos, $M_i \geq 50\,\text{GeV}$. If neutrinos are at the limit, the majoron mass scale must be comparable to the electroweak mass scale. Things could get particularly interesting if there exists a fourth family with a charged lepton up around 50 GeV. In that case, the masses of this doublet neutrino and its singlet counterpart would be comparable, and all sorts of interesting physics would ensue.

The heavier neutrino flavors satisfy the cosmological constraint by means of their invisible decays to lighter neutrinos plus massless majorons ϕ, about which the following results were deduced in refs. [1]and [11]. The requirement that couplings of the scalar mesons to singlet neutrinos be perturbative yields an upper limit to the rate of ν_μ decay,

$$\Gamma(\nu_\mu \to \nu_e \phi) \leq \frac{1}{16} \frac{n_\mu^6 n_e^3}{m_\mu^4 m_e^4},$$

and thus to a lower limit to its lifetime

$$\tau_\mu \geq \left(\frac{250\,\text{KeV}}{n_\mu}\right)^6 \left(\frac{1\,\text{eV}}{n_e}\right)^3 12\,\text{years}.$$

Cosmological consistency becomes a curious *lower* limit to neutrino masses,

$$\left(\frac{n_\mu}{250\,\text{KeV}}\right)^4 \left(\frac{n_e}{1\,\text{eV}}\right)^3 \geq \left(\frac{1}{10}\right).$$

544

Should the direct upper limit on n_μ from pion decay kinematics, or the limit on n_e from double beta decay be significantly improved, this narrow window of neutrino masses at their upper limits can be decisively closed.

The heavier ν_τ decays even more rapidly and is cosmologically safe. Its lifetime may be as short as 10 ns. A fourth neutrino state, if it exists, may have a mass of tens of GeV with a very short lifetime for decay into invisible modes. It would give a phase-space suppressed fractional contribution to LEP's neutrino count. The associated fourth charged lepton, if heavier than its neutrino, would display anomalously large missing visible energy in its decays. If it is lighter than its neutrino and decays by flavor mixing, its lifetime could be long enough to be observable. One can hope that experimenters at LEP will have a ball with such a peculiar fourth family of leptons.

5. Conclusion

Our discipline advances by a sort of punctuated equilibrium wherein bursts of rapid evolution (as in the early 1970's) alternate with periods of consolidation. Now is a quiet time in which the standard model has become generally accepted. We yearn for the deployment of future accelerators and for the bewildering data they may present to us. Theorists who are not satisfied with the status quo have formed two camps: the builders of castles in the sky and the chasers of ambulances. Most have chosen the mathematically sophisticated and less demeaning high road, plunging beyond field theory and hoping to thread their way down from the celestial sphere of the superstring. While they may have found a consistent quantum theory of gravity, they are not yet able to say much about experimentally accessible physics. The low road is equally treacherous but less glamorous and less well travelled. It consists of looking for ways in which future data may stray from the standard route to suggest an improved, even a very slightly improved, new theory. We seek to learn from experimental anomaly just as physiologists learn from human pathology. The trouble is that, up to now, all of the emergencies have turned out to be false alarms and no confirmed discrepancies have appeared. The standard model remains as unchallenged as it is incomplete.

I hope that this situation will soon change, that *something* new will soon be found, most likely at LEP, but perhaps elsewhere, that does not fit into today's big picture. *Be Prepared* is the boy scout motto, and it is in that spirit that I offer these modest suggestions. Nature will no doubt prove to be far more imaginative. Courage, dear colleagues! The search for the ultimate structure of matter has not yet come to an end.

Acknowledgements
Chiral color, a joint production with Paul Frampton, was carried out primarily at the exciting new physics department of Boston University. The remainder of this work is a summary of recent researches carried out with my incomparable team of Harvard graduate students: Eric Carlson, Elizabeth Jenkins, and Uri Sarid. It is to them, who have met the superstring and ignored her, that this paper is dedicated.

References

[1] S.L. Glashow, in *Results and Perspectives in Particle Physics*, p. 539 (Editions Frontières, Gif-Sur-Yvette, 1987).

[2] P.H. Frampton and S.L. Glashow, Phys.Lett. B190 (1987) 157.

[3] F.Cuypers and P.H. Frampton, Preprint IFP-306-UNC (1987).

[4] T.G. Rizzo, Phys. Lett. B197 (1987) 273.

[5] E.D. Carlson, S.L. Glashow and E.E. Jenkins, Preprint HUTP-87/A081.

[6] J. Bagger, S. King and C. Schmidt, Preprint HUTP-87/A056

[7] P.H. Frampton and S.L. Glashow, Phys. Rev. Lett. 58 (1987) 2168.

[8] S.L. Glashow and E.E. Jenkins, Phys. Lett. B196 (1987) 233.

[9] E.D. Carlson, S.L. Glashow, and U. Sarid, Preprint HUTP-87/088.

[10] Edward L. Chupp, Private Communication.

[11] S.L. Glashow, Phys. Lett. B187 (1987) 367.

DISCUSSION

– *Liu:*

Are there any experimental or theoretical bounds on the mass of η^+ or any experiments to verify its existence?

– *Glashow:*

There are no theoretical constraints on the mass of η^+.

The way of finding the particle, if its mass is accessible in Z^0–decay, to see the $\eta^+ + \eta^-$ decay mode. If its mass is less than the mass of the top quark one should observe that the t–quark does not engage in conventional weak decays but decays exclusively to an η^+ plus a down quark.

– *Liu:*

I guess the scale where $SU_L(3) \times SU_R(3)$ breaks to $SU_C(3)$ should be below the scale of $SU(2) \times U(1)$ breaking.

– *Glashow:*

What I have been suggesting is that for some unexplained theoretical reason, the scale of $SU(3)_L \times SU(3)_R$ breaking is similar to the scale of $SU(2) \times U(1)$ breaking.

Because the Higgs responsible for quark masses also breaks chiral color, I do not expect the axigluons to be very light, certainly no lighter than 10 GeV. Such a light axigluon is not excluded experimentally. There is also no upper bound of $SU(3) \times SU(3)$ symmetry breaking. If you put the $SU(3)^2$ breaking scale very high (say 1 TeV) the effective low energy theory is essentially identical to the conventional one. There is only new physics if the scale of chiral color breaking is low.

– *Liu:*

You choose $(3,0)_L + (0,3)_R$ because you want to keep up with $\pi^0 \to 2\gamma$?

– *Glashow:*

The reason why I have restricted to the case where the left handed quark transforms under $SU_L(3)$ and the right handed quark under $SU_R(3)$ is that if you do not make this assignment run the risk of violating the GIM mechanism and generating flavor-changing neutral currents.

– Miele:

My question concerns the majoron particle. Can it also couple to charged leptons?

– Glashow:

Majorons are introduced for the purpose of violating lepton number by two and thus generating neutrino masses. The neutral majoron therefore couples only to neutrinos at tree level. Radiative corrections will generate majoron couplings to charged leptons, but these will be exceedingly small and entirely negligible. For example, the branching ratio for $\mu^{\pm} \to e^{\pm} + \gamma$ or majoron is expected to be smaller than 10^{-20}.

– Singh:

In your unification scheme (B-L) is not conserved; what are the consequences for the proton decay?

– Glashow:

There are none.

– Kiritsis:

Why do you need to embed $SU_L(3) \times SU(3)_C$ in such a big group as $(SU(4))^6$?

– Glashow:

The key point is the cancellation of anomalies. First you must find a fermion representation which is anomaly free in $SU(3)_L \times SU(3)_R \times SU(2) \times U(1)$. Essentially there is only one which makes sense. This is the model which I presented in my lecture and which includes five families.

Then you ask for the simplest unifying group. The answer is $(SU(4))^6$. You could now ask what is the unification mass and what is the value of $sin^2\theta_W$. However, these answers depend strongly on the details of the scalar sector. This is a weak point in comparison to the $SU(5)$ theory. We cannot make a prediction for $sin^2\theta_W$ or the unification scale. On the other hand, we do not predict the proton to decay a hundred times faster than it does.

– Iacovacci:

Does PETRA give you any test for your theories?

– Glashow:

You mean those 7 peculiar events that were observed by MARK J together with the 5 events of JADE? Those are events which include an isolated muon.

They are indeed compatible with a light top quark which decays into a tau which subsequently decays into a muon. However, if that is true we should soon get confirmation from TRISTAN.

– Iacovacci:

Are there any inclusive distributions for which you look at PETRA?

– Glashow:

Electron–positron accelerators are not the best machines to search for axigluons. At hadron colliders, they can show up as two–jet resonances, or as Jacobian peaks in single jet distributions.

CLOSING CEREMONY

The closing cerimony took place on Thursday, 13 August 1987. The Director of the School presented the prizes and scholarships to the winners as specified below.

PRIZES AND SCHOLARSHIPS

Prize for Best Student
> awarded to John QUACKENBUSH, University of California,
> Los Angeles, CA, USA

The Scholarships were open for competition among the participants. They have been awarded as follows:

Patrick M.S. Blanckett Scholarship
> awarded to Jun LIU, University of Texas at Austin, USA.

James Chadwick Scholarship
> awarded to David LEWELLEN, Cornell University, Ithaca, NY, USA.

Amos De-Shalit Scholarship
> awarded to Janos BALOG, Central Research Institut,
> Budapest, Hungary.

Paul A.M. Dirac Scholarship
> awarded to Patricia McBRIDE, CERN, Geneva, Switzerland.

Gunner Kallen Scholarship
> awarded to Elias KIRITSIS, California Institute of Technology,
> Pasadena, CA, USA.

André Lagarrigue Scholarship
> awarded to Gerhard SCHULER, DESY, Hamburg, FRG.

Ettore Majorana Scholarship
> awarded to Tim BOLTON, SLAC, Stanford, CA, USA.

Giulio Racah Scholarship
> awarded to Dale Pitman, SLAC, Stanford, CA, USA.

Jun John Sakurai Scholarship
> awarded to Alvaro DIAZ, ICPT, Trieste, Italy.

Antonio Stanghellini Scholarship

        awarded to Gennaro MIELE, University of Naples, Italy.

Prize for Best Scientific Secretary

        awarded to John QUACKENBUSH, University of California
Los Angeles, CA, USA.

The following participants gave their collaboration in the Scientific Secretarial work:

Janos BALOG	Attilio MORELLI
Tim BOLTON	Gilbert MOUTTAKA
Ana Cristina CADAVID	Oreste NICROSINI
Alvaro DIAZ	Andrea PELISSETTO
Alexander GANCHEV	Dale PITMAN
Andrea GIULIANI	John QUACKENBUSH
Michael JONKER	Jan REYNOLDSON
Elias KIRITSIS	Gerhard SCHULER
David LEWELLEN	N. Nimai SINGH
Jun LIU	Martin STIERLE
Patricia McBRIDE	Francesco TOPPAN
Gennaro MIELE	Eric WICKLUND

Three EPS Scholarship were awarded as follows:

Janos BALOG
Peter BANTAY
Andrej SZCERBA

PARTICIPANTS

Carlo ARTEMI — Università di Perugia
Dipartimento di Fisica
Via Elce di Sotto, 10
06100 PERUGIA, Italy

Fabio BAGARELLO — Istituto di Fisica
Università di Palermo Via Archirafi, 36
90100 PALERMO, Italy

Herbert BALASIN — Institute für Theoretische Physik
Karlsplatz 13
A-1040 WIEN, Austria

Janos BALOG — Central Research Institute for Physics
P.O. Box 49
1525 BUDAPEST, Hungary

Peter BANTAY — Institute for Theoretical Physics
Etvös University
1088 Puskin u. 5-7
BUDAPEST, Hungary

Vincenzo BARONE — Dipartimento di Fisica Teorica
Via P. Giuria, 1
10125 TORINO, Italy

Alice BEAN — Department of Physics
University of California
SANTA BARBARA, CA 93106, USA

Gerald BLAZEY — Department of Physics and Astronomy
The University of Rochester
ROCHESTER, NY 14627, USA

Tim BOLTON — SLAC
Bin 65
STANFORD, CA 94305, USA

Ana Cristina CADAVID — University of California
Department of Physics
405 Hilgard Avenue
LOS ANGELES, CA 90024, USA

Massimo CAMPOSTRINI — Dipartimento di Fisica
Piazza Torricelli, 2
56100 PISA, Italy

Tony CASS — Department of Physics
Liverpool University
LIVERPOOL, L69 3SX, UK

Pietro COLANGELO — INFN
Via Amendola, 173
70126 BARI, Italy

Marc DEJARDIN — DPHPE
CEN Saclay
91191 GIF-SUR-YVETTE, France

Martin DEUTSCH — Massachusetts Institute of Technology
Physics Department
CAMBRIDGE, MA 02139, USA

Alvaro DIAZ — ICTP
P.O. Box 586
34100 TRIESTE, Italy

Michael DUFF — CERN
TH/SP Division
1211 GENEVE 23, Switzerland

Sergio FERRARA — Physics Department
University of California
LOS ANGELES, CA 90024, USA
and
CERN
1211 GENEVE 23, Switzerland

Herman FESHBACH — Massachusetts Institute of Technology
Physics Department
CAMBRIDGE, MA 02139, USA

Maria Luisa FRAU — Dipartimento di Fisica Teorica
Corso Massimo d'Azeglio, 46
10125 TORINO, Italy

Emidio GABRIELLI — Dipartimento di Fisica
Università La Sapienza
Piazzale Aldo Moro, 2
00185 ROMA, Italy

Alexander GANCHEZ — Institute for Nuclear Research
and Nuclear Energy
Bulgarian Academy of Sciences
Blvd. Lenin, 72
SOFIA 1184, Bulgaria

Neil GEDDES — Rutherford Appleton Laboratory
Chilton
DIDCOT, Oxon OX11 0QX, UK

Andrea GIULIANI — Dipartimento di Fisica
Via Celoria, 16
20133 MILANO, Italy

Sheldon L. GLASHOW — Harvard University
Physics Department
CAMBRIDGE, MA 02138, USA

Michel GOURDIN — Université Pierre et Marie Curie
T16 E1
4 Place Jussieu
75230 PARIS, France

Marc T. GRISARU — Brandeis University
Physics Department
WALTHAM, MA 02254, USA

Johan GRUNDBERG — NORDITA
Blegdamsvej, 17
DK-2100 COPENHAGEN, Denmark

Klaus HEIN — Institut für Theoretische Physik
der Universität Heidelberg
Philosophenweg, 16
6900 HEIDELBERG, FRG

Clemens A. HEUSCH — University of California
High Energy Physics
Natural Sciences II
SANTA CRUZ, CA 95060, USA and
CERN
1211 GENEVE 23, Switzerland

Andrew HOCH — Department of Physics
University of Manchester
MANCHESTER, M13 9PL, UK

Michele IACOVACCI — Dipartimento di Fisica
Piazzale Aldo Moro, 2
00185 ROMA, Italy

Mark Mitsuo ITO — Physics Department
Building 510A
Brookhaven National Laboratory
UPTON, NY 11973, USA

Michael JONKER — CERN
EP Division
1211 GENEVE 23, Switzerland

Elias KIRITSIS — California Institute of Technology
452-48
PASADENA, CA 91125, USA

Albrecht Otto KLEMM — Institut für Theoretische Physik
Universität Heidelberg
Philosophenweg 16
D-6900 HEIDELBERG, FRG

Costas KOUNNAS — Lawrence Berkeley Laboratory
Physics Division
1 Cyclotron Road
BERKELEY, CA 94720, USA

Taichiro KUGO — Physics Department
Kyoto University
KYOTO, Japan

Rupert LEITNER — Joint Institute for Nuclear Research
Dubna, Head Post Office
P.O. Box 79
101 000 MOSCOW, USSR

David LEWELLEN — Floyd R. Newman Laboratory
Cornell University
ITHACA, NY 14853, USA

Li Jia LIN — Department of Physics
University of Georgia
ATHENS, GA 30602, USA

Jun LIU — Theory Group
Department of Physics
University of Texas at Austin
AUSTIN, TX 78712, USA

Nicodemo MAGNOLI — INFN
Via Dodecaneso, 33
16146 GENOVA. Italy

Guido MARTINELLI — CERN
TH Division
1211, GENEVA 23, Switzerland

Alberto MASONI — Dipartimento di Scienze Fisiche
Via Ospedale, 76
09100 CAGLIARI, Italy

Patricia McBRIDE — CERN
EP Division
1211 GENEVA 23, Switzerland

Gennaro MIELE — Dipartimento di Fisica
Mostra d'Oltremare - Pad. 19
80125 NAPOLI, Italy

Attilio MORELLI — Dipartimento di Fisica
Università di Trieste
Via A. Valerio, 2
34127 TRIESTE. Italy

Gilbert MOULTAKA — Laboratoire de Physique Mathematique
Université de Montpellier
Place E. Bataillon
34060 MONTPELLIER CEDEX, France

Ramon MUNOZ — Departamento de Fisica Teorica
Universidad Autonoma de Barcelona
BELLATERRA (Barcelona), Spain

Dimitri NANOPOULOS — University of Wisconsin
Physics Department
MADISON, WI 53706, USA

Giuseppe NARDELLI — Dipartimento di Fisica
Università di Trento
38050 POVO (Trento), Italy

Harvey NEWMAN — California Institute of Technology
Physics Department
PASADENA, CA 91125, USA

Oreste NICROSINI — Istituto di Fisica
Via A. Bassi, 6
27100 PAVIA, Italy

Sonia PABAN — Institute of Theoretical Physics
University of Barcelona
08028, BARCELONA, Spain

Andrea PELISSETTO — Scuola Normale Superiore
Piazza dei Cavalieri, 7
56100 PISA, Italy

Oreste PICCIONI — University of California, San Diego
Department of Physics
LA JOLLA, CA 92093, USA

Dale PITMAN SLAC
Bin 65
P.O. Box 4345
STANFORD, CA 94305, USA

Irwin PLESS Massachusetts Institute of Technology
Physics Department
CAMBRIDGE, MA 02139, USA

Martin POPPE CERN
EP Division
1211 GENEVE 23, Switzerland

John QUACKENBUSH University of California
Department of Physics
405 Hilgard Avenue
LOS ANGELES, CA 90024, USA

Martin REUTER DESY
Notkestrasse 85
2000 HAMBURG 52, FRG

Jan REYNOLDSON INFN
Laboratori Nazionali
Casella Postale 13
00044 FRASCATI (Roma), Italy

Martin SCHMIDT Universität Siegen
Fachbereich Physik
Postfach 210 209
5900 SIEGEN 1, FRG

Marcus SCHOLL Institut für Theoretische Physik
Universität Karlsrhue
Kaiserstrasse 12
7500 KARLSRHUE, FRG

Gerhard SCHULER DESY
Notkestrasse 85
2000 HAMBURG 52, FRG

N. Nimai SINGH Department of Physics
University of Delhi
DELHI 110007, India

Volker SOERGEL DESY
Notkestrasse 85
D-2000 HAMBURG 52, FRG

Martin STIERLE Institut für Theoretische Physik
Technische Universität
Karlsplatz 13
1040 WIEN, Austria

Andrej SZCERBA Institute of Physics
Jagellonian University
Reymonta, 4
KRAKOW, Poland

Sam C.C. TING Massachusetts Institute of Technology
Building 44, 51 Vassar Street
Laboratory for Nuclear Science
CAMBRIDGE, MA 02139, USA
and
CERN
1211 GENEVE 23, Switzerland

Francesco TOPPAN SISSA
Strada Costiera - Grignano
Miramare
34100 TRIESTE, Italy

Manuel VILLASANTE University of California
Department of Physics
405 Hilgard Avenue
LOS ANGELES, CA 90024, USA

Eric WICKLUND California Institute of Technology
High Energy Physics
PASADENA, CA 91125, USA

Paul WINDEY Department of Physics
Lawrence Berkeley Laboratory
University of California
BERKELEY, CA 94720, USA

Chien-Shiung WU Columbia University
Department of Physics
NEW YORK, NY 10027, USA